# THE BIOLOGY OF ISLAND FLORAS

Oceanic islands offer biologists unparalleled opportunities to study evolutionary processes and ecological phenomena. However, human activity threatens to alter or destroy many of these fragile ecosystems, with recent estimates suggesting that nearly half of the world's insular endemics are threatened with extinction.

Bringing together researchers from around the world, this book illustrates how modern research methods and new concepts have challenged accepted theories and changed our understanding of island floras. Particular attention is given to the impact of molecular studies and the insights that they provide into topics such as colonisation, radiation, diversification and hybridisation. Examples are drawn from around the world, including the Hawaiian archipelago, Galapagos Islands, Madagascar and the Macaronesian region. Conservation issues are also highlighted, with coverage of alien species and the role of *ex situ* conservation providing valuable information that will aid the formulation of management strategies and genetic rescue programmes.

DAVID BRAMWELL is Director of the Jardín Botánico Canario 'Viera y Clavijo'- Unidad Asociada CSIC. His current research interests include the monitoring and mapping of wild populations of threatened species, working on the *Flora of Macaronesia* project, mapping species decline worldwide, molecular studies of genetic diversity in small populations and the effects of climate change on island plants.

JULI CAUJAPÉ-CASTELLS is Head of the Department of Molecular Biodiversity and DNA bank at the Jardín Botánico Canario 'Viera y Clavijo'- Unidad Asociada CSIC. His research uses molecular data to investigate problems related to the origins, taxonomic identification, microevolution and conservation of terrestrial vascular plants endemic to, or occurring in, Macaronesia.

# THE BIOLOGY OF ISLAND FLORAS

*Edited by*

DAVID BRAMWELL
*Jardín Botánico Canario 'Viera y Clavijo' (Unidad Asociada CSIC), Las Palmas de Gran Canaria, Spain*

and

JULI CAUJAPÉ-CASTELLS
*Jardín Botánico Canario 'Viera y Clavijo' (Unidad Asociada CSIC), Las Palmas de Gran Canaria, Spain*

CAMBRIDGE
UNIVERSITY PRESS

**CAMBRIDGE**
**UNIVERSITY PRESS**

University Printing House, Cambridge CB2 8BS, United Kingdom

One Liberty Plaza, 20th Floor, New York, NY 10006, USA

477 Williamstown Road, Port Melbourne, VIC 3207, Australia

314-321, 3rd Floor, Plot 3, Splendor Forum, Jasola District Centre, New Delhi-110025, India

79 Anson Road, #06-04/06, Singapore 079906

Cambridge University Press is part of the University of Cambridge.

It furthers the University's mission by disseminating knowledge in the pursuit of education, learning and research at the highest international levels of excellence.

www.cambridge.org
Information on this title: www.cambridge.org/9780521118088



First published 2011

*A catalogue record for this publication is available from the British Library*

*Library of Congress Cataloging in Publication data*
The biology of island floras / [edited by] David Bramwell and Juli Caujapé-Castells.
p. cm.
ISBN 978-0-521-11808-8 (hardback)
1. Island plants. I. Bramwell, David. II. Caujapé-Castells, Juli.
QK938.I84B56 2011
578.75′2–dc22
2011011812

ISBN 978-0-521-11808-8 Hardback

# Contents

# Contributors

**Melissa Abdo**, (1) Center for Tropical Plant Conservation, Fairchild Tropical Botanic Garden, 10901 Old Cutler Road, Coral Gables, FL 33156, USA, (2) Department of Biological Sciences, Florida International University, 11200 SW 8th Street, Miami, FL 33199, USA.

**Gregory J. Anderson**, Department of Ecology and Evolutionary Biology, University of Connecticut, Storrs, CT 06269, USA.

**Lisa M. Banfield**, Royal Botanic Garden Edinburgh, 20A Inverleith Row, Edinburgh EH3 5LR, Scotland, United Kingdom.

**Rosalina Berazain**, Jardín Botánico Nacional, Calabazar, Boyeros, CP 19230 Havana, Cuba.

**Gabriel Bernardello**, Instituto Multidisciplinario de Biología Vegetal, CONICET-Universidad Nacional de Córdoba, CC 495, 5000 Córdoba, Argentina.

**David Bramwell**, Jardín Botánico Canario 'Viera y Clavijo' (UA CSIC), Apartado de Correos 14 de Tafira Alta, 35017 Las Palmas de Gran Canaria, Spain.

**Stuart Cable**, Madagascar Science Team, The Herbarium, Royal Botanic Gardens, Kew, Richmond, Surrey TW9 3AE, United Kingdom.

**Mark A. Carine**, Department of Botany, Natural History Museum, Cromwell Road, London SW7 5BD, United Kingdom.

**Juli Caujapé-Castells**, Departamento de Biodiversidad Molecular y Banco de ADN, Jardín Botánico Canario 'Viera y Clavijo' (UA CSIC), Apartado de Correos 14 de Tafira Alta, 35017 Las Palmas de Gran Canaria, Spain.

**Mark W. Chase**, Molecular Systematics Section, Jodrell Laboratory, Royal Botanic Gardens, Kew, Richmond, Surrey TW9 3DS, United Kingdom.

**Colin Clubbe**, The Herbarium, Royal Botanic Gardens, Kew, Richmond, Surrey TW9 3AB, United Kingdom.

**Daniel J. Crawford**, Department of Ecology and Evolutionary Biology and the Natural History Museum and Biodiversity Research Center, University of Kansas, 1200 Sunnyside Avenue, Lawrence, KS 66045, USA.

**Manuel B. Crespo**, CIBIO, Instituto de la Biodiversidad, Universidad de Alicante, PO Box 99, E-03080, Alicante, Spain.

**Jorge V. Crisci**, Laboratorio de Sistemática y Biología Evolutiva (LASBE), Museo de La Plata, Paseo del Bosque s/n, CP 1900, La Plata, Argentina.

**John Dransfield**, The Herbarium, Royal Botanic Gardens, Kew, Richmond, Surrey TW9 3AE, United Kingdom.

**Michael F. Fay**, Molecular Systematics Section, Jodrell Laboratory, Royal Botanic Gardens, Kew, Richmond, Surrey TW9 3DS, United Kingdom.

**Javier Francisco-Ortega**, (1) Department of Biological Sciences, Florida International University, Miami, FL, USA, (2) Centre for Tropical Plant Conservation, Fairchild Tropical Botanic Garden, Coral Gables, Miami, FL, USA.

**Vicki A. Funk**, Department of Botany MRC-166, Smithsonian Institution, PO Box 37012, Washington DC 20013–7012, USA.

**Rosalina Gabriel**, Azorean Biodiversity Group (CITA-A), Departamento de Ciências Agrárias, Universidade dos Açores, Largo da Igreja, Terra Chã, 9700–851 Angra do Heroísmo, Azores, Portugal.

**Juana M. González-Mancebo**, Departamento de Biología Vegetal, Universidad de la Laguna, C/Astrofisico Francisco Sánchez, s/n, 38071, La Laguna, Tenerife, Islas Canarias, Spain.

**Ole Hamann**, Professor Emeritus, Botanical Garden of the Natural History Museum of Denmark, Baunegardsvej 22, 2820 Gentofte, Denmark.

**Vernon H. Heywood**, Professor Emeritus, School of Plant Sciences, The University of Reading, Reading RG6 6AS, United Kingdom.

**Brett Jestrow**, Center for Tropical Plant Conservation, Fairchild Tropical Botanic Garden, 11935 Old Cutler Road, Coral Gables (Miami), FL 33156, USA.

**Francisco Jiménez**, Jardín Botánico Nacional, Apartado Postal 21–9, Santo Domingo, Dominican Republic.

**Per Ola Karis**, Botaniska Institutionen, Stockholms Univesitet, SE 106 91, Stockholm, Sweden.

**Liliana Katinas**, Division de Plantas Vasculares, Museo de La Plata, Paseo del Bosque s/n, CP1900, La Plata, Argentina.

**Sterling C. Keeley**, Department of Botany, University of Hawaii, Manoa, 3190 Maile Way, Honolulu, HI 96822–2279, USA.

**Herbert C. Kesler**, Folius Consulting, 467 S. Main Street, Camp Hill, AL 36850, USA.

**Michael Kiehn**, Institute of Botany and Botanical Garden, Faculty of Life Sciences, University of Vienna, Rennweg 14, A-1030 Vienna, Austria.

**Naomi Kingston**, National Parks and Wildlife Service, 7 Ely Place, Dublin 2, Ireland.

**Ben Laenen**, University of Liège, Institute of Botany, B22 Sart Tilman, B-4000, Liège, Belgium.

**Peter J. de Lange**, The University of Auckland, Private Bag 92109, Auckland, New Zealand.

**Ángela Leiva**, Jardín Botánico Nacional, Calabazar, Boyeros, C.P. 19230 Havana, Cuba.

**María Dolores Lledó**, Molecular Systematics Section, Jodrell Laboratory, Royal Botanic Gardens, Kew, Richmond, Surrey TW9 3DS, United Kingdom.

**Michael Maunder**, (1) Center for Tropical Plant Conservation, Fairchild Tropical Botanic Garden, 10901 Old Cutler Road, Coral Gables, FL 33156, USA, (2) Collections, Conservation and Education Division, Al Ain Wildlife Park and Resort, PO Box 1204, Al Ain, United Arab Emirates, (3) Department of Biological Sciences, Florida International University, 11200 SW 8th Street, Miami, FL 33199, USA.

**Anthony G. Miller**, Royal Botanic Garden Edinburgh, 20A Inverleith Row, Edinburgh EH3 5LR, Scotland, United Kingdom.

**Brian G. Murray**, The University of Auckland, Private Bag 92109, Auckland, New Zealand.

**Sara Oldfield**, Botanic Gardens Conservation International, Descanso House, 199 Kew Road, Richmond, Surrey TW9 3BW, United Kingdom.

**Paula Posadas**, Laboratorio de Sistemática y Biología Evolutiva (LASBE), Museo de La Plata, Paseo del Bosque s/n, CP1900, La Plata, Argentina.

**Mijoro Rakotoarinivo**, Madagascar Conservation Centre, RBG Kew, Madagascar Office, Lot II J131 B, Ambodivoanjo Ivandry, Antananarivo 101, Madagascar.

**Jorge Rodríguez**, Parque Nacional Isla de Coco, Apartado 11383–1000, San José, Costa Rica.

**Fred J. Rumsey**, Department of Botany, Natural History Museum, Cromwell Road, London SW7 5BD, United Kingdom.

**Jennifer L. Trusty**, Gulf Coast State College, 5230 West US 98, Panama City, FL 32401, USA.

**Alan Tye**, Secretariat of the Pacific Regional Environment Programme, PO Box 240, Apia, Samoa.

**Eugenio Santiago-Valentín**, Jardín Botánico, Universidad de Puerto Rico, Apartado Postal, 364984, San Juan, PR 00936, Puerto Rico, USA.

**Kay Van Damme**, Department of Biology, Ghent University, K.L. Ledeganckstr. 35, B-9000 Ghent, Belgium.

**Alain Vanderpoorten**, University of Liège, Institute of Botany, B22 Sart Tilman, B-4000, Liège, Belgium.

**Steve Waldren**, Department of Botany, Trinity College, Dublin 2, Ireland.

# Preface

The diversity of island endemic floras always reminds us of the importance of these enclaves in understanding the origins and evolution of life on our planet. Islands represent roughly 5% of the Earth's land surface, but their plant endemics have unique characteristics which have arisen (mainly) through isolation, and make up about one-quarter of all extant terrestrial plant species. Island floras are thus not only distinctive, but also strikingly prolific.

Despite centuries of botanical exploration, constant new findings remind us how much we still have to learn about plant biodiversity on islands. In some archipelagos, palaeo-botanical data convincingly show that the vegetation makeup may have been radically different barely a few thousand years ago, new insular species are frequently discovered, and some others that were feared extinct reappear.

This book brings together a number of relevant researchers who provide updates and insight into the evolution, diversity and conservation of island plants. Many of its chapters also introduce novel hypotheses that challenge the currently accepted mainstream knowledge on the origins and fates of insular endemics. Islands emerge as burgeoning melting pots of speciation that often are significant sources of biodiversity for other areas, as opposed to earlier thoughts that predominantly conceived them as evolutionary sinks where the colonising organisms could either thrive or die off, but never escape.

Much to our chagrin, the earlier conception of islands as living laboratories for the study of evolution is being replaced by their increasing interest as testing grounds of extinction. Indeed, the small population sizes and ranges of island endemic plants, their unique features and their generally low genetic diversity make them much more vulnerable (with exceptions ...) to the consequences of the present environmental changes than their congeners from other enclaves.

Unarguably, the survival of all life on our planet depends on an inextricable web of interactions among organisms, which is nearly always sustained by the basic ecosystem services that plants provide, and where the death of even a species in the minority may bring about a deluge of extinctions. That such basic tenets still have to be painstakingly reiterated to many decision makers indicates the magnitude of the task needed to preserve this unique natural legacy. The crucial role that plants play in most ecosystems, and the inherent fragility of both insular plant endemics and ecological webs, should suffice to motivate the protection of all island biodiversity. We all know that this is not the case, and about one-quarter of all known plant species are at risk in the world's islands; urgent action is therefore needed to offset the processes that are undermining their intrinsic capabilities for survival, and driving them into an extinction whirlpool.

Evolutionary and ecological considerations notwithstanding, most of our medicines, fabrics, food, building materials and fuel come straight from plant biodiversity, and cannot be synthesised in a laboratory. The relentless alterations that we inflict on the global climate are also impacting on these components of biodiversity, thus backfiring onto us and undermining our well-being. To give just one simple instance, extinction of the wild relatives of common crops would entail the loss of genetic resources that could help such plants overcome the present climatic shifts. Certainly, past climate changes had a serious impact on life on Earth, but none of them was as fast and devastating as today's, and none of them contained so many gloomy prospects for the fate of humankind.

A recent biodiversity audit of 53 archipelagos estimates that at least 7000 insular plant endemic species worldwide may be highly threatened, and *c*.3000 of them are in critical danger of extinction. That survey was complemented by a detailed comparative assessment of the impacts of 13 main drivers of biodiversity loss on 10 focal archipelagos, which revealed the dramatic situation of most island floras, identified emerging factors that possibly will accelerate plant biodiversity loss on these enclaves in the short term, and emphasised the urgent need for a global island plant conservation network.

Botanic gardens conduct much relevant research on plant conservation in every country. Consistent with this mission, the organisation that groups them (Botanic Gardens Conservation International, BGCI) promoted the Global Strategy for Plant Conservation that was unanimously adopted by the United Nations Convention on Biological Diversity in April 2002, and further redefined it to respond more effectively to the dramatic impacts of climate change on plant diversity.

On many islands, the botanic garden is simply the only institution that implements effective conservation measures for the endemic flora, based on the knowledge that its staff generates and synthesises. This book on the biology of island floras contains highly significant inputs from scientists working in renowned research institutions worldwide but, not surprisingly, it was conceived in a botanic garden.

The next chapter in the book of efforts to save our planet's plant diversity will also be supported by botanic gardens, among many other relevant institutions. Barely a few days before writing these lines, the nuts and bolts of the proposed global island plant conservation network were discussed under the auspices of BGCI at the 4th Global Botanic Gardens Congress in Dublin, which hosted a symposium on oceanic island plant conservation. If this unprecedented global research effort delivers, botanic gardens will keep playing a crucial role in preserving our common natural legacy.

Now this book is published and you are reading its preface, so you are probably one of the people who could contribute to the global network, which is already operative at BGCI's website (www.bgci.org/ourwork/islands/). Therefore, let me invite you to join in and give suggestions, share your knowledge and expertise, discuss common challenges, and participate in the formulation of multidisciplinary conservation objectives through the network. The radical threats faced by insular endemic floras are not local, but global, and only a coordinated international effort of the same magnitude may provide the critical mass of knowledge needed to help take the right political decisions.

*Juli Caujapé-Castells*

# 1

# Introduction: islands and plants

DAVID BRAMWELL

The complexity of the modern way of life has meant that humankind is not simply one member of a natural ecosystem but is a user, consumer, dominator and potential destroyer of all the Earth's ecosystems. Unfortunately nowhere is this more apparent than in the fragile ecosystems of the world's oceanic islands.

Recent estimates of the number of flowering plant species on the world's islands suggest that of the 50 000 or so insular endemics, some 20 000 are threatened with extinction. The same is true for other island organisms, and Johnson and Stattersfield (1990) estimate that the extinction rate among island birds in historical times is about 40 times greater than in continental species.

On the other hand, the consensus among biologists is that island studies contribute important ideas, theories, models and tests into mainstream biology (Whittaker, 1998) and island ecosystems are now considered environments in which to seek answers to complex questions about the patterns and pathways of evolution. The possible loss of 40% of these unique plant species, many of them still known only in an alpha-taxonomic state (Bramwell, 2006) is, therefore, a source of concern. Even though there has been considerable progress in its study, the challenge of understanding 'island biology' remains but it is, itself, now threatened by the probability of ever rising extinction rates accelerated by new factors such as global change, in particular the disruption of previously stable climates.

## Some recent publications on island biology

Major developments in island biogeography have taken place, however, over the past 30 years or so and are characterised by the application of new concepts and techniques in molecular systematics as well as more profound studies of reproductive

*The Biology of Island Floras*, eds. David Bramwell and Juli Caujapé-Castells. Published by Cambridge University Press. 

biology and of the role of hybridisation in insular speciation. We should, perhaps, describe these developments as 'helping to explain Darwin' by using modern methods to elaborate and expound the mechanisms, pathways and results of evolution by natural selection on islands.

During this period, a significant number of new publications have appeared, each, in its own right, making a significant contribution to island biology. Among the most important are those written or edited by Williamson (1981), Grant (1998), Whittaker (1998) and Stuessy and Ono (1998). Williamson takes what he refers to as a modern look at McArthur and Wilson's Theory of Island Biology and especially its strengths and weaknesses in relation to species numbers on islands. He concludes, after an extensive review of published data, that the hypothesis of immigration and extinction being in balance is generally correct. It is, however, perhaps not as dynamic a process as envisaged by the original authors, as many immigrants do not establish as long-term colonisers and so the equilibrium requires rather more arrivals than extinctions for species numbers to remain stable. Williamson also considers island ecosystems and is emphatic about the need to relate the structure of island communities and their ecological processes to the evolutionary processes, which form the distinct island biota.

Grant, in 1998, compiled and published an indispensable volume under the title *Evolution on Islands*. In this book, he poses the two fundamental questions in evolutionary biology – 'how does evolution occur?' and 'why does evolution occur?' – and he emphasises the contribution in answering these questions of modern studies of what he terms 'evolution in the relative simplicity of island environments'. In this respect Barratt (1998), in the same volume, considers that many of the problems of evolutionary biology can be studied by using island plants as model systems for investigating the genetic basis of adaptation and speciation. He warns, however, that as islands vary considerably in area, altitude, climate and isolation, broad generalisations about insular evolution may not always apply.

The phylogeny of island plants, a controversial subject over the past 30 years, is also considered and Grant himself comments on the need to reconsider some rather neglected factors such as introgression and hybridisation on speciation and, indeed, on the reconstruction of phylogenies. Thorpe and Malhotra (1998) in their chapter also emphasise the need for caution in phylogenetic interpretation, noting that 'adaptation to current ecological conditions appears to be a primary force influencing morphological population differentiation irrespective of phylogenetic history'. In a further chapter, Clarke *et al.* (1998) reinforce Grant's views on hybridisation and introduce the concept of 'molecular leakage' to explain the genetic similarity between species with widely diverse morphology and ecology. They consider that the results of even very low rates of hybridisation or introgression in some insular organisms can be 'misleading in the interpretation of phylogenetic history'.

Whittaker (1998), in his synthesis of recent developments in the subject, presents an excellent general survey of island biogeography up to the end of the pre-molecular age, though with a considerable emphasis on fauna. On the other hand he devotes only a single page to phylogenetic systematics and cladistic biogeography, and the major contribution made to modern biogeographical concepts by Leon Croizat (1958, 1964), whose alternative ideas of space, time and form and his attempts at re-evaluation of Darwinian interpretation do not even warrant a mention. Cladistic biogeography and its more recent offshoot phylogeography are currently, however, important tools in the interpretation of molecular data.

Apart from the above general texts, several more specifically regional books have appeared. These include Stuessy and Ono (1998) *Evolution and Speciation of Island Plants*, which places a strong emphasis on the Pacific Ocean islands. Wagner and Funk (1995) and Ziegler (2002) both treat distinct aspects of biogeography, ecology and evolution in the Hawaiian archipelago. In addition to these, a number of original ideas on island biology have been presented in individual scientific papers and, among these, Silvertown (2004), for example, has offered a very plausible hypothesis of the role of competitive exclusion and 'niche pre-emption' to explain the inability of new colonisers to establish on many islands despite frequent dispersal events from continental sources. Vargas (2007) has also developed an interesting hypothesis for the interpretation of Macaronesian phytogeography (see below) which may also be applicable to many other island situations throughout the world.

## The biology of island plants

During the past 30 years, the renewed interest in island plants in the light of new data from molecular biology has resulted in the formulation of various new concepts. These are reviewed and discussed in the chapters in this book and are briefly summarised below.

### *Concepts*

In Chapter 2, Daniel Crawford *et al.*, building on the earlier contributions of Carlquist (1974) and Ehrendorfer (1979), present a new overview of the diversity of reproductive biology systems in insular lineages and emphasise the need for future research to integrate traditional studies with new approaches such as the use of molecular markers to estimate outcrossing rates. They specifically compare reproductive biology, including a review of mating systems, in different island groups and attempt to explain similarities and differences. The post-colonisation changes in the reproductive biology of insular plants compared with their nearest continental relatives are also considered and the authors emphasise the importance of reproductive biology for conservation and the need to include conservation implications in its study.

In Chapter 3, Paula Posadas *et al.* consider the need to combine molecular phylogenetics with an innovative spatial methodology involving historical biogeographic approaches such as panbiogeography and track compatibility analysis as well as analysis of parsimony and biota similarity to explain patterns of distribution and to formulate testable biogeographical hypotheses.

## *Modern studies of island diversity*

In the field of insular diversity, Sterling Keeley and Vicky Funk (Chapter 4) take a new look at the origin and evolution of the flora of the Hawaiian archipelago. By comparing pre-molecular age studies carried out by Carlquist (1965, 1974) and Wagner and Funk (1995) with molecular data, they re-evaluate the biogeographic affinities of the flora including probable areas of origin, dispersal modes, age of dispersal and probable number of colonisation events for most of the major groups of Hawaiian endemics. The molecular data provide considerable insight into the direction and timing of colonisations, patterns of adaptive radiation in the archipelago and the role of hybridisation in the evolution of Hawaiian endemics.

The Galapagos Islands have been, since Darwin drew the world's attention to them in the nineteenth century, at the forefront of island biology. In Chapter 5, Alan Tye and Javier Francisco-Ortega review and revise current ideas on the origin and evolution of their unique flora in the light of both molecular studies and recent research using traditional methods. They consider how new information affects conclusions on source areas and colonisation dates and indicates the former existence of now submerged islands. Evolutionary processes within the archipelago are also examined and the role of human activity in altering the flora addressed. In conclusion, the authors make a number of recommendations for the direction of future research.

Michael Maunder *et al.*, in Chapter 6, emphasise the importance of the Caribbean islands as biodiversity centres or 'hotspots' and review the taxonomic and ecological diversity of the region. They also consider the phylogenetics and phytogeography of a number of endemic genera and present a new review of Caribbean conservation issues and challenges.

Madagascar has been one of the islands of special attention during the past 30 years. As John Dransfield and Mijoro Rakotoarinivo demonstrate in Chapter 7, the palm flora, for example, is an astonishing one with 174 described species in 14 genera and with many more still to be documented. The authors comment on the remarkable diversity of relationships of the Madagascar palms with their nearest relatives in regions as close as South West Asia and Africa or as far away as Australia, Indo-China, New Guinea and even South America. They also discuss the diversity of palm ecology on the island and its implications for conservation.

Lisa Banfield *et al.* in Chapter 8 consider the phytogeography of the fascinating island of Socotra in the light of both the classical view of the origins of the flora and more recent molecular studies. Earlier studies suggest that the island acted as a refuge for relictual vicariants, but recent molecular work indicates that later long-distance dispersal played a role in colonisation and establishment of Socotra's flora. The authors consider both aspects in a detailed critique and defend the relictual hypothesis for various taxa. In their conclusions, they include comments on the likely effects of climate change on the conservation of the flora.

One of the least known insular floras is that of New Caledonia with about 75% endemism and a large number of ancient, relict lineages including gymnosperms and basal angiosperm groups. On the other hand, geological evidence of submergence of the island in the Eocene is difficult to reconcile with a relict Gondwana hypothesis for the origin of the flora. In Chapter 9 David Bramwell suggests that there is, indeed, a Gondwana element in the flora but at the same time there are some derived groups of more recent arrival and that both have undergone radiation and diversification on the island. He also comments on the current state of the island's unique vegetation types and human influence on them, suggesting that many plant species on New Caledonia are under threat and formal assessment of their conservation status should be completed.

In Chapter 10, a study of one of the world's most inaccessible island groups, the Pitcairn Islands (Pitcairn, Henderson, Oeno and Ducie), Steve Waldren and Naomi Kingston analyse the biogeographical affinities of the flora and evaluate the endemic taxa and their origins including molecular studies of the largest genus, *Peperomia*. They conclude that insular endemics have originated through geographical isolation rather than by radiation.

Brian Murray and Peter de Lange, in Chapter 11, also consider the biogeography of a Southern Hemisphere insular flora, that of New Zealand. They do it, however, from a more classical approach based on chromosome analysis, especially of the ancient gymnosperms of the islands. Molecular evidence suggests a rather recent origin for many of the endemic angiosperms and their evolution is evaluated using data on polyploidy, aneuploidy and interspecific hybridisation.

## *The Macaronesian model*

The Macaronesian region has been a major focal point for the molecular study of island floras and four chapters here summarise the 'Macaronesian model' for molecular systematics. Chapter 12 by Juli Caujapé-Castells provides a broad overview of the wealth of data generated over the past two decades and their impact on the present understanding of the region's flora. At the same time, he highlights some deficiencies and gaps in the data, which indicate as many new problems as questions answered, and suggests fields for future research. He emphasises the need for testable biogeographic hypotheses. In this context, of recent published works on the region, the paper by Vargas (2007) is an important one from a molecular phylogenetic point of view. In an excellent summary of the evolutionary history of Macaronesian plants, Vargas revives the ideas presented in their day by authors such as Sunding (1979) and Bramwell (1972, 1976, 1985) on the relict nature of the insular floras. He introduces the concept of lineage relictualism (ancient lineages) in terms of absolute time to describe those groups originally colonising Macaronesia before the start of the Quaternary period. His views are supported by Francisco Ortega *et al.* (1997) who suggest, based on molecular evidence, that the exclusively Macaronesian genus *Argyranthemum* of the subtribe Chrysantheminae 'is old relative to continental genera of the subtribe' and is sister to other modern Mediterranean members of the tribe. Estimating divergence times using a molecular clock, these authors indicate that *Argyranthemum* diverged from the rest of the Chrysantheminae, possibly from a now-extinct common ancestor, between 2.5 and 3 million years ago (Ma), i.e. in the Tertiary period. They conclude that the results of three independent sets of molecular data are consistent with a previous hypothesis that some members of the Macaronesian flora may be ancient in relation to their continental relatives. Vargas (2007), reviewing 34 molecular phylogenies and molecular clocks of Macaronesian plant groups, suggests that the majority of these may be considered as lineage relicts with their origin in the Mediterranean region during the Tertiary period. On the other hand, some scepticism remains about the value of molecular clocks in general and, in particular, in situations of rapid radiation. Thomas *et al.* (2006), on the basis of observations that substitution rates vary widely between lineages, question the existence of a universal molecular clock, suggesting that the results of extrapolation from one situation to others should be treated with caution.

Dolores Lledo *et al.* in Chapter 13 consider the biogeographical relationships between Macaronesian endemics and their Mediterranean relatives. The relationships between continental and insular groups, however, still remain partially unresolved. In the case of *Limonium* several colonisation events are postulated and *L. dendroides* is considered to be an isolated relict taxon. The principal group in the Canary Islands, subgenus *Pterocládos* appears to be monophyletic, the result of a single colonisation which probably took place about 7 million years ago and may be considered to be an ancient lineage (Vargas, 2007).

Alain Vanderpoorten *et al.* in Chapter 14 present a review of recent research on the under-studied cryptogam flora of the Macaronesian region. They look at patterns of endemicity and phylogenetic relationships and suggest that the differences in patterns of diversity and relationships between the angiosperm and cryptogam floras may be indicative of very different evolutionary processes. Finally, possible explanations for observed differences are considered and compared, shedding new light on the evolution of insular diversity.

## *Ecology and conservation of island floras*

One of the principal conservation problems facing island plants is competition from invasive alien species. Generally, due to the influence of humans, almost all islands have been colonised by introduced species, often displacing their natural vegetation and competing for niches with individual endemic species. Michael Kiehn, in Chapter 15 on invasive, non-native species, outlines a number of examples and reviews the possible strategies and measures to be taken for native flora conservation and the eradication of aliens.

Ole Hamann, in Chapter 16 on conservation needs in the Galapagos Islands, presents a review of the ecology and demography of the native flora particularly in relation to volcanic eruptions and extreme climate events such as El Niño. He describes a singular example of the negative influence of humankind on one of the scientifically most important archipelagos and outlines the measures needed to restore and protect the flora.

Madagascar is the fourth largest island in the world and has between 10 000 and 12 000 species of plants, of which some 90% are endemic. Unfortunately human settlement of the island almost 2000 years ago has caused enormous destruction and over 80% of the original forest has been cleared. Recent developments outlined by Stuart Cable in Chapter 17 on conservation, however, permit a degree of optimism for the future of the Madagascan flora with the creation of new protected areas through 'debt for nature' exchanges and the involvement of local communities in conservation.

David Bramwell in Chapter 18 gives an account of some of the future problems for island biodiversity. He considers the potential effects of climate change on insular floras, especially in relation to the particular conditions found in island communities. Almost 35 years ago, Sherwin Carlquist, in his book *Island Biology* (Carlquist, 1974) defined a series of characteristics common to many organisms from islands throughout the world as the 'island syndrome'. This syndrome includes such phenomena as woodiness and longevity in generally herbaceous groups of plants and gigantism in animals, reduction of dispersal capacity through enlarged seeds, diminished efficiency of dispersal mechanisms such as a depauperate pappus in plants and the loss of flight capacity in birds and insects. The characteristics of the 'island syndrome' do not seem to aid either adaptation or migration. Insular woodiness and the tendency towards longevity, implying longer generation times, is not conducive to rapid adaptation to changing climatic and ecological conditions. In general terms, therefore, it can be said that the island syndrome creates a critical balance between insular endemics and their environment that can easily be disrupted by climate change and does not help the prospects for their *in situ* conservation in the face of the rapidly changing conditions brought on by climate change.

Jennifer Trusty *et al.* (Chapter 19) consider all the native species including 36 endemics to be threatened with extinction due to the degradation of habitat on the Costa Rican island of Isla del Coco. Applying the minimum area requirements of the IUCN Red List criteria they found, however, that all the species should be included in the CR (critically endangered) category even though they have widely different population sizes and habitat availability. They sensibly suggest that population size should be given priority over area occupied on small oceanic islands. In their contribution, the authors also examine all the threats to Isla de Coco species and make recommendations for their conservation.

Sara Oldfield in Chapter 20 revises the situation of various insular floras and provides a number of examples of their vulnerability to pressures from tourism and development, exotic predators and competitor species. She indicates the need for future conservation planning to take into account a new factor, global climate change. As a possible response to the requirements, she emphasises the need for additional measures to support *in situ* conservation and outlines the future role for botanic gardens and seed banks in securely storing wild plant material on a long-term basis and making it available for conservation projects.

In Chapter 21, a final summing up of the contributions to the book, Vernon Heywood notes that more than just climate change is affecting insular floras and considers this to be one more, very important, factor in the series of problems they face. He emphasises the enormous quantity and quality of biological diversity still surviving on the world's islands but considers the future of island biodiversity to be very hazardous indeed.

## Epilogue on island floras

Ever-expanding agricultural areas, urbanisation and other types of human land use surround protected natural areas. They, therefore, are becoming ever more isolated from each other, creating islands of biodiversity in an 'ocean' of habitat destruction and restriction of gene flow. Eventually such isolation will mean that individual protected areas will be subject to the principles of island biogeography and insular evolutionary processes. Therefore, our knowledge of island biology will become increasingly important in the management of conserved areas, especially in the face of the increasing ecological pressures resulting from climate change, which may well alter both the rates of evolution and extinction in protected areas not only on islands but in general.

## References


Barrett, S. C. H. (1998). The reproductive biology and genetics of island plants. In P. Grant (ed.), *Evolution on Islands*, pp. 18–34. Oxford, Oxford University Press.

Bramwell, D. (1972). Endemism in the flora of the Canary Islands. In D. H. Valentine (ed.), *Taxonomy, Phytogeography and Evolution*, pp. 141–159. London & New York, Academic Press.

Bramwell, D. (1976). The endemic flora of the Canary Islands. In G. Kunkel (ed.), *Biogeography and Ecology in the Canary Islands*, pp. 207–240. The Hague, Junk.

Bramwell, D. (1985). Contribución a la biogeografía de las Islas Canarias. *Bot. Mac.* 14: 3–34.

Bramwell, D. (2006). 'The business of a poet': taxonomy and the conservation of island floras. In E. Leadlay & S. Jury (eds.), *Taxonomy and Plant Conservation*, pp. 205–211. Cambridge, Cambridge University Press.

Carlquist, S. (1965). *Island Life: A Natural History of the Islands of the World.* New York, Natural History Press.

Carlquist, S. (1974). *Island Biology.* New York and London, Columbia University Press.

Clarke, B., Johnson, M. S. & Murray, J. (1998). How 'molecular leakage' can mislead us about island speciation. In P. R. Grant (ed.), *Evolution on Islands*, pp. 181–195. Oxford, Oxford University Press.

Croizat, L. (1958). *Panbiogeography or An Introductory Synthesis of Zoogeography, Phytogeography, Geology; with notes on evolution, systematics, ecology, anthropology, etc.* Published by the author, Caracas.

Croizat, L. (1964). *Space, Time, Form: The Biological Synthesis.* Published by the author, Caracas.

Ehrendorfer, F. (1979). Reproductive biology in island plants. In D. Bramwell (ed.), *Plants and Islands*, pp. 293–306. London & New York, Academic Press.

Francisco-Ortega J., Crawford, D. J., Santos-Guerra A. & Jansen R. K. (1997). Origin and evolution of *Argyranthemum* (Asteraceae: Anthemideae) in Macaronesia. In T. J. Givnish & K. J. Sytsma (eds.), *Molecular Evolution and Adaptive Radiation*, pp. 407–431. Cambridge, Cambridge University Press.

Grant, P. R. (ed.) (1998). *Evolution on Islands: Originating from Contributions to a Discussion Meeting of the Royal Society of London.* Oxford, Oxford University Press.

Johnson, T. H. & Stattersfield, A. J. (1990). A global review of island endemic birds. *Ibis* 132: 167–180.

Silvertown, J. (2004). The ghost of competition past in the phylogeny of island endemic plants. *J. Ecol.* 92: 168–173.

Stuessy, T. F. & Ono, M. (eds.) (1998). *Evolution and Speciation of Island Plants.* Cambridge, Cambridge University Press.

Sunding, P. (1979). Origins of the Macaronesian flora. In D. Bramwell (ed.), *Plants and Islands*, pp. 13–40. London & New York, Academic Press.

Thomas, J. A., Welch, J. J., Wolfit, M. & Bromham, L. (2006). There is no universal molecular clock for invertebrates, but rate variation does not scale with body size. *Proc. Natl. Acad. Sci. USA* 103 (19): 7366–7371.

Thorpe, R. S. & Malhotra, A. (1998). Molecular and morphological evolution within small islands. In P. R. Grant (ed.), *Evolution on Islands*, pp. 67–82. Oxford, Oxford University Press.

Vargas, P. (2007). Are Macaronesian islands refugia of relict plant lineages?: A molecular survey. In S. Weiss, & N. Ferrand, (eds.), *Phylogeography of Southern European Refugia: Evolutionary Perspectives on the Origins and Conservation of European Biodiversity*, pp. 297–341. Dordrecht, Netherlands, Springer.

Wagner, W. L. & Funk, V. L. (eds.) (1995). *Hawaiian Biogeography: Evolution on a Hot Spot Archipelago.* Washington, DC, Smithsonian.

Whittaker, R. J. (1998). *Island Biogeography: Ecology, Evolution, and Conservation.* Oxford, Oxford University Press.

Williamson, M. H. (1981). *Island Populations.* Oxford, Oxford University Press.

Ziegler, A. C. (2002). *Hawaiian Natural History, Ecology and Evolution.* Honolulu, University of Hawaii Press.

# 2

# The reproductive biology of island plants

DANIEL J. CRAWFORD, GREGORY J. ANDERSON AND GABRIEL BERNARDELLO

Since the review of the reproductive biology of island plants by Ehrendorfer (1979) three decades ago, there has been a veritable explosion of studies on plants of oceanic islands. Although the primary emphasis, particularly in the past two decades, has been on molecular phylogenetics of island lineages (Baldwin *et al.*, 1998; Emerson, 2002), there also has been progress in understanding their reproductive biology. However, the comment by Ehrendorfer (1979: 293) about reproductive biology of island plants still applies today: 'This is an extensive, fascinating, and complex field of inquiry, still insufficiently covered by relevant work and factual evidence.' And, of course, there is the exhortation by Herbert Baker from his classic paper in *Evolution* from 1967: 'Clearly, the greatest need in this area of biology is for further field and experimental studies (by as many workers as can spare their time) on the reproductive biology of the plants which make up floras of all kinds.'

The present review uses Ehrendorfer's (1979) chapter as a general framework for highlighting progress since that contribution. Also, Barrett (1996) presented a concise review of the reproductive biology and genetics of insular plants, and incorporated several topics (e.g. phylogenetic analyses) not featured in Ehrendorfer (1979). We will use two general approaches in discussing the evolution of the reproductive and pollination biology of island lineages as compared with their presumptive colonising ancestors. One will be to look at specific islands, and make comparisons between and among them to derive patterns. Another approach will be to discuss specific examples illustrating how plants have adapted, however perfectly or imperfectly, to the situation in the island setting. We will also consider the influence of invasive plants and pollinators on the reproductive biology and conservation of island lineages. Lastly, important areas for future studies will be considered.

*The Biology of Island Floras*, eds. David Bramwell and Juli Caujapé-Castells. Published by Cambridge University Press. 

## Dispersal of colonising ancestors

The attributes of a successful coloniser fall into two general categories, one concerned with dispersal to and establishment on an island, and the other with the radiation and diversification of their progeny in the insular setting. Carlquist (1981) pointed out that migration of colonisers is chiefly a matter of chance, but once they arrive selection operates on the propagules and is the determinate of whether there is successful establishment. Carlquist (1974: 12–13) suggested that chance dispersal was a lesser hurdle than establishment for colonisers. It is, of course, difficult to gather direct evidence on this issue. However, results from molecular phylogenetic studies indicate that, while in most instances the nearest continental source areas are the origins of colonising ancestors (e.g. compilation in Carine *et al.*, 2004 for Canary Islands), there are some exceptions.

Due to space constraints, extensive considerations of dispersal mechanisms and loss of dispersability in island lineages will not be considered in detail. The topics have been the subject of earlier classical reviews (Carlquist, 1966a, 1966b, 1974). Carlquist (1980) also provided an overview for Hawaii and he considered the possibilities for dispersal of Compositae in the Pacific islands (Carlquist, 1966d). These reviews, as well as the more recent one for the Robinson Crusoe (or Juan Fernández) Islands, Chile, by Bernardello *et al.* (2006), provide broad overviews of the probable geographic sources of the island ancestors, the dispersal units of the colonisers and their most likely dispersers to the islands. As with the generally held proposition for plants from other archipelagos such as Hawaii (Carlquist, 1974, 1980), Galapagos (Porter, 1976, 1983), and Bonin (Ogasawara) Islands (Ono, 1991), Bernardello *et al.* (2006) indicate that birds account for the vast majority of dispersal to the Robinson Crusoe Islands. Using a phylogenetic approach, Lowrey *et al.* (2005) investigated the fascinating distribution of the genus *Tetramolopium* in Hawaii, one of the Cook Islands and New Guinea. The data indicate an origin in New Guinea, and while not completely conclusive, the evidence favours dispersal first to Hawaii and then to the Cook Islands, rather than the other way around. Lowrey *et al.* (2005) suggested the bird, the Pacific golden plover, as the dispersal agent for the long trip (over 3000 km) from Hawaii to the Cook Islands. Likewise, Ballard and Sytsma (2000) suggested that one or more of the 50 species of birds that winter in Hawaii and breed in the Arctic likely dispersed the ancestor of the woody Hawaiian violets to the archipelago.

While it appears that birds are common dispersal agents of propagules to islands, Carlquist (1974) discussed the role of oceanic drift in dispersal to islands. Green (1999) determined the identity of disseminules (seeds and fruits) washed up on a beach on Christmas Island in the Indian Ocean. About half of the more than 60 species represented by the stranded disseminules do not grow on the island. Green (1999) discusses the possible distant sources of the seeds and fruits (he suggested that those fruits and seeds of species growing on the island did not originate from plants on the island), and cites prevailing ocean currents and surface winds that could transport disseminules from the source areas to Christmas Island. These and other data from phylogenetic and comparative studies of island plants over the past three decades are perhaps sufficient to provide general support for Carlquist's (1974) hypothesis that dispersal is likely not the limiting process in establishment on oceanic islands.

Phylogenetic and comparative studies indicate that most colonising ancestors come from nearby source areas; however, recent studies suggest a number of instances of exceptionally long-range dispersal. Two of the more notable examples indicating distant source areas include an African origin for the Hawaiian endemic genus *Hesperomannia* (Asteraceae) (Kim *et al.*, 1998) and an Arctic origin for the ancestor of the woody violets in Hawaii (Ballard & Sytsma, 2000). In addition to Lowrey *et al.* (2005), there are other phylogenetic studies showing examples of islands not only being recipients of propagules, but sources of them as well (Harbaugh & Baldwin, 2007; Harbaugh, 2008; Harbaugh *et al.*, 2009). While loss of dispersability occurs in some island groups (Carlquist, 1974; Bernardello *et al.*, 2006), there are situations where it is clear that insular plants are capable of dispersal to other islands (Harbaugh & Baldwin, 2007; Harbaugh, 2008; Harbaugh *et al.*, 2009).

## Breeding systems of colonisers

One of the most important traits of colonising ancestors of island plants is breeding system, in particular whether the invading colonisers are self-compatible (SC) or self-incompatible (SI). It has been more than half a century since Baker (1955) proposed that SC colonisers would have an advantage following long-distance dispersal because one propagule could establish a sexually reproducing population. By contrast, two or more propagules would be required to establish a sexually reproducing population with SI plants. Given that dispersal to a distant small island is not common (even if not the limiting factor in establishment, as above), multiple dispersal and establishment events would be necessary to bring in congeners with compatible breeding groups; this is in all probability especially unlikely. Stebbins (1957) was so convinced of the generality of Baker's (1955) hypothesis that he called it Baker's Law. Carlquist (1974), however, made the point that SC colonists would not provide sufficient genetic variation for the radiation and diversification of their progeny. Both Ehrendorfer (1979) and Barrett (1996) noted that available evidence supports the Baker (1955) hypothesis because the majority of endemic flowering plants that have been studied on islands are SC. Compilations subsequent to Ehrendorfer (1979) for the Galapagos (McMullen, 1987, 1990) and New

Zealand (Webb & Kelly, 1993) and two papers published after Barrett (1996) on the Robinson Crusoe Islands (Anderson *et al.*, 2001a; Bernardello *et al.*, 2001) support their generalisations on the prevalence of SC plants on islands. While the compatibility of the current plants on an island does not confirm the state of the ancestral coloniser, there is little reason to think that initial, successful colonisers in an extant SC group were not also SC because of the aforementioned advantages that SC, and particularly self-pollinating plants, would have in establishing sexual populations. For example, Schueller (2004) demonstrated higher self-pollination in island versus mainland populations of *Nicotiana glauca*. The island populations (which are introduced) were visited as frequently as mainland populations by their hummingbird pollinators, indicating lack of selection for self-pollination in the island setting and suggesting instead that the higher selfing in the islands is due to its advantages in establishment of populations.

Although the preponderance of data indicates that most colonisers are SC, there are a number of notable exceptions to Baker's Law, particularly within Compositae, the family with the most species endemic to oceanic islands. The silversword alliance in Hawaii, one of the most spectacular radiations of island plants, has been shown to be ancestrally SI, perhaps with some leakiness, that is, some level of pseudo-self-compatibility (PSC) (Carr *et al.*, 1986; Barrier *et al.*, 1999; Baldwin, 2003). Other examples of SI or PSC colonisers include *Argyranthemum* (Francisco-Ortega *et al.*, 1997), the woody *Sonchus* alliance (S.-C. Kim, pers. com.), and *Tolpis* (Crawford *et al.*, 2008) from the Canary Islands, *Commidendrum* from St Helena (Eastwood *et al.*, 2004; A. Eastwood, pers. com.), *Dendroseris* from the Robinson Crusoe Islands (Anderson *et al.*, 2001a; Bernardello *et al.*, 2001) and *Scalesia*, which is the largest endemic genus in the Galapagos Islands (Nielsen *et al.*, 2000, 2003). As emphasised by Levin (1996) and Crawford *et al.* (2008), among others, a PSC coloniser would have the advantage of establishing a sexually reproducing population from a single propagule, and because the propagule originated from a largely outcrossing population, it would carry more genetic diversity than a seed from a highly selfing population. More information derived from careful studies of the reproductive biology and mating systems of island plants and their close relatives is needed to determine the frequency of PSC colonising ancestors of island lineages, and to document fully the level of SC versus SI in island species and their relatives.

## The general syndrome for island flowers and their pollinators

Earlier surveys of island plants, particularly in Hawaii, indicated a generalised floral morphology with small, shallow, open, dull coloured 'inconspicuous' flowers (Carlquist, 1966d, 1974: 540). Subsequent studies have supported the

(a)

(b)

(c)

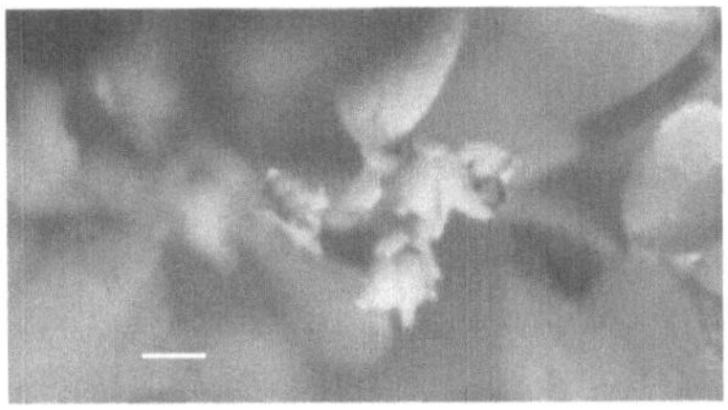

Figure 2.1 (a) *Lactoris fernandeziana* plant in field on Masatierra, Juan Fernández Islands, Chile. Photo by G. J. Anderson. (b) Branch with the small, greenish flowers of *Lactoris fernandeziana*. The three carpels are the conspicuous part of the flowers. (c) Sepals visible at base of flowers. Scale lines: 4.0 mm in both photos. Photos by D. J. Crawford.

earlier observations for flowers of the Galapagos (McMullen, 1989, 1999: 26), the Ogasawara (Bonin) Islands (Abe, 2006), and the Robinson Crusoe Islands (Bernardello *et al.*, 2001). Bernardello *et al.* (2001) showed a very strong correlation among the features for small, inconspicuous, and green flowers (Fig. 2.1a,b,c). Another especially interesting and unusual outcome was the recognition that the floral morphology of the probable colonising ancestors and current species of the Robinson Crusoe Islands, particularly in terms of size and colour, are quite similar. This suggests that, in some cases, perhaps dependent on the age of the archipelago, the generalised floral type on islands can be related to dispersal or establishment rather than to *in situ* selection (Bernardello *et al.*, 2001), and/or that the particular system under study is relatively young in evolutionary time.

Consider next some possible reasons for the prevalence of small, open, colourless, hermaphroditic flowers in island floras. One of the major challenges faced by the colonising ancestors of island lineages is the paucity of biotic pollinators, especially insects (Ehrendorfer, 1979; McMullen, 1987, 1990; Sakai *et al.*, 1995a; Barrett, 1996; Anderson *et al.*, 2001a; Bernardello *et al.*, 2001). The lack of pollinators on islands can also pose major limitations on successful establishment of colonists, especially those that have specialised pollination systems. Broad comparative studies, such as Anderson *et al.* (2001a), Bernardello *et al.* (2001) and Price and Wagner (2004) provide insights into possible pollinators of island colonisers as well as current pollinators of island plants. In addition to autogamy (pollination within the same flower without a pollen vector), it has been generally accepted that plants have coped with the paucity-of-pollinators problem by using either wind or generalist animal pollinators; a brief review of more recent studies will be compared to 'conventional wisdom' on the topic.

Barrett (1996) emphasised that, while there is the general perception that wind pollination is prevalent on oceanic islands, it is not clear that it is more frequent than on continental source areas. The possible advantages of wind pollination on islands include the assurance of pollination without reliance on the often-scarce biotic pollinators, atmospheric conditions on islands favouring wind over animal pollinators, and wind dispersal of pollen being more effective than animals for outcrossing (Barrett, 1996). Bernardello *et al.* (2001) contrasted the most likely pollinators of the ancestral colonisers with the current pollinators of plants of the Robinson Crusoe Islands. Pollinators are unknown for about 45% of the total flora. Bernardello *et al.* (2001) estimated that about half of the colonisers were insect pollinated and 40% wind pollinated; bird-pollinated colonisers and those with some sort of mixed pollination system were very rare. In contrast, studies on the islands suggest that the current island species are: 47% wind pollinated, 9% bird pollinated, with possibly one insect-pollinated species. One element of uncertainty in inferring the pollinators of some Robinson Crusoe species is that their flowers show certain features associated with wind pollination and other attributes suggesting biotic pollination (Bernardello *et al.*, 2001). Successful colonisers may move into new niches, and in these new environments with fewer competitors, elements of ancestral morphology may persist because of lack of selection against them. As a consequence, the recognition of wind pollination may be obscured by retention of zoophilous features, and detailed studies of pollination biology of island floras may reveal anemophily for species where morphology would not predict it.

An unusual and extreme example illustrating the impact of wind transferring pollen was identified in the species *Wahlenbergia berteroi* (Campanulaceae) from Robinson Crusoe Island (Anderson *et al.*, 2000). In other *Wahlenbergia* in the archipelago (and indeed the family as a whole; Yeo, 1993), autogamy occurs at the end of the life of a flower when the stigmatic branches recurve nearly 360° and contact pollen on the style. In *W. berteroi* the degree of stigmatic curvature is insufficient to reach the style and facilitate autogamy, yet most flowers have pollen on

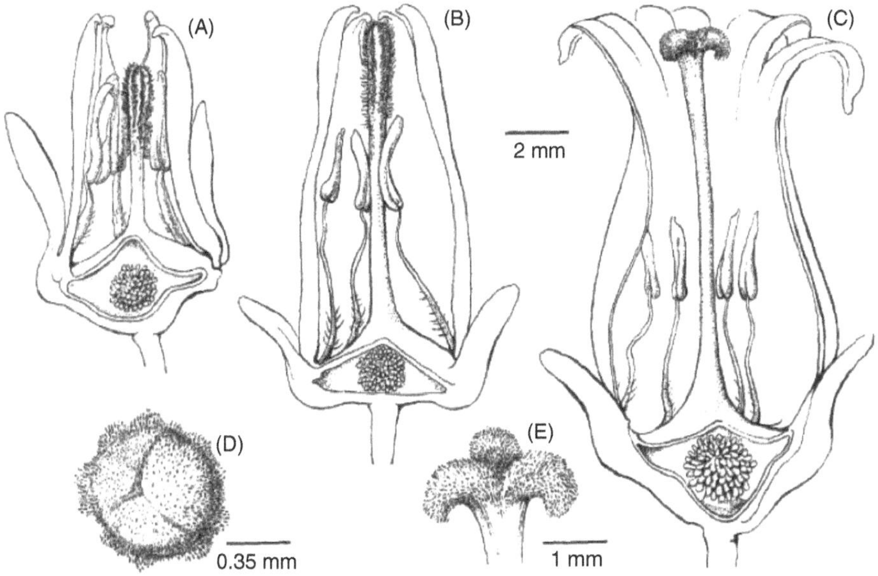


Figure 2.2 Drawings of flowers of *Whalenbergia berteroi* showing different stages of maturation. A–C, longitudinal sections of flowers; D, E, stigmatic lobes. The flowers are protandrous, and after the anthers dehisce, pollen is collected on the dense hairs on the upper part of the styles, and on the outer surface of the stigmatic lobes, which have not yet been exposed (A, D). The style elongates and carries the pollen upward near the throat of the corolla tube (B). The stigmatic lobes expand (E), the hairs on the styles retract and the anthers shrivel (C). Some pollen is deposited on the throat of the corolla at about the level of the expanded stigmatic surfaces, and self-pollination (autogamy) presumably can occur when flowers are shaken by the persistent wind, bringing the stigmatic lobes and pollen on the inner sides of the corolla in contact. See text for additional explanation. From Anderson *et al.* (2000), *Pl. Syst. Evol.* 223: 109–123. Used with permission from Springer-Verlag, Wien.

the stigmas. As anthers dehisce in bud stage (protandry), the corolla throat closely surrounds the anther cylinder and becomes coated with pollen and the stigma lobes have not expanded (Fig. 2.2 A, D). The pollen is presented by a stylar brush as the style elongates and pushes the pollen up near the open mouth of the now-open corolla, and against the inner faces of the corolla tube near its apex (Fig. 2.2 B, C). The stigma is then exposed by the reflexing of the lobes (Fig. 2.2 E). Because pollen stays viable throughout the receptive stage of the stigma, transfer from corolla to stigmata occurs when flowers are shaken by the ever present wind, thus promoting contact between the stigmata and corolla. This, and other studies on Robinson Crusoe plants, suggests that the frequency of wind pollination in the islands is slightly higher than the continental source areas but also indicates that it could, indeed, be much higher. Even by 'island standards', insect pollinators are very scarce on the Robinson Crusoe Islands (i.e. there were no bee species reported until a very recent documentation by Engel, 2000), and there is essentially no evidence for generalist insects as effective pollinators (Anderson *et al.* 2001a; Bernardello *et al.*, 2001). If estimates of the frequency of insect pollinators of island ancestors are close to being accurate, then there has been a clear shift from insect to other means of pollination for plants native to the Robinson Crusoe Islands. While autogamy is quite rare in the species studied (Anderson *et al.*, 2001a; Bernardello *et al.*, 2001), over 40% of the Robinson Crusoe flora has not been examined, and the importance of autogamy remains an open question.

Available data for the Galapagos Islands, including morphology of flowers and airborne pollen, suggest that wind pollination, both by the colonising ancestors and the present flora of the archipelago, is rarer than in the Robinson Crusoe Islands (McMullen, 1987, 1990, 1993). However, Bernardello *et al.* (2001) caution that only 10% to 15% of the native flora of the Galapagos has been examined in detail, and suggest that flowers of some Galapagos species that otherwise do not appear to be anemophilous could in fact be wind pollinated. The endemic Galapagos carpenter bee, the only native bee (Nielsen *et al.*, 2002), is the primary insect pollinator (it has been at least observed at flowers of 79 taxa including 25 endemics) in the Galapagos (McMullen, 1987, 1990, 1993; McMullen & Close, 1993). A variety of other insects have been seen on Galapagos plants (McMullen, 1993; Philipp *et al.*, 2004), but their importance as effective transmitters of pollen versus simply random visitors is questionable. Other studies indicate that a variety of insects visit flowers of the same species, and visitors may vary by island, or the visitors may be limited and still vary by island (e.g. McMullen & Viderman, 1994). Available evidence for the Galapagos indicates that autogamy is quite common (Rick, 1966, McMullen, 2007, 2009, but additional study is needed), while generalist insect pollinators and wind pollination are relatively uncommon.

Sakai *et al.* (1995b) inferred the pollinators of the colonists of the native flora of Hawaii, and suggested that the majority were insect pollinated, with about half as many wind pollinated; much less frequent were bird pollination and self-pollination. Price and Wagner (2004) listed insects as the most frequent mode of pollination in native angiosperms, with abiotic and bird pollination equally frequent but much rarer than insects. As with any archipelago, the more ancient the initial colonisation of an archipelago the more difficult it becomes to infer with certainty the original pollinators of the ancestors of current lineages.

A wide variety of generalist insect pollinators has been observed in the relatively insect-rich Canary Islands (Forfang & Olesen, 1998; Garcia, 2000; Olesen *et al.*, 2002a; Dupont & Skov, 2004; Dupont *et al.*, 2003). Other than possible wind pollination of the Canary Island date palm (Meekijjaroenroj & Anstett, 2003), we are unaware of reports of wind pollination in the Canaries. However, despite the

Figure 2.3 Hermaphroditic flower of *Solanum vespertilio* subsp. *vespertilio* from plant on Tenerife, Canary Islands, Spain. Scale line = 4.0 cm. Photo by G. J. Anderson.

lack of known reports, members of families such as Chenopodiaceae, Cyperaceae, Myricaceae, Poaceae and Urticaceae (Bramwell & Bramwell, 2001) almost certainly are wind pollinated, and there are no doubt many other wind-pollinated plants in the archipelago. Bird pollination is apparently limited to fewer than 10 species in the Canaries (Olesen, 1985; Dupont *et al.*, 2004; Valido *et al.*, 2004), with pollination by non-specialist opportunistic nectar feeders common (Rodríguez-Rodríguez & Valido, 2008; Ollerton *et al.*, 2009). In addition to generalist pollinators, there are also instances of dedicated bee pollination, one example being the endemic *Solanum vespertilio* (Anderson *et al.*, 2006a; Prohens *et al.*, 2007; Fig. 2.3).

The Ogasawara (Bonin) Islands consist of a group of small (largest less than 25 km$^2$) volcanic islands situated some 1000 km south of the southern tip of Japan (Abe, 2006). The oldest islands have been above sea level for several million years. The most abundant native pollinators are insects (over 70% of the species), followed by 26% of the species (mostly grasses and sedges) pollinated by wind (Abe, 2006).

A plausible explanation for this combination of features – small, dull-coloured, open, inconspicuous, hermaphroditic flowers on islands – is that they can be pollinated by the few small, generalist insects (e.g. solitary bees, flies) present, and known to visit flowers. As indicated by Carlquist (1966c), the lack of long-tongued bees and larger Hymenoptera on islands ‘disfavors many angiosperms with larger and more colorful flowers’ other than bird-pollinated species. Bernardello *et al.* (2001) suggest that the open shallow flowers may be suitable for wind pollination. In addition, autogamy could function as a pollination mechanism. While the hypothesis of 'typical' floral morphology of island plants being related to pollination seems reasonable and available data support it, more careful field studies are needed to provide more rigorous tests of the hypothesis. The question of whether wind pollination is more frequent on islands remains open. For example, perhaps for 'young' islands with depauperate pollinator faunas like the Robinson Crusoe Islands wind pollination may be frequent but, even here, a substantial 40% of the flora is unknown in terms of field pollination studies (Bernardello *et al.*, 2001).

## Sex of flowers and separation of sexes

As indicated above, there is considerable monotony among many island flowers but there has also been floral diversification in some lineages. The majority of species native or endemic to islands have hermaphroditic flowers: estimated values include over 60% for Hawaii (Wagner *et al.*, 1990), over 70% for the Robinson Crusoe Islands (Bernardello *et al.*, 2001), 75% for the Ogasawara Islands (Abe, 2006) and over 80% for the Canary Islands (Francisco-Ortega, unpubl., cited in Helfgott *et al.*, 2000) and for the Galapagos (McMullen, 1990, 1999). Among species with hermaphroditic flowers available evidence, meagre as it is, indicates that the majority are SC (e.g. Carlquist, 1974: 516–519; McMullen, 1990; Calero & Santos, 1993; Bernardello *et al.*, 2001). An important and interesting question for which there is little information is how these two floral features (hermaphroditic and SC) translate into the actual mating systems in island plants. On the one hand, autogamy would result in largely selfing populations, as appears to be the case in the Galapagos (McMullen, 1990), but on the other hand there are floral mechanisms that could promote outcrossing in SC hermaphroditic plants.

Consider some mechanisms for promoting outcrossing in SC hermaphroditic island plants. One such mechanism is dichogamy (protandry and protogyny). While dichogamy could greatly reduce autogamy, that does not necessarily mean that it promotes outcrossing because pollen could be transferred between different flowers on the same plant (geitonogamy), as appears to be the case in some plants of the Robinson Crusoe Islands (Anderson *et al.*, 2001a; Bernardello *et al.*, 2001). Barrett (2003) questioned the adaptive significance of dichogamy for promoting outcrossing because it frequently occurs in SI plants. However, in such instances, dichogamy (or herkogamy) may be beneficial because it reduces pollen 'clogging' by self pollen, in essence, interference that could reduce the occurrence or effectiveness of outcrossing. Whether this is the general situation in island plants is not known, and is in need of further study; however, the compilation found in Bernardello *et al.* (2001, table 1) shows that for species for which information is available, dichogamy and SC occur together. While the level of autogamy can be determined by bagging experiments, when a pollinating agent is required molecular markers (usually allozymes or microsatellites) of progeny arrays provide the best estimates of the levels of selfing and outcrossing in natural populations. Unfortunately, such studies are essentially non-existent for island plants (but see discussion of *Bidens* below).

The most efficient mechanism for promoting outcrossing in hermaphroditic flowers is total SI, which precludes selfing. As indicated above, SC appears to be the common condition in island plants, but examples of SI are known, and emerging evidence indicates that the situation may not be as simple as has generally been reported or envisioned. Earlier workers such as Baker (1955, 1967), Stebbins (1957) and Carlquist (1974) envisioned the SC system as a logical feature of the establishment of plants, perhaps single plants, but at least a few individuals, on remote islands with little flora and perhaps few or likely no appropriate pollinators. Two elements of this argument will be considered. First, just conceptually, the tacit assumption in making the wholly logical proposition of SC being 'a' or 'the' reasonable system for early establishment is that the whole flora arrived at the same 'ecological' time in the sense that the islands were 'bare' or lacking other plant species, or pollinators. That certainly must be the case for early-arriving species. But, island floras are built over time, not all at once, and clearly many species arrive later in the geological evolutionary history of islands. The later arrivals may still have to deal with the issues of few propagules, and SC is advantageous. However, those later arrivals may have available a more diverse pollinator fauna than some of the early arrivals, and thus be able to have the pollen moved among their progeny at a level that was not available to the progeny of the first arrivals. Second, we will consider several examples showing that the SC system is not always clear cut. Studies by Rick (1966), McMullen (1987) and McMullen and Naranjo (1994) indicated that four of the species of *Scalesia*, the largest endemic genus in the Galapagos, are SC and do not require a vector for pollination. However, one of those species, *S. affinis*, was examined by Nielsen *et al.* (2003) and found to be partially SI and a second species, *S. divisa*, was highly SI (Nielsen *et al.*, 2000). Studies of additional island endemics such as the silversword alliance in Hawaii (Carr *et al.*, 1986) and *Tolpis* in the Canary Islands (Crawford *et al.*, 2008) show that, while apparently fundamentally SI, there can be considerable inter-populational variation in compatibility, with significant 'leakiness' in the system. In one species of *Tolpis* in the Canaries there has been total breakdown of SI, even accompanied by some floral characters associated with SC. Studies using molecular markers show that some of the 'SI' populations have mixed mating systems, with about a 1:1 mix of outcrossing and selfing (Crawford *et al.*, unpubl.).

Of course, an additional advantage of a partially SI breeding system in islands is reproductive assurance via SC when pollinators are limiting. Populations could be established from dispersal of single propagules into new areas during radiation and diversification and, because the single dispersal unit is from a highly outcrossed population, it would presumably carry more genetic diversity than if it were from an SC population. More work is needed to elucidate the mating systems of island plants that are basically SI but display various levels of leakiness. It appears that some of the most successful insular lineages, especially in Compositae, are SI or with 'leaky' SI (Crawford *et al.*, 2009).

While hermaphroditic flowers are most common in islands, a variety of other sex expression patterns are known. The more common conditions, among others, are monoecy, andromonoecy, gynomonoecy, dioecy, subdioecy and gynodioecy (e.g. Sakai *et al.*, 1995b; Helfgott *et al.*, 2000; Bernardello *et al.*, 2001). The first three, along with hermaphroditism, are sometimes designated monomorphic systems while the latter three are termed dimorphic (Lloyd, 1980; Sakai *et al.*, 1995b). One question of interest about sex expression in island plants is whether a given condition was present in the coloniser or evolved *in situ* in the island setting. Estimates for Hawaiian endemic species indicate that 11% of the colonists were dimorphic and gave rise to only dimorphic species. By contrast, nearly one-third of dimorphic species had monomorphic colonising ancestors (Sakai *et al.*, 1995b). In the Robinson Crusoe Islands, four dimorphic genera appear to have originated from dimorphic colonists whereas three genera ostensibly arose from monomorphic colonists (Bernardello *et al.*, 2001). Of all the systems, only strict dioecy (like complete SI in hermaphroditic flowers) results in obligate outcrossing, and so in a sense it is different from other forms of sex expression. An important question is how the different forms of sex expression influence mating systems in populations of island plants. Unfortunately, empirical studies of the question are largely lacking.

A study by Sun and Ganders (1988) is an example of the influences and complexities of breeding system, sex expression and dichogamy (protandry) on the mating system in Hawaiian *Bidens.* The approximately 20 endemic Hawaiian *Bidens* originated from a single introduction (Ganders *et al.*, 2000) but the extant species are diverse morphologically and ecologically (Ganders & Nagata, 1984). The clade has capitula with sterile ray florets, but there is variation in sex expression of the disk florets. Some species have hermaphroditic, SC, strongly protandrous disk florets whereas female (totally male sterile) disk florets and totally hermaphroditic disk florets occur in separate flowers on separate plants in other species (they are gynodioecious). Allozyme markers of progeny arrays and seed set without pollinators were used to estimate mating system in populations. Low seed set without pollinators indicates that protandry is effective in preventing autogamy. Gynodioecy is normally viewed as a mechanism for promoting outcrossing in SC plants, and it evolved *in situ* in Hawaiian *Bidens* (Sun & Ganders, 1986, 1988). Sun and Ganders (1986) found a positive correlation between frequencies of females and selfing rates of hermaphrodites, evidence supporting the contention that females are established and survive in part because their outcrossed progeny are more fit than the selfed progeny of hermaphrodites. Sun and Ganders (1988) explored the components of inbreeding (autogamy, geitonogamy and biparental inbreeding) in populations with gynodioecious and hermaphroditic individuals. Autogamy can be estimated directly by seed set without pollinators, and apparent selfing estimated from allozyme markers must be due to biparental inbreeding because neither of the other two forms of inbreeding is possible with female plants. It follows then that selfing via geitonogamy can be estimated indirectly by subtracting estimates of the other two forms of inbreeding from the estimate of effective selfing in hermaphroditic flowers. In Hawaiian *Bidens*, mean effective selfing is 0.43, biparental inbreeding is 0.15 and autogamy is 0.04, which indicates that the contribution of geitonogamy to selfing is 0.24. Species of *Bidens* with capitula in compact compound capitulescences had higher levels of geitonogamy than those species with solitary capitula, which is what might be expected (Sun & Ganders, 1988). Hawaiian *Bidens* illustrates the potential complexities of mating systems in an island lineage.

The origin and evolution of dioecy has been one of the most discussed aspects of the reproductive biology of island plants. A central issue is whether dioecy was present in the colonising ancestors or the condition originated in the island setting (e.g. Baker & Cox, 1984; Sakai *et al.*, 1995b; Bernardello *et al.*, 2001). If the colonising ancestors were strictly dioecious, then obviously one propagule could not establish a sexually reproducing population. On the other hand, if the ancestors of dioecious island plants were frequently hermaphroditic, then it would be of interest to know what factor(s) are associated with the evolution of dioecy on oceanic islands. In Hawaii, the present relatively high level of dioecy appears to be the result of both dimorphic colonists and evolution from hermaphroditic colonists in the archipelago (Sakai *et al.*, 1995b). On the Robinson Crusoe Islands, four of the dioecious genera appear to have evolved from colonists that were either dioecious or manifested incipient dioecy while the largest dioecious genus, *Robinsonia*, ostensibly evolved from hermaphroditic ancestors (Bernardello *et al.*, 2001). In the Canary Islands, dioecy in the *Bencomia* alliance (Rosaceae) has apparently arisen twice within the archipelago, either from monoecious or gynomonoecious ancestors (Helfgott *et al.*, 2000) and the functionally dioecious *Withania aristata* (Solanaceae) (Fig. 2.4) reputedly originated from an SC colonist (Anderson *et al.*, 2006b). Thus, accumulating data indicate that dioecy on islands does not have one simple explanation; dioecious taxa have originated both from dimorphic colonists and the autochthonous evolution of dioecy in the island setting.

For those lineages originating from dimorphic colonists, it is instructive to examine such features as the probable dispersal unit of their colonising ancestors and possible 'leakiness' in dioecy because either (or both) could facilitate the establishment of sexual populations from a single colonising event. One dispersal

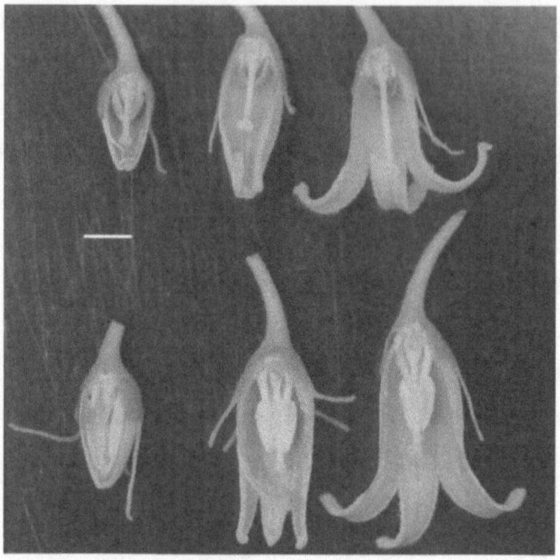

Figure 2.4 Female (top row) and male (bottom row) flowers of *Withania aristata* on Tenerife, Canary Islands, Spain. Scale line = 3.0 cm. Photo by G. Bernardello.

event could be effective in establishing a sexual, dioecious population if the dispersal unit were a fleshy fruit and a bird (the most common dispersal agents, see above) ingested a single multi-seeded fruit or several single-seeded fruits. Sakai *et al.* (1995a) found an association between fleshy fruits and dimorphism for genera in Hawaii. The fleshy fruits were presumably dispersed via bird ingestion. In the Robinson Crusoe Islands, three of the four dioecious genera that presumably originated from dimorphic colonists have fleshy fruits, and likely were ingested and dispersed by birds (Bernardello *et al.*, 2001, 2006). Over 70% of the endemic or native dioecious species in the Canary Islands have fleshy fruits (Bramwell & Bramwell, 2001), and a similar or higher value characterises the Azores (Schäfer, 2005). Lastly, over half of the dioecious natives or endemics in the Galapagos have fleshy fruits (McMullen, 1999).

Consider, next, 'leaky' dioecy (rare occurrence of functionally hermaphroditic flowers in a population) in island colonists as a means of facilitating establishment after long-distance dispersal. If the species is SC, then seed could be produced on the rare hermaphroditic flowers even though a coloniser is largely dioecious. This topic was reviewed by Baker and Cox (1984). Anderson *et al.* (2006b) report the rare occurrence of fruits on male plants of the Canary Island endemic *Withania aristata*, and cite additional examples of the phenomenon from the more recent literature. The frequency of leaky dioecy and its significance in the establishment of dioecious colonists in oceanic archipelagos are largely unknown.

The stage has now been reached, or perhaps was reached much earlier, when there is little point in discussing the relative frequency with which insular dioecious plants originated from dioecious colonisers versus the autochthonous origin of dioecy in the island setting. It is now evident that both can occur, and detailed phylogenetic-evolutionary-ecological studies of individual island lineages are needed to provide additional insight into the origin and evolution of dioecy. One of the most extensive series of studies of the phylogeny, evolution and ecology of breeding systems (particularly dimorphic ones) in an island lineage is on the genus *Schiedea*, a monophyletic group of some 35 species that originated from a hermaphroditic colonising ancestor (e.g. Sakai *et al*., 1997, 2006; Culley *et al*., 1999, 2006; Weller & Sakai, 1991; Weller *et al*., 1998, 2005). A phylogeny for the lineage (Wagner *et al*., 2005) indicates that dimorphic breeding systems originated twice and autogamy evolved three times. Dimorphic species usually occur in dry windy habitats and are pollinated by wind whereas autogamous species are generally found in more moist areas (Sakai *et al*., 2006). The hypothesis is that with shifts to drier, more open habitats there were higher selfing rates due to pollinator loss, resulting in high inbreeding depression. The high inbreeding depression provided selection pressure for a dimorphic breeding system, and the open dry habitat would facilitate wind pollination in areas where biotic pollinators are lacking (Sakai *et al*., 2006). In contrast, autogamy may have evolved in the small populations of very wet areas, and the small, pendent flowers would be 'sheltered' from heavy precipitation. Similar comprehensive studies on other island lineages and other archipelagos would be most helpful in achieving a deeper understanding of the evolution of complex breeding systems and the factors that may shape them.

## Conservation of island plants: reproductive biology and the impact of invasive plants and pollinators

It almost goes without saying that reproductive biology is among the most important aspects of the conservation of island plants. Any study of reproductive biology should include its conservation implications (Holsinger, 1991; Anderson, 1995; Anderson *et al*., 2001a, 2001b; Bernardello *et al*., 1999, 2001, 2004; Dupont & Skov, 2004; Abe, 2006). Conservation issues may be particularly complex when both plants and pollinators are involved (Inoue, 1993; Kearns *et al*., 1998). Important factors in formulating conservation strategies include breeding and mating systems, and cross-compatibility and interfertility among congeneric species. Lastly, the influence of invasive species, both plants and pollinators, may have an impact on the reproductive biology of island plants (Anderson *et al*., 2001b; Dupont *et al*., 2003; Abe, 2006). In many instances, it is the rarity of island species that makes reproductive biology of such critical concern in conservation; Weller (1994) reviewed some of the important relationships between rarity and reproductive biology, and many of the issues he raised are relevant to island plants. Basically, the challenge for a rare plant on an island (or anywhere) is finding a compatible mate of the same species, with the fitness of their resulting progeny sufficient for survival and reproduction. Conservation strategies are aimed at increasing the likelihood of this happening, considered in the context of what is known about the reproductive biology of the species.

A case study from the Robinson Crusoe Islands illustrates some of the conservation implications of reproductive biology mentioned above. Bernardello *et al.* (2004) studied *Sophora fernandeziana* (Leguminosae), which has been classified as endangered because of the small size and number of extant populations (Ricci, 2006). The results of Bernardello *et al.* (2004) were not totally conclusive, but indicate strongly that the species exhibits ovarian SI, which is known for other legumes. If indeed the species is SI, then the lack of compatible mates in the very small populations, and especially if individual plants become isolated, would be a serious impediment to seed set. Genetic (allozyme) diversity in the species is quite low, even by island standards (Crawford *et al.*, 2001). The pollinators of the plant are two species of hummingbird, one endemic and the other native, and they obtain nectar from *S. fernandeziana*. When it is blooming there are few other species from which the birds can get nectar (Bernardello *et al.*, 2004). This system is an example of a mutualism where a decline in either the plant or the pollinator would have a negative effect on the other. Another possible negative factor for *Sophora* on the Robinson Crusoe Islands is invasive plant species such as *Rubus ulmifolius* (Rosaceae), which can outcompete native species (Stuessy *et al.*, 1998; Greimler *et al.*, 2002). In addition, if an invasive species could provide nectar for the hummingbirds at the same time that *S. fernandeziana* is blooming and the invasive also occurs in greater density, which is a certainty, then *Sophora* would suffer from lower bird visitation and pollination. The low seed set observed by Bernardello *et al.* (2004) in a natural population of *S. fernandeziana* is a clear indication that one or more of the factors mentioned above, and possibly others, act in concert to reduce greatly the production of viable seeds.

The impact of invasive plants and pollinators on endemic and native plants and animals in insular biotas is receiving particular attention by island biologists. Efforts to understand the role of invasives in the reproductive biology of island plants and their pollinators are particularly important because the more frequent travel to many islands by tourists and other visitors will increase the introduction of alien elements into insular biotas. *A priori* predictions or broad generalisations about the impact of invasives may not be warranted for many individual situations. Rather, detailed observations, especially over time, are needed to elucidate the effects of invasive species on individual endemics, both plant and pollinator, or the impact on the dynamics of plant-pollinator networks (Kearns & Inouye, 1997; Kearns *et al.*, 1998; Cox & Elqvist, 2000; Olesen *et al.*, 2002b; Dupont *et al.*, 2003; Abe, 2006).

The impact of introduction of pollinators could be positive or negative, or perhaps both, in particular situations. For example, the impacts of the presence (or at least documented occurrence) of bees, possibly a new species (Engel, 2000), on Robinson Crusoe Island during the past 30 years are not yet apparent (Anderson *et al.*, 2001b). The bees are found almost exclusively around the single village, and have not yet been detected on native-endemic plants in native habitats (Anderson *et al.*, 2001a, 2001b; Bernardello *et al.*, 2001). Given the essentially total lack of bee pollinators on Robinson Crusoe, it is conceivable that the bees could be effective pollinators of some endemic largely (by necessity) wind pollinated species that retain features such as colourful corollas and nectar rewards from their presumed biotically-pollinated ancestors (Anderson *et al.*, 2001a, 2001b; Bernardello *et al.*, 2001). On the negative side, if the bees were to take nectar from species pollinated by the one native and one endemic hummingbird species, then the hummingbirds could suffer from loss of their limited food sources. Or, if the bees prefer the often more abundant floral rewards of the invasives, they may indirectly facilitate the spread of the invasives at a cost to the natives.

In the Bonin Islands there are observations suggesting that introduced honeybees are displacing endemic visitors to plants (Abe, 2006). For example, on the two main islands of the archipelago, honeybees visited about 60% of the observed plant species. By contrast, on smaller satellite islands lacking honeybees, endemic small bees were the most frequent visitors to the majority (about 65%) of the native species. It is clear that the honeybees, and other introduced pollinators such as wasps and butterflies, could have a major impact on pollinator networks in the Bonin Islands, and on the native bee fauna. More recently, Abe *et al.* (2008) have shown that an invasive lizard is a threat to native pollinators on the Bonin Islands.

Dupont *et al.* (2003) studied the network (interactions among different plants and animals) of plants and their flower visitors in the subalpine high desert on Tenerife in the Canary Islands. This habitat has a short growing season, and the plants and animals are nearly all endemic or native to the Canaries. One exception is the introduced honeybee, which is very abundant in the subalpine zone. Dupont *et al.* (2003) state that they cannot at present assess the impact, if any, of the honeybees on the interactions among native plants and animals. However, because the honeybees visit more than half of the plant species, and they are very efficient at collecting resources from the plants, there is the potential for the bees to outcompete native visitors.

Dupont and Skov (2004) studied bee visitors to members (20 endemic and one introduced species) of the plant genus *Echium*, one of the largest and most diverse radiations in the Canary Islands (Bramwell & Bramwell, 2001). They found that the only good predictor of the richness (number of species) of bees visiting species of *Echium* was the distribution range of the plant species. That is, more widely distributed plant species were visited by more bee species. A species introduced into the Canary Islands, *Echium plantagineum*, has spread over several islands; it occurs with endemic congeners, is more widespread than any of the endemic species and attracts the same suite of bee visitors as the endemic species. In fact, the introduced species attracted a higher number of bee species than any of the endemics. One of the conservation concerns is whether the introduced plant species will compete with the endemic taxa for pollinators and, since most species of *Echium* are obligately outcrossing, seed set could be reduced in the endemic taxa. Another potential conservation implication of the introduced plant species is its hybridisation with the less abundant endemic species; the results would vary depending on the fertility and vigour of the hybrids (Levin *et al.*, 1996; Francisco-Ortega *et al.*, 2000). For example, if hybrids are vigorous and fertile, then the rarer endemics could be assimilated by the alien species.

It is evident from the above discussion of specific examples that an accurate assessment or prediction of the impact of invasives, whether plant or animal, in any given situation, requires basic information on many aspects of the reproductive biology of both native and endemic plants.

## Future studies

Regardless of the area of biological research, it is always fashionable, and probably valid, to say that what is needed in the future is more of the same kinds of studies that have already been done, as the quote in the opening paragraph from H. G. Baker advocated. In addition, we suggest that more basic, meticulous studies of the reproductive biology of island plants such as breeding systems, pollinators and mating systems (Neal & Anderson, 2005) are needed. This is true because very few species have been critically examined, and in more than a few cases such studies have shown that the situation is more complex or subtle than it appears from casual observation. Several examples include: flowers appearing to be hermaphroditic may be functionally dioecious; the role of biotic pollinators versus wind in the pollination of ‘typical’ (small, shallow, colourless) island flowers; and the significance of ‘leaky’ SI and ‘leaky’ dioecy in colonisation and the subsequent diversification of island lineages. Superficial observations of island plants may lead to interpretations of reproductive biology that are heavily influenced by results from continental species. While island plants are often viewed, and correctly so, as wonderful bizarre organisms in general habit and vegetative features, it is evident that in some instances evolution to the island habitat has involved very subtle or cryptic changes in reproductive biology. Additional studies may call into question, or at least reveal many additional exceptions to, some of the more popular generalisations about reproductive biology of island plants.

It would also be desirable to have more highly integrated studies in which more traditional investigations are combined with additional approaches. For example, molecular markers (e.g. allozymes, microsatellites, etc.) could be employed to estimate outcrossing rates; data from field observations of pollinators, floral morphology (separation of sexes) and breeding system studies would provide insight into the factors yielding the observed mating system.

As indicated in the introduction, molecular phylogenetic studies of island plants have flourished over the past 20 years, but with the notable exceptions of several investigations cited earlier, there have been relatively few attempts to examine reproductive biology within a phylogenetic context. This applies both to the colonising ancestors of insular lineages as well as within the lineages themselves.

In closing, it is appropriate to place future research on islands within a global perspective. Islands are scattered throughout the oceans of the world, and are under the political jurisdiction of many different countries. These geographical and political factors create challenges to those wishing to do research on oceanic islands, including the logistics of getting to the islands and of obtaining access to facilities and personnel for research in the islands. Acquiring the necessary permissions to work in an archipelago can be a daunting task. Studies of plant reproductive biology of island plants may be especially challenging because of the time, effort and local assistance required for locating, observing and manipulating plants in natural populations. In this regard, there is a need for a research coordination network in which basic information on the logistics of getting to and working on islands is available.

Another challenge for those working on oceanic islands is being aware of similar projects or efforts occurring in different archipelagos around the world so that information on methodology, etc., could be shared. Island researchers often work in geographic isolation, and local researchers in smaller archipelagos may not have easy access to current developments in theory and practices related to ecological, evolutionary and conservation issues of common concern. As discussed earlier, information on reproductive biology is critical for formulating conservation strategies. Thus, the sharing of basic information and observations, some of which have not and perhaps never will be published, on breeding systems, pollinators and mating systems of both native and invasive species on a global scale is an especially important component of any plan to conserve the vulnerable plant treasures often teetering on the brink of extinction on small specks of lava in the vast oceans of the world. Clearly, it is past time for those interested in preserving these treasures to develop a vehicle for sharing information and coordinating efforts in this important enterprise.

## Acknowledgements

We thank Professor David Bramwell for the invitation to contribute to this volume. The authors' research on the Juan Fernández and Canary Islands has been supported by: the National Science Foundation (DJC, GJA), Connecticut Research Foundation and Department of Ecology and Evolutionary Biology, University of

Connecticut (GJA); the Department of Ecology and Evolutionary Biology, the Natural History Museum and Biodiversity Research Center, the University of Kansas, and a Kansas National Science Foundation (NSF) EPSCoR (DJC); and CONICET, SECYT-Universidad Nacional de Córdoba (GB).

This paper is dedicated to Professor Tod F. Stuessy. He first introduced us to the endemic plants of the Robinson Crusoe Islands, and facilitated and inspired our studies of the fascinating flora of the archipelago.

## References

Abe, T. (2006). Threatened pollination systems in native flora of the Ogasawara (Bonin) Islands. *Ann. Bot.* 98: 317–334.

Abe, T., Makino, S. & Okochi, I. (2008). Why have endemic pollinators declined on the Ogasawara Islands? *Biodivers. Conserv.* 17: 1465–1473.

Anderson, G. J. (1995). Systematics and reproductive biology. In P. C. Hoch & A. G. Stephenson (eds.), *Experimental and Molecular Approaches to Plant Biosystematics*, pp. 263–272. Monographs in Systematic Botany 53. St Louis, Missouri Botanical Garden.

Anderson, G. J., Bernardello, G., Crawford, D. J. & Stuessy, T. F. (2000). Reproductive biology of *Wahlenbergia* (Campanulaceae) endemic to Robinson Crusoe Island (Chile). *Pl. Syst. Evol.* 223: 109–123.

Anderson, G. J., Bernardello, G., Stuessy, T. F. & Crawford, D. J. (2001a). Breeding system and pollination of selected plants endemic to Juan Fernández Islands. *Am. J. Bot.* 88: 220–233.

Anderson, G. J., Bernardello, G. & Engel, M. S. (2001b). Conservation implications of a newly discovered bee species on Isla Robinson Crusoe, Chile. *Conserv. Biol.* 15: 803–805.

Anderson, G. J., Bernardello, G., Bohs, L., Weese, T. & Santos-Guerra, A. (2006a). Phylogeny and biogeography of the Canarian *Solanum vespertillo* and *S. lidii* (Solanaceae). *Anales Jard. Bot. Madrid* 63: 159–167.

Anderson, G. J., Bernardello, G., Opel, M. R., Santos-Guerra, A. & Anderson, M. (2006b). Reproductive biology of the dioecious Canary Islands endemic *Withania aristata* (Solanaceae). *Am. J. Bot.* 93: 1295–1305.

Baker, H. G. (1955). Self-compatibility and establishment after 'long-distance' dispersal. *Evolution* 9: 347–348.

Baker, H. G. (1967). Support for Baker's Law: as a rule. *Evolution* 21: 853–856.

Baker, H. G. & Cox, P. A. (1984). Further thoughts on dioecism and islands. *Ann. Mo. Bot. Gard.* 71: 244–253.

Baldwin, B. G. (2003). A phylogenetic perspective on the origin and evolution of Madiinae. In S. Carlquist, B. G. Baldwin & G. D. Carr (eds.), *Tarweeds and Silverswords: Evolution of the Madiinae (Asteraceae)*, pp. 193–228. St Louis, Missouri Botanical Garden Press.

Baldwin, B. G., Crawford, D. J., Francisco-Ortega, J., Kim, S.-C., Sang, T. & Stuessy, T. F. (1998). Molecular phylogenetic insights on the origin and evolution of oceanic island plants. In D. Soltis, P. Soltis & J. Doyle, J. (eds.), *Molecular Systematics of Plants II. DNA Sequencing*, pp. 410–444. New York, Kluwer Academic Publishers.

Ballard, H. E. & Sytsma, K. J. (2000). Evolution and biogeography of the woody violets (*Viola*, Violaceae): arctic origins, herbaceous ancestry and bird dispersal. *Evolution* 54: 1521–1532.

Barrett, S. C. H. (1996). The reproductive biology and genetics of island plants. *Phil. Trans. R. Soc. Lond. B.* 351: 725–733.

Barrett, S. C. H. (2003). Mating strategies in flowering plants: the outcrossing-selfing paradigm and beyond. *Phil. Trans. R. Soc. Lond. B.* 358: 991–1004.

Barrier, M., Baldwin, B. G., Robichaux, R. H. & Purugganan, M. D. (1999). Interspecific hybrid ancestry of a plant adaptive radiation: allopolyploidy of the Hawaiian silversword alliance (Asteraceae) inferred from floral homeotic gene duplications. *Molec. Phylogen. Evol.* 16: 1105–1113.

Bernardello, G., Anderson, G. J., Lopez S. P., Cleland, M. A., Stuessy, T. F. & Crawford, D. J. (1999). Reproductive biology of *Lactoris fernandeziana* (Lactoridaceae). *Am. J. Bot.* 86: 829–840.

Bernardello, G., Anderson, G. J., Stuessy, T. F. & Crawford, D. J. (2001). A survey of floral traits, breeding systems, floral visitors, and pollination systems of the angiosperms of the Juan Fernandez Islands (Chile). *Bot. Rev. (Lancaster)* 67: 255–308.

Bernardello, G., Aguilar, R. & Anderson, G. J. (2004). The reproductive biology of *Sophora fernandeziana* (Leguminosae), a vulnerable endemic species from Isla Robinson Crusoe. *Am. J. Bot.* 91: 198–206.

Bernardello, G., Anderson, G. J., Stuessy, T. F. & Crawford, D. J. (2006). The angiosperm flora of the archipelago Juan Fernandez (Chile): origin and dispersal. *Can. J. Bot.* 84: 1266–1281.

Bramwell, D. & Bramwell, Z. (2001). *Wild Flowers of the Canary Islands*, 2nd edition, Madrid, Editorial Rueda.

Calero, A. & Santos, A. (1993). Reproductive biology of the high altitude Canarian flora. *Proceedings of the Fifth Meeting of OPTIMA (Organization for the Phyto-Taxonomic Investigation of the Mediterranean Area)*, pp. 497–502.

Carine, M. A., Russell, S. J., Santos-Guerra, A. & Francisco-Ortega, J. (2004). Relationships of the Macaronesian and Mediterranean floras: molecular evidence for multiple colonizations into Macaronesia and back-colonization of the continent in *Convolvulus* (Convolvulaceae). *Am. J. Bot.* 91: 1070–1085.

Carlquist, S. (1966a). The biota of long-distance dispersal. I. Principles of dispersal and evolution. *Q. Rev. Biol.* 41: 247–270.

Carlquist, S. (1966b). The biota of long-distance dispersal. III. Loss of dispersability in the Hawaiian flora. *Brittonia* 18: 310–335.

Carlquist, S. (1966c). The biota of long-distance dispersal. IV. Genetic systems in the floras of oceanic islands. *Evolution* 20: 433–455.

Carlquist, S. (1966d). The biota of long-distance dispersal. II. Loss of dispersability in Pacific Compositae. *Evolution* 20: 30–48.

Carlquist, S. (1974). *Island Biology*. New York, Columbia University Press.

Carlquist, S. (1980). *Hawaii: A Natural History*. Lawai, Hawaii, National Tropical Botanical Garden.

Carlquist, S. (1981). Chance dispersal. *Am. Sci.* 69: 509–516.

Carr, G. D., Powell, E. A. & Kyhos, D. W. (1986). Self-incompatibility in the Hawaiian Madiinae (Compositae): an exception to Baker's rule. *Evolution* 40: 430–434.

Cox, P. A. & Elqvist, T. (2000). Pollinator extinction in the Pacific Islands. *Conserv. Biol.* 14: 237–239.

Crawford, D. J., Ruiz, E., Stuessy, T. F., Tepe, E., Aqeveque, P., Gonzalez, F. et al. (2001). Allozyme diversity in endemic flowering plant species of the Juan Fernandez archipelago, Chile: ecological and historical factors with implications for conservation. *Am. J. Bot.* 88: 2195–2203.

Crawford, D. J., Archibald, J. K., Mort, M. E., Kelly, J. K. & Santos-Guerra, A. (2008). A test of Baker's law: the radiation of *Tolpis* (Asteraceae) in the Canary Islands. *Int. J. Plant Sci.* 169: 782–791.

Crawford, D. J., Lowrey, T. K., Anderson, G. J., Bernardello, G., Santos-Guerra, A. & Stuessy, T. F. (2009). Genetic diversity in the colonizing ancestors of Asteraceae endemic to oceanic islands: Baker's Law and polyploidy. In V. A. Funk, A. Susanna, T. Stuessy & R. Bayer (eds.), 139–151 *Systematics, Evolution, and Biogeography of the Compositae*. Vienna, International Association of Plant Taxonomy.

Culley, T. M., Weller, S. G., Sakai, A. K. & Rankin, A. E. (1999). Inbreeding depression and selfing rates in a self-compatible hermaphroditic species *Schiedea membranacea* (Caryophyllaceae). *Am. J. Bot.* 86: 980–987.

Culley, T. M., Dunbar-Wallis, A. K., Sakai, A. K., Weller, S. G. & Campbell, D. R. (2006). Genetic variation of ecophysiological traits in two gynodioecious species of *Schiedea* (Caryophyllaceae). *New Phytol.* 169: 589–601.

Dupont, Y. L., Hansen, D. M. & Olesen, J. M. (2003). Structure of a plant-flower-visitor network in the high-altitude sub-alpine desert on Tenerife, Canary Islands. *Ecography* 26: 301–310.

Dupont, Y. L., Hansen, D. M., Rasmussen, T. J. & Olesen J. M. (2004). Evolutionary changes in nectar sugar composition associated with switches between bird and insect pollination: the Canarian bird-flower element revisted. *Funct. Ecol.* 18: 67–676.

Dupont, Y. L. & Skov, C. (2004). Influence of geographical distribution and floral traits on species richness of bees (Hymenoptera: Apoidea) visiting *Echium* species (Boraginaceae) of the Canary Islands. *Int. J. Plant Sci.* 165: 377–386.

Eastwood, A., Gibby, M. & Cronk, Q. C. B. (2004). Evolution of St. Helena (Asteraceae): relationships of the genera *Commidendrum* and *Melanodendron*. *Bot. J. Linn. Soc.* 144: 69–83.

Ehrendorfer, E. (1979). Reproductive biology of island plants. In D. Bramwell (ed.), *Plants and Islands*, pp. 293–306. London & New York, Academic Press.

Emerson, B. C. (2002). Evolution on oceanic islands: molecular phylogenetic approaches to understanding pattern and process. *Molec. Ecol.* 11: 951–966.

Engel, M. S. (2000). A new *Lasioglossum* from the Juan Fernandez Islands (Hymenoptera: Halictidae). *Rev. Chil. Entomol.* 27: 5–10.

Forfang, A.-S. & Olesen, J. M. (1998). Male-biased sex ratio and promiscuous pollination in the dioecious island tree *Laurus azorica* (Lauraceae). *Plant Syst. Evol.* 212: 143–157.

Francisco-Ortega, J., Crawford, D. J., Santos-Guerra, A. & Jansen, R. K. (1997). Origin and evolution of *Argyranthemum* (Asteraceae: Anthemideae) in Macaronesia. In

T. J. Givnish & K. J. Sytsma (eds.), *Molecular Evolution and Adaptive Radiation*, pp. 407–431. Cambridge, Cambridge University Press.

Francisco-Ortega, J., Santo-Guerra, A., Kim, S.-C. & Crawford, D. J. (2000). Plant genetic diversity in the Canary Islands: a conservation perspective. *Am. J. Bot.* 87: 909–919.

Ganders, F. R., Berbee, M. & Pirseyedi, M. (2000). ITS base sequence phylogeny in *Bidens* (Asteraceae): evidence for the continental relatives of Hawaiian and Marquesan *Bidens*. *Syst. Bot.* 25: 122–133.

Ganders, F. R. & Nagata, K. M. (1984). The role of hybridization in the evolution of *Bidens* on the Hawaiian islands. In W. F. Grant (ed.), *Plant Biosystematics*, pp. 179–194. Toronto & New York, Academic Press.

Garcia, J. D. D. (2000). Patterns of insect flower visitation in *Lavandula buchii* Webb (Lamiaceae), an endemic shrub of Tenerife (Canary Islands). *J. Nat. Hist. (London)* 34: 2145–2155.

Green, P. T. (1999). Greta's Garbo: stranded seeds and fruits from Greta Beach, Christmas Island, Indian Ocean. *J. Biogeogr.* 26: 937–946.

Greimler, J., Stuessy, T. F., Swenson, U., Baeza, C. M. & Matthei, O. (2002). Plant invasions on an oceanic archipelago. *Biol. Invas.* 4: 73–85.

Harbaugh, D. T. (2008). Polyploid and hybrid origins of Pacific island sandalwoods (*Santalum*, Santalaceae) inferred from low-copy nuclear and flow cytometery data. *Int. J. Plant Sci.* 169: 677–685.

Harbaugh, D. T. & Baldwin, B. G. (2007). Phylogeny and biogeography of the sandalwoods (*Santalum*, Santalaceae): repeated dispersals throughout the Pacific. *Am. J. Bot.* 94: 1028–1040.

Harbaugh, D. T., Wagner, W. L., Allen, G. J. & Zimmer, E. A. (2009). The Hawaiian Archipelago is a stepping-stone for dispersal in the Pacific: an example from the plant genus *Melicope* (Rutaceae). *J. Biogeogr.* 36: 230–241.

Helfgott, D. M., Francisco-Ortega, J., Santos-Guerra, A., Jansen, R. K. & Simpson, B. B. (2000). Biogeography and breeding system evolution of the woody *Bencomia* alliance (Rosaceae) in Macaronesia based on ITS sequence data. *Syst. Bot.* 25: 82–97.

Holsinger, K. E. (1991). Conservation of genetic diversity in rare and endangered plants. In E. C. Dudley (ed.), *The Unity of Evolutionary Biology*, pp. 626–633. Portland, OR, Dioscorides Press.

Inoue, K. (1993). Evolution of mutualism in plant-pollinator interactions on islands. *J. Biosci.* 18: 525–536.

Kearns, C. A. & Inouye, D. W. (1997). Pollinators, flowering plants, and conservation biology. *BioScience* 47: 297–307.

Kearns, C. A., Inouye, D. W. & Wasser, N. M. (1998). Endangered mutualisms: the conservation of plant-pollinator interactions. *Annu. Rev. Ecol. Syst.* 29: 83–112.

Kim, H.-G., Keely, S. C., Vroom, P. S. & Jansen, R. K. (1998). Molecular evidence for an African origin of the Hawaiian endemic *Heperomannia* (Asteraceae). *Proc. Natl. Acad. Sci. USA* 95: 15440–15445.

Levin, D. A. (1996). The evolutionary significance of pseudo-self-fertility. *Am. Nat.* 148: 321–332.

Levin, D. A., Francisco-Ortega, J. & Jansen, R. K. (1996). Hybridization and the extinction of rare plant species. *Conserv. Biol.* 10: 10–16.

Lloyd, D. G. (1980). Sexual strategies in plants III. A quantitative method for describing the gender of plants. *New Zeal. J. Bot.* 18: 103–108.

Lowrey, T. K., Whitkus, R. & Sykes, W. R. (2005). A new species of *Tetramolopium* (Asteraceae) from Mitiaro, Cook Islands: biogeography, phylogenetic relationships, and dispersal. *Syst. Bot.* 30: 448–455.

McMullen, C. K. (1987). Breeding systems of selected Galapagos Islands angiosperms. *Am. J. Bot.* 74: 1694–1705.

McMullen, C. K. (1989). Flowering colonizers of the Galapagos Islands: drab but not dull. *Plants Today* 2: 192–196.

McMullen, C. K. (1990). Reproductive biology of Galapagos Islands angiosperms. In J. E. Lawesson, O. Hamann, G. Rogers, G. Reck & H. Ochoa (eds.), *Botanical Research and Management in the Galapagos*, pp. 35–45. St Louis, Missouri Botanical Garden.

McMullen, C. K. (1993). Flower-visiting insects of the Galapagos Islands. *Pan-Pac. Entomol.* 69: 95–106.

McMullen, C. K. (1999). *Flowering Plants of the Galapagos*. Ithaca, NY, Cornell University Press.

McMullen, C. K. (2007). Pollination biology of the Galapagos endemic, *Tournefortia rufo-sericea* (Boraginaceae). *Bot. J. Linn. Soc.* 153: 21–31.

McMullen, C. K. (2009). Pollination biology of a night-flowering Galapagos endemic, *Ipomoea habeliana* (Convolvulaceae). *Bot. J. Linn. Soc.* 160: 11–20.

McMullen, C. K. & Close, D. D. (1993). Wind pollination in the Galapagos Islands. *Not. Galápagos* 52: 12–17.

McMullen, C. K. & Naranjo, S. J. (1994). Pollination of *Scalesia baurii* ssp. *hopkinsii* (Asteraceae) on Pinta Island. *Not. Galápagos* 53: 25–28.

McMullen, C. K. & Viderman, D. M. (1994). Comparative studies on the pollination biology of *Darwiniothamnus tenuifolius* (Asteraceae) and *Plumbago scandens* (Plumbaginaceae) on Pinta Island and Santa Cruz Island, Galápagos. *Phytologia* 76: 30–38.

Meekijjaroenroj, A. & Anstett, M.-C. (2003). A weevil pollinating the Canary Islands date palm: between parasitism and mutualism. *Naturwissenchaften* 90: 452–455.

Neal, P. R. & Anderson, G. J. (2005). Are 'mating systems' 'breeding systems' of inconsistent terminology in plant reproductive biology? Or, is it the other way around? *Plant Syst. Evol.* 250: 173–185.

Nielsen, L. R., Philipp, M., Adsersen, H. & Siegismund, H. R. (2000). Breeding system of *Scalesia divisa* Andersson, an endemic Asteraceae from the Galápagos Islands. *Det Norske Videnskaps-Akademi I. Matematisk-Naturvidenskapelige Klasse, Shrifter, Ny Serie* 39: 127–138.

Nielsen, L. R., Philipp, M. & Siegismund, H. S. (2002). Selective advantage of ray florets in *Scalesia affinis* and *S. pedunculata* (Asteraceae), two endemic species from the Galápagos. *Evol. Ecol.* 16: 139–153.

Nielsen, L. R., Siegismund, H. S. & Philipp, M. (2003). Partial self-incompatibility in the polyploid endemic species *Scalesia affinis* (Asteraceae) from the Galapagos: remnants of a self-incompatibility system? *Bot. J. Linn. Soc.* 142: 93–101.

Olesen, J. M. (1985). The Macaronesian bird-flower element and its relation to bird and bee opportunists. *Bot. J. Linn. Soc.* 91: 395–414.

Olesen, J. M., Valido, A. & Dupont, Y. L. (2002a). Polinización de plantas Canarias. *El Indiferente* 2: 19–29.

Olesen, J. M., Eskildsen, L. I. & Venkatasamy, S. (2002b). Invasion of pollination networks on oceanic islands: importance of invader complexes and endemic super generalists. *Diversity & Distrib.* 8: 181–192.

Ollerton, J., Cranmer L., Stelzer, R. J. & Sullivan, S. (2009). Bird pollination of Canary Island endemic plants. *Naturwissenchaften* 96: 221–232.

Ono, M. (1991). The flora of the Bonin (Ogasawara) Islands: endemism and dispersal modes. *Aliso* 13: 95–105.

Philipp, M., Hansen, L. B., Adsersen, H. & Siegismund, H. R. (2004). Reproductive ecology of the endemic *Lecocarpus pinnatifidus* (Asteraceae) in an isolated population in the Galapagos Islands. *Bot. J. Linn. Soc.* 146: 171–180.

Porter, D. M. (1976). Geography and dispersal of Galapagos Islands vascular plants. *Nature* 264: 745–746.

Porter, D. M. (1983). Vascular plants of the Galapagos: origins and dispersal. In R. I. Bowman, M. Berson & A. E. Leviton (eds.), *Patterns of Evolution in Galapagos Organisms*, pp. 33–96. San Francisco, Pacific Division AAAS.

Price, J. P. & Wagner, W. L. (2004). Speciation in Hawaiian angiosperm lineages: cause, consequence, and mode. *Evolution* 58: 2185–2200.

Prohens, J., Anderson, G. J., Herraiz, J., Bernardello, G., Santos-Guerra, A. & Crawford, D. J. (2007). Genetic diversity and conservation of two endangered eggplant relatives (*Solanum vespertilio* Aiton and *S. lidii* Sunding) endemic to the Canary Islands. *Genet. Resour. Crop Ev.* 54: 451–464.

Ricci, M. (2006). Conservation *status* and *ex situ* cultivation efforts of endemic flora of the Juan Fernández Archipelago. *Biodivers. Conserv.* 15: 3111–3130.

Rick, C. M. (1966). Some plant-animal interactions on the Galápagos Islands. In R. I. Bowman (ed.), *The Galápagos*, pp. 97–106. Berkeley, University of California Press.

Rodríguez-Rodríguez, M. C. & Valido, A. (2008). Opportunistic nectar-feeding birds are effective pollinators of bird-flowers from Canary Islands: experimental evidence from *Isoplexis canariensis* (Scrophulariaceae). *Am. J. Bot.* 95: 1408–1415.

Sakai, A. K., Wagner, W. L., Ferguson, D. M. & Herbst, D. R. (1995a). Biogeographical and ecological correlates of dioecy in the Hawaiian flora. *Ecology* 76: 2530–2543.

Sakai, A. K., Wagner, W. L., Ferguson, D. M. & Herbst, D. R. (1995b). Origins of dioecy in the Hawaiian flora. *Ecology* 76: 2517–2529.

Sakai, A. K., Weller, S. G., Chen, M.-L., Chou, S.-Y. & Tasanont, C. (1997). Evolution of gynodioecy and maintenance of females: the role of inbreeding depression, outcrossing rates, and resource allocation in *Schiedea adamantis* (Caryophyllaceae). *Evolution* 51: 724–736.

Sakai, A. K., Weller, S. G., Wagner, W. L., Nepokroeff, M. & Culley, T. M. (2006). Adaptive radiation and evolution of breeding systems in *Schiedea* (Caryophyllaceae), an endemic Hawaiian genus. *Ann. Mo. Bot. Gard.* 93: 49–63.

Schäfer, H. (2005). *Flora of the Azores: A Field Guide*, 2nd edition. Wilkersheim, Margraf Publishers.

Schueller, S. K. (2004). Self-pollination in island and mainland populations of the introduced humming-bird pollinated plant *Nicotiana glauca* (Solanaceae). *Am. J. Bot.* 91: 672–681.

Stebbins, G. L. (1957). Self-fertilization and population variability in the higher plants. *Am. Nat.* 91: 337–354.

Stuessy, T. F., Swenson, U., Crawford, D. J., Anderson, G. & Silva O. M. (1998). Plant conservation in the Juan Fernandez archipelago, Chile. *Aliso* 16: 89–101.

Sun, M. & Ganders, F. R. (1986). Female frequencies in gynodioecious populations correlated with selfing rates in hermanphrodites. *Am. J. Bot.* 73: 1645–1648.

Sun, M. & Ganders, F. R. (1988). Mixed mating systems in Hawaiian *Bidens* (Asteraceae). *Evolution* 42: 516–527.

Valido, A., Dupont, Y. L. & Olesen, J. M. (2004). Bird–flower interactions in the Macaronesian islands. *J. Biogeogr.* 31: 1945–1953.

Wagner, W. L., Herbst, D. R. & Sohmer, S. H. (1990). *Manual of the Flowering Plants of Hawaii.* Honolulu, University of Hawaii Press & Bishop Museum Press.

Wagner, W. L., Weller, S. G. & Sakai, A. K. (2005). Monograph of *Schiedea* (Caryophyllaceae-Alsinoideae). *Syst. Bot. Monogr.* 72: 1–169.

Webb, C. J. & Kelly, D. (1993). The reproductive biology of the New Zealand flora. *Trends Ecol. Evol.* 8: 442–447.

Weller, S. G. (1994). The relationship of rarity to plant reproductive biology. In M. L. Bowles & C. J. Whelen (eds.), *Restoration of Endangered Species: Conceptual Issues, Planning, and Implementation*, pp. 90–117. Cambridge, Cambridge University Press.

Weller, S. G. & Sakai, A. K. (1991). The genetic basis of male sterility in *Schiedea* (Caryophyllaceae), an endemic Hawaiian genus. *Heredity* 67: 265–273.

Weller, S. G., Sakai, A. K., Rankin, A. E., Golonka, A., Kutcher, B. & Ashby, K. E. (1998). Dioecy and the evolution of wind pollination in *Schiedea* and *Alsinidendron* (Caryophyllaceae: Alsinoideae) in the Hawaiian Islands. *Am. J. Bot.* 85: 1377–1388.

Weller, S. G., Sakai, A. K., Thai, D. A., Tom, J. & Rankin, A. E. (2005). Inbreeding depression and heterosis in populations of *Schiedea viscosa*, a highly selfing species. *J. Evol. Biol.* 18: 1434–1444.

Yeo, P. F. (1993). Secondary pollen presentation. *Plant Syst. Evol.* Suppl. 6, pp. 1–268. New York, Springer Verlag.

# 3

# Spatial methodologies in historical biogeography of islands

PAULA POSADAS, JORGE V. CRISCI AND LILIANA KATINAS

The study of insular biotas has influenced biology for a considerable period of time. The uniqueness of many biotic assemblages on islands or isolated ecosystems (e.g. mountaintops, caves or lakes), which act as habitat islands has prompted the development of biological fields such as evolutionary biology and ecology.

It is well known that the ideas of Darwin and Wallace on evolution were rooted in a historical approach to the geographical distribution of organisms (Crisci & Katinas, 2009). As Darwin himself put it: 'In considering the distribution of organic beings over the face of the globe, the first great fact which strikes us is, that neither the similarity nor the dissimilarity of the inhabitants of various regions can be accounted for by their climatal and other physical conditions' (Darwin, 1859: 346, quoted in Crisci & Katinas, 2009). Notably, the experiences of Darwin on the Galapagos and Wallace in the East Indies profoundly influenced their thinking (Brown & Lomolino, 1998) through the observed patterns of organism distributions in those archipelagos. So, island biogeography has been the root of major paradigm shifts which affect the entire field of biology.

Heaney (2007) postulates that current island biogeographic research is framed in two opposite paradigms that have existed side by side during the last four decades. These two paradigms are MacArthur and Wilson's (1963, 1967) equilibrium theory of island biogeography, and vicariance biogeography (Nelson & Platnick, 1981). Alternatively, equilibrium theory and vicariance biogeography may be better placed as two different approaches to biogeography: ecology vs history (see Crisci *et al.*, 2003).

While MacArthur and Wilson's (1963, 1967) equilibrium theory is an approach to ecological biogeography, vicariance biogeography is an approach to historical biogeography. Ecological and historical biogeography are two research traditions

*The Biology of Island Floras*, eds. David Bramwell and Juli Caujapé-Castells. Published by Cambridge University Press. 

which have been distinguished since the early nineteenth century (de Candolle, 1820, as 'stations' and 'habitations', respectively). Each subdiscipline uses different epistemological approaches in addition to focusing on different spatio-temporal scales – i.e. evolutionary processes occurring over millions of years on a large, often global, scale in historical biogeography versus ecological processes occurring over a short temporal and small spatial scale in ecological biogeography (Myers & Giller, 1988; Crisci *et al.*, 2006).

Within a historical biogeographic framework, the emergence of phylogenetic and molecular systematics over the past years has been widely employed to infer evolutionary and biogeographic hypotheses on islands. For example, molecular information has been applied to study colonisation scenarios in such well-known islands and archipelagos as Hawaii (Desalle, 1995; Baldwin & Robichaux, 1995), Macaronesia (Helfgott *et al.*, 2000; Francisco-Ortega *et al.*, 1997), Caribbean islands (Calsbeek & Smith, 2003; Vidal *et al.*, 2008) and Madagascar (Yoder & Nowak, 2006).

Historical biogeographic, non-phylogeny-based, approaches exist which are also important tools for providing testable hypotheses to explain the spatial arrangement of island and continental organisms. To Darwin, the distribution of organisms was sufficiently historically informative that it became central in his evolutionary theory. This belief prevailed until the arrival of modern methods of phylogenetic reconstruction, especially those using molecular data. Hence, the empirical data on geographical distribution of organisms that were so useful to Darwin and others are in many current analyses considered historically uninformative if they are not combined with a phylogenetic analysis. Thus, geographical distributions become informative, understandable and comparable only when combined with a genealogical hypothesis and expressed in an area cladogram. The construction of area cladograms assumes that the nodes of the cladograms, with their information about relationships among taxa, are the medium through which geographical distributions are conveyed in a historically informative way.

There are, however, authors who support the idea that the geographical distribution of organisms, in itself, is historically informative. Craw *et al.* (1999) stressed that many aspects of current distribution (geographical structure, distributional limits, spatial geometry, boundaries and range, etc.) are potentially informative of evolutionary aspects. They emphasised the importance of the spatial or geographical dimension of life's diversity in the study of biogeography. This attempt is based on the panbiogeographical approach of Croizat (1958, 1964). Another method that focuses on the distribution of organisms is parsimony analysis of endemicity or PAE (Rosen, 1988), which classifies areas (or localities or quadrats) according to their shared taxa, resulting in a hierarchical classification of the geographical units. According to Rosen (1988) parsimony analysis of endemicity generates historical (hence geological) hypotheses even from modern distributions (Crisci & Katinas, 2009).

In the last few years, there has been an extraordinary development of what is called 'spatial thinking' (National Research Council, 2006). Spatial thinking is not a content-based discipline in the way that physics, biology or economics are disciplines: it is not a stand-alone subject in its own right. It is a way of thinking that permeates those disciplines and virtually all subject matter disciplines. Spatial thinking is one form of thinking based on three elements: concepts of space, tools of representation and processes of reasoning. It is the concept of space that makes spatial thinking a distinctive form of thinking. By understanding the meaning of space, its properties (e.g. dimensionality, continuity, proximity and separation) can be used as a vehicle for structuring problems, finding answers and expressing solutions. Expressing relationships within spatial structures (e.g. maps) enables perception, recognition and analysis of the static and, via transformations, the dynamic properties of objects and the relationships between objects. This is clearly the way that Darwin, through the analysis of geographic distribution of South American living and fossil organisms, justified the reality of biological evolution, and established natural selection as the promoting process of evolution. It is also the underlying philosophy of panbiogeography and PAE. These two approaches can be referred to as 'spatial methodologies'.

In this chapter we will briefly describe parsimony analysis of endemicity (PAE) and panbiogeography track compatibility analysis; and present examples of their application to insular systems of southern South America, including habitat and continental and oceanic island systems.

## Parsimony analysis of endemicity

Parsimony analysis of endemicity or PAE is a tool of historical biogeography that allows discovery of the natural patterns of organism distribution (Rosen, 1988; Rosen & Smith, 1988). PAE classifies geographic units (e.g. localities, quadrats or areas) according to their shared taxa by means of the most parsimonious solution (parsimony criterion) (Crisci *et al.*, 2003). PAE is one of the most widely applied techniques described in the historical biogeographic literature (Posadas & Donato, 2007). PAE was developed based on an analogy between systematics and biogeography. This analogy is established on the basis that taxa distribution in geographical units is analogous to character distribution in taxa. For example, one character could be unique to a taxon (autapomorphy) or it could be shared by two or more taxa; similarly, one taxon could be unique to one locality or it could inhabits two or more localities. Then, it is possible to procede to analyse distributional data of taxa using the same methodological tools as to analyse character distribution among taxa. Rosen (1988) proposed that parsimony analysis, as in cladistic systematics, could be usefully applied to the distribution of taxa. Thus, PAE application consists in the construction of a data matrix in which terminals are represented by area units and characters are represented by the presence (scored with 1) or absence (scored with 0) of taxa. The data matrix is then analysed using a parsimony algorithm resulting in an area cladogram, which reflects relationships between the area units considered.

PAE has been criticised because it ignores cladistic relationships among taxa, considering only their distributions (Humphries, 1989). Garzón-Orduña *et al.* (2008), by comparing PAE results with that of other historical biogeographical methodologies (e.g. Brooks parsimony analysis and dispersal-vicariance analysis), concluded that PAE aims to describe but does not explain the current distribution of organisms in terms of historical biogeography. These authors disregard PAE as a historical biogeographical method. They propose that PAE is a useful tool in other biogeographical or ecological analyses for exploring the distribution of taxa or for establishing hypotheses of primary homology between areas. Nonetheless, the search for primary homology between areas is a necessary first step in historical biogeography. According to Rosen (1988), a PAE cladogram represents a hypothesis of contemporaneous ecological conditions, where sister localities are then those that are most closely related ecologically. In addition, PAE cladograms represent schemes of successive emergences of biotas, that is, relative recency of biotic interchange between sister localities or, conversely, historical sequences of biotic divergences and isolations.

## Panbiogeography track compatibility analysis

Panbiogeography is an approach to historical biogeography that focuses on the role of place in the history of life. It was originally proposed by Léon Croizat (1952, 1958, 1964, 1981). Panbiogeographers acknowledge that an understanding of locality is a fundamental precondition to any adequate analysis of the patterns and processes of evolutionary change (Crisci *et al.*, 2003).

Croizat (1964) posed the idea that barriers evolve together with biotas when he postulated that 'earth and life evolve together'. From this seminal statement grew the concept of panbiogeography. Croizat's method was basically graphic, consisting of plotting distributions of organisms on maps and connecting the disjunct distribution areas (or collection localities) together with lines called tracks. If superimposing of individual tracks of unrelated groups of organisms showed coincidence, the resulting summary lines were considered generalised tracks. According to panbiogeographic thinking, generalised tracks indicate the preexistence of ancestral biotas, which subsequently become fragmented by tectonic and/or climatic change. Intersection of two or more generalised tracks in a given area results in a panbiogeographic node. A node is interpreted as a composite area in which different ancestral biotic and geologic fragments interrelate in space-time. Generalised tracks commonly stretch across the major oceanic barriers (e.g. between New Zealand and South America through the Pacific Ocean). Croizat thus developed the concept of considering the ocean basins, not continents, as the major focus for biogeographic attention (Keast, 1991).

Several authors developed quantitative approaches to panbiogeography. Probably the most accepted one has been the track compatibility analysis developed by Craw (1988, 1989). Craw's method is based on the concept of distributional compatibility. In track compatibility analysis, two or more individual tracks are considered compatible with each other when tracks are either included within or replicated by one another (Crisci *et al.*, 2003; Katinas *et al.*, 2004). Thus, in panbiogeography, compatibility is applied in a more restricted way compared with its original use in taxonomy (see Meacham, 1984).

A data matrix of areas vs tracks is constructed to apply track compatibility analysis. In this matrix an entry is 1 if the track is present in the area and 0 if it is absent. The matrix is analysed by means of a compatibility algorithm to find the largest cliques of compatible tracks. If more than one large clique or several cliques of considerable size are found, it could be hypothesised that: (1) several generalised tracks link the localities or distribution areas in different ways, or (2) those tracks common to all the largest cliques can be identified as a generalised track (Craw, 1989). According to the first hypothesis, a given locality or distribution area could be part of more than one generalised track, which is not allowed in the second hypothesis.

## Example 1: Tierra del Fuego

The Antarctic is a continental mass which is isolated from the other continents by a wide oceanic belt (from *c.*66° S to 40° S). The subantarctic islands, southern New Zealand and Tierra del Fuego are the only terrestrial areas that penetrate this oceanic zone.

The Tierra del Fuego archipelago constitutes the southernmost tip of Patagonia in South America. It has a surface area of 66 000 km$^2$. From a biogeographic viewpoint, Tierra del Fuego belongs to the Subantarctic subregion of the Andean region (Morrone, 2001). However, floristically, southern South America (including Tierra del Fuego) has been a controversial area belonging either to the Antarctic or to the Neotropical floristic kingdoms (Tuhkanen *et al.*, 1990).

The archipelago is characterised by the presence of the Andean cordillera, which generates a remarkable topographic and climatic gradient over short distances.

Thus, the phytogeographical zones of the archipelago form a gradient of vegetation formation from semiarid steppes (northeast) to the extremely humid Magellanic forest and oceanic heath in the west and south (Niemelä, 1990). In southeastern Patagonia and northern Tierra del Fuego the Antarctic floristic element of moist and cool climatic conditions (as cushion plants and soft herbs) is mixed with spiny dwarf shrubs, annual grasses and tussocks (typical of Patagonian semiarid flora). Such mixing constitutes a transitional belt of humid steppe, the sub-Andean floristic province described by Cabrera (1978). According to Tuhkanen *et al.* (1990), several Antarctic genera are conspicuous in this area (e.g. *Acaena*, *Nothofagus*), representing the ancient Gondwanan flora. These Antarctic elements occupied southern South America, New Zealand and Tasmania, and fossil evidence of their presence has been found in the Kerguelen Islands and Antarctic Peninsula. Summarising, the flora of Tierra del Fuego could be interpreted as an extension of that of southern South America and it shows conection with other austral areas, including the subantarctic islands.

The subantarctic islands (Fig. 3.1) form a loose ring within the latitudinal belt of 54° to 48° S clustered around the major climatic and oceanic boundary, the Antarctic Polar Frontal Zone (Whinan *et al.*, 2005). The main islands of this group are: Islas Malvinas (Falkland Islands), South Georgia, Diego Ramírez, Tristan da Cunha–Gough in the South Atlanctic Ocean; Macquarie, Campbell, Auckland and Chatham in the South Pacific Ocean; and Marion, Crozet, Prince Edward, Amsterdam, Kerguelen and Heard in the South Indian Ocean. Bergstrom and Chown (1999) classified subantarctic islands into two groups according to their history: older islands with a continental origin (e.g. Islas Malvinas, South Georgia, Campbell and Auckland); and younger islands, of volcanic origin (e.g. most of the subantarctic islands of the Indian Ocean).

The objective of this analysis is to apply a spatial-based biogeographic methodology to infer the relationships of the flora of Tierra del Fuego with that of other areas, especially subantarctic islands.

## *Material and methods*

### *Delimitation of areas*

We included all areas that share vascular plant species with Tierra del Fuego. A total of 20 areas were considered, which included continental areas and islands. Among the continental areas are North America, Eurasia, Africa, the Antarctic Peninsula, South East Australia and South America. The latter was divided in its main biogeographic components, Andean and Neotropical regions, since South America is a hybrid area according to previous studies (e.g. Crisci *et al.*, 1991; Katinas *et al.*, 1999). Among islands, we included the subantarctic group and other islands. In the

Figure 3.1 Map showing the main areas considered in the parsimony analysis of endemicity of Tierra del Fuego's flora.

first group we included Islas Malvinas, South Georgia, Diego Ramírez, Tristan da Cunha–Gough, Crozet–Prince Edward–Marion, Kerguelen–Heard–Amsterdam, Macquarie–Campbell, Chatham and New Zealand. In the second group we included Islas Juan Fernández and Tasmania.

### *Selection of taxa*

Distributional data of the 386 species and infraspecific taxa of vascular plants inhabiting Tierra del Fuego were considered. Data were taken from *Flora of Tierra del Fuego* (Moore, 1983). Nomenclatural changes and current systematic status of each taxon were checked with recent literature (e.g. Zuloaga *et al.*, 2008). The

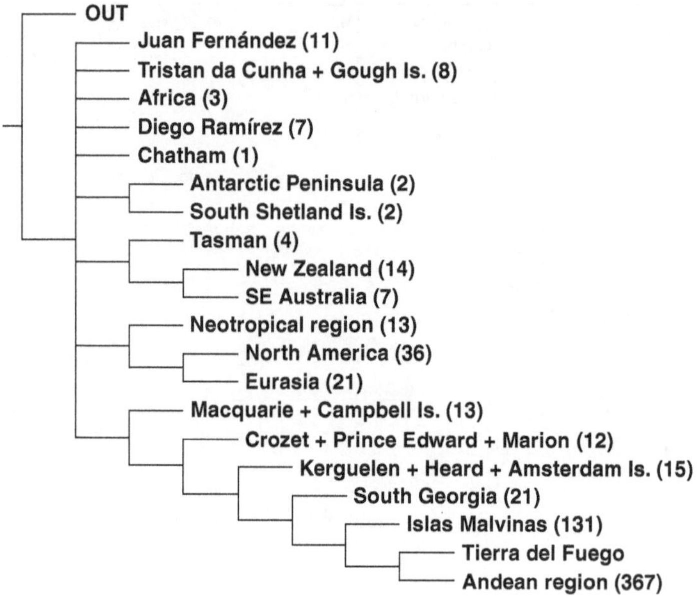


Figure 3.2 Strict consensus cladogram resulting from parsimony analysis of endemicity applied to the flora from Tierra del Fuego. Numbers indicate the quantity of species in each area also present in Tierra del Fuego.

distribution of each taxon beyond Tierra del Fuego was confirmed from the literature in checklists, floristic treatments and revisionary studies (e.g. Dawson, 1958; Greene & Groves, 1963; Lindsay, 1971; Allison & Smith, 1973; Meurk *et al.*, 1994; Convey, 1996; Convey *et al.*, 2000; Bergstrom *et al.*, 2006). Adventitious, cultivated and naturalised species were excluded from the analysis.

### *Parsimony analysis of endemicity*

A data matrix of 21 areas (the 20 aforementioned plus a hypothetical outgroup with all species absent) by 386 taxa (presence in a given area coded as 1 and absence as 0) was analysed using TNT ver. 1.1 (Goloboff *et al.*, 2003). An exhaustive search was applied using implicit enumeration option.

## *Results and discussion*

Parsimony analysis of the data matrix yielded five equally most parsimonious cladograms (L = 493). The strict consensus cladogram is shown in Fig. 3.2. Since our analysis focuses on Tierra del Fuego archipelago relationships, the main group of interest in our results is that containing Tierra del Fuego. The archipelago is part of a pectinate clade which contains a continental Andean region and several subantarctic islands. The most basal area of this clade is occupied by the South Pacific

Campbell and Macquarie islands, which share 13 plant species with Tierra del Fuego. The next two groups of islands in the clade are the South Indian Ocean subantarctic islands: first, the group composed of Crozet, Prince Edward and Marion islands (sharing 12 plant species with Tierra del Fuego); second, the group containing Kerguelen, Heard and Amsterdam islands (this group shares 15 plant species with Tierra del Fuego). Then, the South Atlantic subantarctic islands are depicted in the cladogram: first, South Georgia (which shares 21 species with Tierra del Fuego), followed by Islas Malvinas (131 species in common with Tierra del Fuego). The Andean region and Tierra del Fuego are sister areas with 367 species in common, of the 386 considered. The relationships of Tierra del Fuego's flora can be summarised as follows: (Pacific Ocean subantarctic islands (Indian Ocean subantarctic islands (Atlantic Ocean subantarctic islands (continental Andean region, Tierra del Fuego)))). This means that Tierra del Fuego flora shows a movement from West to East across the southern oceans to other subantarctic islands. This sequence is congruent with the direction of the Antarctic Circumpolar Current (West Wind Drift).

When considering the number of Tierra del Fuego's plant species present in the other areas according to the depiction sequence of the clade, we observed an increasing quantity of shared species in those areas more closely related to Tierra del Fuego (see Fig. 3.2). However, the number of Fuegian species in the other areas did not affect the topology of the cladogram, since distant areas such as North America, which has 35 species in common with Tierra del Fuego, appears in a separate clade (see Fig. 3.2).

The sister relation of Tierra del Fuego with the continental Andean region and the relationships of this pair with Islas Malvinas have been established for other taxonomic groups. Morrone (1998) found these same relationships when applying PAE to analyse the weevil fauna (Insecta: Coleoptera: Curculionidae) of subantarctic islands. In fact, as pointed out above, Tierra del Fuego's flora includes part of the Magellanic forest and moorland and the semiarid Patagonian steppe of southernmost South America. Also, the closest relation between Tierra del Fuego and the Andean region with Islas Malvinas is expected, since Islas Malvinas' flora is very similar to that of Tierra del Fuego, except for the lack of arboreal species and a diminished number of other species (Posadas, 2008). Posadas and Morrone (2004) have hypothesised that the geographical isolation of Islas Malvinas from Tierra del Fuego was broken more than once during sea-level changes in Tertiary and Quaternary times. The link of Tierra del Fuego flora with other subantarctic islands following the sequence South Atlantic–South Indian–South Pacific Ocean, in accordance with the Antarctic Circumpolar Current, could indicate a transoceanic dispersal origin. However, it is important to note that the final breakup of southern South America and the Antarctic Peninsula occurred around 30 million years ago

(Ma). This created the Drake Passage and generated the Antarctic Circumpolar Current (Ortiz-Jaureguizar & Cladera, 2006), which led to climatic changes in Antarctica, accompanying the marked fall in marine temperatures (Zachos *et al.*, 2001). By this time, most of the 'younger' subantarctic islands had already emerged. For example, the Kerguelen Plateau, in which the Kerguelen, Heard and McDonald archipelagos are placed, is dated as older than 110 Ma (Duncan, 2002). The Kerguelen archipelago exhibits volcanism dated from 29 Ma to the Pleistocene (Wallace *et al.*, 2002). Thus, an alternative hypothesis to long-distance dispersal must be considered: at least part of the flora of subantarctic islands could belong to a widespread Antarctic-subantarctic flora fragmented by tectonic and climatic events. Antarctica could hold the key to testing these alternative hypotheses through its fossil record.

These two hypotheses require testing through new geological, palaeontological and phylogenetic information.

### Example 2: the Tandilia and Ventania mountain ranges

Since Darwin's journey in Argentina in 1835 the geology of Tandilia and Ventania and their inhabiting biota have attracted several generations of geologists and biologists (Crisci *et al.*, 2001). Tandilia and Ventania are the only systems of mountain ranges situated in a grassy steppe or 'pampas' in the political province of Buenos Aires, in central eastern Argentina. The pampas are a plain situated at sea level, or even below sea level (e.g. the big depression of the Salado river), with the exception of two major systems of mountain ranges or 'sierras', namely Tandilia and Ventania. From a geological point of view the two areas do not appear to be closely related. Tandilia (500 m high), in the southeast of Buenos Aires province, originated in the Proterozoic Age, being the oldest nucleus in Argentina (2200 Ma). Ventania (900–1000 m high) is located in the southwest of the province and has a more recent origin dating from the Palaeozoic (490 Ma). Gondwanaland and pre-Gondwanaland events seem to have played a major role in the building of these mountain ranges.

Biologically, Tandilia and Ventania exhibit a high taxon diversity and endemicity. These mountain ranges not only have a characteristic petrophilous flora, but are the only regions in the province with endemic plant taxa. Within the diverse plant families inhabiting Tandilia and Ventania, Asteraceae hold second place (after Poaceae, the grass family). Approximately 30% of the Asteraceae known in Buenos Aires province occur in Ventania and Tandilia. In addition, there are four species of Asteraceae endemic to Tandilia, three species and two varieties endemic to Ventania, and one species endemic to both. Based on their particular biodiversity, and isolation within grassy ecosystems, these ranges are considered orographic islands (Kristensen & Frangi, 1995) or habitat islands. According to MacArthur (1984), the chief difference between habitat islands on the mainland and oceanic islands is that the latter exist in a vacuum insofar as land birds or insects are concerned whereas mainland islands are adjacent to other habitats filled with birds and insects. Consequently, provided that the mainland islands are not too different from their surroundings, birds from the surrounding areas are able to migrate onto mainland islands more frequently and with more success than for immigration onto oceanic islands. Crisci *et al.* (2001) wondered if the same principle could be applied to plants, exemplified by Asteraceae.

A parsimony analysis of endemicity and a track compatibility analysis were carried out (Crisci *et al.*, 2001) to seek historical explanations for the high diversity and endemicity of Asteraceae in these island-like mountain ranges and the relationships of the biota inhabiting Tandilia and Ventania with other areas.

## Material and methods

### Delimitation of areas

Thirteen areas were delimited for this study on the basis of several overlapping taxon distributions and phytogeographical and geological criteria: Central Chile (CC), Central and North America (CNA), Chaco (CHA), Mahuidas (MA), northern Andes (NA), Pampa (PAM), Patagonia (PAT), Sierras Pampeanas (SPA), Sierras Subandinas (SA), southern Brazil (SB), Tandilia (TA), Uruguay (UR) and Ventania (VE). Mahuidas, Pampeanas and Subandinas are mountain ranges west of Tandilia and Ventania.

### Selection of taxa

Distributional data of the 112 species and infraspecific taxa of Asteraceae inhabiting Tandilia and Ventania were analysed. Data were obtained from the labels of specimens (*c.*2500, correctly identified prior to data extraction) at the Herbarium of La Plata, Argentina (LP) and from the literature (checklists, floristic treatments and revisionary studies). Adventitious, cultivated, naturalised and cosmopolitan species were excluded from the analysis.

### Track compatibility analysis

Distributions of the 112 taxa were plotted on maps and their separate distribution localities were connected with lines (individual tracks) according to their geographic proximity. The constructed data matrix (13 areas × 112 tracks) was analysed using SECANT 2.2 (Salisbury, 1999).

### *Parsimony analysis of endemicity*

The same areas and matrix used in the track compatibility analysis were used and analysed for PAE, using PAUP* version 4.0 beta 2 (Swofford, 1999) and applying the branch-and-bound option. A strict consensus tree was constructed based on the results generated from PAUP*. The cladogram was rooted with a hypothetical area coded all zeros.

## ***Results***

### *Track compatibility analysis*

Analysis of the data matrix resulted in four largest cliques (generalised tracks) of 30 individual tracks each (Fig. 3.3A–D). The intersection of 27 individual tracks common to the four largest cliques is identified as the fifth clique or generalised track (Fig. 3.3E). In this way, six individual tracks are combined in groups of three with the 27 individual tracks to form the four cliques of 30 individual tracks. The main distributional pattern obtained connects Tandilia and Ventania with southern Brazil, Pampa, Uruguay and Sierras Pampeanas.

### *Parsimony analysis of endemicity*

Analysis of the data matrix generated four area cladograms (Fig. 3.4A–D). The strict consensus of the four trees (Fig. 3.4E) shows the following area relationships:

(Mahuidas (Patagonia (Chile (Sierras Subandinas (Sierras Pampeanas (Ventania (Pampa (southern Brazil-Tandilia-Uruguay))))))).

### *Comparison of methods*

Comparison of the results obtained applying track compatibility analysis and parsimony analysis of endemicity to the areas under study shows a coincidence in the relationships of Sierras Pampeanas (SPA), Ventania (VE), Pampa (PAM), southern Brazil (SB), Tandilia (TA) and Uruguay (UR). The areas relationship PAM-SB-SPA-TA-UR-VE forms a main distributional pattern strongly supported by the two approaches, and represented in Fig. 3.5 as a generalised track. Both methods show that Tandilia is more closely related to Uruguay, southern Brazil and Pampa than to Ventania or other austral areas. In addition, PAE and track compatibility analysis resulted in different relationships of the remaining areas with the main distributional pattern or generalised track.

## ***Discussion***

The main distributional pattern connecting Tandilia and Ventania with southern Brazil, Pampa, Uruguay and Sierras Pampeanas agrees with the previous

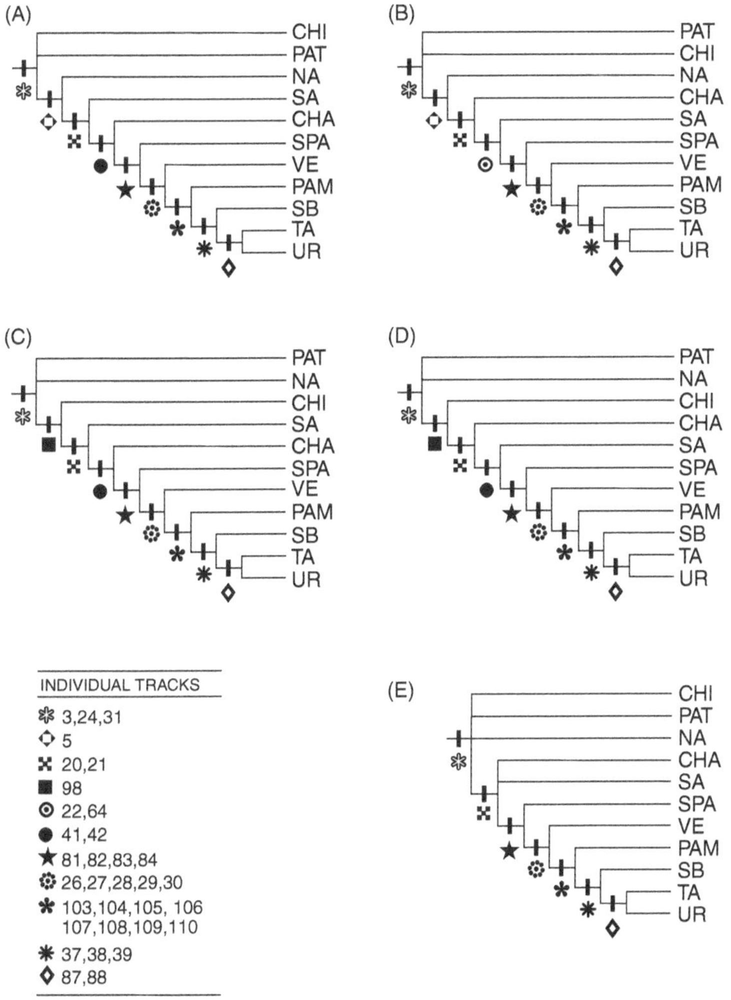


Figure 3.3 Largest cliques resulting from track compatibility analysis of Asteraceae from Tandilia and Ventania. A–D: Four largest cliques (generalised tracks) of 30 individual tracks each. E: Intersection of 27 individual tracks common to the four largest cliques. Abbreviations as in text. Each number corresponds to a taxon name as cited in the original list of taxa (see Crisci *et al.*, 2001).

hypothesis. A biotic migratorial route from the Andes to Brazil, through Sierras Pampeanas, the sierras of Buenos Aires, the sierras of Uruguay and Planalto and the sierras of Brazil was postulated many years ago (e.g. Hicken, 1918–1919). Frenguelli (1950) gave the name of peripampasic orogenic arc ('arco peripampásico serrano') to a biotic corridor constituting Sierras Pampeanas, Mahuidas and Tandilia. De la Sota (1967) concluded that the pteridological flora of Tandilia and Ventania have four origins – Austral-Antarctic, Andean-Pampean, Austrobrazilian and endemic – and that these mountain ranges constitutes an

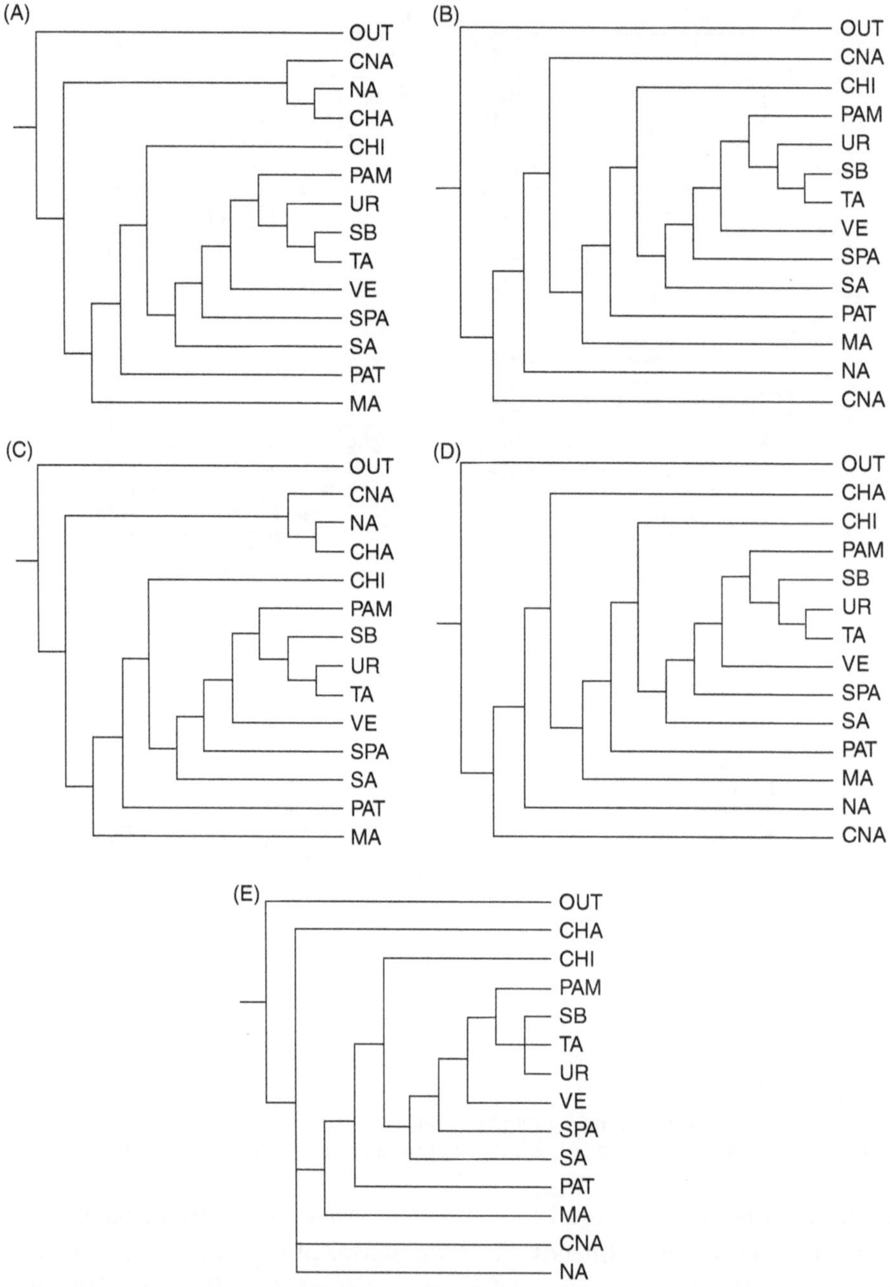


Figure 3.4 Cladograms resulting from parsimony analysis of endemicity of Asteraceae from *Tandilia* and Ventania. A–D: Four equally parsimonious cladograms. E: Strict consensus cladogram. Abbreviations as in text. Modified from Crisci *et al.* (2001).

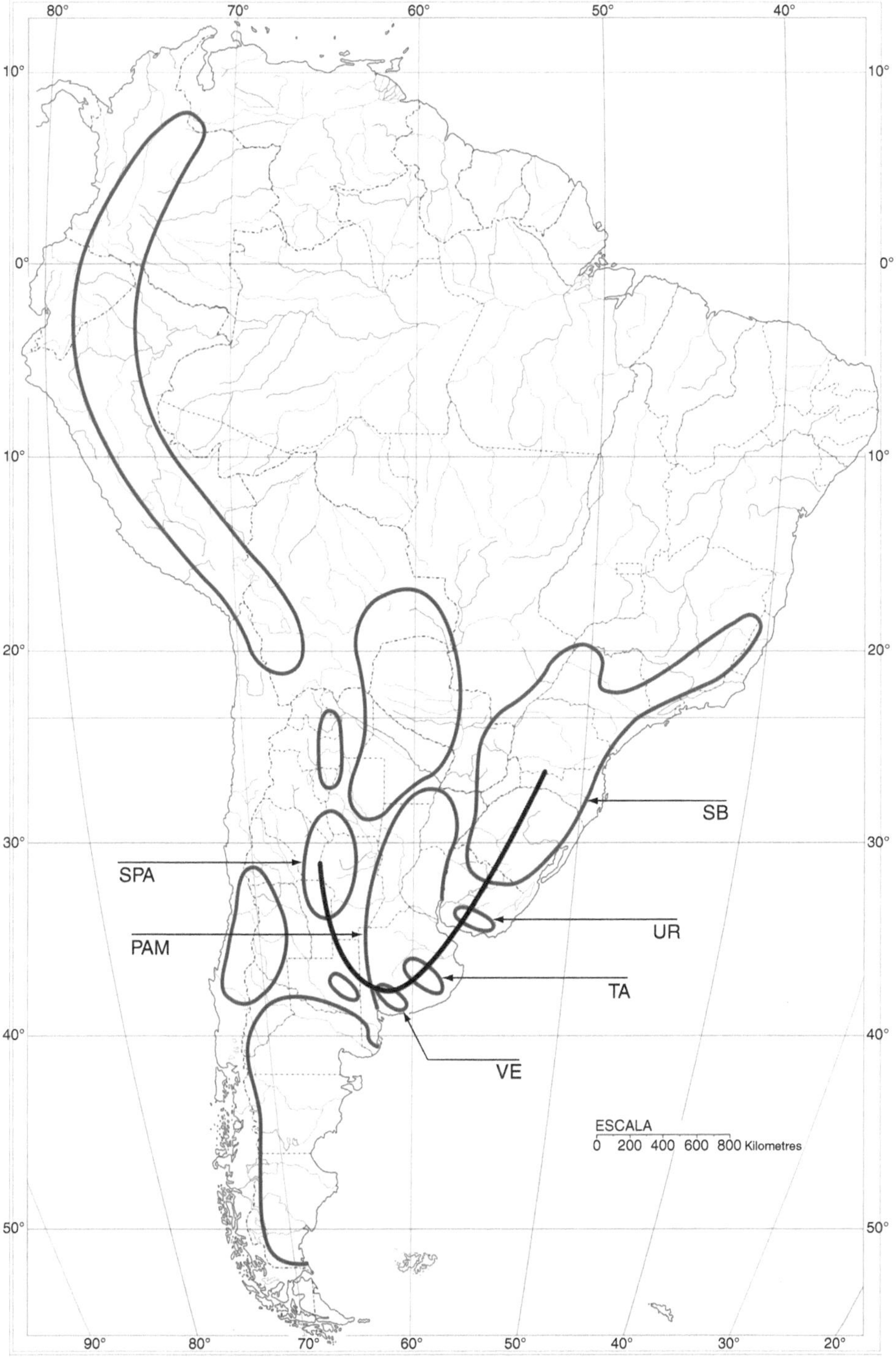


Figure 3.5 Map showing the main distributional pattern of Asteraceae from Tandilia and Ventania represented as a generalised track according to track compatibility analysis and parsimony analysis of endemicity. Abbreviations as in text. Modified from Crisci *et al.* (2001).

intermediate orophilous station on a migratorial route. Therefore, the study of different plant and animal taxa inhabiting Tandilia and Ventania, including this example of Asteraceae, coincide with the idea of these mountain ranges being part of a biotic corridor.

The relationships of the Asteraceae inhabiting Tandilia with Uruguay and southern Brazil rather than with Ventania coincide with the different origins of these mountain ranges. Tandilia was a part of the Rio de la Plata craton together with northeastern Argentina, east Paraguay, Uruguay and southeast Brazil. Ventania, on the other hand, was the result of the collision of a Patagonian block that drifted from the south. The family Asteraceae was hypothesised to have originated in southern South America during the Oligocene (Stuessy *et al.*, 1996), long before Tandilia and Ventania were raised. The uplift of the Andes and the Pleistocene glaciations may have significantly affected the evolution of this family in South America. The former event resulted in drier climatic conditions beginning in the Miocene (25.5 Ma) and in additional uplift of several mountain systems. The Quaternary glaciations resulted in dry and wet cycles that caused fragmentation and differentiation of populations (Simpson Vuilleumier, 1971; Simpson, 1975). The main distributional pattern found in this study and the high endemicity of Asteraceae now found in Tandilia and Ventania could be explained by these two events that resulted in eventual fragmentation of populations in more elevated regions, restricted now to rocky and loose soils.

## Conclusions

The two examples presented here, applying PAE and panbiogeography, illustrate the importance of geographic information as spatial data to infer the biogeographic history of organisms. The use of phylogenies in most historical biogeographic studies, which has become mandatory in the last years, has obscured the great relevance of the use of space information on biogeography.

Biogeography is fundamentally based in spatial thinking and it would be an error to disregard spatial information as visualised by spatial thinking, i.e. the use of space as a framework for scientific understanding of the nature, structure and function of evolutionary phenomena. Panbiogeography and PAE are historical biogeographic approaches which apply spatial thinking, since both of them use the spatial information of taxa as the main source to uncover their distributional histories.

Islands (oceanic, continental or habitat islands) are perfect subjects for biogeographical analysis, since they have clear limits which separate each from its surrounding areas. Their isolation generates questions about the relationship of their biotas, questions which can only be answered in terms of spatial thinking.

## Acknowledgements

The authors would like to thank David Bramwell for his kind invitation to participate on this book. The authors are affiliated to Consejo Nacional de Investigaciones Científicas y Técnicas (CONICET) and to Universidad Nacional de la Plata (UNLP), Argentina, whose continuous support they acknowledge. This work was financed by CONICET grant PIP 0729 and ANPCyT grant PICT 2007–01977.

## References

Allison, J. S. & Smith, F. I. L. (1973). The vegetation of Elephant Island, South Shetland Islands. *Br. Antarct. Surv. Bull.* 33 & 34: 185–212.

Baldwin, B. G. & Robichaux, R. H. (1995). Historical biogeography and ecology of the Hawaiian silversword alliance (Asteraceae). In W. L. Wagner & V. A. Funk (eds.), *Hawaiian Biogeography: Evolution on a Hot Spot Archipelago*, pp. 259–287. Washington, DC, Smithsonian Institution Press.

Bergstrom, D. N. & Chown, S. L. (1999). Life at the front: history, ecology and change on Southern Ocean islands. *Trends Ecol. Evol.* 14: 472–477.

Bergstrom, D. N., Turner, P. A. M., Scott, J., Copson, G. & Shaw, J. (2006). Restricted plant species on sub-Antarctic Macquarie and Heard Islands. *Polar Biol.* 29: 532–539.

Brown, J. H. & Lomolino, M. B. (1998). *Biogeography*, 2nd edition. Sunderland, MA, Sinauer.

Cabrera, A. L. (1978). La vegetación de Patagonia y sus relaciones con la vegetación altoandina y puneña. In C. Troll & W. Lauer (eds.), *Geoecological Relations between the Southern Temperate and the Tropical Mountains*, pp. 329–343. Wiesbaden, Franz Steiner.

Calsbeek, R. & Smith, T. B. (2003). Ocean currents mediate evolution in island lizards. *Nature* 426: 552–555.

Convey, P. (1996). Reproduction of Antarctic flowering plants. *Antarct. Sci.* 8: 127–134.

Convey, P., Lewis Smith, R. I., Hodgson, D. A. & Peat, H. J. (2000). The flora of South Sandwich Islands, with particular reference to the influence of geothermal heating. *J. Biogeogr.* 27: 1279–1295.

Craw, R. C. (1988). Continuing the synthesis between panbiogeography, phylogenetic systematics and geology as illustrated by empirical studies on the biogeography of New Zealand and the Chatham Islands. *Syst. Zool.* 37: 291–310.

Craw, R. C. (1989). Quantitative panbiogeography: introduction to methods. *New Zeal. J. Zool.* 16: 485–494.

Craw, R. C., Grehan, J. R. & Heads, M. J. (1999). *Panbiogeography: Tracking the History of Life*. New York, Oxford University Press.

Crisci, J. V., Cigliano, M. M., Morrone, J. J. & Roig–Juñent, S. (1991). Historical biogeography of southern South America. *Syst. Zool.* 40: 152–171.

Crisci, J. V., Freire, S. E., Sancho, G. & Katinas, L. (2001). Historical biogeography of the Asteraceae from Tandilia and Ventania mountain ranges (Buenos Aires, Argentina). *Caldasia* 23 (1): 21–41.

Crisci, J. V. & Katinas, L. (2009). Darwin, historical biogeography, and the need of overcoming binary opposites. *J. Biogeogr.* 36: 1027–1031.

Crisci, J. V., Katinas, L. & Posadas, P. (2003). *Historical Biogeography: An Introduction.* Cambridge, MA, Harvard University Press.

Crisci, J. V., Sala, O., Katinas, L. & Posadas, P. (2006). Bridging historical and ecological approaches in biogeography. *Aust. Syst. Bot.* 19: 1–10.

Croizat, L. (1952). *Manual of Phytogeography.* The Hague, Junk.

Croizat, L. (1958). *Panbiogeography,* Vols. I, IIa, and IIb. Caracas, Venezuela, published by the author.

Croizat, L. (1964). *Space, Time, Form: The Biological Synthesis.* Caracas, Venezuela, published by the author.

Croizat, L. (1981). Biogeography: past, present, and future. In G. Nelson & D. E. Rosen (eds.), *Vicariance Biogeography: A Critique,* pp. 501–523. New York, Columbia University Press.

Darwin, C. (1859). *On the Origin of Species by Means of Natural Selection or the Preservation of Favoured Races in the Struggle for Life.* London, John Murray.

Dawson, J. W. (1958). Interrelationships of the Australasian and South American floras. *Tuatara* 7: pp.6. Electronic version.

de Candolle, A. P. (1820). Géographie botanique. *Dict. Sci. Nat.* 18: 359–422.

de la Sota, E. A. (1967). Composición, origen y vinculación de la flora pteridológica de la sierras de Buenos Aires (Argentina). *Bol. Soc. Argentina Bot.* 11: 105–128.

Desalle, R. (1995). Molecular approaches to biogeographic analysis of Hawaiian Drosophilidae. In W. L. Wagner & V. A. Funk (eds.), *Hawaiian Biogeography: Evolution on a Hot Spot Archipelago,* pp. 72–89. Washington, DC, Smithsonian Institution Press.

Duncan, R. A. (2002). A time frame for construction of the Kerguelen Plateau and Broken Ridge. *J. Petrol.* 43: 1109–1119.

Francisco-Ortega, J., Crawford, D. J., Santos-Guerra, A. & Jansen, R. K. (1997). Origin and evolution of *Argyranthemum* (Asteraceae: Anthemidae) in Macaronesia. In T. J. Givinish & K. J. Sytsma (eds.), *Molecular Evolution and Adaptive Radiation,* pp. 407–431. Cambridge, Cambridge University Press.

Frenguelli, J. (1950). Rasgos generales de la morfología y la geología de la provincia de Buenos Aires. La Plata, Argentina: L.E.M.I.T. Serie 2, n. 33, pp. 72.

Garzón-Orduña, I. J., Miranda-Esquivel, D. R. & Donato, M. (2008). Parsimony analysis of endemicity describes but does not explain: an illustrated critique. *J. Biogeogr.* 35: 903–913.

Goloboff, P., Farris, J. & Nixon, K. (2003). T.N.T. Tree Analysis using New Technology. Program and documentation, available from the authors and www.zmuc.dk/public/phylogeny.

Greene, S. W. & Groves, E. W. (1963). The W. N. Bonner (1955–61) collection of plants from South Georgia. *Br. Antarct. Surv. Bull.* 2: 93–96.

Heaney, L. R. (2007). Is a new paradigm emerging for oceanic island biogeography? *J. Biogeogr.* 34: 753–757.

Helfgott, D. M., Francisco-Ortega, J., Santos-Guerra, A., Jansen, R. K. & Simpson, B. B. (2000). Biogeography and breeding system evolution of the woody *Bencomia* alliance (Rosaceae) in Macaronesia based on ITS sequence data. *Syst. Bot.* 25: 82–97.

Hicken, C. M. (1918–1919). La migración de los helechos en la flora de Tucumán. Pp. 187–209. *Primera Reunión Nacional de la Sociedad Científica Argentina de Ciencias Naturales*, Tucumán, Argentina.

Humphries, C. J. (1989). Any advance in assumption 2? *J. Biogeogr.* 16: 101–102.

Katinas, L., Morrone, J. J. & Crisci, J. V. (1999). Track analysis reveals the composite nature of the Andean biota. *Aust. J. Bot.* 47: 111–130.

Katinas, L., Crisci, J. V., Wagner, W. & Hoch, P. C. (2004). Geographical diversification of tribes Epilobieae, Gongylocarpeae, and Onagreae (Onagraceae) in North America, based on parsimony analysis of endemicity and track compatibility analysis. *Ann. Missouri Bot. Gard.* 91: 159–185.

Keast, A. (1991). Panbiogeography: then and now. *Q. Rev. Biol.* 66: 467–472.

Kristensen, M. J. & Frangi, L. (1995). La Sierra de la Ventana: una isla de biodiversidad. *Ciencia Hoy (Argentina)* 5 (30): 25–34.

Lindsay, D. C. (1971). Vegetation of the South Shetland Islands. *Br. Antarct. Surv. Bull.* 25: 59–83.

MacArthur, R. H. (1984). *Geographical Ecology: Patterns in the Distribution of Species.* Princeton, NJ, Princeton University Press.

MacArthur, R. H. & Wilson, E. O. (1963). An equilibrium theory of insular biogeography. *Evolution* 17: 373–387.

MacArthur, R. H. & Wilson, E. O. (1967). *The Theory of Insular Biogeography.* Princeton, NJ, Princeton University Press.

Meacham, C. (1984). Evaluating characters by character compatibility analysis. In T. Duncan & T. F. Stuessy (eds.), *Cladistics: Perspectives on the Reconstruction of Evolutionary History*, pp. 152–165. New York, Columbia University Press.

Meurk, C. D., Foggo, M. N. & Wilson, B. (1994). The vegetation of Subantarctic Campbell Island. *New Zeal. J. Ecol.* 18: 123–168.

Moore, D. M. (1983). *Flora of Tierra del Fuego.* St Louis, Anthony Nelson, Missouri Botanical Garden.

Morrone, J. J. (1998). On Udvardy's Insulantarctic province: a test from the weevils (Coleoptera: Curculionidae). *J. Biogeogr.* 25: 947–955.

Morrone, J. J. (2001). Homology, biogeography and areas of endemism. *Diversity Distrib.* 7: 1–3.

Myers, A. A. & Giller, P. S. (eds.) (1988). *Analytical Biogeography: An Integrated Approach to the Study of Animal and Plant Distributions.* New York, Chapman & Hall.

National Research Council (2006). *Learning to Think Spatially.* Washington, DC, National Academy Press, www.nap.edu.

Nelson, G. & Platnick, N. I. (1981). *Systematics and Biogeography: Cladistics and Vicariance.* New York, Columbia University Press.

Niemelä, J. (1990). Habitat distribution of carabid beetles in Tierra del Fuego, South America. *Entomol. Fennica* 1: 3–16.

Ortiz-Jaureguizar, E. & Cladera, G. A. (2006). Paleoenvironmental evolution of southern South America during the Cenozoic. *J. Arid Environ.* 66: 498–532.

Posadas, P. (2008). A preliminar overview of species composition and geographical distribution of Malvinian weevils (Insecta: Coleoptera: Curculionidae). *Zootaxa* 1704: 1–26.

Posadas, P. & Donato, M. (2007). Everything you always wanted to know about historical bioegography, but were afraid to ask: a preliminary overview based on papers published in *Journal of Biogeography* 2005–2006. *Biogeografía* 2: 1–6.

Posadas, P. & Morrone, J. J. (2004). A new species of *Antarctobius* from Islas Malvinas (Coleoptera: Curculionidae: Cyclominae). *Insect Syst. Evol.* 35: 353–359.

Rosen, B. R. (1988). From fossils to earth history: applied historical biogeography. In A. A. Myers & P. S. Giller (eds.), *Analytical Biogeography: An Integrated Approach to the Study of Animal and Plant Distribution*, pp. 437–481. New York, Chapman & Hall.

Rosen, B. R. & Smith, A. B. (1988). Tectonics from fossils? Analysis of reef-coral and sea-urchin distributions from Late Cretaceous to Recent, using a new method. In M. G. Audley-Charles & A. Hallam (eds.), *Gondwana and Tethys*, pp. 275–306. Oxford, Oxford University Press.

Salisbury, B. A. (1999). *SECANT: Strongest Evidence Compatibility Analytic Tool. Version 2.2*. New Haven, CT, Department of Ecology and Evolutionary Biology, Yale University.

Simpson, B. B. (1975) Pleistocene changes in the flora of high tropical Andes. *Paleobiology* 1: 273–294.

Simpson Vuilleumier, B. B. (1971). Pleistocene changes in the fauna and flora of South America. *Science* 173: 771–774.

Stuessy, T. F., Sang, T. & de Vore, M. L. (1996). Phylogeny and biogeography of the subfamily Barnadesioideae with implications for early evolution of the Compositae. In D. J. N. Hind & H. J. Beentje (eds.), *Compositae: Systematics. Proceedings of the International Compositae Conference Kew, 1994*, Vol. 1, pp. 463–490. London, Royal Botanical Gardens Kew.

Swofford, D. L. (1999). *PAUP*: Phylogenetic Analysis Using Parsimony. Version 4.0 beta 2*. Sunderland, MA, Sinauer Associates.

Tuhkanen, S., Kuokka, I., Hyvönen, J., Stenroos, S. & Niemelä, J. (1990). Tierra del Fuego as a target for biogeographical research in the past and present. *Anal. Inst. Patagonia* 19 (2): 1–107.

Vidal, N., Azvolinski, A., Cruaud, C. & Hedges, S. B. (2008). Origin of tropical American burrowing reptiles by transatlantic rafting. *Biol. Lett.* 4: 115–118.

Wallace, P. J., Frey, F. E., Weis, D. & Coffin, M. F. (2002). Origin and evolution of the Kerguelen Plateau, Broken Bridge, and Kerguelen archipelago: Editorial. *J. Petrol.* 43: 1105–1108.

Whinan, J., Chilcott, N. & Bergstrom, D. M. (2005). Subantarctic hitchhikers: expeditioners as vectors for the introduction of alien organisms. *Biol. Conserv.* 121: 207–219.

Yoder, A. D. & Nowak, M. D. (2006). Has vicariance or dispersal been the predominant biogeographic force in Madagascar? Only time will tell. *Annu. Rev. Ecol. Evol. Syst.* 37: 405–431.

Zachos, J., Pagani, M., Sloan, L., Thomas, E. & Billups, K. (2001). Trends, rhythms, and aberrations in global climate 65 Ma to Present. *Science* 292: 686–693.

Zuloaga, F. O., Morrone, O. & Belgrano, M. J. (eds.) (2008). Catálogo de las plantas vasculares del cono sur (Argentina, sur de Brasil, Chile, Paraguay y Uruguay). *Monogr. Syst. Bot. Mo. Bot. Gard.* 107, 3 vols.

# 4

# Origin and evolution of Hawaiian endemics: new patterns revealed by molecular phylogenetic studies

STERLING C. KEELEY AND VICKI A. FUNK

The current high islands of the Hawaiian archipelago are among the most remote land masses in the world. They lie 3500 km from California, the nearest continental source, and approximately 2300 km from the Marquesas, the nearest islands (Fig. 4.1). They are the southernmost islands in the Hawaiian Ridge, formed successively over a 'hot spot' that has allowed magma to penetrate the Pacific Plate. The plate has moved gradually north and northwestwards over the past 85 Ma, leaving the previously formed islands to gradually erode and subside (Clague, 1996). The current high islands (Fig. 4.1, inset) range in age from Kauai/Niihau (5.1–4.9 Ma), to Oahu (3.7–2.6 Ma), to Maui Nui (2.2–1.2 Ma), during the Pleistocene comprising several islands – West Maui (1.3 Ma), East Maui (0.75 Ma), Molokai (1.76–1.90 Ma), Lanai (1.28 Ma) and Kaho'olawe (1.03 Ma) – and Hawaii (0.5 Ma to present) (Price & Clague, 2002).

Important for the establishment and evolution of the extant Hawaiian flora is the historic pattern of island formation within the archipelago. For example, islands with elevations greater than 1000 m did not exist from 30 to 23 Ma and from *c.* 8 to 5 Ma when the current high islands began to emerge (Clague, 1996; Price & Clague, 2002; Clague *et al.*, 2010). These gaps likely precluded colonisation of the current high islands by taxa on the older islands, although some may have persisted long enough to disperse to the current high islands (Clague *et al.*, 2010). As a result of the history of the islands' formation and data from recent molecular studies, the processes of colonisation and diversification are generally assumed to have taken place within the past *c.* 5 Ma, as a result of long-distance dispersal (Price & Clague, 2002; Price & Wagner, 2004; Cowie & Holland, 2006).

Hawaii's flowering plants are *c.* 90% endemic, and its fern flora is 70% endemic, one of the highest percentages in the world. They arose from an estimated 263–270

*The Biology of Island Floras*, eds. David Bramwell and Juli Caujapé-Castells. Published by Cambridge University Press. 

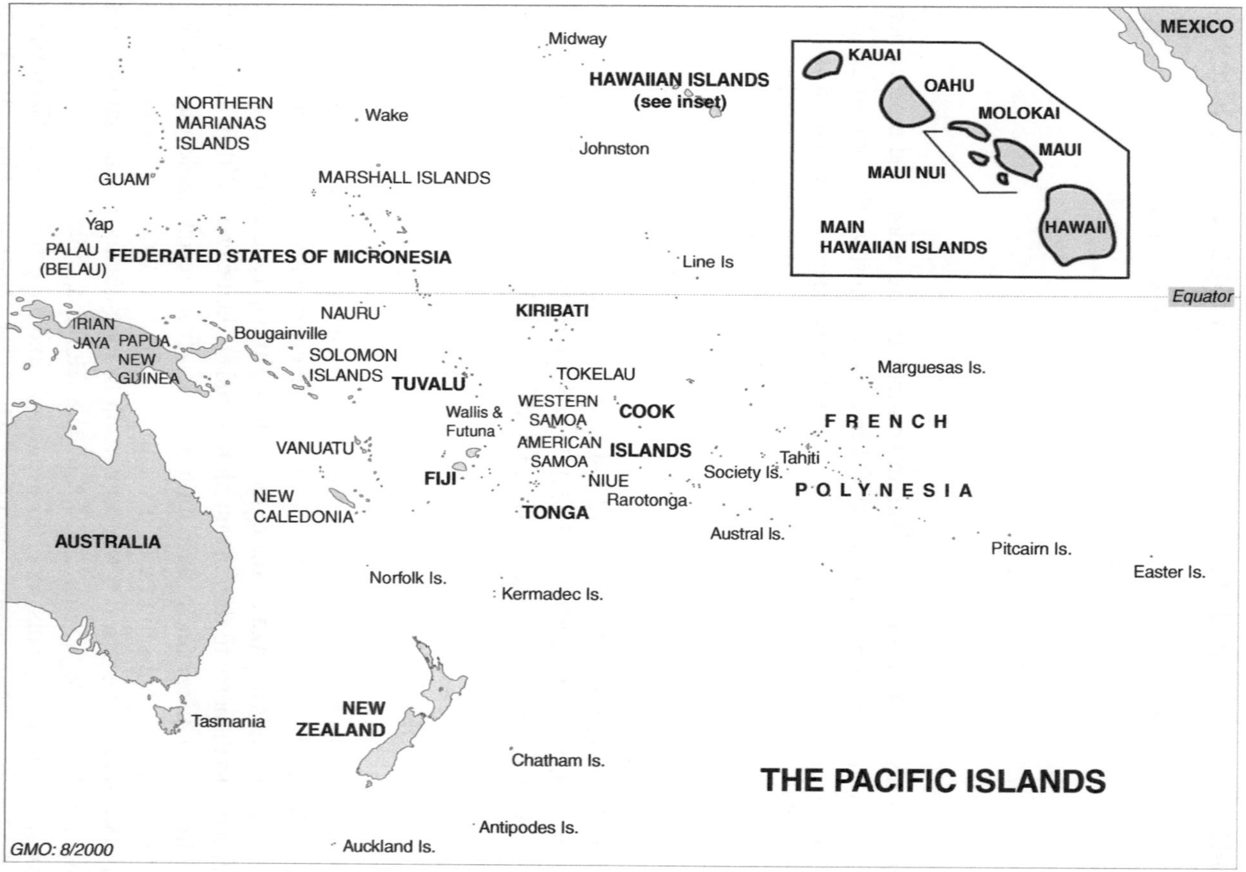


Figure 4.1 The Pacific islands, Australia, New Guinea, New Zealand and Mexico. Inset: the main Hawaiian Islands. Used by permission, C. Gemmill.

original colonists which have given rise to approximately 1030 species of flowering plants and 162 species of ferns for a total of 1192 species of vascular plants (Palmer, 2003; Price & Wagner, 2004). As a result the diversification and spread of genera within and between islands, the factors influencing their radiations, the sources of colonists and the morphological and genetic diversity of Hawaiian plants have been of interest from both evolutionary and biogeographic perspectives as the islands offer an unusual opportunity for separating these factors from the more complex interactions responsible for the creation of continental floras (Fosberg, 1948; Carlquist, 1970, 1974; Wagner *et al.*, 1995).

Wagner *et al.* (1990, 1999) listed the predominate source of the Hawaiian flora as Malesia (the region including the Malay Peninsula, Indonesia, the Philippines and New Guinea) with other relatives in Australia, New Zealand and the Pacific. However, recent work has shown that boreal, temperate and tropical regions of the Americas, Asia and Africa have also served as the source of Hawaiian genera. Dispersal has been largely by birds, with fruits carried either externally or internally, but also by wind and water (Carlquist, 1970, 1974; Wagner *et al.*, 1995; Ballard & Sytsma, 2000; Price & Wagner, 2004). Wagner and Funk (1995) brought together much of the information available at that time on Hawaiian plant and animal groups. The overarching question they attempted to answer was: Do lineages of plants and animals follow biogeographic patterns similar to the geologic pattern of island formation? That is, are the basal taxa on the oldest islands and do they move in a regular pattern to younger islands (the Progression Rule)? The largely morphological data were analysed using a consistent cladistic approach. Authors of the individual chapters also discussed the speciation and radiation of various lineages, the likely age of colonisations, habitat shifts and back-colonisations.

Molecular studies since Wagner and Funk (1995) have expanded the scope beyond the patterns within the islands themselves to phylogenetic relationships across broad geographical regions. These studies have revealed new sources of colonists, clarified the number of colonisation events, added information on the age of several lineages and revealed a role for the Hawaiian Islands within the Pacific that were not previously known. Additional data on hybridisation and polyploidy have also contributed to our understanding of the origin and evolution of the Hawaiian flora. A selection of these studies is presented below to illustrate the main findings.

## Geographic sources of colonists

The endemic Hawaiian flora is the product of colonisation from a variety of geographic sources (Wagner *et al.*, 1990; 1999). Fosberg (1948) and Carlquist (1970) suggested that the majority of species lineages were derived from ancestors in the tropical Western Pacific, but believed that other areas were to be expected from the Americas and perhaps more distant sources. Contributions by Lowrey, and Baldwin and Robichaux in Wagner and Funk (1995), using molecular data, identified New Guinea and western North America, respectively, as sources of Hawaiian *Tetramolopium* and the silversword alliance, both in the family Compositae, while geographical relationships of other genera often remained unconfirmed pending further study. Molecular studies since that time have revealed that the ancestors of the Hawaiian flora came from other areas separated from the Hawaiian Islands by great over-water and climatic distances. Among the most unusual of these are colonists from Africa and the Arctic or boreal regions. Additional colonisations have also been reported for taxa originating in western North America, Australia, New Zealand and Asia. For some of these taxa the area of origin may change with the addition of more outgroup taxa, as even now differing interpretations are possible in some cases. Regardless, it is clear that the Hawaiian flora is the product of colonisations from geographically diverse ancestors.

## *Africa*

An unexpected source of direct colonisation to the Hawaiian Islands was revealed for the endemic genera *Hesperomannia* and *Kokia*. *Hesperomannia* (Fig. 4.2) was shown, unequivocally, to be derived from an African/Madagascan ancestor (Kim *et al.*, 1998). Although this was unexpected for a Hawaiian group, both the nuclear and chloroplast data confirmed its placement in the Compositae tribe, Vernonieae, rather than in the South American tribe, Mutisieae, where it had previously been placed. Futhermore, phylogenetic study of Old and New World Vernonieae (Keeley *et al.*, 2007) continues to support the placement of *Hesperomannia* near the base of the phylogeny. The three species of *Hesperomannia* form a monophyletic group nested within a clade of African/Madagascan taxa. Kim *et al.* (1998) estimated that the time of divergence between the African ancestor and *Hesperomannia* at 17–26 Ma ± 1–3.43 Ma depending on gene region used for dating. They postulated the ancestor of *Hesperomannia* arrived in the archipelago when one of the northwestern Hawaiian Islands was still above water.

However, it is also possible that the ancestor of *Hesperomannia* was growing somewhere else during this period and arrived in Hawaii more recently. The estimated divergence among the extant species of *Hesperomannia* suggests that their radiation occurred between 1.8 and 4.9 Ma (Kim *et al.*, 1998; Price & Wagner, 2004), within the time frame of the current high islands. The only other taxon currently proposed to have an African origin (Table 4.1) is *Kokia*. Seelanan *et al.* (1997) used nuclear chloroplast sequences in a worldwide study of the cotton genus *Gossypium*. In the course of their study it was discovered that the African/

Figure 4.2 Endemic Hawaiian genera: (a) *Cyrtandra lessoniana* Gard, (b) *Hesperomannia arborescens* A. Gray, (c) *Argyroxiphium sandwicensis* DC., (d) *Clermontia parviflora* Gaud. ex A. Gray, (e) *Melicope clusiifolia* (A. Gray) T. G. Hartley & B. C. Stone, (f) *Phyllostegia ambigua* (A. Gray) Hillebr. All photos by G. Carr except (c) by V. Funk. For colour figure see www.cambridge.org/9780521118088

Madagascan genus *Gossypioides* was sister to the Hawaiian genus *Kokia* and that the monophyletic *Gossypium* was the sister of the *Kokia* + *Gossipioides* clade. Using molecular clocking methods it was estimated that *Gossypium* diverged from its sister clade, *Kokia* + *Gossipioides*, approximately 12.5 Ma. The small amount of sequence difference between *Kokia* and *Gossypioides* resulted in a divergence between these two genera well within the age range of the modern high islands of Hawaii. It was thought that the ancestor of *Kokia* was most likely dispersed over water directly from Africa/Madagascar as there are no known stopping places between these two regions (Seelanan *et al.*, 1997). The origin of *Kokia* may need

Table 4.1 *Genera, number of species, estimated age and area of origin of Hawaiian endemics.*

| Name | No. of species | Age (Ma) | Geographic origin | Type of data | Reference |
|---|---|---|---|---|---|
| Araliads: *Tetraplasandra, Reynoldsia, Munroidendron* (Araliaceae) | 9 | no date | Pacific or palaeotropics | ITS, 5S-NTS, morphology | Costello & Motley, 2007 |
| *Bidens* (Compositae) | 19 | no date | Western North America-Meso-America | ITS | Kimball & Crawford, 2004 |
| *Broussaissia arguta* (Hydrangeaceae) | 1 | no date | Asia | matK, rbcL | Hufford *et al.*, 2001 |
| *Chamaesyce* (Euphorbiaceae) | 27 | no date | North America – Mexico | ITS, trn*L-F*, trn*H*-psb*A*, ndhA intron, nuclear DEADX9 | Yang *et al.*, 2009 |
| *Cheirodendron* (Araliaceae) | 5 | no date | uncertain | ITS | Plunkett & Lowry, 2001 |
| *Cuscuta sandwichiana* (Convolvulaceae) | 1 | no date | North America (possibly) | 26S | Neyland, 2001; Stefanovic *et al.*, 2007 |
| *Cyrtandra* (Gesneriaceae) | 58 | no date | Asia-Pacific | ITS | Cronk *et al.*, 2005 |
| *Euphorbia* (Euphorbiaceae) | 1 | no date | Australia, Melanesia | ITS, *ndhF* | Steinmann & Porter, 2002 |
| *Geranium* (Geraniaceae) | 7 | 2 | Western North America | rbcL | Pax *et al.*, 1997 |
| *Gossypium tomentosum* (Malvaceae) | 1 | 1–2 | Meso-America | DNA restriction site | DeJoode & Wendle, 1992 |
| *Gunnera* (Gunneraceae) | 2 | no date | North America or the Americas | ITS, morphology | Wanntorp & Wanntorp, 2003 |
| *Hedyotis* (Rubiaceae) | 20 | no date | uncertain | ITS | Motley *et al.*, 1998 |
| *Hesperomannia* (Compositae) | 3 | 17–26 (+1–3.43) 20.5 / 3.36 | Africa | ITS, ndhF | Kim *et al.*,1998; Price & Wagner, 2004 |

| | | | | | |
|---|---|---|---|---|---|
| *Hillebrandia sandwicensis* (Begoniaceae) | 1 | 15 | Australasia or the tropics | 18S, rbcL, ITS | Clement *et al.*, 2004 |
| *Kanaloa* (Leguminosae) | 1 | no date | uncertain | nr 5.8S; ITS | Hughes *et al.*, 2003 |
| *Kokia* (Malvaceae) | 4 | 3 | Africa or tropics | ITS, ndhF | Seelanan *et al.*, 1997 |
| *Korthalsella* (Santalaceae) | 6 (or 2) | no date | Australasia | ITS, trnL-F; 18S, rbcL | Molvray *et al.*, 1999 |
| *Labordia* (Loganiaceae) | 16 | no date | Asia-Pacific | morphology, hybridisation studies, chromosome number | Motley & Carr, 1998 |
| Lobeliads: *Brighamia, Clermontia, Cyanea, Delissea, Lobelia, Trematolobelia* (Lobeliaceae) | 126 | 13 | Asia-Polynesia or Africa | rpl16, rbcL, psbA-trnH, trnL-trnF, trmT-TrnL, tnrV-tnrk, atpB-rbcL | Givnish *et al.*, 2009 |
| *Lysimachia* (Myrsinaceae) | 15 | no date | Southeast Asia | trnL-F, ITS | Hao *et al.*, 2004 |
| *Melicope* (Rutaceae) | 47 | no date | Austral region | ITS, trnL-F, trnE-T, trmH-psbA | Harbaugh *et al.*, 2009b |
| *Metrosideros* (Myrtaceae) | 1–3 | 3.9 (1.4–6.3) | Pacific | ITS, ETS; trnS-trnG spacer, trnG intron, trnD-trnE-trn-trnT spacers, rpL16 spacer; microsatellites | Wright *et al.*, 2001; Percy *et al.*, 2008; Harbaugh *et al.*, 2009c |
| Mints: *Stenogyne, Haplostachys, Phyllostegia* (Labiateae) | 60 | 5 (2.6–7.4) | Western North America | rbcL, trnL, ITS 5.8 S | Lindqvist & Albert, 2002; Lindqvist *et al.*, 2003 |
| *Pittosporum* (Pittosporaceae) | 10 | no date | Melanesia | ITS | Gemmil *et al.*, 2002; Cayzer *et al.*, 2000 |
| *Plantago* (Plantaganaceae) | 6 | no date | North America | ITS, ETS, ndhF, rpl32 & rpl32-trnL | Dunbar-Co *et al.*, 2008; Wagner, pers. comm. |

Table 4.1 (*cont.*)

| Name | No. of species | Age (Ma) | Geographic origin | Type of data | Reference |
|---|---|---|---|---|---|
| *Pritchardia* (Palmae) | 23 | no date | South Pacific (New Zealand, Australia, Papua New Guinea and neighbouring islands) | supertree | Baker *et al.*, 2009; Bacon, pers. com. |
| *Psychotria* (Rubiaceae) | 11 | no date | Austral | ITS, ETS | Nepokroeff *et al.*, 1999, 2003 |
| *Rhus sandwicensis* (Anacardiaceae) | 1 | 13.5 + –3 | Asia | ITS, ndhF, trnL-F | Yi *et al.*, 2004 |
| *Rubus hawaiiensis* (Rosaceae) | 1 | no date | Western North America | ITS, ndhF | Alice & Campbell, 1999 |
| *Rubus macraei* (Rosaceae) | 1 | no date | Western North America | ITS, ndhF | Alice & Campbell, 1999 |
| *Santalum* 1st dispersal (Santalaceae) | 2 | 1.0–1.5 | Australia | ITS, ETS, trnK | Harbaugh & Baldwin, 2007; Harbaugh, 2008 |
| *Santalum* 2nd dispersal (Santalaceae) | 2 | 411–616 KY | Australia | ITS, ETS, trnK | Harbaugh & Baldwin, 2007; Harbaugh, 2008 |
| *Scaevola glabra* (Goodeniaceae) endemic tetraploid | 1 | no date | Australia | ITS, LFY, NIA, G3PDH | Howarth *et al.*, 2003; Howarth & Baum, 2005 |
| *Scaevola* (Goodeniaceae) endemic diploid clade | 8 | no date | Australia | ITS, LFY, NIA, G3PDH | Howarth *et al.*, 2003; Howarth & Baum, 2005 |
| *Scaevola taccada* (Goodeniaceae) indigenous | 1 | no date | Australia – Polynesia | ITS, LFY, NIA, G3PDH | Howarth *et al.*, 2003; Howarth & Baum, 2005 |

| | | | | | |
|---|---|---|---|---|---|
| *Schiedea* (Caryophyllaceae) | 34 | no date | Arctic | matK, trnL-F, and rps16 | Nepokroeff *et al.*, 1999, 2003; Harbaugh *et al.*, 2009a |
| *Sicyos* (Cucurbitaceae) | 14 | no date | Australia | nuclear & chloroplast | Sebastian *et al.*, 2009 |
| *Silverswords: Argyroxiphium, Dubautia, Wilkesia* (Compositae) | 30 | 5 (+0.8) | Western North America | ITS, ETS, trnK, matK, APETLA-1, APETALA-3/ TMS6 | Synthesised in Carlquist *et al.*, 2003 |
| *Silene* (Caryophyllaceae) | 7 | no date | uncertain: Americas or Mediterranean | rps 16 intron, ITS, intron 23 of RPB2 | Eggens *et al.*, 2007 |
| *Tetramolopium* (Compositae) | 11 | >1 | Australasia | rbcL | Lowrey *et al.*, 2005; Lowrey, pers. comm. |
| *Vaccinium* (Ericaceae) | 3 | no date | Arctic, Boreal | ITS, matK, ndhF | Powell & Kron, 2002 |
| *Viola* (Violaceae) | 9 | 3.7 (upper limit) | Arctic, Beringia | ITS | Ballard & Sytsma, 2000; Havran *et al.*, 2009 |

further examination, however, as the African connection is found only in the sister group to *Kokia*, while the other outgroups show a more general tropical to subtropical distribution.

### *Arctic-boreal*

Ballard and Sytsma (2000) identified the Arctic (Northern Pacific) *Viola langsdorffii* Fisch. ex Ging. as the sister to Hawaiian *Viola*, contrary to the prior view that they were descended from South American species. Using nuclear and chloroplast data (Table 4.1) to look at the relationships among *Viola* taxa in South and Central America, North America and Mexico, Asia and Europe, as well as the Hawaiian Islands, they also showed that Hawaiian species were derived, rather than basal as had been proposed based on their woody habit. They argued that a mid-Pliocene dispersal would have involved a *V. langsdorffi* already adapted to cold tundra and bogs. Although there is no tundra in Hawaii, Hawaiian violets are restricted to high elevation habitats, commonly bogs and other moist sites. The data provided an upper limit of 3.7 Ma for successful colonisation of Kauai (the oldest high island) and subsequent radiation of species within the islands.

A more recent study (Havran *et al.*, 2009), however, suggests that the original site of colonisation was likely the Maui Nui complex, thus estimating an age of 1.2–2 Ma. Ballard and Sytsma (2000) also documented the routes of migratory birds, identifying 260 species that visit Hawaii with approximately 39% coming from the Arctic and North Pacific in contrast to Australasia and the South Pacific where only about 20% of the migratory birds are found. They further suggested that the low numbers of species from Arctic ancestors may be the result of poor survival of cold-adapted species to tropical conditions or to misidentification of the source of ancestors of Hawaiian taxa.

Similarly, the three Hawaiian *Vaccinium* species (section *Macropelma* (Klotzsch) Hook. f.) were also shown to have a boreal origin (Powell & Kron, 2002). Chloroplast and nuclear DNA phylogenies showed that Hawaiian *Vaccinium* formed a monophyletic group; they had previously been placed in the wrong section. Instead of being South Pacific in origin, as had been hypothesised, the Hawaiian species were shown to be derived from within the Arctic-boreal section *Myrtillus* Dumort. *Vaccinium cereum* (L.f.) Forster (section *Macropelma*), from the Society Islands, showed affinity with the Hawaiian species based on chloroplast DNA, but this was in conflict with relationships determined by morphology and nuclear sequence data which showed Arctic origins. As a result a hybrid origin was hypothesised for this species, but the manner in which such a hybrid could be formed was not resolved.

*Schiedea* is the largest radiation in the Hawaiian Islands (34 species) with a purported Arctic origin (Wagner *et al.*, 2005). This study showed a single introduction with several different radiations within the islands. The most closely related genera are the monotypic *Honckenya* and *Wilhelmina*, which have distributions that are circumboreal and Alaskan to northeastern Asian, respectively. The authors postulated that the first colonist may have been on an island older than the current high islands (Wagner *et al.*, 2005).

### Western North America

The most thoroughly documented case of a Hawaiian species group derived from a western North American ancestor is that of the silversword complex, containing 30 species in the genera *Argyroxiphium, Dubautia* and *Wilkesia*. The origin and evolution of this group has been extensively investigated over the past 30 years by Sherwin Carlquist (Santa Barbara Botanic Garden), Bruce Baldwin (University of California, Berkeley), Gerald Carr (University of Hawaii, Manoa) and Robert Robichaux (University of Arizona) culminating in the publication of *Tarweeds and Silverswords* (Carlquist *et al.*, 2003). Phylogenetic data from nuclear and chloroplast DNA and nuclear homeotic genes (Table 4.1) along with data from crossing experiments, cytology, chemistry and anatomy were synthesised by Baldwin (2003) as part of this extensive treatment. The information from molecular work, as well as that by the other contributors, confirmed earlier hypotheses that the Hawaiian silverswords were monophyletic and derived from a single colonisation event (Carlquist *et al.*, 2003). In addition, the origin of the group from the primarily Californian 'Madia' lineage in the Madiinae (Heliantheae) was confirmed and the ancestors identified from among extant taxa, a most remarkable finding (Baldwin, 2003). The lineage is one of the largest and best understood of the Hawaiian taxa exhibiting adaptive radiation. Hybridisation and polyploidy, discussed below, were also important is the establishment and speciation of the silverswords.

Several other taxa including some with large radiations, i.e. the Hawaiian mints (60 species), *Chamaesyce* (27) and *Bidens* (19), have been identified as having a western North American origin as well (Table 4.1). The Hawaiian mints are found in three genera, *Phyllostegia, Stenogyne* and *Haplostachys*, and were previously thought to be Asian in origin. However, recent studies have shown them to be most closely related to species in the temperate North American genus *Stachys* and documented that they are the result of one colonisation event (Lindqvist & Albert, 2002; Lindqvist *et al.*, 2003). It is also likely that the 19 endemic species of *Bidens* originated from ancestors in North America, Mexico or Central America (Kimball & Crawford, 2004; Funk *et al.*, unpublished) and Hawaiian *Chamaesyce* have recently been found to be monophyletic and to be nested in a group of weedy Mexican species (Yang *et al.*, 2009).

Similarly, the six species of Hawaiian *Plantago* (Dunbar-Co *et al.*, 2008; Wagner, pers. com.), the five *Geranium* species (Pax *et al.*, 1997), and the two species of *Gunnera* (Wanntorp & Wanntorp, 2003) each form monophyletic groups based on a single introduction, most likely from North America (Table 4.1). There are also several other Hawaiian taxa related to American lineages, but these have only one known species each. The Hawaiian endemic species of *Gossypium* (DeJoode & Wendel, 1992) and *Cuscuta* (Stefanovic *et al.*, 2007) and the indigenous species of *Ilex* (Cuenoud *et al.*, 2000) were found to be most closely related to North American or Meso-American taxa. A slightly different scenario has been put forward for *Rubus* whose two endemic species, *R. hawaiensis* A. Gray and *R. macraei* A. Gray are not closely related to each other. They were found to be the result of two separate colonisations from North American ancestors (Howarth *et al.*, 1997; Alice & Campbell, 1999).

The seven species of *Silene* (Eggens *et al.*, 2007) were found to be a single radiation in Hawaii. However, the exact origin of this clade is not clear. The sister taxon is a weedy species from the Americas (*S. antirrhina* L.) but the next three clades are from the Mediterranean as well as the Middle East, Northern Africa, the Arctic and Europe. Sister to these outgroups is one from Asia reaching over into northwest North America (Eggens *et al.*, 2007). At this point it might be best to list the origin of this taxon as 'uncertain' until more North America taxa are included in the analysis.

## *Asia*

Asia has been postulated to be the site of origin of a number of Hawaiian lineages (Table 4.1). Hawaiian *Lysimachia* (Hao *et al.*, 2004) in the subgenus *Lychimachiopsis* (A. Heller) Hand.-Mazz. were suggested to be related to Malesian taxa (Wagner *et al.*, 1999) and to be the result of two colonisation events. However, a worldwide phylogenetic study of the genus using both chloroplast and nuclear sequence data (Table 4.1) strongly support the monophyly of the 15 endemic Hawaiian species and a sister group relationship with taxa in the Asian subgenus *Palladia* (Moench) Hand.-Mazz. (Hao *et al.*, 2004). For the endemic taxa sampled only a single colonisation was supported from a Southeast Asian ancestor. However, the study did not include *L. mauritiana* Lam., an indigenous Hawaiian species also in subgenus *Palladia*, so it remains to be verified that *L. mauritiana* originated from a separate colonisation event. In addition, an Asian origin for the single Hawaiian endemic species of the genus *Rhus*, *R. sandwicensis* A. Gray, was indicated in a study of Northern Hemisphere *Rhus* species employing both chloroplast and nuclear DNA sequence data (Yi *et al.*, 2004). *Rhus sandwicensis* was found to be most closely related to *R. chinensis* Mill., distributed from Asia to Malaysia, and was estimated to have diverged from it 13.5 + 3.0 Ma. Given the limited availability of high islands during this period, it is likely that the immediate ancestor of *R. sandwicensis* was somewhere in Asia or between Asia and Hawaii for a significant period before dispersing to the islands.

In addition to those taxa with a clearly Asian ancestry, two genera, *Cyrtandra* (Fig. 4.2) and *Labordia*, appear to have originated in a somewhat broader Asian-Pacific region. *Cyrtandra* (Cronk *et al.*, 2005) is widely distributed throughout the Pacific islands and Malesia and has *c.* 58 endemic species in the Hawaiian Islands. These species were thought to be the product of four to six independent colonisations (Wagner *et al.*, 1990, 1999). The phylogenetic study of Cronk *et al.* (2005) using nuclear DNA revealed a large Pacific clade that was sister to a species (*C. umbellifera* Merr.) from the Bataan Islands (located between Taiwan and the Philippines). Within this Pacific clade there were two smaller well-supported sister clades, a Hawaiian clade and a clade made up of the rest of the Micronesian (Guam, the Federated States of Micronesa, Kiribati, Marshal Islands and the Northern Mariana Islands) and Polynesian species (Cook Islands, Marquesas, Samoa, Society Islands and Tonga). However, the small sample size of Hawaiian and Pacific species and the sister group relationship between the two clades did not allow the identification of the location of the most recent common ancestor. The limited resolving power of the available data makes it impossible to distinguish between a single dispersal event to the Pacific with further dispersal to Hawaii or two independent dispersal events from Southeast Asia. The data do support the monophyly of the Hawaiian species sampled, and therefore indicate a single colonisation event to the islands. However, more study is needed to better resolve the relationships, as a second colonisation event could not be entirely discounted (Cronk *et al.*, 2005).

Like *Cyrtandra*, the origin of the endemic genus *Labordia* (Motley & Carr, 1998; Wen *et al.*, 2001) is not yet clear. The sister taxon in analyses to date appears to be in the Pacific, but related groups are also in Asia so it seems best at this time to include both as possible areas of origin. Finally, the origin of the Hawaiian lobelioids (Fig. 4.2d) is among the least well understood of those taxa whose ancestry may be Asian. Data indicate the possibility of an Asian-Polynesian origin, like its sister group, or in an altogether different place, Africa, where the closest outgroup is located. Only additional studies will resolve this problem. All six Hawaiian lobelioid genera (126 species) (Table 4.1) are the result of a single introduction (Givnish *et al.*, 2009).

## *Australia, New Zealand and the Western Pacific*

The genus *Scaevola* consists of 130 species widely dispersed throughout the Pacific. In Hawaii there are 10 species, nine endemics and one indigenous pan-Pacific species, *S. taccada* (Gaertn.) Roxb. (often referred to as *S. sericea* Vahl). Of the endemics, there are eight diploid and one tetraploid species (Wagner *et al.*, 1999). Patterson (1995), in a morphological cladistic study, postulated that all diploid Hawaiian species were the result of a single colonising event, and that the tetraploid species, *S. glabra* Hook. & Arnott, originated from a separate colonisation. The widespread *S. taccada* was found by Patterson (1995) to be the sister taxon to the Hawaiian diploid species, and New Caledonia taxa were sister to all the Hawaiian taxa. As a result it was hypothesised that there were at least two separate colonisations of the Hawaiian Islands (Patterson, 1995). Earlier Carlquist (1974) had suggested three possible colonisations, one for the diploid species, one for the tetraploid and a third for *S. taccada*. The closest relatives were previously assumed to be from nearby Pacific islands, but this was confounded by the sister group relationship of the New Caledonia species.

Howarth *et al.* (2003) used nuclear sequence data to examine the biogeography and phylogeny of *Scaevola* throughout its geographical range. The data showed a complex pattern of dispersals from Australia into the Pacific involving three different clades and six dispersal events. Three of these dispersals were to the Hawaiian Islands and each gave rise to a different lineage (Table 4.1). One of these dispersals was found to be responsible for establishment of the Hawaiian diploid clade, supporting earlier hypotheses of diploid monophyly (Patterson, 1995). However, the sister group to the diploid species was not *S. taccada*, as previously thought, but included the Cuban species, *S. wrightii* (Griseb.) M. Gómez, and the Socotran species, *S. socotraensis* St. John, along with the widespread *S. plumiere* Vahl. None of these taxa occur in Hawaii, however. *Scaevola plumiere* is morphologically similar to the Hawaiian diploid species, *S. coriacea* Nutt., and was proposed as a possible ancestor to the Hawaiian diploids (Carlquist, 1969), a finding in agreement with the data of Howarth *et al.* (2003). As a result Howarth *et al.* (2003) concluded that the original diploid colonist was likely from the Americas or Africa where *S. plumiere* is found and not from Tonga, Samoa, the Marquesas or Tahiti, all locations that are more geographically proximate to the Hawaiian Islands (Fig. 4.1) and which were earlier considered the most likely sources. The data also showed that the tetraploid species, *S. glabra*, was most closely related to two Australian species in a different section and subsection from the diploids (Howarth *et al.*, 2003). The widespread Southern Hemisphere *S. taccada* was shown to be related to taxa in Australia, New Caledonia and the South Pacific. It was established independently of other Hawaiian *Scaevola*. The three separate colonisation events responsible for establishing *Scaevola* in Hawaii are the most yet documented for any genus in the flora.

Hawaiian *Santalum* were also shown to result from more than one colonisation event. Harbaugh and Baldwin (2007) and Harbaugh (2008) investigated the relationships among Pacific *Santalum* species using nuclear and the chloroplast DNA (Table 4.1). There are four endemic species in the Hawaiian Islands, all tetraploid (Wagner *et al.*, 1999). These have been placed in sections *Solenantha* Tuyama and *Hawaiiensia* Skottsb. (and sometimes sections *Polyensia* Skottsb. and *Santalum* Skottsb.) (Skottsberg, 1930). The two species in section *Hawaiiensia, S. ellipticum* Gaud. and *S. paniculatum* Hook & Arn., were thought to be the result of a colonisation from Polynesia where one species, *S. insulare* Bertero ex DC., is widely distributed. The species in section *Solenantha, S. freycinetianum* Gaud. and *S. halaekalae* Hillebr., were postulated to be the result of colonisation from Australia or the Austral region (Wagner *et al.*, 1990, 1999). Harbaugh and Baldwin (2007) confirmed that two separate colonisation events from Australia/Papua New Guinea, but not from Polynesia, were responsible for establishing the Hawaiian species. Within their larger study of *Santalum* they found there were at least five, putatively bird-mediated (carrying fleshy fruits), long-distance dispersal events from Australia to explain the current distribution of the 15 species found from Australia across the Pacific islands, north to the Bonin Islands off Japan and south to the Juan Fernandez Islands.

Several Hawaiian endemics that are members of widespread genera do not have specific areas of origin, although they are thought to be from Austral or Australasian/Malesian areas (Table 4.1). Among these are *Melicope/Platydesma* (Harbaugh *et al.*, 2009b), *Pittosporum* (Gemmil *et al.*, 2002), *Psychotria* (Nepokroeff *et al.*, 1999, 2003) and *Sicyos* (Sebastian *et al.*, 2009). Some of these were previously thought to be the result of multiple colonisation events as well, but have since been shown to be monophyletic and therefore the result of single colonisations.

For example, *Melicope* (Fig. 4.2e) has 47 species found across a broad range of habitats on all the high islands. It was hypothesised to be closely related to species of the Hawaiian endemic genus, *Platydesma*, the latter with only four species. It was generally thought that the Hawaiian *Melicope* arose from a single Indo-Pacific introduction (Fosberg, 1948; Hartley, 2000) and that *Platydesma* was the result of a second separate colonisation event. The molecular study of Pacific *Melicope* by Harbaugh *et al.* (2009b) using chloroplast and nuclear regions (Table 4.1) showed that *Platydesma* was nested within *Melicope*, and therefore not the result of a separate colonisation event. The ancestor of the Hawaiian species was found to be from the Austral region although, given the size and distribution of the genus (230 species widely distributed in the Southern Hemisphere from Australia, Malaysia, New Zealand, Madagascar, Southeast Asia, to the Pacific islands) and the number of species sampled, the precise source of the Hawaiian species is not clear. Similarly, *Psychotria*, estimated to have 1000–1650 species in the tropics worldwide, was thought to have originated in Australasia (Nepokroeff *et al.*, 1999, 2003). The 11 Hawaiian species were initially thought to be the result of one (Fosberg, 1962), two or three separate colonisation events (Sohmer, 1977; Wagner *et al.*, 1999). However, Nepokroeff *et al.* (2003), using nuclear DNA sequence data (Table 4.1), showed that the Hawaiian taxa were monophyletic. As with many of the groups in Hawaii, genetic differences were low among species, suggesting rapid radiation within the islands. Nepokroeff *et al.* (2003) also identified several radiations within *Psychotria* from its Australasian origin to Africa with subsequent radiation to the neotropics and to the Pacific, including Melanesia and Hawaii. However, broader taxon sampling will be necessary to determine the origin of the Hawaii ancestor.

The precise origin of Hawaiian *Pittosporum* is also unknown due to the lack of resolution among species from New Caledonia, New Zealand, Fiji, Tonga and eastern Polynesia (Gemmil *et al.*, 2002). The nearest relatives of the 10 Hawaiian endemics were proposed to be among species in Fiji and Tonga, based on the shared presence of unisexual flowers and morphology (Haas, 1977; Wagner *et al.*, 1990; Cayzer *et al.*, 2000). The findings of Gemmil *et al.* (2002) using nuclear DNA sequence data (Table 4.1) confirmed the monophyly of the Hawaiian taxa, but the source of the original colonists remained unclear. The data indicated a close relationship between Hawaiian *Pittosporum* and two species, *P. rhytidocarpum* A. Gray from Fiji and *P. yunckeri* A. C. Smith from Tonga. The New Caledonian species, *P. gatopense* Guillaumin, was sister to the Fijian and Tongan species, although support for this relationship was not strong. The greatest diversity of *Pittosporum* is found in Australia, but a direct connection to Hawaii is lacking. In this case further work will be needed to verify whether or not a single event gave rise to the Fijian, Tongan and then to the Hawaiian species or whether perhaps the Hawaiian colonisation was separate from that giving rise to species in the other two island groups (perhaps directly from Australia or New Zealand). There are 150 species of *Pittosporum* found in Africa, Asia, Australia and New Zealand as well as the Pacific islands, suggesting high dispersibility and the likelihood of multiple introductions in the Pacific.

## Single species colonisation without radiation

Although most radiations of Hawaiian endemics examined so far are the product of a single colonisation event followed by speciation, not all have this history. Some taxa have, as far as we know, failed to develop beyond the original colonisation (Table 4.1). For example, the molecular study of *Korthalsella*, a parasitic mistletoe with few definitive morphological characters, suggests that previous treatments listing multiple Hawaiian species may be in error (Molvray *et al.*, 1999). Wagner *et al.* (1990, 1999) listed four endemic and two indigenous Hawaiian species and postulated three independent introductions. However, Molvray *et al.* (1999) indicate that only one species is likely to exist in Hawaii and suggest that it may not be endemic. They proposed that all the species but *K. cylindrica* (Tieghem) Engler should be considered as part of the widely distributed Pacific species, *K. taenioides* (Commerson ex DC.) Engler. They further suggested that additional study might show that *K. cylindrica* should also be included in *K. taenioides*. If this should prove to be the case then there would be no endemic species of *Korthalsella* in the Hawaiian Islands and its presence there is probably the result of a relatively recent colonisation event. The precise geographical origin of the genus is unknown as it has a patchy distribution, but Molvray *et al.* (1999) proposed Papuasia.

A number of other genera from a variety of different areas appear not to have speciated in the islands although they are widely distributed within the archipelago. Among these taxa with only one species each are *Rhus* (Yi *et al.*, 2004), *Hillebrandia* (Clement *et al.*, 2004) and *Broussasia* of Asian origin (Hufford *et al.*, 2001), and *Cuscuta* putatively an immigrant from western North America (Stefanovic *et al.*, 2007). Additionally, the Hawaiian endemic legume, *Kanaloa kahoolawensis* D. H. Lorence & K. R. Wood, was described from two individuals growing on the 'Ale'ale sea stack off the coast of the small island of Kaho'olawe (Lorence & Wood, 1994) and placed in the *Leucaena* group (Hughes *et al.*, 2003). The precise relationship of *Kanaloa* to *Schleinitzia*, from the Western Pacific, and to *Desmanthus*, from the Americas, is not certain so the origin of this endemic species cannot be resolved at this time.

Finally and perhaps most interestingly, is the case of the signature Hawaiian plant, *Metrosideros polymorpha* Gaud. This is the dominant tree (woody) species in the islands where it grows in multiple ecosystems ranging from bare lava flows to mesic forests and bogs. It is so variable morphologically that species identification(s) have not been resolved despite data from ecological, morphological, physiological and population studies using isozymes and RAPDs (e.g. Rock, 1917; Skottsberg, 1936; Corn, 1979; Stemmermann, 1983; Aradhya *et al.*, 1991; Cordell *et al.*, 1998; Melcher *et al.*, 2000; James *et al.*, 2004). It has been treated as a widespread endemic (Skottsberg, 1936), as five separate species (Wagner *et al.*, 1999) or as part of the widespread taxon, *M. collina* A. Gray (Rock, 1917). Wright *et al.* (2001) and Percy *et al.* (2008) studied the Hawaiian species of *Metrosideros* within the larger context of probable origin, dispersal and biogeographic affinities in the Pacific. The data showed that the origin of the clade that contains the Hawaii species is either New Zealand or the Pacific (Wright *et al.*, 2001; Percy *et al.*, 2008). Percy *et al.* (2008), using 10 non-coding chloroplast DNA regions (Table 4.1), showed that most of the 'species' were nested inside *M. polymorpha* and therefore not distinct from it. They also found that there was no correlation between genetic and morphological diversity as the same haplotype was shared by plants with differing morphologies and that those with the same morphology could have different haplotypes. The data generally supported a single colonisation event and Percy *et al.* (2008) suggested intra-island diversification as the cause of *M. polymorpha*'s extreme morphological variability. Using microsatellites (Harbaugh *et al.*, 2009c) also supported the idea of multiple inter-island dispersals with later parallel evolution as among the major factors explaining the variability of this species. Subsequent investigations (Wagner, pers. com.) indicate that there may be from one to three species in Hawaii. Given the failure of multiple types of evidence to show consistent differences within and among islands and the lack of agreement on taxonomic status, it seems to us that this taxon will continue to be a source of controversy for the foreseeable future.

## Hawaii as a source of colonists

The studies of Hawaiian and Pacific *Santalum* (Harbaugh & Baldwin, 2007; Harbaugh, 2008; Harbaugh *et al.*, 2009b), *Tetramolopium* (Lowrey *et al.*, 2001) and *Melicope* (Harbaugh *et al.*, 2009b) illustrate the importance of the previously unknown role of the Hawaiian Islands as a source of colonising species, as well as a recipient of colonists from other areas.

The phylogenies produced from nuclear and chloroplast data on Pacific *Santalum* (Harbaugh & Baldwin, 2007; Harbaugh, 2008) showed an unexpected relationship to Hawaiian taxa by two species, *S. insulare* Bert. ex DC. from the Marquesas and French Polynesia (Society Islands) and *S. boninensis* (Nakai) Tuyama from the Bonin Islands. These two tetraploid species, as well as hexaploid and octoploid varieties of *S. insulare*, were nested within the Hawaiian clade. The most parsimonious explanation of these findings (Harbaugh & Baldwin, 2007) was that these species were the products of dispersals from Hawaii. Harbaugh and Baldwin reported that this was the first time Hawaii was unequivocally shown to be the source of colonists for other Pacific radiations.

*Tetramolopium* has a disjunct distribution that includes, besides Hawaii and New Guinea, Mitiaro in the Cook Islands (Lowrey, 1995; Okada *et al.*, 1997; Lowrey *et al.*, 2001). The genus was shown to have an origin in New Guinea with a single colonisation of the Hawaiian Islands. The Cook Island species, *T. mitiaroense* Lowrey, R., Whitkus and W. R. Sykes, was initially considered conspecific with the Hawaiian taxon, *T. sylvae* Lowrey; however, it was found to be a distinct species based on nuclear DNA sequence (Table 4.1) and RFLP (restriction fragment length polymorphism) data (Okada *et al.*, 1997; Lowrey *et al.*, 2005). The preferred explanation, in light of the molecular data, is that the Cook Island plants were derived from those in Hawaii and that the dispersal pathway for *Tetramolopium* was New Guinea to Hawaii to the Cook Islands. Final confirmation of this pathway will require additional study, however, to rule out alternative explanations for the observed patterns.

Harbaugh *et al.* (2009b), following up with another Pacific genus, *Melicope*, found a similar pattern to that seen for *Santalum* and *Tetramolopium*. Marquesan

*Melicope* species were found within the Hawaiian clade at two different places in the ITS (internal transcribed spacers) phylogeny. Based on morphological and molecular data, Harbaugh *et al.* (2009b) concluded that Marquesan species were descended from one and perhaps two different colonisations from Hawaii. They further speculate that other genera will also be found to have dispersed from Hawaii to elsewhere in the Pacific (i.e. *Pritchardia* (Arecaceae), *Cheirodendron* (Araliaceae), *Phyllostegia* (Lamiaceae) and *Leptocophylla* (Epacridaceae)). Recent work by M. Knope (pers. com.) indicates that two undescribed species of *Bidens* from the Marquesas are also likely to be derived from Hawaiian species, as previously suggested by Ganders *et al.* (2000), but this remains to be confirmed.

## Polyploidy and hybridisation

Polyploidisation and hybridisation have been invoked as potentially among the most important events in the establishment of taxa on distant islands and in subsequent adaptive radiation (Carr, 1998; Crawford *et al.*, 2009). This follows from the idea that polyploid colonisers would have greater genetic diversity than diploids, thus allowing for more adaptive possibilities in response to the newly encountered environments. Additionally, self-compatibility and low levels of inbreeding depression would be more likely to ensure persistence, once established. The relatively high number of polyploids among those reported for Hawaiian plants by Carr (1998) led him to conclude that a large number of Hawaiian plants originated from polyploid ancestors (palaeopolyploids). The reduced likelihood of polyploids originating from diploids on islands was discussed by Stuessy and Crawford (1998). They reasoned that this would require, first, successful establishment of both parent species, likely from different colonisation events, then hybridisation and doubling, followed by radiation, a much more unlikely set of events as compared to divergence from an original polyploid ancestor. Rapid evolution among natural and synthesised polyploids has also been demonstrated in a number of studies, making these 'excellent ancestors for rapidly diverging lineages' (Crawford *et al.*, 2009: 145). Hybridisation among species originating from both polyploid and diploid ancestors has also occurred, although this appears to be less common, or at least less well documented. Some examples of studies available are given below.

Polyploidy and hybridisation have been shown to be important in the Hawaiian silversword alliance (*Argyroxiphium, Dubautia, Wilkesia*) (Carlquist *et al.*, 2003). Crossing studies and chromosome morphology, along with allozyme and molecular data, support the derivation of the Hawaiian species from an allotetraploid ancestor (n = 14) that arose and diversified on the mainland before dispersing to Hawaii (Carlquist *et al.*, 2003; Carr, 2003). Natural hybridisation within and between genera (i.e. *Dubautia menziesii* (A. Gray) D. Keck × *Argyroxiphium sandwicense* DC.), and artificially produced hybrids between virtually any combination of taxa within the silversword alliance (Carr, 2003) indicated that hybridisation was likely, particularly given the abrupt habitat transitions and relatively frequent disruptions due to volcanic activities that are common in Hawaiian landscapes. Conflicting chloroplast and nuclear DNA phylogenies (Baldwin & Robichaux, 1995; Baldwin, 2003) also suggest that hybridisation has been important in silversword evolution (and perhaps in the entire tribe; Baldwin & Wessa, 2000). Data from floral homeotic gene sequences of the Hawaiian species and the California 'Madia' lineage also showed the allopolyploid nature of the Hawaiian taxa (Barrier *et al.*, 1999, 2001). There were two copies of homeotic genes (Table 4.1) in Hawaiian taxa, but not in mainland species, and each of these copies was most closely related to the mainland taxa (*Carlquistia muirii* (A. Gray) B. G. Baldwin and *Anisocarpus scabridus* (Eastw.) B. G. Baldwin) rather than to other Hawaiian species. Chromosome numbers of silversword taxa are n = 14 and n = 13. Molecular phylogenetic evidence showed that the n = 13 and n = 14 Hawaiian taxa were nested among those diploid Californian species in the 'Madia' alliance with chromosome numbers of n = 6 to n = 9 (Baldwin, 1996). Earlier crossing studies (Carr *et al.*, 1996) also showed that *Dubautia* scabra (DC.) D. Keck (n = 14) could be crossed with *Kyosia bolanderi* (A. Gray) B. G. Baldwin (n = 6) and *Carlquistia muirii* (A. Gray) B. G. Baldwin (n = 8) separately and in combination, and these crossing relationships could be extended to include *Anisocarpus scabridus* (Eastw.) B. G. Baldwin (n = 7). Barrier *et al.* (1999) proposed, on the basis of the close relationship of the genomes of Hawaiian and North American species, that Hawaiian taxa were descended from an extinct allopolyploid formed from among extant taxa.

Howarth and Baum (2005) examined the role of hybridisation in Hawaiian *Scaevola* using four nuclear regions (Table 4.1). Five diploid species (*S. chamissoniana* Gaudich, *S. gaudichaudiana* Cham., *S. mollis* Hook. & Arnott, *S. coriaceae* Nutt. and *S. gaudichaudii* Hook. & Arnott) showed clear monophyly when analysed separately by region and in combined analyses. However, *S. kilaueae* Degner and *S. procera* Hillebr. were placed in different parts of the individual gene trees and were not together in any of them. On the basis of genetic, morphological and ecological data it was inferred that these two diploid species were the result of homoploid hybridisation. *Scaevola kilaueae* is postulated to be a hybrid between *S. coriaceae* and *S. chamissoniana* while *S. procera* is a hybrid between *S. gaudichaudii* and *S. mollis*. The latter two species are known to hybridise when sympatric (Gillett, 1966). Howarth and Baum (2005) proposed that homoploid hybrids could be important in allowing for ecological shifts as well as adaptive radiation in *Scaevola* and perhaps in other island taxa as well. Patterson (1995) hypothesised that species of dry areas were ancestral to those of more mesic sites and that establishment was related to the ability of *Scaevola* to find suitable open habitats. *Scaevola* species are found in different habitats, but may have occurred together at various times in the past before widespread habitat destruction extirpated connecting populations (Patterson, 1995; Howarth & Baum, 2005). *Scaevola procera* is found in a different habitat and at a different elevation from its putative parents and has a distinctly different morphology. *Scaevola kilaueae* is intermediate in morphology and genetic composition to that of its putative parents and it occurs on young lava flows on the youngest island, where the parental species are also found. New combinations generated by hybridisation could potentially increase the ability of species to adapt to new habitats.

Harbaugh (2008) also found that hybridisation and polyploidy were important in *Santalum* species. Using a nuclear gene (Table 4.1) along with flow cytometry, at least six polyploid or hybridisation events in the genus were postulated as it radiated from Australia to Malesia and across the Pacific to Hawaii. All four Hawaiian species are tetraploid (Harbaugh & Baldwin, 2007) but the data indicate that they were produced by at least two different hybridisation events. *Santalum freycinetianum* and *S. halaekalae* (section *Solenantha*) were determined to be the result of autopolyploidy in an Australian ancestor that subsequently led to the first successful colonisation event in Hawaii (Harbaugh & Baldwin, 2007). The second colonisation event was by the lineage that gave rise to *S. ellipticum* and *S. paniculatum* (section *Hawaiiensia*). This lineage resulted from allopolyploidy following hybridisation of Australian diploid species prior to dispersal to Hawaii (Harbaugh, 2008). Disagreement between nuclear and chloroplast phylogenies (Harbaugh & Baldwin, 2007) along with the data of Harbaugh (2008) also suggested that *S. boninensis* (Nakai) Tuyama of the Bonin Islands ($n = 4$) was the result of hybridisation between Hawaiian tetraploid taxa related to *S. freycinetianum* and *S. paniculatum*. Additionally, *S. insulare* Bert. ex DC. var. *marchionense* (Skottsb.) Skottsb. from the Marquesas was putatively determined to be hexaploid and *S. insulare* Bert. ex DC. var. *raiateense* (J. W. Moore) Fosberg & Sachet from the Society Islands a putative octaploid, suggesting a stepwise increase in ploidy from a Hawaiian tetraploid ancestor. Harbaugh (2008) found that there were more than twice as many successful long-distance colonisation events to Pacific Islands by polyploid as diploid *Santalum* species.

Other cases of presumed hybridisation involve the Hawaiian mints (Lindqvist & Albert, 2002; Lindqvist *et al.*, 2003) and *Cyrtandra* (Cronk *et al.*, 2005) (Fig. 4.2a); experimental hybrids have also been formed between *Portulaca* species and among *Plantago* species (Carr, unpubl.) and many species in the Hawaiian flora are presumed to be polyploids as well (Wagner *et al.*, 1990, 1999).

## Morphological diversity and genetic uniformity

Among endemic Hawaiian taxa, morphological and genetic diversity do not necessarily go hand in hand. For example, the genus *Bidens* (Compositae) is one of the most morphologically diverse in the Hawaiian Islands (Carlquist, 1970; Ganders

& Nagata, 1984; Helenurm & Ganders, 1985). Species vary extensively in leaf shape, growth form, inflorescence structure, head and achene morphology, the latter unique to the Hawaiian species (Carlquist, 1974; Helenurm & Ganders, 1985). The range of morphological variation is greater among the Hawaiian taxa than that found in the rest of the 200–300 species in the genus that are distributed across five continents (Ganders *et al.*, 2000). The 19 endemic Hawaiian species (and eight subspecies) also grow in a wide variety of habitats which is reflected in their morphologies. These habitats include coastal dunes, cinder cones, rainforests, montane ridges over 2000 m and bog edges. One species, *B. cosmoides* (A. Gray) Sherff, has significantly different floral morphology from the other species. It has large, pendent and nectar-rich heads and is presumed to be adapted to bird pollination. It was proposed that this species was the result of a different colonisation event from that which gave rise to the other Hawaiian species (Ganders & Nagata, 1983). They do hybridise readily, however, under artificial and natural conditions when they come into contact (Ganders & Nagata, 1984; Ganders *et al.*, 2000). Despite the great variation in morphology and habitat, Hawaiian *Bidens* are remarkably similar genetically (Ganders, 1989; Ganders *et al.*, 2000). For example, earlier data from isozymes, flavonoids and polyacetylenes showed little differentiation, supporting a single colonisation event, but failing to differentiate species (Ganders, 1989). This same low level of genetic difference was also found by Ganders *et al.* (2000) using nuclear DNA (Table 4.1). All Hawaiian species were genetically identical, confirming the monophyly of the genus in the islands, including *B. cosmoides*.

The silversword alliance is also morphologically diverse and species are found in a wide variety of habitats. There are 30 species, five in *Argyroxiphium* (Fig. 4.2c), two in *Wilkesia* and 23 in *Dubautia* that range from rosette plants, to prostrate and erect shrubs, trees, cushion plants and lianas. Species grow in a wide variety of habitats from lava and cinders, to dry cliff sides, in dry and mesic forests and bogs, and from low elevations to 3750 m with annual precipitation varying from 40 cm to 1300 cm (Carr, 1985). Molecular studies using nuclear DNA (Baldwin & Robichaux, 1995; Baldwin, 1996, 1997, 2003) showed little sequence variation among the *Argyroxiphium* species, however, and while identifying the *Dubautia*/*Wikesia* sister lineage, again showed only small levels of difference within clades of *Dubautia*. This lack of variation was attributed to recent and rapid radiation (Baldwin & Sanderson, 1998) as well as to possible extinction of close relatives of *Argyroxiphium*.

The largest radiation within the Hawaiian Islands occurs within the lobeliads (*Brighamia, Clermontia, Cyanea, Delissia, Lobelia, Trematolobelia;* Table 4.1) (Givnish *et al.*, 1995, 2004). There are 126 species distributed among these genera, all derived from a single common ancestor (Givnish *et al.*, 2009). Taxa range from shrubs and trees to cliff-dwelling succulents, to rosette plants of high elevation bogs and even vines. The variation in leaf morphology is notable, especially in *Clermontia* and *Cyanea*, the largest genera (Givnish *et al.*, 1995; Lammers, 1995; Wagner *et al.*, 1999). Chloroplast DNA (Table 4.1) has been used to establish relationships, including that of *Brighamia* with *Delissia* and these with *Cyanea* and *Clermontia*; however, genetic differences among taxa remain small.

Other genera that have radiated into different habitats and express morphological and ecological diversity but have relatively low genetic differentiation include *Schiedea* (now including *Alsinodendron*) (Soltis *et al.*, 1996; Sakai *et al.*, 1997; Wagner *et al.*, 2005; Harbaugh *et al.*, 2009a), the Hawaiian mints (*Haplostachys, Phyllostegia, Stenogyne*) (Lindqvist & Albert, 2002; Lindqvist *et al.*, 2003), *Tetramolopium* (Lowrey, 1995; Okada *et al.*, 1997, 2000) and *Plantago* (Dunbar-Co *et al.*, 2008).

## Conclusions

Estimated dates for the time of divergence of Hawaiian lineages from their sister group have been published for 14 genera and monophyletic groups (Table 4.1). These dates reveal several different patterns of geographical origin and spread. Among those dated > 5 Ma, for instance, all appear to have been in the Hawaiian Islands *c.* 13–15 Ma, indicating they may have once inhabited Gardner Pinnacles in the Northwest Hawaiian Islands which formed a large high island at that time but has since eroded. *Hesperomannia*, derived from African ancestors, may be the oldest lineage in the islands with an estimated age ranging using ndhF from 24.0 to 17.14 ± 3.43 Ma. While a date around 14 Ma could include residence on Gardner Pinnacles, the older age estimates would mean that the ancestor was likely somewhere else between tropical Africa and the Pacific until at least 15 Ma. Age estimates for various taxa of < 5 Ma place them within the age of the current high islands (5.1–0.5 Ma). This group includes more taxa and a wider range of source areas than found for the older genera, which is not surprising. Among these are some of the best-known examples of adaptive radiation, i.e. the silverswords (30 species), mints (60 species), lobeliads (126 species) and others. Smaller radiations within this same time frame are known for *Viola* (7 species), *Geranium* (7 species), *Kokia* (4 species), and one of the arrivals of *Santalum* (4 species) (Table 4.1). The lack of detectable genetic divergence among species within genera indicates that intra- and inter-island radiations have likely been rapid as well.

The geographical sources of Hawaiian taxa vary widely and include Arctic/Boreal regions, western North America, Central America, Africa, Asia, Malesia and Australasia (Table 4.1). Among the 42 studies listed in Table 4.1, 37 have restricted areas of origin that can be estimated. These 37 groups have a total of 568 species, over half the native flora. Based on these data there are 22 groups

(60% of the lineages and 64% of the species) are derived from Asia, Australasia, Malesia or the South Pacific (a loosely used term that often includes New Zealand, Australia, Papua New Guinea and neighbouring islands). In comparison, 14 groups (38% of the lineages and 35% of the species) are estimated to be from the Americas + the Arctic; of these, 10 (27%) are proposed to be from North America (including Mexico). The remaining 2% is accounted for by the one confirmed taxon with an African origin. Correlated with the geographic origin of the lineages is the size of radiations within the islands. These radiations fall into four categories: very large (58–126 species), large (14–32 species), medium (6–11 species) and small (2–5 species). Of the very large radiations, two (Lobeloides and *Cyrtandra*) are Asian-Pacific in origin and one is American (Hawaiian mints). Within the large category there are seven groups, four from the Americas and three from the Asia-Pacific region. There are nine medium-size groups (six from the Asia-Pacific and three from the America-Arctic regions) and finally there are six small clades with two from the Asia-Pacific region, three from the Americas and one from Africa. There are 12 lineages that have only one taxon so they have either failed to radiate or extinction has left only one taxon from a previously larger group. These numbers indicate a much greater contribution by North American taxa to the Hawaiian flora than previously suggested (Fosberg, 1948; Wagner *et al.*, 1990; 1999). Australasian, Asian, Malesian and South Pacific groups are still in the majority. Work is under way in many additional groups and these figures are bound to change.

Adaptive radiation has been 'explosive' in many groups, giving rise to great morphological divergence while at the same time showing little genetic differentiation. This is likely a product of the young age and the often allopatric distribution of species on different islands (many single-island endemics) and habitat differences. As a result of close genetic similarity within monophyletic groups, hybridisation is frequent, and it is likely that hybridisation followed by polyploidy was important in their establishment and diversification. Polyploidy, whether by allo- or autopolyploidisation, also appears to have been important as more than a quarter of all endemic taxa thus far investigated (Carr, 1998) are polyploids. Intra-island as well as inter-island divergence has been important in species diversification within many genera in the islands. Diversification has also resulted in changes in habit and breeding system (e.g. *Schiedea*) which in some cases can be correlated with genetic divergence.

The original plant colonists arrived in Hawaii as a result of long-distance dispersal, largely facilitated by migratory birds, but also by wind and water. Among the taxa investigated in molecular studies, many had been suggested to be the product of more than one colonisation event. The available phylogenies, however, show monophyletic lineages resulting from a single original coloniser, in all but two cases cases, i.e. *Scaevola* and *Santalum*. The Hawaiian Islands connect floras from the Arctic of North America south to New Zealand, and from the eastern to the western sides of the Pacific, Southeast Asia and even Africa. As the histories of more Hawaiian taxa become known it is likely that this 'laboratory of evolution' will be revealed to have a significant role in the biogeography and evolution of plant groups now found throughout the Pacific and beyond.

## Acknowledgements

We would like to thank David Clague, Tim Lowrey and Warren Wagner for providing additional information, Chrissen Gemmill for permission to use her map, Carol Kelloff for production of Fig. 4.1 and Gerrald Carr for photographs. We also thank the Departments of Botany at the University of Hawaii and the Smithsonian Institution for their support and, last but not least, we thank the Editors for their patience and understanding during the preparation of this manuscript.

## References

Alice, L. A. & Campbell, C. S. (1999). Phylogeny of *Rubus* (Rosaceae) based on nuclear ribosomal DNA Internal Transcribed Spacer region sequences. *Am. J. Bot.* 86: 81–97.

Aradhya, K. M., Mueller- Dombois, D. & Ranker, T. A. (1991). Genetic evidence for recent and incipient speciation in the evoltuion of Hawaii *Metrosideros* (Myrtaceae). *Heredity* 67: 129–138.

Baker, W. J., Savolainen, V., Asmussen-Lange, C. B., Chase, M. W., Dransfield, J., Forest, F. *et al.* (2009). Complete generic-level phylogenetic analyses of Palms (Arecaceae) with comparisons of supertree and supermatrix approaches. *Syst. Biol.* 58: 240–256.

Baldwin, B. G. (1996). Phylogenetics of the California Tarweeds and the Hawaiian Silversword Alliance (Madiinae; Heliantheae *sensu lato*). In D. J. N. Hind & H. J. Beentje (eds.), *Compositae: Systematics. Proceedings of the International Compositae Conference.* pp. 377–391. London, Royal Botanic Gardens Kew.

Baldwin, B. G. (1997). Adaptive radiation of the Hawaiian Siversword Alliance: congruence and conflict of phylogenetic evidence from molecular and non-molecular investigations. In T. J. Givnish & K. J. Sytsma (eds.), *Molecular Evolution and Adaptive Radiation*, pp. 104–128. Cambridge, Cambridge University Press.

Baldwin, B. G. (2003). A phylogenetic perspective on the origin and evolution of the Madiinae. In S. Carlquist, B. G. Baldwin & G. D. Carr (eds.), *Tarweeds & Silverswords*, pp. 193–228. St Louis, Missouri Botanical Garden Press.

Baldwin, B. G. & Robichaux, R. H. (1995). Historical biogeography and ecology of the Hawaiian silversword alliance (Asteraceae): new molecular phylogenetic perspectives. In W. L. Wagner & V. A. Funk (eds.), *Hawaiian Biogeography: Evolution on a Hot Spot Archipelago*, pp. 259–287. Washington, DC, Smithsonian Institution Press.

Baldwin, B. G. & Sanderson, M. J. (1998). Age and rate of diversification of the Hawaiian silversword alliance (Compositae). *Proc. Natl. Acad. Sci. U.S.A.* 95: 9402–9406.

Baldwin, B. G. & Wessa, B. L. (2000). Origin and relationships of the Tarweed-Silversword lineage (Compositae – Madiinae). *Am. J. Bot.* 87: 1890–1908.

Ballard, H. E. & Sytsma, K. J. (2000). Evolution and biogeography of the woody Hawaiian violets (*Viola*, Violaceae): arctic origins, herbaceous ancestry and bird dispersal *Evolution* 54: 1521–1532.

Barrier, M., Baldwin, B. G., Robichaux, R. H. & Puruggaganan, M. D. (1999). Interspecific hybrid ancestry of a plant adaptive radiation: allopolyploidy of the Hawaiian Silversword Alliance (Asteraceae) inferred from floral homeotic gene duplications *Molec. Biol. Evol.* 16: 1105–1113.

Barrier, M., Robichaux, R. H. & Purugganan, M. D. (2001). Accelerated regulatory gene evolution in an adaptive radiation. *Proc. Natl. Acad. Sci. USA* 98: 10208–10213.

Carlquist, S. (1969). Wood anatomy of the Goodeniaceae and the problem of insular woodiness. *Ann. Mo. Bot. Gard.* 56: 358–390.

Carlquist, S. (1970). *Hawaii: A Natural History.* Lawai, Hawaii, Pacific Tropical Botanical Garden (reprinted 1980).

Carlquist, S. (1974). *Island Biology.* New York, Columbia University Press.

Carlquist, S., Baldwin, B. G. & Carr, G. D. (2003). *Tarweeds and Silverswords: Evolution of the Madiinae (Asteraceae).* St Louis, Missouri Botanical Garden Press.

Carr, G. D. (1985). Monograph of the Hawaiian Madiinae (Asteraceae): *Argyroxiphium, Dubautia*, and *Wilkesia. Allertonia* 4: 1–123.

Carr, G. D. (1998). Chromosome evolution and speciation in Hawaiian flowering plants. In T. F. Stuessy & M. Ono (eds.), *Evolution and Speciation of Island Plants*, pp. 5–47. Cambridge, Cambridge University Press.

Carr, G. D. (2003). Hybridization in Madiinae. In S. Carlquist, B. G. Baldwin & G. D. Carr (eds.), *Tarweeds and Silverswords: Evolution of the Madiinae (Asteraceae)*, pp. 79–104. St Louis, Missouri Botanical Garden Press.

Carr, G. D., Baldwin, B. G. & Khyos, D. W. (1996). Cytogenetic implications of artificial hybrids between the Hawaiian silversword alliance and North American tarweeds (Asteraceae: Heliantheae – Madiinae). *Am. J. Bot.* 83: 653–660.

Cayzer, L. W., Crisp, M. D. & Telford, J. R. H. (2000). Revision of *Pittosporum* (Pittosporaceae) in Australia. *Aust. Syst. Bot.* 13: 845–902.

Clague, D. A. (1996). Growth and subsidence of the Hawaiian-Emperor volcanic chain. In A. Keast & S. E. Miller (eds.), *The Origin and Evolution of Pacific island Biotas, New Guinea to Eastern Polynesia: Patterns and Processes*, pp. 35–50. Amsterdam, SPB Academic Press.

Clague, D. A., Braga, J. C., Bassi, D., Fullagar, P. D., Renema, W. & Webster, J. M. (2010). Early Oligocene to early Miocene carbonate bank deposits on Koko Seamount, Emperor Seamount Chain: implications for the biogeography of the Hawaiian Islands. *J. Biogeogr.* 37: 1022–1033.

Clement, W. L., Tebbitt, M. C., Forrest, L. L., Blair, J. E., Brouillet, L., Eriksson, T. & Swensen, S. (2004). Phylogenetic position and biogeography of *Hillebrandia sandwicensis* (Begoniaceae): a rare Hawaiian relict. *Am. J. Bot.* 91: 905–917.

Cordell, S., Goldstein, G., Mueller- Dombois, D., Webb, D. & Vitousek, P. M. (1998). Physiological and morphological variation in *Metrosideros polymorpha*, a dominant Hawaiian tree species, along an altitudinal gradient; the role of phenotypic plasticity. *Oecologia* 113: 381–387.

Corn, C. (1979). Variation in Hawaiian *Metrosideros.* PhD thesis: Department of Botany, University of Hawaii, Honolulu.

Costello, A. & Motley, T. J. (2001). Molecular systematics of *Tetraplasandra*, *Munroidendron* and *Reynoldsia sandwicensis* (Araliaceae) and the evolution of superior ovaries in *Tetraplasandra*. *Edinb. J. Bot.* 58 (2): 229–242.

Cowie, R. H. & Holland, B. S. (2006). Dispersal is fundamental to biogeography and the evolution of biodiversity on oceanic islands. *J. Biogeogr.* 33: 193–198.

Crawford, D. J., Lowrey, T. K., Anderson, G. J., Bernardello, G., Santos-Guerra, A. & Stuessy, T. F. (2009). Genetic diversity in Asteraceae endemic to oceanic islands: Baker's Law and polyploidy. In V. A. Funk, A. Susanna, T. F. Stuessy & R. J. Bayer (eds.), *Systematics, Evolution, and Biogeography of Compositae*, p. 963. Vienna, International Association for Plant Taxonomy.

Cronk, Q., Kiehn, M., Wagner, W. L. & Smith, J. F. (2005). Evolution of *Cyrtandra* (Gesneriaceae) in the Pacific Ocean: the origin of a supertramp clade. *Am. J. Bot.* 92 (6): 1017–1024.

Cuenoud, P., Del Pero Martinez, M. A., Loizeau, P. A., Spichiger, R., Andrews, S. & Manen, J. F. (2000). Molecular phylogeny and biogeography of the genus *Ilex* L. (Aquifoliaceae). *Ann. Bot.* 85: 111–122.

DeJoode, D. R. & Wendel, J. F. (1992). Genetic diversity and origin of the Hawaiian Islands cotton, *Gossypium tomentosum*. *Am. J. Bot.* 79: 1311–1319.

Dunbar-Co, S., Wieczorek, A. M. & Morden, C. W. (2008). Molecular phylogeny and adaptive radiation of the endemic Hawaiian *Plantago* species (Plantaginaceae). *Am. J. Bot.* 95: 1177–1188.

Eggens, F., Popp, M., Neopkroeff, M., Wagner, W. L. & Oxelman, B. (2007). The origin and number of introductions of the Hawaiian endemic *Silene* species (Caryophyllaceae). *Am. J. Bot.* 94: 210–218.

Fosberg, F. R. (1948). Derivation of the Hawaiian Flora. In E. C. Zimmerman (ed.), *Insects of Hawaii*, pp. 107–119. Honolulu, University of Hawaii Press.

Fosberg, F. R. (1962). Miscellaneous notes on Hawaiian plants–3. *Occas. Pap. Bishop Mus.* 23: 42–44.

Ganders, F. R. (1989). Adaptive radiation in Hawaiian *Bidens*. In L. V. Giddings, W. W. Anderson & K. Y. Kaneshiro (eds.), *Genetics, Speciation, and the Principle*, pp. 99–112. New York & Oxford, Oxford University Press.

Ganders, F. R. & Nagata, K. M. (1983). Relationships and floral biology of *Bidens cosmoides* (Asteraceae). *Lyonia* 2: 21–31.

Ganders, F. R. & Nagata, K. M. (1984). The role of hybridization in the evolution of *Bidens* on the Hawaiian Islands. In: W. F. Grant (ed.), *Plant Biosystematics*. pp. 179–194. Don Mills, Ontario: Academic Press Canada.

Ganders, F. R., Berbee, M. & Pirseyede, M. (2000). ITS base sequence phylogeny in *Bidens* (Asteraceae): evidence for the continental relatives of Hawaiian and Marquesan *Bidens* *Syst. Bot.* 25 (1): 122–133.

Gemmil, C. E. C., Allan, G. J., Wagner, W. L. & Zimmer, E. A. (2002). Evolution of insular Pacific *Pittosporum* (Pittosporaceae): origin of the Hawaiian radiation. *Molec. Phylogenet. Evol.* 22: 31–42.

Gillett, G. W. (1966). Hybridization and its taxonomic implications in the Scaevola gaudichaudiana complex of the Hawaiian Islands. *Evolution* 20: 505–516.

Givnish, T. J., Sytsma, K. J., Smith, J. F., & Hahn, W. J. (1995). Molecular evolution, adaptive radiation and geographic speciation in *Cyanea* (Campanulaceae), the largest plant genus endemic to Hawaii. In W. J. Wagner & V. L. Funk (eds.), *Hawaiian Biogeography: Evolution on a Hot Spot Archipelago*, pp. 288–337. Washington, DC, Smithsonian Institution Press.

Givnish, T. J., Montgomery, R. A. & Goldstein, G. (2004). Adaptive radiation of photosynthetic phisiology in the Hawaiian Lobeliads: light regimes, static light responses, and whole-plant compensation points. *Am. J. Bot.* 91: 228–246.

Givnish, T. J., Millam, K. C., Mast, A. R., Paterson, T. B., Theim, T. J., Hipp, A. L. *et al.* (2009). Origin, adaptive radiation and diversification of the Hawaiian lobeliads (Asterales: Campanulaceae). *Proc. R. Soc. Lond. B Biol. Sci.* 276 (1656): 407–416.

Haas, J. E. (1977). The pacific species of *Pittosporum* Banks ex Gaertn. (Pittosporaceae). *Allertonia* 1: 73–167.

Hao, G., Yuan, Y.-M., Hu, C.-M., Ge, X.-J. & Zhao, N.-X. (2004). Molecular phylogeny of *Lysimachia* (Myrsinaceae) based on chloroplast trnL-Fand nuclear ribosomal ITS sequences. *Molec. Phylogenet. Evol.* 31: 323–339.

Harbaugh, D. T. (2008). Polyploid and hybrid origins of Pacific island sandalwoods (Santalum, Santalaceae) inferred from low-copy nuclear and flow cytometry data. *Int. J. Plant Sci.* 169: 677–685.

Harbaugh, D. & Baldwin, B. G. (2007). Phylogeny and biogeography of the sandalwoods (*Santalum*, Santalaceae): repeated dispersals throughout the Pacific. *Am. J. Bot.* 94(6): 1028–1040.

Harbaugh, D. T., Nepokroeff, M., Rabeler, R. K., Zimmer, E. & Wagner, W. L. (2009a). Classification of the Caryophyllaceae and colonization of a divergent Hawaiian lineage. In *Botany and Mycology 2009 Participating Societies Abstract Book*, p. 170. Snowbird, UT.

Harbaugh, D. T., Wagner, W. L., Allan, G. J. & Zimmer, E. A. (2009b). The Hawaiian Archipelago is a stepping stone for dispersal in the Pacific: an example from the plant genus *Melicope* (Rutaceae) *J. Biogeogr.* 36: 230–241.

Harbaugh, D. T., Wagner, W. L., Percy, D. M., James, H. F. & Fleischer, R. C. (2009c). Genetic structure of the polymorphic *Metrosideros* (Myrtaceae) complex in the Hawaiian Islands using nuclear microsatellit data. *PLoS ONE* 4 (3): e4698. doi:4610.1371/journal.pone.0004698

Hartley, T. G. (2000). On the taxonomy and biogeography of *Euodia* and *Melicope* (Rutaceae). *Allertonia* 8: 1–319.

Havran, J. C., Sytsma, K. J. & Ballard, H. E., Jr. (2009). Evollutionary relationships, interisland biogeography, and molecular evolution in the Hawaiian violets (*Viola*: Violaceae). *Am. J. Bot.* 96: 1–14.

Helenurm, K. & Ganders, F. R. (1985). Adaptive radiation and genetic differentiaion in Hawaiian *Bidens*. *Evolution* 39: 753–769.

Howarth, D. G. & Baum, D. A. (2005). Genealogical evidence of homoploid hybrid speciation in an adaptive radiation of *Scaevola* (Goodeniaceae) in the Hawaiian Islands. *Evolution* 59: 948–961.

Howarth, D. G., Gardner, D. E. & Morden, C. W. (1997). Phylogeny of *Rubus* subgenus *Idaeobatus* (Rosaceae) and its implications toward colonization of the Hawaiian Islands. *Syst. Bot.* 22: 433–441.

Howarth, D. G., Gustafsson, M. H. G., Baum, D. A. & Motley, T. J. (2003). Phylogenetics of the genus *Scaevola* (Goodeniaceae): implication for dispersal patterns across the Pacific Basin and colonization of the Hawaiian Islands. *Am. J. Bot.* 90: 915–923.

Hufford, L., Moody, M. L. & Soltis, D. E. (2001). A phylogenetic analysis of Hydrangeaceae based on sequences of the plastid gene *matk* and their combination with rbcl and morphological data. *Int. J. Plant Sci.* 162: 835–846.

Hughes, E. E., Bailey, C. D., Krosnick, S. & Luckow, M. A. (2003). Relationships among genera of the informal *Dichrostachys* and *Leucaena* groups (Mimosoideae) inferred from nuclear ribosomal ITS sequences. In B. B. Klitgaard & A. Bruneau (eds.), *Advances in Legume Systematics, part 10, Higher Level Systematics*, pp. 221–238. London, Royal Botanic Gardens Kew.

James, S. A., Puttock, C. F., Cordell, S. & Adams, R. P. (2004). Morphological and genetic variation within *Metrosideros polymorpha* on Hawaii. *New Zeal. J. Bot.* 42: 263–270.

Keeley, S., Forsman, Z. H. & Chan, R. (2007). A phylogeny of the 'evil tribe' (Vernonieae: Compositae) reveals Old/New World long distance dispersal: support from separate and combined congruent datasets (trnL-F, ndhF, ITS). *Molec. Phylogenet. Evol.* 44 (1): 89–103.

Kim, H. -G., Keeley, S., Vroom, P. S. & Jansen, R. K. (1998). Molecular evidence for an African origin of the Hawaiian endemic *Hesperomannia* (Asteraceae). *Proc. Natl. Acad. Sci. USA* 95: 15440–15445.

Kimball, R. T. & Crawford, D. J. (2004). Phylogeny of the Coreopsideae (Asteraceae) usiing ITS sequences suggest lability in reproductive characters. *Molec. Phylogenet. Evol.* 33: 127–139.

Lammers, T. (1995). Patterns of speciation and biogeography in *Clermonita* (campanulaceae, Lobelioideae). In W. J. Wagner & V. L. Funk (eds.), *Hawaiian Biogeography: Evolution on a Hot Spot Archipelago*, pp. 338–362. Washington, DC, Smithsonian Institution Press.

Lindqvist, C. & Albert, V. A. (2002). Origin of the Hawaiian endemic mints within North American *Stachys* (Lamiaceae). *Am. J. Bot.* 89: 1709–1724.

Lindqvist, C., Motley, T. J., Jeffrey, J. J. & Albert, V. A. (2003). Cladogenesis and reticulation in the Hawaiian endemic mints (Lamiaceae). *Cladistics* 19: 480–495.

Lorence, D. H. & Wood, K. R. (1994). *Kanaloa*, a new genus of Fabaceae (Mimosoideae) from Hawaii. *Novon* 4: 137–145.

Lowrey, T. K. (1995). Phylogeny, adaptive radiation, and biogeography of Hawaiian *Tetramolopium*. In W. L. Wagner & V. A. Funk (eds.), *Hawaiian Biogeography: Evolution on a Hot Spot Archipelago*, pp. 195–220. Washington, DC, Smithsonian Institution Press.

Lowrey, T. K., Quinn, C. J., Taylor, R.K., Chan, R., Kimball, R. T. & de Nardi, J. (2001). Molecular and morphological reassessment of relationships within the *Vittandinia* group of Astereae (Asteraceae). *Am. J. Bot.* 88: 1279–1289.

Lowrey, T. K., Whitkus, R. & Sykes, W. R. (2005). A new species of *Tetramolopium* (Asteraceae) from Mitiaro, Cook Islands: biogeography, phylogenetic relationships and dispersal. *Syst. Bot.* 30 (2): 448–455.

Melcher, P. J., Cordell, S., Jones, T. J., Scowcroft, P. G., Niemczura, W., Gimabelluca, T. W. & Goldstein, G. (2000). Supercooling capacity increases from sea level to tree line in the Hawaiian tree species *Metrosideros polymorpha*. *Int. J. Plant Sci.* 161: 369–379.

Molvray, M., Kores, P. J. & Chase, M. W. (1999). Phylogenetic relationships within *Korthalsella* (Viscaceae) based on nuclear ITS and plastid Trnl-F sequence data. *Am. J. Bot.* 86: 249–260.

Motley, T. J. & Carr, G. D. (1998). Artificial hybridization in the Hawaiian endemic genus Labordia (Loganiaceae). *Am. J. Bot.* 85: 654–660.

Nepokroeff, M., Bremer, B. & Sytsma, K. J. (1999). Reorganization of the genus *Psychotria* and tribe Psychotrieae (Rubiaceae) inferred from ITS and rbcL sequence data. *Syst. Bot.* 24: 5–27.

Nepokroeff, M., Sytsma, K. J., Wagner, W. L. & Zimmer, E. A. (2003). Reconstructing ancestral patterns of colonization and dispersal in the Hawaiian understory tree genus Psychotria (Rubiaceae): a comparison of parsimony and likelihood approaches. *Syst. Biol.* 52: 820–838.

Neyland, R. (2001). A phylogeny inferred from large ribosomal subunit (26S) rDNA sequences suggests that *Cuscuta* is a derived member of Convolvulaceae. *Brittonia* 53: 108–115.

Okada, M., Whitkus, R. & Lowrey, T. K. (1997). Genetics and adaptive radiation in Hawaiian and Cook Island species of *Tetramolopium* (Asteraceae: Astereae) I. nuclear RFLP analysis. *Am. J. Bot.* 84: 1236–1246.

Okada, M., Lowrey, T. K. & Whitkus, R. (2000). Quantitative morphological variationin *Tetramolopium* (Asteraceae) in Hawaii and the Cook Islands. *Plant Syst. Evol.* 22: 1–13.

Palmer, C. (2003). *Hawaii's Ferns and Fern Allies*. Honolulu, University of Hawaii Press.

Patterson, R. (1995). Phylogenetic analysis of Hawaiian and other Pacific species of *Scaevola* (Goodeniaceae). In W. L. Wagner & V. A. Funk (eds.), *Hawaiian Biogeography: Evolution on a Hot Spot Archipelago*, pp. 363–378. Washington, DC, Smithsonian Institution Press.

Pax, D., Price, R. A. & Michaels, H. J. (1997). Phylogenetic position of the Hawaiian geraniums based on rbcL sequences. *Am. J. Bot.* 84: 72–78.

Percy, D. M., Garver, A. M., Wagner, W. L., James, H. F., Cunningham, C. W., Miller, S. E. *et al.* (2008). Progressive island colonization and ancient origin of Hawaiian Metrosideros (Myrtaceae). *Proc. R. Soc. Lond. B Biol. Sci.* 275: 1479–1490.

Plunkett, G. M. & Lowry, P. P. (2001). Relationships among "Ancient Araliads" and their significance for the systematics of Apiales. *Molec. Phylogenet. Evol.* 19 (2): 259–276.

Powell, A. E. & Kron, K. A. (2002). Hawaiian blueberries and their relatives – a phylogenetic analysis of *Vaccinium* sections Macropelma, Myrtillus, and Hemimyrtillus (Ericaceae). *Syst. Bot.* 27: 768–779.

Price, J. P. & Clague, D. A. (2002). How old is the Hawaiian biota? Geology and phylogeny suggest recent divergence. *Proc. R. Soc. Lond. B Biol. Sci.* 269: 2429–2435.

Price, J. P. & Wagner, W. L. (2004). Speciation in Hawaiian angiosperm lineages: cause, consequence and mode. *Evolution* 58 (10): 2185–2200.

Rock, J. F. (1917). The *ohia lehua* trees of Hawaii. A revision of the Hawaiian species of the genus *Metrosideros* Banks, with special reference to the varieties and forms of

*Metrosideros collina* (forster) A. Gray subspecies *polymorpha* (Gaud.) Rock. *Hawaii Board of Agriculture, Division of Forestry Botanical Bulletin* 4: 1–76.

Sakai, A. K., Weller, S. G., Wagner, W. L., Soltis, P. S. & Soltis, D. E. (1997). Phylogenetic perspectives on the evolution of dioecy: adaptive radiation in the endemic Hawaiian genera *Schiedea* and *Alsinidendron* (Caryophyllaceae: Alsinoideae). In T. J. Givnish & K. J. Sytsma (eds.), *Molecular Evolution and Adaptive Radiation*, pp. 455–473. Cambridge, Cambridge University Press.

Sebastian, P. M., Lira, R., Cross, H. B., Telford, I. R. H., Motley, T. J. & Renner, S. S. (2009). Biogeography of the *Sicycos* clade (Cucurbitaceae): from the New World to Australia, New Zealand, and Hawaii. In *Botany and Mycology 2009 Abstract Book*, p. 221. Snowbird, UT.

Seelanan, T., Schnabel, A. & Wendel, J. F. (1997). Congruence and consensus in the cotton tribe (Malvaceae). *Syst. Bot.* 22: 259–290.

Skottsberg, C. (1930). Further notes on the Pacific Sandalwoods. *Acta Horti Gothob.* 5: 135–145.

Skottsberg, C. (1936). Vascular plants from the Hawaiian Islands II. *Acta Horti Gothob.* 10: 97–193.

Sohmer, S. H. (1977). *Psychotria* L. (Rubiaceae) in the Hawaiian islands. *Lyonia* 1: 103–186.

Soltis, P. S., Soltis, D. E., Weller, S. G., Sakai, A. & Wagner, W. (1996). Molecular phylogenetic analysis of the Hawaiian endemics *Shiedea* and *Alisinidondron* (Caryphyllaceae). *Syst. Bot.* 21: 365–379.

Stefanovic, S., Kuzmina, M. & Costea, M. (2007). Delimitation of major lineages within *Cuscuta* subgenus *Grammica* (Convolvulaceae) using plastid and nuclear DNA sequences. *Am. J. Bot.* 94: 568–589.

Steinmann, V. W. & Porter, J. M. (2002). Phylogenetic relationships in Euphorbieae (Euphorbiaceae) based on ITS and ndhF sequence data. *Ann. Mo. Bot. Gard.* 89 (4): 453–490.

Stemmermann, L. (1983). Ecological studies of Hawaiian *Metrosideros* in a successional context. *Pac. Sci.* 37: 361–373.

Stuessy, T. F. & Crawford, D J. (1998). Chromosomal stasis during speciation in angiosperms of oceanic islands. In T. F. Stuessy & M. Ono (eds.), *Evolution and Speciation of Island Plants*, pp. 307–324. Cambridge, Cambridge University Press.

Wagner, W. L. & Funk, V. A. (eds.) (1995). *Hawaiian Biogeography:Evolution on a Hot Spot Archipelago.* Washington, DC, Smithonsian Institution Press.

Wagner, W. L., Herbst, D. R. & Sohmer, S. H. (1990). *Manual of the Flowering Plants of Hawaii.* Honolulu, Bishop Museum Press.

Wagner, W. L., Herbst, D. R. & Sohmer, S. H. (1999). *Manual of the Flowering Plants of Hawaii*, revised edition. Honolulu, University of Hawaii Press.

Wagner, W. L., Weller, S. G. & Sakai, A. K. (2005). Monograph of Schiedea (Caryophyllaceae-Alsinoideae). *Syst. Bot. Monog.* 72: 35.

Wanntorp, L. & Wanntorp, H. -E. (2003). The biogeography of *Gunnera* L.: vicariance and dispersal. *J. Biogeogr.* 30: 979–987.

Wen, J., Plunkett, G. M., Mitchell, A. D. & Wagstaff, S. J. (2001). The evolution of Araliaceae: a phylogenetic analysis based on ITS sequences of nuclear ribosomal DNA. *Syst. Bot.* 26: 144–167.

Wright, S. D., Yong, C. G., Wichman, S. R., Dawson, J. W. & Gardner, R. C. (2001). Stepping stones to Hawaii: a trans-equatorial dispersal pathway for *Metrosideros* (Myrtaceae) inferred from nrDNA (ITS + ETS). *J. Biogeogr.* 28: 769–774.

Yang, Y., Morden, C. W., Sack, L., Sporck, M. J. & Berry, P. E. (2009). Phylogeny and adaptive radiation of woody Hawaiian Chamaesyce from herbaceous and annual ancestors in subtropical North America (Euphorbia-Euphorbiaceae). In *Botany and Mycology 2009 Participating Societies Abstract Book*, p. 182. Snowbird, UT.

Yi, T., Miller, A. J. & Wen, J. (2004). Phylogenetic and biogeographic diversification of *Rhus* (Anacardiaceae) in the Northern Hemisphere. *Molec. Phylogenet. Evol.* 33: 861–879.

# 5

# Origins and evolution of Galapagos endemic vascular plants

ALAN TYE AND JAVIER FRANCISCO-ORTEGA

Thirty years ago, Porter (1979) summarised what was then known of the origins and evolution of Galapagos endemic vascular plants. He concluded that 43% of indigenous taxa (231 out of 543 species, subspecies and varieties) were endemic, although this figure included six questionably endemic and 14 questionably distinct taxa plus 13 others known only from the type or not collected for at least 45 years. Removing those led to 186 endemics in a total vascular flora of 498 taxa (37% endemic). This percentage endemism is higher than some published estimates, partly because Porter (1976, 1979, 1983) excluded from the indigenous flora 47 species that he regarded as introduced by people, but which have been treated as native by others (e.g. Lawesson *et al.*, 1987). Porter (1979) recognised 192 naturalised introduced taxa (including the 47 that were treated as native by Lawesson *et al.*, 1987). Porter (1979) recognised seven endemic genera, and radiations that had produced at least three endemic taxa in '19' genera (actually 21 according to his systematic list, including two about which some doubt was expressed), plus '10' more genera (12 in his systematic list including two less certain ones) which had produced two endemics by radiation. He suggested that birds had played the most important role in dispersing plants to the islands.

Porter (1976) had earlier concluded that, contrary to previous assumptions, the Galapagos flora had no affinities with North America and the Caribbean, with 99% of its non-endemic taxa being found in South America and the other 1% in Mexico and Central America, and that the endemics had their main affinities in South America too. He considered that, with only eight exceptions, the endemics and non-endemic natives all had their origins in the Andean region. Porter later (1979) amended these figures to 92% of the non-endemic native plants and 88% of the endemics having their closest relatives in South America (including those with

*The Biology of Island Floras*, eds. David Bramwell and Juli Caujapé-Castells. Published by Cambridge University Press. 

pantropical relatives), with the remainder originating from Central America, the Caribbean or North America.

Porter's (1979) review was carried out when the maximum age of the archipelago was considered to be about 3 million years (Ma) (Bailey, 1976), during which the endemics were assumed to have evolved. Porter (1979) drew attention to the fact that most evolutionary radiation had apparently taken place in the drier lowland vegetation zones, in which most of the endemics are found. In 1979, it was believed that the present climate of Galapagos, which supports mesic habitats above a certain altitude that depends on position and aspect, had persisted for the past 10 000 years, but that the preceding 14 000+ years had been drier (Colinvaux, 1972). This probably contributed to the differences in evolutionary history and levels of endemism between the present-day Dry and Humid vegetation zones.

Porter (1979) listed all endemic vascular plant taxa, with their known distribution by major island, habitats and chromosome numbers, and discussed the relationships and origins of each. In this chapter we review advances in our knowledge of the origins and evolution of the Galapagos endemic flora since Porter's 1979 review. Authorities for scientific names of endemic taxa are given in Table 5.1, others at first mention in the text.

## The Galapagos archipelago

The Galapagos Islands straddle the equator about 1000 km west of mainland Ecuador. Different authors recognise between 12 and 20 'main' islands in the archipelago. Here, we define 14 (named in Fig. 5.1) based on the biogeographical criterion of continued separation by sea through the Pleistocene ice ages (see glacial maximum map of the archipelago in Geist, 1996). There is uncertainty about the separation of Fernandina, the youngest island among these 14, which may have been connected to Isabela during the latest period of low sea level 17 000 years ago, although its recent origin (possibly emerged within the past 60 000 years: Geist 1996) suggests that it may have been much lower, smaller and perhaps isolated by sea even then. The islands other than these 14 (over 120 islets and rocks capable of supporting at least one species of terrestrial plant excluding mangroves: see Snell *et al.*, 1996) are treated as satellites of the main island to which they were linked by land during Pleistocene sea level falls or of the nearest main island in the case of a few small emergent rocks that are separated from the nearest main island by deep water (e.g. Roca Redonda off northern Isabela); these satellite islets do not have high levels of independent endemism. The largest island, Isabela (4700 $km^2$), makes up over half of the total land area of the archipelago (*c.* 7900 $km^2$), and is composed of six volcanoes that are treated as 'satellite islets' in this scheme; they have probably been separated from each other during periods of high sea level and

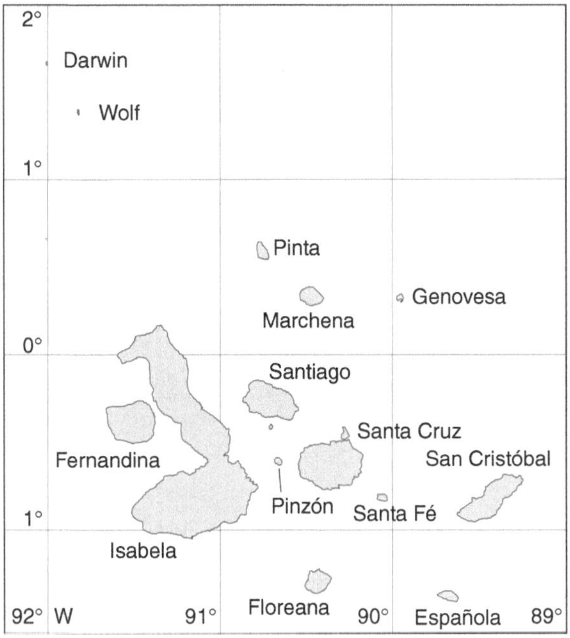


Figure 5.1 The Galapagos archipelago with named major islands. See text for biogeographical definition of 'major'.

certainly when they were younger, lower and smaller than they are now, and they demonstrate levels of independent endemism similar to or greater than some of the true satellite islets.

## Description of the vegetation

Porter (1979) followed Wiggins and Porter (1971) in defining seven principal vegetation zones in the archipelago (Fig. 5.2, top). However, this was based on detailed study of Santa Cruz island, with relatively less attention paid to the other islands; this scheme therefore applies well only to Santa Cruz, especially as regards the higher altitude, more humid zones. For example, of Wiggins and Porter's (1971) highland zones, the Brown Zone is only evident on Santa Cruz and Pinta, and the Miconia Zone only on Santa Cruz and San Cristóbal. We therefore provide a more generally applicable scheme in Fig. 5.2 (bottom). The main differences from the scheme of Wiggins and Porter (1971) are the relegation of the Scalesia, Brown, Miconia and Fern-Sedge Zones to subdivisions (vegetative formations) of the Humid Zone and the formal recognition of a High-Altitude Dry Zone, the existence of which was noted by Porter (1979), on the highest volcanoes of the western islands, above the inversion layer at which clouds commonly form and compounded by rain-shadow effects (Colinvaux, 1984). The smaller and lower islands have only the Littoral and

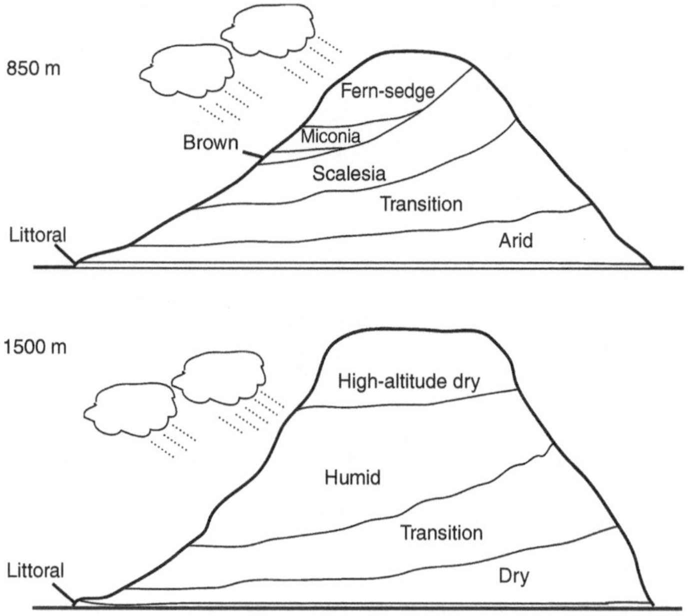


Figure 5.2 Above: major vegetation zones according to Wiggins and Porter (1971), applicable in full only to Santa Cruz island. Below: new scheme of major climatic-vegetation zones, applicable throughout the archipelago. Clouds indicate the normal level of the inversion layer (see text).

Dry Zones, and only seven islands are high enough to carry Humid Zone vegetation. The five zones recognised under this scheme are therefore more explicitly identified as climatic zones (or in the case of Littoral, otherwise abiotically determined), which interact with substrate type and hydrology to determine the major vegetation formations at any given locality. The climatic factors that produce this zonation are more fully discussed by Trueman and d'Ozouville (2010).

## Advances in methods and knowledge since 1979

Several new techniques have become available since 1979 (e.g. DNA sequencing), while new uses for older techniques have been applied to Galapagos plants only since then (e.g. pollen analysis for determining species arrival dates). Here we review how these and more traditional methods have added to and changed our understanding of the Galapagos flora in the past three decades.

### *Advances in knowledge of the origins and phylogeny of the flora*

The earliest, widely used molecular taxonomic method was allozyme analysis. This was first applied to Galapagos plants just before Porter's (1979) review, to examine the population genetics and origins of *Solanum cheesmaniae* (Rick & Fobes, 1975), and shortly afterwards to show that some of the morphological characteristics of forma *minor* of this species, including its 3–4 pinnate leaf structure, are controlled by single loci (Rick, 1980, 1983). Some 10 years later, enzyme analysis was used again to determine the ancestors, geographic origins, divergence dates and genetic variability of the two Galapagos endemic cottons *Gossypium* spp. (Wendel & Percival, 1990; Wendel & Percy 1990) and *Galvezia leucantha* (Elisens, 1989, 1992).

Allozyme-based techniques were soon supplemented by DNA analyses, commencing with chloroplast DNA (cpDNA) fragment hybridisation and culminating in increasingly sophisticated sequencing methods using a variety of marker regions of cp and nuclear DNA. Several authors have combined allozyme and DNA techniques, including Wendel and Percival (1990), who used both enzyme analysis and the first cpDNA fragment hybridisation experiments on a Galapagos plant, to investigate the origins of the Galapagos endemic diploid *Gossypium klotzschianum*, finding its closest relative to be the Baja Californian *G. davidsonii* Kell.

The first DNA sequencing study to have included a Galapagos endemic plant seems to have been that of Small and Wendel (2000), who used a nuclear gene to reinvestigate the phylogeny of New World diploid *Gossypium* spp. (once again), and confirmed the close relationship between *G. klotzschianum* and *G. davidsonii*. Later analysis of three other nuclear genes and amplified fragment-length polymorphisms (AFLPs) further confirmed most of these results, including the *klotzschianum-davidsonii* link and Wendel and Percival's (1990) allozyme-based finding of low genetic variability in *G. klotzschianum* (Alvarez *et al.*, 2005; Alvarez & Wendel, 2006; Guo *et al.*, 2007).

Since the early 1990s, DNA techniques have been used to investigate many other Galapagos plant taxa, especially endemics, and including: the origins of the endemic genus *Scalesia* (Schilling *et al.*, 1994); the age, origin and relationships of the allotetraploid cotton *Gossypium darwinii* (Small *et al.*, 1998; Lacape *et al.*, 2007); the phylogeny of *Phaseolus* (Delgado-Salinas *et al.*, 1999); the relationships of *Macraea* (Panero *et al.*, 1999); the origin, time of divergence and intraspecific variation within *Solanum* sec. *Lycopersicon* (Peralta & Spooner, 2001; Nesbitt & Tanksley, 2002; Nuez *et al.*, 2004) and the identity and origin of introduced weedy tomato taxa (Nuez *et al.*, 2004); the relationships of *Capsicum galapagoense* (Walsh & Hoot, 2001); the relationships of *Euphorbia equisetiformis* (Steinmann & Porter, 2002); the relationships of *Lecocarpus* (Rauscher, 2002); the population genetics of *Scalesia affinis* (Nielsen, 2004); the relationships of *Acnistus/Iochroma* (Smith & Baum, 2006); the origins of *Tiquilia* spp. (Moore *et al.*, 2006); relationships of *Brachycereus* and *Jasminocereus* cacti (Lendel *et al.*, 2006); the diverse origins of the two *Cuscuta* spp. (Stefanović *et al.*, 2007); population genetics and phylogeny

Figure 5.3 *Alternanthera filifolia pintensis* on Pinta Island. The colonisation history and area of origin of Galapagos *Alternanthera* have recently been clarified by Sánchez del Pino (2007). In the background stands a specimen of *Opuntia g. galapageia*; recent evidence (Helsen *et al.*, 2009a) suggests that the Galapagos *Opuntia* comrpise a single large radiation resulting from a single colonisation event, whereas they were formerly suggested to comprise two separate radiations from two colonisation events.

of *Opuntia* cacti (Helsen *et al.*, 2007, 2009a, 2009b; Griffith & Porter, 2009); the relationships of *Nolana galapagensis* (Dillon *et al.*, 2007; Tu *et al.*, 2008); the phylogeny of *Alternanthera* (Sánchez-del Pino, 2007; Fig. 5.3), *Encelia* (Fehlberg & Ranker, 2007), *Polygala* (Forest *et al.*, 2007), *Lycium* and *Grabowskia* (Levin *et al.*, 2007); the diverse origins of species formerly included in the endemic genus *Darwiniothamnus* (Andrus *et al.*, 2009; Tye, 2010); the status of *Bursera* taxa (Weeks & Tye, 2009); the origins and evolution of endemic *Varronia* (Weeks *et al.* 2010) and the relationships of *Sicyos* and *Sicyocaulis* (Sebastian *et al.*, 2010).

Meanwhile, 'classical' taxonomic work based primarily on comparative morphology has continued, as exemplified by the studies of: Adsersen (1980) and Brok and Adsersen (2007) on species limits in *Lecocarpus*; Eliasson (1984) on the affinity of *Macraea*; Seberg (1984) on taxon limits in *Acalypha*; Jansen (1985) on relationships in *Acmella*; Eliasson (1985, 2004) on endemic Amaranthaceae including *Alternanthera*, *Pleuropetalum*, *Amaranthus*, *Lithophila* and *Blutaparon*; Huft and van der Werff (1985) on taxon limits in *Chamaesyce*; Bernardello (1986) on *Lycium*; Lawesson and Adsersen (1987) on *Darwiniothamnus* with the description of a new species; Lawesson (1988) and K. Porter-Utley (pers. comm.) on the status of *Passiflora* taxa; Eliasson (1996) on *Sesuvium*; Harley (1999) and Kearns and Wunderlin (1999) (in Jørgensen & León-Yánez 1999) on the status of *Salvia* taxa and *Sicyocaulis*; Darwin *et al.* (2003) on *Solanum* sec. *Lycopersicon*; Tye (2008) on *Urochloa*; and Blaschke and Sanders (2009) on *Scalesia* phylogeny.

Innovative combinations of classical morphological techniques combined with molecular evidence have also been used, such as to investigate hybridisation between various *Scalesia* spp. (Nielsen *et al.*, 2003b; Lindhardt *et al.* 2009), and the evolution of *Phaseolus* (Delgado-Salinas *et al.*, 1999) and *Opuntia* (Helsen *et al.*, 2009a).

Similarly, allozyme analysis has not been entirely supplanted by DNA work, with Browne *et al.* (2003) having used allozymes to study genetic variability in *Opuntia* cacti. Alternative biochemical techniques have also been used in a few cases, such as high-performance liquid chromatography (HPLC) analysis of leaf trichome aromatics to investigate taxon limits in *Scalesia* and the origins of the genus (Spring *et al.*, 1999).

### *Contributions to knowledge of the evolution of the 'island syndrome'*

In parallel with the explicit phylogenetic and population genetic studies mentioned above, investigations of the evolution of 'island syndrome' characteristics (Carlquist, 1974) have also continued in Galapagos, such as Adsersen and Adsersen's (1993) study of the evolution of cyanogenic compounds. Most such studies have focused on breeding systems, including: McMullen (1987) on the proportions of different breeding systems in the flora; McMullen and Viderman (1994) on *Darwiniothamnus* and *Plumbago*; Mauchamp (1997), who found that the normally dioecious *Croton* may be evolving monoecy; Nielsen (2004), Nielsen *et al.* (2000, 2003a, 2003b, 2007) and Philipp and Nielsen (2010) on *Scalesia*; Philipp *et al.* (2004) on *Lecocarpus*; McMullen (2007, 2009) on *Tournefortia* and *Ipomoea*; Smith and Baum (2007) and Smith *et al.* (2008a, 2008b) on Iochrominae; and Jaramillo *et al.* (2010) on *Jasminocereus*.

### *Palaeobotanical determinations of status*

Palaeobotanical techniques are beginning to be used to determine the status of some of the 47 species that were classed as introduced by Porter (1983) but accepted as native by Lawesson *et al.* (1987) plus other 'doubtful natives' (Tye, 2006). These studies (van Leeuwen *et al.*, 2008) have resulted in the confirmation as truly native of four doubtful natives: *Ageratum conyzoides* L.; *Brickellia diffusa* (Vahl) A. Gray; *Spermacoce remota* Lam. (includes the taxa formerly identified as

*Borreria laevis* (Lam.) Griseb. and *Diodia radula* (Willd. & Hoffmanns. ex Roem. & Schult.) Cham. & Schltdl.); *Ranunculus flagelliformis* Sm. The first three of these are pantropical weedy species. Investigations in progress suggest that several additional doubtfully native taxa are truly native (E. Coffey, pers. comm.). Further, this technique has also identified as native *Cuphea carthagenensis* (Jacq.) G. F. MacBr. and *Hibiscus diversifolius* Jacq. (van Leeuwen *et al.*, 2008), both considered by all former authors as introduced. Such determinations have the effect of reducing the level of endemicity in the Galapagos flora, by increasing the number of non-endemic natives. These findings have led to changes in conservation policy too, as *H. diversifolius* had previously been considered invasive.

### *New geophysical research*

In parallel with botanical research, new insights into the climate and geology of Galapagos have provided improved knowledge of the environmental background to plant evolution in the archipelago. The precise times of emergence of individual islands are still not known, but Geist (1996) set minimum and maximum ages for each of the main islands, based respectively on their oldest known subaerial lavas, and their distance from Fernandina compared with the rate of movement of the Nazca Plate, which is currently about 7 cm per year (Kellogg & Vega, 1995). These calculations extend the possible age of the existing eastern islands to 6 Ma, while individual islands might be expected to remain emergent for 3–5 Ma, as indicated by the ages of surface lavas and the likely period of activity of each shield volcano (Bailey, 1976; Geist, 1996; O'Connor *et al.*, 2007). Further, seamounts presently found to the east of the modern archipelago have been found to carry subaerial lavas, extending the age of the archipelago to at least 14 Ma (Christie *et al.*, 1992; Werner *et al.*, 1999; Harpp *et al.*, 2005), while evidence of hotspot activity further extends it, possibly to as much as 90 Ma (Duncan & Hargreaves, 1984). This provides a long period for colonisation and evolution on the archipelago, with species able to hop from older to newer islands as they emerge. Further, evidence for the formation of the Caribbean Plate over the Galapagos hotspot (Duncan & Hargreaves, 1984) raises the possibility of biotic vicariance relationships between Galapagos and the Caribbean, with descendents of taxa that evolved on early hotspot islands drifting into the Caribbean while persisting on newer islands in the Galapagos (Grehan, 2001; Iturralde-Vinent, 2006). Later, dispersal from the Caribbean to Galapagos could have been facilitated by the Circumtropical Palaeocurrent, which ran westwards through the Tropical American Seaway from the Proto-Caribbean Sea into the Pacific Ocean, until the closure of the Panama Isthmus about 3.5 Ma (Collins, 1996).

Floral interchange between islands within the archipelago would have been facilitated by falls in sea level associated with the Pleistocene glaciations. Sea level is known to have been about 120 m below its present level at the time of the last glacial maximum around 17 000 years ago (Fairbanks, 1989). This would have increased the size and altitude of all 14 major islands, reduced most inter-island distances, and connected almost all of the present satellite islets to their neighbouring major island, including connecting Rábida (the largest satellite 'islet') with Santiago. It would also have exposed additional islands between Floreana and Santa Fe, between Santiago and Santa Cruz, and to the east of Española and San Cristóbal (Geist, 1996).

Climatic research has recently been stimulated by concern over global warming, and many studies have refined ideas about climatic history and climate cycles. Hartley and Chong (2002) showed that the climate of western South America became more arid in the late Pliocene, around 6 Ma, and that there had been prior arid phases coinciding with global cooling periods. It is not clear how Galapagos would have been affected by these shifts. In more recent times, Seltzer *et al.* (2002) showed that the climate of the equatorial Andean region shifted at the end of the last glacial period from cold and wet to hotter and drier about 20 000 BP. The timing of this change was widespread in the tropical Pacific, including around the Galapagos (Lea *et al.*, 2000, 2006), and occurred at least 5000 years before major warming in the Northern Hemisphere. So if the climatic shift (from dry to wetter) that occurred in Galapagos at the end of the last ice age coincided with this, then the current Galapagos climate might have existed for much longer than the 10 000 years that Colinvaux (1972) suggested. The discrepancy in timing could be explained by the fact that Colinvaux's results came from a lake close to the atmospheric inversion layer over Galapagos, above which conditions are much drier since cloud level commonly coincides with the inversion, and that major sea level rise came later, when deglaciation occurred at higher latitudes. If the Galapagos inversion were correspondingly lower when sea level was lower, then drier conditions at the lake site up to 10 000 BP need not indicate that drier conditions persisted until then throughout the archipelago or at contemporary altitudes equivalent to that of the present lake (H. Adsersen in Loope, 1989).

The timing and extent of immediate post-glacial changes in Galapagos thus remain incompletely resolved, but it is at least clear that during the past 135 000 years the Galapagos climate shifted several times between conditions cooler and drier than at present and conditions similar to the present or perhaps wetter (Colinvaux, 1972; Lea *et al.*, 2000, 2006). During drier periods, the Humid Zone vegetation would probably have been more restricted than at present, although the fact that the islands would have been cooler, and also higher then due to sea level fall, could have compensated to some extent for this. Whether many humid zone species would have been driven extinct is not clear; the long evolutionary history in Galapagos of some Humid-zone restricted genera such as *Cyathea* shows that humid conditions were not entirely absent (Colinvaux & Schofield, 1976a, 1976b). Nor is it clear how such changes would have impacted the evolutionary history of species that survived them, but some would certainly have suffered significant population reductions and consequent genetic bottlenecks, which may have favoured rapid evolutionary change by genetic drift.

On a finer scale, climatic shifts between drier periods or periods with fewer or weaker El Niño events, and wetter periods or periods with increased frequency of El Niño events, have now been elucidated for most of the Holocene (Conroy *et al.*, 2008). Shifts of this nature may not have caused major changes in the Galapagos flora as a whole but would certainly have affected species ranges in detail, especially aquatic plants (e.g. Colinvaux, 1972), whose habitats may have disappeared entirely for long intervals in the period since the last glaciation. Recent climatic research in Galapagos is reviewed by Bush *et al.* (2010).

## Analysis and review of current state of knowledge

### *Systematic list of endemic taxa, with latest data on origins and relationships*

Table 5.1 lists all taxa that were considered endemic by Porter (1979) and those that are considered endemic now, including six that are currently considered questionably endemic to Galapagos. For each, their area of origin, determined as precisely as possible in 1979 and now, is given, with references to any changes in status and origin.

### *Endemicity of the flora*

The data in Table 5.1 reveal that 233 taxa were considered endemic in 1979, vs 236 now. This apparent similarity masks some changes in status that have occurred since 1979, with 14 endemics that were recognised by Porter being no longer considered so or reduced to synonymy, but with 16 other endemic taxa having been resurrected or newly described. Endemicity levels have also been affected by the confirmation as native of taxa regarded as introduced by Porter (1979, 1983) and the discovery of other, probably native, non-endemics.

Porter (1979) noted that the percentage endemism had fallen in recent years (i.e. up to 1979), with better study of the mainland South American flora. This trend has not continued to any great extent: Porter (1979) included among his endemics only one taxon, *Amaranthus squamulatus*, that has since been unequivocally found outside the archipelago (Eliasson, 1985). Four others have supposedly been reported outside Galapagos, but we maintain them as Galapagos endemics pending further study. *Sesuvium edmondstonei* was reported from the southern Caribbean

Table 5.1 *Status and source areas for Galapagos endemic vascular plant taxa recognised by Porter (1979) and in this paper.*

| Family | Genus, species and infraspecies | Name used by Porter (1979) if different | Status in Porter (1979) | Present status (changes indicated in bold or by a dash) | Source in Porter (1979)[1] | Present source (changes indicated in bold or by a dash)[1] | References to changes |
|---|---|---|---|---|---|---|---|
| **Pteridophyta** | | | | | | | |
| Aspleniaceae | *Asplenium formosum* Willd. var. *carolinum* (Maxon) C. V. Morton | | En var | En var | Pantrop | Pantrop | |
| Cyatheaceae | *Cyathea weatherbyana* (C. V. Morton) C. V. Morton | | En Sp | En Sp | Trop Am | Trop Am | |
| Dryopteridaceae | *Megalastrum pleiosoros* (Hook. f.) A. R. Sm. & R. C. Moran | *Ctenitis pleiosoros* | En Sp | En Sp | Trop Am | Trop Am | |
| Lycopodiaceae | *Huperzia galapagensis* (O. Hamann) Holub | *Lycopodium setaceum galapagense* | En Ssp | **En Sp** | Pantrop | Pantrop | Holub, 1985 |
| Polypodiaceae | *Polypodium insularum* (C. V. Morton) de la Sota | | En Sp | En Sp | Andes | Andes | |
| Polypodiaceae | *Polypodium tridens* Kunze | | En Sp | En Sp | C Am | C Am | |
| Pteridaceae | *Notholaena galapagensis* Weath. & Svenson | | En Sp | En Sp | C Am | C Am | |

Table 5.1 (*cont.*)

| Family | Genus, species and infraspecies | Name used by Porter (1979) if different | Status in Porter (1979) | Present status (changes indicated in bold or by a dash) | Source in Porter (1979)[1] | Present source (changes indicated in bold or by a dash)[1] | References to changes |
|---|---|---|---|---|---|---|---|
| Thelypteridaceae | *Thelypteris tetragona* (Sw.) Small ssp. *aberrans* Morton | | En Ssp | En Ssp | Trop Am | Trop Am | |
| **Magnoliophyta** | | | | | | | |
| Acanthaceae | *Justicia galapagana* Lindau | | En Sp | En Sp | Andes | Andes | |
| Aizoaceae | *Sesuvium edmonstonei* Hook. f.[2] | | En Sp | En Sp | Pantrop | **Carib** | Eliasson, 1996 |
| Amaranthaceae | *Alternanthera filifolia* (Hook. f.) Howell ssp. *filifolia* | | En Ssp | En Ssp | Andes | **Trop Am** | Sánchez-del Pino, 2007 |
| Amaranthaceae | *Alternanthera filifolia* (Hook. f.) Howell ssp. *glauca* Howell | | En Ssp | En Ssp | Andes | **Trop Am** | Sánchez-del Pino, 2007 |
| Amaranthaceae | *Alternanthera filifolia* (Hook. f.) Howell ssp. *glaucescens* Eliasson | | En Ssp | En Ssp | Andes | **Trop Am** | Sánchez-del Pino, 2007 |

| | | | | | | |
|---|---|---|---|---|---|---|
| Amaranthaceae | *Alternanthera filifolia* (Hook. f.) Howell ssp. *microcephala* Eliasson | En Ssp | En Ssp | Andes | **Trop Am** | Sánchez-del Pino, 2007 |
| Amaranthaceae | *Alternanthera filifolia* (Hook. f.) Howell ssp. *nudicaulis* (Hook. f.) Eliasson | En Ssp | En Ssp | Andes | **Trop Am** | Sánchez-del Pino, 2007 |
| Amaranthaceae | *Alternanthera filifolia* (Hook. f.) Howell ssp. *pintensis* Eliasson | En Ssp | En Ssp | Andes | **Trop Am** | Sánchez-del Pino, 2007 |
| Amaranthaceae | *Alternanthera filifolia* (Hook. f.) Howell ssp. *rabidensis* Eliasson | En Ssp | En Ssp | Andes | **Trop Am** | Sánchez-del Pino, 2007 |
| Amaranthaceae | *Alternanthera flavicoma* (Andersson) Howell | En Sp | En Sp | Andes | **Trop Am** | Sánchez-del Pino, 2007 |
| Amaranthaceae | *Alternanthera galapagensis* (A. Stewart) Howell | En Sp | En Sp | Andes | **Trop Am** | Sánchez-del Pino, 2007 |
| Amaranthaceae | *Alternanthera helleri* (B. L. Rob.) Howell | En Sp | En Sp | Andes | **Trop Am** | inference |
| Amaranthaceae | *Alternanthera nesiotes* Johnst. | En Sp | En Sp | Andes | **Trop Am** | Sánchez-del Pino, 2007 |
| Amaranthaceae | *Alternanthera snodgrassii* (B. L. Rob.) Howell | En Sp | En Sp | Andes | **Trop Am** | Sánchez-del Pino, 2007 |

Table 5.1 (*cont.*)

| Family | Genus, species and infraspecies | Name used by Porter (1979) if different | Status in Porter (1979) | Present status (changes indicated in bold or by a dash) | Source in Porter (1979)[1] | Present source (changes indicated in bold or by a dash)[1] | References to changes |
|---|---|---|---|---|---|---|---|
| Amaranthaceae | *Amaranthus anderssonii* Howell | | En Sp | En Sp | Andes | **Ec-Peru** | Eliasson, 1985 |
| Amaranthaceae | *Amaranthus furcatus* Howell | | En Sp | En Sp | Andes | **Ec-Peru** | inference |
| Amaranthaceae | *Amaranthus sclerantoides* (Andersson) Andersson | | En Sp | En Sp | Andes | **Ec-Peru** | inference |
| Amaranthaceae | *Amaranthus squamulatus* (Andersson) B. L. Rob. | | En Sp | **Na** | Andes | — | Eliasson, 1985 |
| Amaranthaceae | *Blutaparon rigidum* (B. L. Rob. & Greenm.) Mears | *Philoxerus rigidus* | En Sp | En Sp | Pantrop | **Trop Am** | Eliasson, 1985 |
| Amaranthaceae | *Froelichia juncea* B. L. Rob. & Greenm. ssp. *alata* Howell | | En Ssp | En Ssp | Trop Am | Trop Am | |
| Amaranthaceae | *Froelichia juncea* B.L. Rob. & Greenm. ssp. *juncea* | | En Ssp | En Ssp | Trop Am | Trop Am | |

| | | | | | | | |
|---|---|---|---|---|---|---|---|
| Amaranthaceae | *Froelichia nudicaulis* Hook. f. ssp. *curta* Howell | | En Ssp | En Ssp | Trop Am | Trop Am | |
| Amaranthaceae | *Froelichia nudicaulis* Hook. f. ssp. *lanigera (Andersson)* Eliasson | | En Ssp | En Ssp | Trop Am | Trop Am | |
| Amaranthaceae | *Froelichia nudicaulis* Hook. f. ssp. *nudicaulis* | | En Ssp | En Ssp | Trop Am | Trop Am | |
| Amaranthaceae | *Lithophila radicata* (Hook. f.) Standl. | | En Sp | En Sp | Carib | Carib | |
| Amaranthaceae | *Lithophila (radicata)* (Hook. f.) Standl. *scirpoides* (Hook. f.) | *L. radicata* | — | **En Ssp?** | — | **Carib** | Eliasson, 1985 |
| Amaranthaceae | *Lithophila subscaposa* (Hook. f.) Standl. | | En Sp | En Sp | Carib | Carib | |
| Amaranthaceae | *Pleuropetalum darwinii* Hook. f. | | En Sp | En Sp | Trop Am | Trop Am | |
| Apiaceae | *Hydrocotyle galapagensis* B.L. Rob. | | En Sp | En Sp | Americas | Americas | |
| Apocynaceae | *Vallesia glabra* (Cav.) Link var. *pubescens* (Andersson) Wiggins | | En var | En var | Trop Am | Trop Am | |
| Asclepiadaceae | *Sarcostemma angustissimum* (Andersson) R.W. Holm | | En Sp | En Sp | S Am | S Am | |

Table 5.1 (*cont.*)

| Family | Genus, species and infraspecies | Name used by Porter (1979) if different | Status in Porter (1979) | Present status (changes indicated in bold or by a dash) | Source in Porter (1979)[1] | Present source (changes indicated in bold or by a dash)[1] | References to changes |
|---|---|---|---|---|---|---|---|
| Asteraceae | *Acmella darwinii* (D.M. Porter) R.K. Jansen | *Spilanthes darwinii* | En Sp | En Sp | Trop Am | **Ec-Peru** | Jansen, 1985 |
| Asteraceae | *Baccharis steetzii* Andersson | | En Sp | En Sp | Andes | Andes | |
| Asteraceae | *Chrysanthellum fagerlindii* Eliasson | — | — | **En Sp?** | — | **C Am** | Lawesson *et al.*, 1987 |
| Asteraceae | *Chrysanthellum pusillum* Hook. f. | | En Sp | En Sp | C Am | C Am | |
| Asteraceae | *Darwiniothamnus lancifolius* (Hook. f.) Harling ssp. *glabriusculus* (A. Stewart) Lawesson & Adsersen | *D. l. lancifolius* | En Ssp | En Ssp | Chile | **Carib** | Andrus, 2002; Andrus *et al.*, 2009 |
| Asteraceae | *Darwiniothamnus lancifolius* (Hook. f.) Harling ssp. *glandulosus* Harling | | En Ssp | En Ssp | Chile | **Carib** | Andrus, 2002; Andrus *et al.*, 2009 |
| Asteraceae | *Darwiniothamnus lancifolius (Hook. f.) Harling ssp. lancifolius* | — | — | **En Ssp** | | **Carib** | Andrus, 2002; Andrus *et al.*, 2009 |

| | | | | | | | |
|---|---|---|---|---|---|---|---|
| Asteraceae | *Darwiniothamnus tenuifolius* (Hook. f.) Harling *santacruzianus* Harling | | En Ssp | En Ssp | Chile | **Carib** | Andrus, 2002; Andrus *et al.*, 2009 |
| Asteraceae | *Darwiniothamnus tenuifolius* (Hook. f.) Harling var. *tenuifolius* | | En Ssp | En Ssp | Chile | **Carib** | Andrus, 2002; Andrus *et al.*, 2009 |
| Asteraceae | *Delilia inelegans* (Hook. f.) Kuntze | | En Sp | En Sp | Trop Am | Trop Am | |
| Asteraceae | *Delilia repens* (Hook. f.) Kuntze | | En Sp | En Sp | Trop Am | Trop Am | |
| Asteraceae | *Encelia hispida* Andersson | | En Sp | En Sp | Peru-Chile | Peru-Chile | Fehlberg & Ranker, 2007 |
| Asteraceae | *Erigeron alternifolius* (Lawesson & Adsersen) Tye & Andrus | — | — | **En Sp** | — | **Chile** | Lawesson & Adsersen, 1987; Andrus *et al.*, 2009; Tye, 2010 |
| Asteraceae | *Jaegeria gracilis* Hook. f. | | En Sp | En Sp | Trop Am | Trop Am | |
| Asteraceae | *Lecocarpus darwinii* Adsersen | — | — | **En Sp** | — | **N Andes** | Adsersen, 1980; Rauscher, 2002; Andrus *et al.*, 2009 |
| Asteraceae | *Lecocarpus lecocarpoides* (B.L. Rob. & Greenm.) Cronquist & Stuessy | | En Sp | En Sp | Trop Am | **N Andes** | Rauscher, 2002; Andrus *et al.*, 2009 |

Table 5.1 (*cont.*)

| Family | Genus, species and infraspecies | Name used by Porter (1979) if different | Status in Porter (1979) | Present status (changes indicated in bold or by a dash) | Source in Porter (1979)[1] | Present source (changes indicated in bold or by a dash)[1] | References to changes |
|---|---|---|---|---|---|---|---|
| Asteraceae | *Lecocarpus leptolobus* (S.F. Blake) Cronquist & Stuessy | | En Sp | En Sp | Trop Am | **N Andes** | Rauscher, 2002; Andrus *et al.*, 2009; pers. obs., CDS |
| Asteraceae | *Lecocarpus pinnatifidus* Decne. | | En Sp | En Sp | Trop Am | **N Andes** | Rauscher, 2002; Andrus *et al.*, 2009 |
| Asteraceae | *Macraea laricifolia* Hook. f. | | En Sp | En Sp | Trop Am | Trop Am | |
| Asteraceae | *Pectis subsquarrosa* (Hook. f.) Sch. Bip. | | En Sp | En Sp | S Am | S Am | |
| Asteraceae | *Pectis tenuifolia* (DC.) Sch. Bip. | | En Sp | En Sp | S Am | S Am | |
| Asteraceae | *Scalesia affinis* Hook. f. ssp. *affinis* | | En Ssp | En Ssp | Andes | Andes | |
| Asteraceae | *Scalesia affinis* Hook. f. ssp. *brachyloba* Harling | | En Ssp | En Ssp | Andes | Andes | |

| | | | | | | | |
|---|---|---|---|---|---|---|---|
| Asteraceae | *Scalesia affinis* Hook. f. ssp. *gummifera* (Hook. f.) Harling | | En Ssp | En Ssp | Andes | Andes | |
| Asteraceae | *Scalesia aspera* Andersson | | En Sp | En Sp | Andes | Andes | |
| Asteraceae | *Scalesia atractyloides* Arn. var. *atractyloides* | | En var | En var | Andes | Andes | |
| Asteraceae | *Scalesia atractyloides* Arn. var. *darwinii* (Hook. f.) Eliasson | | En var | En var | Andes | Andes | |
| Asteraceae | *Scalesia bauri* B.L. Rob. & Greenm. ssp. *bauri* | | En Ssp | En Ssp | Andes | Andes | |
| Asteraceae | *Scalesia bauri* B.L. Rob. & Greenm. ssp. *hopkinsii* (B.L. Rob.) Eliasson | | En Ssp | En Ssp | Andes | Andes | |
| Asteraceae | *Scalesia cordata* A. Stewart | | En Sp | En Sp | Andes | Andes | |
| Asteraceae | *Scalesia crockeri* Howell | | En Sp | En Sp | Andes | Andes | |
| Asteraceae | *Scalesia divisa* Andersson | | En Sp | En Sp | Andes | Andes | |
| Asteraceae | *Scalesia gordilloi* O. Hamann & Wium-And. | — | — | **En Sp** | — | **Andes** | Hamann & Wium-Andersen, 1986 |

Table 5.1 (*cont.*)

| Family | Genus, species and infraspecies | Name used by Porter (1979) if different | Status in Porter (1979) | Present status (changes indicated in bold or by a dash) | Source in Porter (1979)[1] | Present source (changes indicated in bold or by a dash)[1] | References to changes |
|---|---|---|---|---|---|---|---|
| Asteraceae | *Scalesia helleri* B.L. Rob. ssp. *helleri* | | En Ssp | En Ssp | Andes | Andes | |
| Asteraceae | *Scalesia helleri* B.L. Rob. ssp. *santacruziana* Harling | | En Ssp | En Ssp | Andes | Andes | |
| Asteraceae | *Scalesia incisa* Hook. f. | | En Sp | En Sp | Andes | Andes | |
| Asteraceae | *Scalesia micro-cephala* B.L. Rob. var. *cordifolia* Eliasson | | En Ssp | En Ssp | Andes | Andes | |
| Asteraceae | *Scalesia micro-cephala* B.L. Rob. var. *microcephala* | | En Ssp | En Ssp | Andes | Andes | |
| Asteraceae | *Scalesia peduncu-lata* Hook. f. | | En Sp | En Sp | Andes | Andes | |
| Asteraceae | *Scalesia retroflexa* Hemsl. | | En Sp | En Sp | Andes | Andes | |
| Asteraceae | *Scalesia stewartii* Riley | | En Sp | En Sp | Andes | Andes | |
| Asteraceae | *Scalesia villosa* A. Stewart | | En Sp | En Sp | Andes | Andes | |

| | | | | | | | |
|---|---|---|---|---|---|---|---|
| Boraginaceae | *Tiquilia darwinii* (Hook. f.) A.T. Richardson | | En Sp | En Sp | Ec-Chile | **SW N Am or N Andes** | Moore *et al.*, 2006 |
| Boraginaceae | *Tiquilia fusca* (Hook. f.) A.T. Richardson | | — | **En Sp** | — | **SW N Am or N Andes** | Moore *et al.*, 2006 |
| Boraginaceae | *Tiquilia galapagoa* (Howell) A.T. Richardson | | En Sp | En Sp | Ec-Chile | **SW N Am or N Andes** | Moore *et al.*, 2006 |
| Boraginaceae | *Tiquilia nesiotica* (Howell) A.T. Richardson | | En Sp | En Sp | Ec-Chile | **SW N Am or N Andes** | Moore *et al.*, 2006 |
| Boraginaceae | *Tournefortia pubescens* Hook. f. | | En Sp | En Sp | Trop Am | Trop Am | |
| Boraginaceae | *Tournefortia rufo-sericea* Hook. f. | | En Sp | En Sp | Trop Am | Trop Am | |
| Boraginaceae | *Heliotropium anderssonii* B.L. Rob. | | En Sp | En Sp | Andes | Andes | Weeks *et al.*, 2010 |
| Boraginaceae | *Varronia anderssonii* (Kuntze) *Borhidi* | *Cordia anderssonii* | En Sp | En Sp | Ec-Peru | **Trop Am** | Weeks *et al.*, 2010 |
| Boraginaceae | *Varronia leucophlyctis (Hook. f.) Andersson* | *Cordia leucophlyctis* | En Sp | En Sp | Ec-Peru | **Trop Am** | Weeks *et al.*, 2010 |
| Boraginaceae | *Varronia revoluta (Hook. f.) Borhidi* | *Cordia revolute* | En Sp | En Sp | Ec-Peru | **Trop Am** | Weeks *et al.*, 2010 |
| Boraginaceae | *Varronia scouleri (Hook. f.) Borhidi* | *Cordia scouleri* | En Sp | En Sp | Ec-Peru | **Trop Am** | Weeks *et al.*, 2010 |

Table 5.1 (*cont.*)

| Family | Genus, species and infraspecies | Name used by Porter (1979) if different | Status in Porter (1979) | Present status (changes indicated in bold or by a dash) | Source in Porter (1979)[1] | Present source (changes indicated in bold or by a dash)[1] | References to changes |
|---|---|---|---|---|---|---|---|
| Burseraceae | *Bursera graveolens* (Kunth) Triana & Planch. ssp. *malacophylla* (B.L. Rob.) A. Weeks & Tye | *Bursera malacophylla* | En Sp | **En Ssp** | N Andes | N Andes | Weeks & Tye 2009 |
| Cactaceae | *Brachycereus nesioticus* (K. Schum.) Backeb. | | En Sp | En Sp | Ecuador | Ecuador | Lendel *et al.* 2006 |
| Cactaceae | *Jasminocereus thouarsii* (F.A.C. Weber) Backeb. var. *delicatus* (E.Y. Dawson) E.F. Anderson & Walk. | | En var | En var | Ecuador | Ecuador | Lendel *et al.* 2006 |
| Cactaceae | *Jasminocereus thouarsii* (F.A.C. Weber) Backeb. var. *sclerocarpus* (K. Schum.) E.F. Anderson & Walk. | | En var | En var | Ecuador | Ecuador | Lendel *et al.* 2006 |
| Cactaceae | *Jasminocereus thouarsii* (F.A.C. Weber) Backeb. var. *thouarsii* | | En var | En var | Ecuador | Ecuador | Lendel *et al.* 2006 |

| | | | | | |
|---|---|---|---|---|---|
| Cactaceae | *Opuntia echios* Howell var. *barringtonensis* E.Y. Dawson | En var | En var | Andes | Andes |
| Cactaceae | *Opuntia echios* Howell var. *echios* | En var | En var | Andes | Andes |
| Cactaceae | *Opuntia echios* Howell var. *gigantea* Howell | En var | En var | Andes | Andes |
| Cactaceae | *Opuntia echios* Howell var. *inermis* E.Y. Dawson | En var | En var | Andes | Andes |
| Cactaceae | *Opuntia echios* Howell var. *zacana* (Howell) E.F. Anderson & Walk. | En var | En var | Andes | Andes |
| Cactaceae | *Opuntia galapageia* Hemsl. var. *galapageia* | En var | En var | Andes | Andes |
| Cactaceae | *Opuntia galapageia* Hemsl. var. *macrocarpa* E.Y. Dawson | En var | En var | Andes | Andes |
| Cactaceae | *Opuntia galapageia* Hemsl. var. *profusa* E.F. Anderson & Walk. | En var | En var | Andes | Andes |

Table 5.1 (*cont.*)

| Family | Genus, species and infraspecies | Name used by Porter (1979) if different | Status in Porter (1979) | Present status (changes indicated in bold or by a dash) | Source in Porter (1979)[1] | Present source (changes indicated in bold or by a dash)[1] | References to changes |
|---|---|---|---|---|---|---|---|
| Cactaceae | *Opuntia helleri* K. Schum. | | En Sp | En Sp | Andes | Andes | |
| Cactaceae | *Opuntia insularis* A. Stewart | | En Sp | En Sp | Andes | Andes | |
| Cactaceae | *Opuntia megasp-erma* Howell var. *megasperma* | | En var | En var | Andes | Andes | |
| Cactaceae | *Opuntia megasp-erma* Howell var. *mesophytica* J. Lundh | | En var | En var | Andes | Andes | |
| Cactaceae | *Opuntia megasp-erma* Howell var. *orientalis* Howell | | En var | En var | Andes | Andes | |
| Cactaceae | *Opuntia saxicola* Howell | | En Sp | En Sp | Andes | Andes | |
| Caryophyllaceae | *Drymaria monticola* Howell | | En Sp | En Sp | Andes | Andes | |
| Convolvulaceae | *Cuscuta acuta* Engelm. | | En Sp | En Sp | Peru-Chile | **SW N Am** | Stefanovic *et al.* 2007 |
| Convolvulaceae | *Cuscuta gymno-carpa* Engelm. | | En Sp | En Sp | Argentina | **SW N Am** | Stefanovic *et al.* 2007 |

| | | | | | | | |
|---|---|---|---|---|---|---|---|
| Convolvulaceae | *Ipomoea habeliana* Oliv. | | En Sp | En Sp | Trop Am | Trop Am | |
| Convolvulaceae | *Ipomoea linearifolia* Hook. f. | | En Sp | En Sp | Andes | Andes | |
| Convolvulaceae | *Ipomoea tubiflora* Hook. f. | | En Sp | En Sp | Trop Am | Trop Am | |
| Cucurbitaceae | *Rytidostylis carthaginensis* (Jacq.) Kuntze var. *cordatum* (Hook. f.) Svenson | *Elaterium carthagenense cordatum* | En var | En var | Ec-Peru | Ec-Peru | |
| Cucurbitaceae | *Sicyos pentagonus* ('Wiggins) | *Sicyocaulis pentagonus* | En Gen | **En Sp** | Trop Am | **Ec-Peru** | Sebastian *et al.* 2010 |
| Cucurbitaceae | *Sicyos villosus* Hook. f. | *S. villosa* | En Sp | En Sp | Peru-Chile | **SW N Am** | Sebastian *et al.* 2010 |
| Cyperaceae | *Cyperus anderssonii* Boeck. | | En Sp | En Sp | Trop Am | Trop Am | |
| Cyperaceae | *Cyperus elegans* L. ssp. *rubiginosus* (Hook. f.) Eliasson | | En Ssp | En Ssp | Trop Am | Trop Am | |
| Cyperaceae | *Cyperus grandifolius* Andersson | | En Sp | En Sp | Trop Am | Trop Am | |
| Ericaceae | *Pernettya howellii* Sleumer | | En Sp | En Sp | Chile | Chile | |
| Euphorbiaceae | *Acalypha abingdonii* Seberg | *A. s. sericea* | En var | **En Sp** | S Am | **Trop Am** | Seberg 1984 |
| Euphorbiaceae | *Acalypha baurii* B.L. Rob. & Greenm. | *A. s. baurii* | En var | **En Sp** | S Am | **Trop Am** | Seberg 1984 |
| Euphorbiaceae | *Acalypha flaccida* Hook. f. | | En Sp | — | S Am | — | Seberg 1984 |

Table 5.1 (*cont.*)

| Family | Genus, species and infraspecies | Name used by Porter (1979) if different | Status in Porter (1979) | Present status (changes indicated in bold or by a dash) | Source in Porter (1979)[1] | Present source (changes indicated in bold or by a dash)[1] | References to changes |
|---|---|---|---|---|---|---|---|
| Euphorbiaceae | *Acalypha parvula* Hook. f. | *A. flaccida, A. p. chathamensis, A. p. reniformis, A. p. strobilifera, A. velutina* | 2 En Spp, 3 En vars, listed separately | **En Sp** | – | **Trop Am** | Seberg 1984 |
| Euphorbiaceae | *Acalypha parvula* Hook. f. var. *chathamensis* (Robins.) Webster | | En var | — | S Am | — | Seberg 1984 |
| Euphorbiaceae | *Acalypha parvula* Hook. f. var. *reniformis* (Hook. f.) Mueller-Arg. | | En var | — | S Am | — | Seberg 1984 |
| Euphorbiaceae | *Acalypha parvula* Hook. f. var. *strobilifera* (Hook. f.) Mueller-Arg. | | En var | — | S Am | — | Seberg 1984 |
| Euphorbiaceae | *Acalypha velutina* Hook. f. | | En Sp | — | S Am | — | Seberg 1984 |

| | | | | | | | |
|---|---|---|---|---|---|---|---|
| Euphorbiaceae | *Acalypha wigginsii* G.L. Webster | | En Sp | En Sp | S Am | **Trop Am** | Seberg 1984 |
| Euphorbiaceae | *Chamaesyce abdita* D.G. Burch | | En Sp | En Sp | Trop Am | Trop Am | Huft & van der Werff 1985 |
| Euphorbiaceae | *Chamaesyce amplexicaulis* (Hook. f.) D.G. Burch | | En Sp | En Sp | Trop Am | Trop Am | Huft & van der Werff 1985 |
| Euphorbiaceae | *Chamaesyce bind-loensis* (Stewart) D.G. Burch | | En Sp | — | Trop Am | — | Huft & van der Werff 1985 |
| Euphorbiaceae | *Chamaesyce gala-pageia* (B.L. Rob. & Greenm.) D.G. Burch | | En Sp | En Sp | Trop Am | Trop Am | Huft & van der Werff 1985 |
| Euphorbiaceae | *Chamaesyce num-mularia* (Hook. f.) D.G. Burch var. *glabra* D.G. Burch | | En var | En var | Trop Am | Trop Am | Huft & van der Werff 1985 |
| Euphorbiaceae | *Chamaesyce num-mularia* (Hook. f.) D.G. Burch var. *nummularia* | | En var | En var | Trop Am | Trop Am | Huft & van der Werff 1985 |
| Euphorbiaceae | *Chamaesyce punc-tulata* (Andersson) D.G. Burch | *C. bindloensis, C. punctulata* | En Sp | En Sp | Trop Am | Trop Am | Huft & van der Werff 1985 |
| Euphorbiaceae | *Chamaesyce recurva* (Hook. f.) D.G. Burch | | En Sp | En Sp | Trop Am | Trop Am | Huft & van der Werff 1985 |
| Euphorbiaceae | *Chamaesyce viminea* (Hook. f.) D.G. Burch | | En Sp | En Sp | Carib | Carib | Huft & van der Werff 1985 |

Table 5.1 (*cont.*)

| Family | Genus, species and infraspecies | Name used by Porter (1979) if different | Status in Porter (1979) | Present status (changes indicated in bold or by a dash) | Source in Porter (1979)[1] | Present source (changes indicated in bold or by a dash)[1] | References to changes |
|---|---|---|---|---|---|---|---|
| Euphorbiaceae | *Croton scouleri* Hook. f. var. *brevifolius* (Andersson) Müll. Arg. | | En var | En var | Ecuador | Ecuador | |
| Euphorbiaceae | *Croton scouleri* Hook. f. var. *darwinii* G.L. Webster | | En var | En var | Ecuador | Ecuador | |
| Euphorbiaceae | *Croton scouleri* Hook. f. var. *grandifolius* Müll. Arg. | | En var | En var | Ecuador | Ecuador | |
| Euphorbiaceae | *Croton scouleri* Hook. f. var. *scouleri* | | En var | En var | Ecuador | Ecuador | |
| Euphorbiaceae | *Euphorbia equiseti-formis* A. Stewart | | En Sp | En Sp | Trop Am | **Carib** | Steinmann & Porter, 2002 |
| Fabaceae | *Dalea tenuicaulis* Hook. f. | | En Sp | En Sp | N Andes | N Andes | |
| Fabaceae | *Phaseolus mollis* Hook. f. | | En Sp | En Sp | Trop Am | Trop Am | |

| | | | | | | | |
|---|---|---|---|---|---|---|---|
| Iridaceae | *Sisyrinchium galapagense* Ravenna | *S. macrocephalum* | Na | **En Sp** | — | **Andes** | Ravenna, 1979 |
| Lamiaceae | *Hyptis gymnocaulos* Epling | | En Sp | En Sp | Andes | Andes | |
| Lamiaceae | *Salvia insularum* Epling | | En Sp | — | Carib | — | Harley, 1999 |
| Lamiaceae | *Salvia prostrata* Hook. f. | | En Sp | En Sp | Carib | Carib | |
| Lamiaceae | *Salvia pseudoserotina* Epling | *S. insularum, S. pseudoserotina* | En Sp | En Sp | Carib | Carib | Harley, 1999 |
| Linaceae | *Linum cratericola* Eliasson | | En Sp | En Sp | Peru | Peru | |
| Linaceae | *Linum harlingii* Eliasson | | En Sp | En Sp | Peru | Peru | |
| Malvaceae | *Abutilon depauperatum* (Hook. f.) Andersson ex B.L. Rob. | | En Sp | En Sp | Trop Am | Trop Am | |
| Malvaceae | *Fuertesimalva insularis* (Kearney) Fryxell | *Urocarpidium insulare* | En Sp | En Sp | Argentina | Argentina | |
| Malvaceae | *Gossypium darwinii* G. Watt | *G. barbadense* var *darwinii* | En var | **En Sp** | Ec-Peru | **Trop Am** | Small *et al.* 1998; Lacape *et al.*, 2007 |
| Malvaceae | *Gossypium klotzschianum* Andersson | | En Sp | En Sp | SW N Am | SW N Am | |
| Melastomataceae | *Miconia robinsoniana* Cogn. | | En Sp | En Sp | Col-Ven | Col-Ven | |

Table 5.1 (*cont.*)

| Family | Genus, species and infraspecies | Name used by Porter (1979) if different | Status in Porter (1979) | Present status (changes indicated in bold or by a dash) | Source in Porter (1979)[1] | Present source (changes indicated in bold or by a dash)[1] | References to changes |
|---|---|---|---|---|---|---|---|
| Mimosaceae | *Vachellia rorudiana* (Christoph.) Seigler & Ebinger | *Acacia rorudiana* | En Sp | En Sp | Ec-Peru | Ec-Peru | Seigler & Ebinger 2005 |
| Molluginaceae | *Mollugo crockeri* Howell | | En Sp | En Sp | Americas | Americas | |
| Molluginaceae | *Mollugo flavescens* Andersson ssp. *flavescens* | | En Ssp | En Ssp | Americas | Americas | |
| Molluginaceae | *Mollugo flavescens* Andersson ssp. *gracillima* (Andersson) Eliasson | | En Ssp | En Ssp | Americas | Americas | |
| Molluginaceae | *Mollugo flavescens* Andersson ssp. *insularis* (Howell) Eliasson | | En Ssp | En Ssp | Americas | Americas | |
| Molluginaceae | *Mollugo flavescens* Andersson ssp. *striata* (Howell) Eliasson | | En Ssp | En Ssp | Americas | Americas | |
| Molluginaceae | *Mollugo floriana* (B.L. Rob.) Howell ssp. *floriana* | | En Ssp | En Ssp | Americas | Americas | |

| | | | | | | | |
|---|---|---|---|---|---|---|---|
| Molluginaceae | *Mollugo floriana* (B.L. Rob.) Howell ssp. *gypsophiloides* Howell | | En Ssp | En Ssp | Americas | Americas | |
| Molluginaceae | *Mollugo floriana* (B.L. Rob.) Howell ssp. *santacruziana* (Christoph.) Eliasson | | En Ssp | En Ssp | Americas | Americas | |
| Molluginaceae | *Mollugo snodgrassii* B.L. Rob. | | En Sp | En Sp | Americas | Americas | |
| Myrtaceae | *Psidium galapageium* Hook. f. var. *galapageium* | | En var | En var | C Am | C Am | |
| Myrtaceae | *Psidium galapageium* Hook. f. var. *howellii* Porter | | En var | En var | C Am | C Am | |
| Nyctaginaceae | *Pisonia floribunda* Hook. f. | | En Sp | En Sp | Pantrop | Pantrop | |
| Orchidaceae | *Cranichis lichenophila* D. Weber | | En Sp | En Sp | Carib | Carib | |
| Orchidaceae | *Cyclopogon werffii* Dodson | — | — | **En Sp** | — | **Andes** | Dodson, 1993 |
| Orchidaceae | *Epidendrum spicatum* Hook. f. | | En Sp | En Sp | Andes | Andes | |
| Passifloraceae | *Passiflora colinvauxii* Wiggins | | En Sp | En Sp | S Am | S Am | |
| Passifloraceae | *Passiflora foetida* L. var. *galapagensis* Killip. | | En var | En var | Trop Am | Trop Am | |

Table 5.1 (*cont.*)

| Family | Genus, species and infraspecies | Name used by Porter (1979) if different | Status in Porter (1979) | Present status (changes indicated in bold or by a dash) | Source in Porter (1979)[1] | Present source (changes indicated in bold or by a dash)[1] | References to changes |
|---|---|---|---|---|---|---|---|
| Passifloraceae | *Passiflora tridactylites* Hook. f. | *P. suberosa* (syn. of) | Na | **En Sp** | — | **Trop Am** | Lawesson 1988; K. Porter-Utley pers. comm. |
| Piperaceae | *Peperomia galapagensis* Hook. f. ex Mig. var. *galapagensis* | | En var | En var | S Am | S Am | |
| Piperaceae | *Peperomia galapagensis* Hook. f. ex Mig. var. *ramulosa* (Andersson) Yunck. | | En var | En var | S Am | S Am | |
| Piperaceae | *Peperomia obtusilimba* C. DC. | | En Sp | En Sp | S Am | S Am | |
| Piperaceae | *Peperomia petiolata* Hook. f. | | En Sp | En Sp | S Am | S Am | |
| Plantaginaceae | *Plantago galapagensis* Rahn | | En Sp | En Sp | SW N Am | SW N Am | |
| Poaceae | *Aristida divulsa* Andersson | | En Sp | En Sp | Brazil | Brazil | |

| | | | | | |
|---|---|---|---|---|---|
| Poaceae | *Aristida repens* Trin. | En Sp | En Sp | Brazil | Brazil |
| Poaceae | *Aristida subspicata* Trin. & Rupr. | En Sp | En Sp | Brazil | Brazil |
| Poaceae | *Aristida villosa* B.L. Rob. & Greenm. | En Sp | En Sp | Brazil | Brazil |
| Poaceae | *Cenchrus platyacanthus* Andersson | En Sp | En Sp | Pantrop | Pantrop |
| Poaceae | *Paspalum galapageium* Chase var. *galapageium* | En Ssp | En Ssp | Andes | Andes |
| Poaceae | *Paspalum galapageium* Chase var. *minoratum* Chase | En Ssp | En Ssp | Andes | Andes |
| Poaceae | *Paspalum redundans* Chase | En Sp | En Sp | Andes | Andes |
| Poaceae | *Pennisetum pauperum* Nees ex Steud. | En Sp | En Sp | Andes | Andes |
| Poaceae | *Trichoneura lindleyana* (Kunth) Ekman var. *albemarlensis* Reeder | En Ssp | En Ssp | Andes | Andes |
| Poaceae | *Trichoneura lindleyana* (Kunth) Ekman var. *lindleyana* | En Ssp | En Ssp | Andes | Andes |
| Poaceae | *Trisetum howellii* Hitchc. | En Sp | En Sp | Trop Am | Trop Am |

Table 5.1 (*cont.*)

| Family | Genus, species and infraspecies | Name used by Porter (1979) if different | Status in Porter (1979) | Present status (changes indicated in bold or by a dash) | Source in Porter (1979)[1] | Present source (changes indicated in bold or by a dash)[1] | References to changes |
|---|---|---|---|---|---|---|---|
| Poaceae | *Urochloa multiculma* (Andersson) Morrone & Zuloaga | *Panicum fasciculatum* | Na | **En Sp** | — | **Trop Am** | Tye 2008 |
| Polygalaceae | *Polygala anderssonii* B.L. Rob. | | En Sp | En Sp | Trop Am | **C Am** | Forest *et al.*, 2007 |
| Polygalaceae | *Polygala galapageia* Hook. f. var. *galapageia* | | En var | En var | Trop Am | **C Am** | Forest *et al.*, 2007 |
| Polygalaceae | *Polygala galapageia* Hook. f. var. *insularis* (A.W. Benn.) B.L. Rob. | | En var | En var | Trop Am | **C Am** | Forest *et al.*, 2007 |
| Polygalaceae | *Polygala sancti-georgii* Riley var. *oblanceolata* Howell | | En var | En var | Trop Am | **C Am** | Forest *et al.*, 2007 |
| Polygalaceae | *Polygala sancti-georgii* Riley var. *sancti-georgii* | | En var | En var | Trop Am | **C Am** | Forest *et al.*, 2007 |
| Polygonaceae | *Polygonum galapagense* Caruel | | En Sp | En Sp | Americas | Americas | |

| | | | | | |
|---|---|---|---|---|---|
| Portulacaceae | *Calandrinia galapagosa* H. St. John | En Sp | En Sp | Chile | Chile |
| Portulacaceae | *Portulaca howellii* (D. Legrand) Eliasson | En Sp | En Sp | Argentina | Argentina |
| Rhamnaceae | *Scutia spicata* (Humb.& Bonpl. ex Willd.) Weberb. var. *pauciflora* (Hook. f.) M.C. Johnst. | En var | En var | Ec-Peru | Ec-Peru |
| Rubiaceae | *Borreria dispersa* Hook. f. | En Sp | En Sp | Ecuador | Ecuador |
| Rubiaceae | *Borreria ericaefolia* Hook. f. | En Sp | En Sp | Ecuador | Ecuador |
| Rubiaceae | *Borreria linearifolia* Hook. f. | En Sp | En Sp | Ecuador | Ecuador |
| Rubiaceae | *Borreria perpusilla* Hook. f. | En Sp | En Sp | Ecuador | Ecuador |
| Rubiaceae | *Borreria rotundifolia* Andersson | En Sp | En Sp | Ecuador | Ecuador |
| Rubiaceae | *Borreria suberecta* Hook. f. | En Sp | En Sp | Ecuador | Ecuador |
| Rubiaceae | *Galium galapagoense* Wiggins | En Sp | En Sp | S Am | S Am |
| Rubiaceae | *Psychotria angustata* Andersson | En Sp | En Sp | Andes | Andes |
| Rubiaceae | *Psychotria rufipes* Hook. f. | En Sp | En Sp | Andes | Andes |

Table 5.1 (*cont.*)

| Family | Genus, species and infraspecies | Name used by Porter (1979) if different | Status in Porter (1979) | Present status (changes indicated in bold or by a dash) | Source in Porter (1979)[1] | Present source (changes indicated in bold or by a dash)[1] | References to changes |
|---|---|---|---|---|---|---|---|
| Sapindaceae | *Cardiospermum galapageium* B.L. Rob. & Greenm. | | En Sp | En Sp | Pantrop | Pantrop | |
| Sapindaceae | *Dodonaea viscosa* Jacq. var. *galapagensis* (Scherff) Porter | | En var | En var | Trop Am | Trop Am | |
| Scrophulariaceae | *Galvezia leucantha* Wiggins ssp. *leucantha* | | En var | En var | Peru | Peru | |
| Scrophulariaceae | *Galvezia leucantha* Wiggins ssp. *porphyrantha* Tye & H. Jäger | — | — | **En Ssp** | — | **Peru** | Tye & Jaeger, 2000 |
| Scrophulariaceae | *Galvezia leucantha* Wiggins ssp. *pubescens* Wiggins | | En var | En var | Peru | Peru | |
| Simaroubaceae | *Castela galapageia* Hook. f. | | En Sp | En Sp | Carib | Carib | |
| Solanaceae | *Capsicum galapagoense* Hunz. | | En Sp | En Sp | Andes | **Trop Am** | Walsh & Hoot, 2001 |
| Solanaceae | *Exodeconus miersii* (Hook. f.) D'Arcy | | En Sp | En Sp | Ec-Peru | Ec-Peru | |

| | | | | | | | |
|---|---|---|---|---|---|---|---|
| Solanaceae | *Iochroma ellipticum* (Hook. f.) Hunz. | *Acnistus ellipticus* | En Sp | En Sp | Andes | **N Andes** | Smith & Baum, 2006 |
| Solanaceae | *Jaltomata werffii* D'Arcy | — | — | **En Sp** | — | **Trop Am** | D'Arcy, 1982a, 1982b |
| Solanaceae | *Lycium minimum* C.L. Hitchc. | | En Sp | En Sp | SW N Am | **E Argentina** | Bernardello, 1986; Levin *et al.*, 2007 |
| Solanaceae | *Nolana galapagensis* (Christoph.) Johnst. | | En Sp | En Sp | Chile | **Peru** | Dillon *et al.*, 2007; Tu *et al.*, 2008, M. Dillon pers. comm. |
| Solanaceae | *Physalis galapagoensis* Waterf. | | En Sp | En Sp | C Am | C Am | |
| Solanaceae | *Solanum cheesmaniae* (Riley) Fosberg var. *cheesmaniae* | *Lycopersicon cheesmanii* var. *cheesmanii* | En var | En var | Ec-Peru | Ec-Peru | Darwin *et al.*, 2003, Nuez *et al.*, 2004 |
| Solanaceae | *Solanum cheesmaniae* (Riley) Fosberg var. *minor* | *Lycopersicon cheesmanii* var. *minor* | En var | En var | Ec-Peru | Ec-Peru | Darwin *et al.*, 2003, Nuez *et al.*, 2004 |
| Solanaceae | *Solanum edmonstonei* Hook. f.[2] | | En Sp? | **Not from Galapagos** | S Am | — | |
| Urticaceae | *Pilea baurii* B.L. Rob. | | En Sp | En Sp | Andes | Andes | |

Table 5.1 (*cont.*)

| Family | Genus, species and infraspecies | Name used by Porter (1979) if different | Status in Porter (1979) | Present status (changes indicated in bold or by a dash) | Source in Porter (1979)[1] | Present source (changes indicated in bold or by a dash)[1] | References to changes |
|---|---|---|---|---|---|---|---|
| Verbenaceae | *Clerodendrum molle* Kunth var. *glabrescens* Svenson | | En var | En var | Trop Am | Trop Am | |
| Verbenaceae | *Lantana peduncularis* Andersson | | — | En Sp | — | Ec-Peru | |
| Verbenaceae | *Lantana peduncularis* Andersson var. *macrophylla* Moldenke | | En var | **—** | Ec-Peru | **—** | AT pers. obs. |
| Verbenaceae | *Lantana peduncularis* Andersson var. *peduncularis* | | En var | **—** | Ec-Peru | **—** | AT pers. obs. |
| Verbenaceae | *Lippia rosmarinifolia* Andersson var. *latifolia* Moldenke | | En var | En var | Andes | Andes | AT pers. obs. |
| Verbenaceae | *Lippia rosmarinifolia* Andersson var. *rosmarinifolia* | | En var | En var | Andes | Andes | |
| Verbenaceae | *Lippia rosmarinifolia* Andersson var. *stewartii* Moldenke | | — | **En var** | — | **Andes** | New records |
| Verbenaceae | *Lippia salicifolia* Andersson | | En Sp | En Sp | Andes | Andes | |

| | | | | | | |
|---|---|---|---|---|---|---|
| Verbenaceae | *Verbena grisea* B.L. Rob. & Greenm. | En Sp | En Sp | Americas | Americas | |
| Verbenaceae | *Verbena sedula* Moldenke | — | En Sp? | — | Americas | |
| Verbenaceae | *Verbena sedula* Moldenke var. *darwinii* Moldenke | En var | — | Americas | — | AT pers. obs. |
| Verbenaceae | *Verbena sedula* Moldenke var. *fournieri* Moldenke | En var | — | Americas | — | AT pers. obs. |
| Verbenaceae | *Verbena sedula* Moldenke var. *sedula* | En var | — | Americas | — | AT pers. obs. |
| Verbenaceae | *Verbena townsendii* Svenss. | En Sp | En Sp? | Americas | Americas | |
| Viscaceae | *Phoradendron henslovii* (Hook. f.) B.L. Rob. | En Sp | En Sp | Carib | Carib | |
| Zygophyllaceae | *Kallstroemia adscendens* (Andersson) B.L. Rob. | En Sp | En Sp | SW N Am | SW N Am | |

[1] 'Source' means range of closest presumed relative outside Galapagos. Source abbreviations: C Am = Central America (Panama isthmus to S Mexico); Carib = Caribbean; Col-Ven = Colombia to Venezuela; E Argentina = Eastern Argentina; Ec-Chile = Ecuador to Chile; Ec-Peru = Ecuador to Peru; N Andes = northern Andes; Pantrop = Pantropical; S Am = South America; SW N Am = Southwest North America (including NW Mexico); Trop Am = Tropical Americas.

[2] Eliasson (1996) pointed out that Hooker made two orthographic errors in each name when naming these plants '*edmonstonei*' after Edmondston and that he should have used the form '*edmondstonii*'. However, the names have not been changed by decision of the ICBN and so remain as published by Hooker.

but, since there are consistent though slight morphological differences between the Galapagos and Caribbean populations (Eliasson 1996), we maintain it as a Galapagos endemic pending molecular study. *Cuscuta acuta*, *Euphorbia equisetiformis* and *Vachellia rorudiana* (as *Acacia rorudiana*) have been reported from coastal mainland Ecuador (Jørgensen & León-Yánez, 1999; Rico-Arce, 2007) but mainly based on specimens from the same collector, which have not been critically determined; the first two were maintained as Galapagos endemics in more recent studies (Stefanović *et al.*, 2007; Steinmann & Porter, 2002). *Vachellia rorudiana* has also been reported from Chile (Wiggins & Porter, 1971) but no evidence to support this has been found (D. Neill, pers. comm.).

Botanical survey since 1979 has also resolved questions regarding some endemic taxa. Porter (1979) mentioned 'nine' (actually 10 according to his systematic list, or 11 including *Solanum edmonstonei*) taxa known only from the Type, plus seven more that had not been seen since 1932 or earlier. Of these 18 (Table 5.2), *Solanum edmonstonei* was probably not actually collected in Galapagos (Porter, 1983); it was based on an Edmondston specimen and his collections were mixed up after his death on the mainland coast of Ecuador. It has never been found in Galapagos since, and specimens from coastal Peru have been referred to it. Of the remaining 17, three are currently regarded as extinct, nine have been rediscovered (including two that are questionably distinct), three have been synonymised, and two, both also questionably distinct, are still known only from the Type (Table 5.2).

Porter (1979) also included nine endemic taxa that are now synonymised or regarded as questionably distinct: *Acalypha flaccida*, *A. parvula chathamensis*, *A. p. reniformis*, *A. p. strobilifera* and *A. velutina* have all been synonymised within *A. parvula* (Seberg, 1984), *Chamaesyce bindloensis* within *C. punctulata* (Huft & van der Werff, 1985) and *Salvia insularum* within *S. pseudoserotina* (Harley, 1999), while *Lantana peduncularis macrophylla* and *L. p. peduncularis* are not recognised here, as the species shows a continuous range of morphology (AT, pers. obs. of specimens at CDS), and varieties of *Verbena sedula* are similarly not recognised.

Balancing these losses, endemicity level has been increased by the description of the following eight new endemic taxa since Porter's (1979) review: *Sisyrinchium galapagense* Ravenna (1979); *Jaltomata werffii* D'Arcy (1982a, 1982b); *Acalypha abingdonii* Seberg (1984); *Scalesia gordilloi* Hamann & Wium-Andersen (1986); *Darwiniothamnus alternifolius* Lawesson & Adsersen (1987), now *Erigeron alternifolius* (Tye, 2010); *Cyclopogon werffii* Dodson (1993); *Galvezia leucantha porphyrantha* Tye & H. Jäger (2000). In addition, *Lecocarpus darwinii* was erected by Adsersen (1980) not as a new taxon but as a replacement name for populations that had previously been called *L. leptolobus*; but Adsersen's reasoning for suppressing *L. leptolobus* was faulty (AT, unpublished) and two taxa seem to be involved (see Fig. 5.4).

Table 5.2 *Galapagos endemic taxa known only from the Type or not collected since 1932 (or before), and therefore considered possibly extinct by Porter (1979), with their present status.*

| | Porter (1979) | Present status |
|---|---|---|
| *Delilia inelegans* (Asteraceae) | Known only from Type, 1835. | Extinct |
| *Lecocarpus leptolobus* (Asteraceae) | Known only from Type, 1906. Questionably distinct. | Rediscovered (CDS) |
| *Ipomoea tubiflora* (Convolvulaceae) | Known only from Type, 1835. | Rediscovered (CDS) |
| *Sicyos villosus* (Cucurbitaceae) | Known only from Type, 1835. | Extinct |
| *Acalypha flaccida* (Euphorbiaceae) | Known only from Type, 1835. | Synonymised in *A. parvula* (Seberg, 1984) |
| *Salvia insularum* (Lamiaceae) | Known only from Type, 1846. | Synonymised in *S. pseudoserotina* (Harley, 1999) |
| *Mollugo flavescens striata* (Molluginaceae) | Known only from Type, 1906. | Rediscovered (CDS: Adsersen) |
| *Mollugo floriana santacruziana* (Molluginaceae) | Known only from Type, 1927. | Rediscovered (CDS: many collections) |
| *Borreria rotundifolia* (Rubiaceae) | Known only from Type, 1852. | Known only from Type, questionably distinct. |
| *Solanum edmonstonei* (Solanaceae) | Known only from Type, 1846. Possibly not collected in Galapagos. | Probably not from Galapagos. |
| *Verbena sedula darwinii* (Verbenaceae) | Known only from Type, 1835. Questionably distinct. | Known only from Type, not regarded here as distinct. |
| *Blutaparon rigidum* (Amaranthaceae) | Last collected 1906. | Extinct |
| *Acalypha velutina* (Euphorbiaceae) | Last collected 1891. | Synonymised in *A. parvula* (Seberg, 1984) |
| *Mollugo floriana gypsophiloides* (Molluginaceae) | Last collected 1932. | Rediscovered (C: Hamann) |
| *Paspalum redundans* (Poaceae) | Last collected 1932. | Rediscovered (CDS; Adsersen 1987) |
| *Borreria linearifolia* (Rubiaceae) | Last collected 1932. | Rediscovered? Questionably distinct (CDS: many) |
| *Borreria suberecta* (Rubiaceae) | Last collected 1932. | Rediscovered? Questionably distinct (CDS: many) |
| *Psychotria angustata* (Rubiaceae) | Last collected 1852. | Rediscovered (CDS: many) |

Figure 5.4 Charles Darwin's collection of *Lecocarpus* from San Cristóbal Island. The specimens to the right are of *L. darwinii*, while that to the left corresponds to *L. leptolobus*, not to *L. lecocarpoides* from Española Island. *L. leptolobus* has recently been rediscovered close to where Darwin probably collected his plant (A. Tye, unpubl.). Photo A. Tye, of specimen at CGE courtesy of the University of Cambridge Herbarium.

Further, three endemic taxa that were synonymised by Porter (1979) have since been reinstated following additional study: *Passiflora tridactylites* (Lawesson, 1988; K. Porter-Utley, pers. comm.), *Urochloa multiculma* (Tye, 2008) and *Lippia rosmarinifolia stewartii* (specimens at CDS: AT, pers. obs.). We also tentatively recognise the following three more, pending further study: *Lithophila (radicata) scirpoides* (Eliasson, 1985); *Chrysanthellum fagerlindii* (Lawesson *et al.*, 1987); *Darwiniothamnus l. lancifolius* (Andrus *et al.*, 2009).

The other factor that affects percentage endemicity is the size of the non-endemic native flora. Most recently discovered non-endemic species are known or can safely be assumed to have been introduced by people. However, of species discovered or recognised since 1979, the following may be native, having escaped earlier collection: *Galium canescens* Kunth (= '*Rubia* sp.' of Hooker, 1847; see Porter, 1980, *G. reynoldsii* of Lawesson 1988: it was not the new discovery then that Lawesson stated it to be); *Annona glabra* L. (first record 1906 but ignored until recent confirmation); *Eleocharis geniculata* (L.) Roem. & Schult. (CDS: first record 1906 but not recognised until 2003); *Nephrolepis undulata* (Afzel. ex Sw.) J. Sm. (CDS: first record 1966 but not recognised until 2004); *Buddleja americana* L. (Lawésson, 1987, 1988: first record 1970); *Soliva anthemifolia* (Juss.) Sweet, *Rorippa nana* (Schltdl.) J. F. Macbr., *Crassula aquatica* (L.) Schönl. and *Juncus pallescens* Lam. (CDS: first records 1974 but unknown to Porter 1983); *Varronia polycephala* Lam. and *Cressa truxillensis* Kunth (CDS: first records 1977 but unknown to Porter, 1983); *Cranichis werffii* Garay (Garay, 1978) and *Sida veronicaefolia* Lam. (Hamann 1979), neither included by Porter (1983); *Eragrostis pastoensis* (Kunth) Trin. (CDS: first record 1981); *Oplismenus compositus* (L.) P. Beauv. (CDS: 1986); *Sisyrinchium micranthum* Cav. (CDS: 1999); *Rhynchospora radicans* (Schltdl. & Cham.) Pfeiffer (CDS: 2003); *Ctenitis microchlaena* (Fee) Stolze (CDS: 2004); *Ombrophytum subterraneum* (Aspl.) B. Hansen (CDS: 2006). Further, the recent palaeobotanical work, which has shown that at least six species formerly regarded as introduced or doubtfully native are truly native, has reduced the proportion of endemic taxa further.

Balancing this to some extent are reductions in the number of non-endemic natives due to the correction of some earlier erroneous records. Examples of non-endemic taxa included in the native flora by Porter (1983) but now considered erroneous include *Calystegia (Convolvulus) soldanella* (L.) R. Br. (Edmondston specimen probably not collected in Galapagos, plus later specimens at CDS now re-identified as other species) and *Diodia radula* (together with '*Borreria laevis*' now considered to be *Spermacoce remota*: C. Taylor, pers. comm.).

The number of non-endemic native vascular plant taxa is now regarded as 319, plus the 236 endemics, resulting in 43% endemism, a level unchanged from Porter's (1979) estimate. If all 58 remaining doubtful natives were to be included within the native flora, endemism would reduce to 38%.

## *Areas of origin of the flora and evidence for dispersal events*

Molecular techniques have generated more changes to presumed area of origin of the endemics (79 in total, including changes that follow from status changes) than to endemic status. Fewer endemic taxa are now thought to have evolved from colonisers from the southern Andes, and more from the northern Andes, Central America, the Caribbean and southwest North America, as well as 'Tropical America' (Table 5.3). Some changes are due to increased precision in our knowledge of area of origin resulting from the identification of sister taxa: fewer taxa are now referred to the general regions 'Andes' and 'South America' and more can be referred to individual countries or to a group of two neighbouring countries, with (for example) more taxa now recognised as having their closest relatives in Ecuador–Peru (14% of colonisations). These changes thus result in an overall weakening of the importance of southern source areas, and a strengthening of the links with the Americas from Ecuador northward.

Following the work of Levin and Moran (1989) and Andrus *et al.* (2009) there is no longer thought to be any direct floristic connection between Galapagos and the Revillagigedos Islands off the Pacific coast of Mexico. Possible links were identified by Johnston (1931), but none has withstood subsequent examination. Floral links between Galapagos and Baja California (north of the Revillagigedos) have however been confirmed (e.g. Wendel & Percival, 1990), and it is no longer possible to claim an absence of floral linkage to North America (Porter, 1976, 1979).

However, overall, South America (45% of colonisations) still appears to be the primary source for the Galapagos endemic flora, as would be expected given its proximity and the fact that birds are still considered to have been the most important dispersal mechanism (Porter, 1983). The other major source areas are Central America and the Caribbean (12% of colonisations) and North America (5%), with the source areas of the remainder being less precisely known than this (Table 5.3). The strong connection now recognised with the Caribbean and Central America corresponds with the recent geological evidence linking the Galapagos hotspot with that region.

In some cases, changes to ideas about closest relative have not meant changes to known area of origin, when both postulated taxa are found in the same area. For example, the closest relative of *Solanum cheesmaniae* is now suggested to be *S. lycopersicum* L. (Nesbitt & Tanksley, 2002) rather than *S. pimpinellifolium* L. (Rick & Fobes, 1975). Although results of different studies do not completely concur regarding the origins and taxonomy of this group, probably in part due to continuing gene flow between *S. lycopersicum* varieties and *S. pimpinellifolium* (Peralta & Spooner, 2001; Nesbitt & Tanksley, 2002), both of these possible sister taxa have similar original ranges.

Phylogenetic studies often refer to 'area of origin' of a Galapagos taxon, but this usually simply means 'area where the sister taxon is found', and actually tells us nothing about the direction of migration. However, it does imply that a dispersal event (in either direction) took place, as no evidence has yet been found to support the Caribbean vicariance championed by Grehan (2001). In most cases one can assume that migration was from continent to islands, but not always; and if reverse

Table 5.3 *Source areas of the Galapagos endemic vascular flora. Data for 1979 are recalculated from taxon information in the systematic list of Porter (1979).*

| Source area (as precise as possible) | | No. of colonisations (Porter 1979) | No. of colonisations now (change) | No. of taxa (Porter 1979) | No. of taxa now (change) | Overall source region | | No. of colonisations (Porter 1979) | No. of colonisations now (change) | No. of taxa (Porter 1979) | No. of taxa now (change) |
|---|---|---|---|---|---|---|---|---|---|---|---|
| Area number | Area name | | | | | Area numbers | Region name | | | | |
| 1 | Ecuador | 4 | 3 (–1) | 14 | 14 (0) | | | | | | |
| 2 | Peru | 2 | 3 (+1) | 4 | 6 (+2) | | | | | | |
| 3 | Chile | 4 | 3 (–1) | 7 | 3 (–4) | | | | | | |
| 4 | Andean Argentina | 3 | 2 (–1) | 3 | 2 (–1) | | | | | | |
| 5 | E Argentina | 0 | 1 (+1) | 0 | 1 (+1) | | | | | | |
| 6 | Brazil | 1 | 1 (0) | 4 | 4 (0) | | | | | | |
| 7 | Colombia–Venezuela | 1 | 1 (0) | 1 | 1 (0) | | | | | | |
| 8 | Ecuador–Peru | 9 | 10 (+1) | 13 | 12 (–1) | 8(+1, 2) | Ecuador–Peru | 15 | 16 (+1) | 31 | 32 (+1) |
| 9 | Peru–Chile | 3 | 2 (–1) | 3 | 2 (–1) | 9(+2+3) | Peru–Chile | 9 | 8 (–1) | 14 | 11 (–3) |
| 10 | N Andes | 2 | 4 (+2) | 2 | 7 (+5) | 10(+1+2+7+8) | N Andes | 18 | 21 (+3) | 34 | 40 (+6) |
| 11 | Ecuador–Chile | 1 | 0 (–1) | 3 | 0 (–3) | 11(+1–4,8,9) | S Andes | 26 | 22 (–4) | 47 | 36 (–9) |
| 12 | Andes | 27 | 19 (–8) | 73 | 59 (–14) | 12(+1,2,3,4,7, 8,9,10,11) | Andean S America | 56 | 46 (–10) | 123 | 105 (–18) |
| 13 | Central America | 5 | 6 (+1) | 6 | 12 (+6) | | | | | | |
| 14 | Caribbean | 5 | 8 (+3) | 9 | 16 (+7) | | | | | | |
| 15 | SW N America | 4 | 5 (+1) | 4 | 5 (+1) | | | | | | |
| 16 | S America | 7 | 5 (–2) | 18 | 9 (–9) | 16(+1 to 12) | S America | 64 | 53 (–11) | 145 | 119 (–26) |
| 17 | Tropical America | 27 | 33 (+6) | 46 | 60 (+14) | 17(+1 to 14) | Tropical America | 94 | 95 (+1) | 188 | 198 (+10) |
| 18 | Americas | 5 | 4 (–1) | 16 | 14 (–2) | 18(+1 to 17,20) | Americas | 110 | 112 (+2) | 226 | 231 (+5) |
| 19 | Pantropical | 7 | 5 (–2) | 7 | 5 (–2) | 19(+1 to 14,16, 17,18) | Tropics | 113 | 109 (–4) | 229 | 225 (–3) |
| 20 | SW N America or N Andes | 0 | 2 (+2) | 0 | 4 (+4) | | | | | | |
| | **Totals** | **117** | **117** | **233** | **236** | | | | | | |

migration has occurred, as may be the case in *Alternanthera* and *Tiquilia* (Moore *et al.*, 2006; Sánchez-del Pino, 2007), then the continental taxon which first gave rise to the Galapagos taxon will be an apparently more distant (= more basal) relative, which may however still be possible to identify (see Moore *et al.*, 2006).

In some cases, endemic taxa within a single genus were suggested (by Porter, 1979, or by predecessors cited therein) to have arisen from more than one separate colonisation. New data have modified this view in some cases, such as *Opuntia*, where all Galapagos taxa seem to be very closely related to one another (Helsen *et al.*, 2009a, see Fig. 5.3), thereby implying a single colonisation rather than the two previously suggested. However, in other cases, endemic groups formerly presumed to be monophyletic, such as *Darwiniothamnus* (Lawesson & Adsersen, 1987), have been shown to result from separate introductions, sometimes from widely separated areas of origin (Andrus *et al.*, 2009). *Darwiniothamnus* is nested within *Erigeron*, with its original two species having a Caribbean origin, but with the third species derived from a different ancestor from Chile (Andrus *et al.*, 2009; Tye, 2010). *Alternanthera* appears to have colonised Galapagos at least three times, with one clade radiating in the archipelago (see Fig. 5.3) and the other two leading to *A. nesiotes* and the non-endemic *A. echinocephala* (Hook. f.) Christoph. (Sánchez-del Pino, 2007). Galapagos *Alternanthera* spp. may include affinities with Mexico (the origin of the main Galapagos endemic clade), Peru (*A. echinocephala*) and Chile (back-migrations from Galapagos of *A. halimifolia* (Lam.) Standl. ex Pittier and *A. vestita* (Andersson) Howell) (Sánchez-del Pino, 2007). Several other genera (*Cuscuta*, *Cyperus*, *Gossypium*, *Ipomoea*, *Passiflora*, *Polypodium* and *Sicyos*) contain endemic species that result from more than one colonisation, with each arrival having produced only a single endemic taxon.

## Times of colonisation and divergence

Few attempts have been made, using molecular clock techniques, to date the times of divergence of Galapagos endemic plant taxa from their closest relatives (Table 5.4). The data are imprecise (large ranges in estimates) and doubt remains about their accuracy owing to uncertainties associated with the techniques and to rates of evolutionary change in the various markers used.

Consequently we have no clear idea so far of the process and sequence of the colonisation of Galapagos by plants. However, evidence of hybridisation among some of the endemics (e.g. Nielsen *et al.*, 2003b; Lindhardt *et al.*, 2009) or of an ability to hybridise with their mainland sister taxa (e.g. Wendel & Percival, 1990; Wendel & Percy, 1990; Tye, 2008), together with the quite high proportion of infraspecific relative to specific differentiation in the Galapagos endemic flora (33% of 263 differentiation events: data from Table 5.1 column 5 plus the 27 full species that comprise more

Table 5.4 *Estimates of divergence times of Galapagos endemic plant taxa.*

| Taxon | Divergence date (M yr ago) Min | Max | Range | Reference |
|---|---|---|---|---|
| *Gossypium darwinii* | <0.5 | <2 | 1.5 | Small *et al.*, 1998 |
| *Lecocarpus* | | <4 | | Rauscher, 2002 |
| *Scalesia* | 1.9 | 6.3 | 4.4 | Schilling *et al.*, 1994 |
| *Sicyos pentagonus* | 0.2 | 2.6 | 2.4 | Sebastian *et al.*, 2010 |
| *Sicyos villosus* | 2 | 6 | 4 | Sebastian *et al.*,2010 |
| *Solanum cheesmaniae* | 0.9 | 1.3 | 0.4[1] | Nesbitt & Tanksley, 2002 |
| *Tiquilia* | 2.5 | 6 | 3.5 | Moore *et al.*, 2006 |
| *Varronia* | 1.2 | 9 | 7.8 | Weeks *et al.*, 2010 |

[1] Estimate was 1.1 ± 0.2 (S.E.) Ma.

than one infraspecific taxon), might suggest either that many colonisations occurred quite recently or that rates of evolutionary change are low, or both.

### *New information on evolutionary processes*

Although the percentage endemism at the species level remains much as Porter (1979) estimated, the number of valid endemic genera has been (or will probably be) reduced from seven to four as a result of molecular investigations (Table 5.5).

Porter (1979) recognised radiations that had produced at least three endemic taxa in '19' genera (21 in his systematic list), plus '10' more genera (12 in his systematic list) that had produced two endemics. This situation remains little changed in terms of the number of genera involved: there are still 20 genera with radiations having produced three or more taxa, plus 12 that contain radiations having produced two taxa, but the numbers of taxa produced by most of the radiations that have been investigated have changed following discoveries of new taxa or phylogenetic revisions (Table 5.6). One of these radiations from a single colonisation includes what have so far been regarded as two endemic genera (*Brachycereus* and *Jasminocereus*).

Aside from the 33 radiations currently recognised in Table 5.6, 83 or 84 colonisations (72% of colonisations) have produced a single endemic taxon each by phyletic evolution, with 35.5% of the endemic taxa having resulted from phyletic evolution and the remainder forming part of radiations.

### *New information on 'island syndrome' characteristics*

As in other island endemic plants (DeJoode & Wendel, 1992; Barrett, 1998), many molecular investigations have revealed very low levels of genetic variability in

Table 5.5 *Plant genera considered endemic to Galapagos in 1979 and now.*

| Endemic genus recognised by Porter (1979) | Status now |
|---|---|
| *Darwiniothamnus* (Asteraceae) | Still named, pending revision of *Erigeron*, within which it is nested in Sect. *Microcephalum* (Andrus *et al.*, 2009) |
| *Lecocarpus* (Asteraceae) | Endemic genus |
| *Macraea* (Asteraceae) | Endemic genus |
| *Scalesia* (Asteraceae) | Endemic genus |
| *Brachycereus* (Cactaceae) | Endemic genus but sister to *Jasminocereus* (Lendel *et al.*, 2006) |
| *Jasminocereus* (Cactaceae) | Endemic genus |
| *Sicyocaulis* (Cucurbitaceae) | Nested within *Sicyos* (Kearns & Wunderlin, 1999; Sebastian *et al.*, 2010) |

Galapagos endemics (e.g. *Solanum*, Rick & Fobes, 1975; *Galvezia*, Elisens, 1989; *Gossypium*, Wendel & Percival, 1990, Alvarez & Wendel, 2006; *Opuntia*, Browne *et al.*, 2003). One exception, with high variability, seems to be due to introgression with a related species that has recently been introduced to the islands by people (Wendel & Percy, 1990). However, in other cases, high levels of intra- and inter-population variability have been found in rapidly evolving markers such as AFLPs and microsatellites (e.g. Nielsen *et al.*, 2003b; Nielsen, 2004; Helsen *et al.*, 2009b), in one such case (*Opuntia*) after a number of investigations had found very little variation in more slowly evolving markers (e.g. Browne *et al.*, 2003; Helsen *et al.*, 2009a).

Many of the more recent studies have cast doubt on some earlier assumptions about the evolution of the Galapagos flora, or revealed that the true situation is more complex than earlier supposed. Aside from the advantage of self-compatibility on arrival, when a single plant may be sufficient to establish a new population (Baker, 1955), it had been suggested that there would be strong selection in Galapagos for the evolution of autogamy in insect-pollinated species, because of a paucity of pollinator insects (e.g. Rick, 1966). Rick (1983) considered that the existence of only one native bee species would have severely limited insect cross-pollination in the ancestor of *Solanum cheesmaniae*. Similarly, the normally dioecious, wind-pollinated *Croton scouleri* (whose mainland relatives are also dioecious) is occasionally monoecious, which could favour its establishment on islands (Mauchamp, 1997). In contrast, it has been suggested that, once established, island plants should be selected in the opposite direction, for outcrossing (Carlquist, 1974).

The first genetic and breeding system studies showed that some Galapagos plants appear to be almost completely autogamous (Rick, 1966, 1983; Rick & Fobes, 1975).

However, later studies have revealed that numerous smaller insects in Galapagos, including many species that are clearly native (or endemic) visit and probably cross-pollinate a variety of Galapagos flowers (e.g. McMullen, 1993; McMullen & Viderman, 1994; Philipp *et al.* 2006; Simbaña & Tye, 2009; Jaramillo *et al.*, 2010), as they do on many other oceanic archipelagos (Barrett, 1998). Philipp *et al.* (2006) found that only one of the plant species they studied appeared to be pollen-limited for outcrossing, suggesting low selection pressure for selfing. The results of Philipp *et al.* (2006) also suggest that nocturnal pollination may be important for a wider range of plants than formerly assumed.

Table 5.6 *Genera containing Galapagos radiations that have produced two or more endemic taxa. An asterisk marks groups that have not received any molecular evolutionary study to date.*

| Genus | Endemic taxa in the radiation recognised by Porter (1979) | Current status of radiation | Reference to change |
|---|---|---|---|
| *Scalesia* (Asteraceae) | 20 | 21 | Hamann & Wium-Andersen (1986) |
| *Opuntia* (Cactaceae) | 5 + 9 (from 2 colonisations) | 14 (single colonisation) | Helsen *et al.* (2009a) |
| *Alternanthera* (Amaranthaceae) | 8 (+ 4 endemic taxa from 4 additional colonisations) | 11 (+ 1 from an additional colonisation) | Sánchez-del Pino (2007) |
| **Mollugo* (Molluginaceae) | 9 | 9 *(no study, no change)* | |
| **Chamaesyce* (Euphorbiaceae) | 9 | 8 | Huft & van der Werff (1985) |
| **Borreria* (Rubiaceae) | 6 | 6 *(no study, no change)* | |
| **Acalypha* (Euphorbiaceae) | 6 | 4 | Seberg (1984) |
| **Froelichia* (Amaranthaceae) | 5 | 5 *(no study, no change)* | |
| **Polygala* (Polygalaceae) | 5 | 5 *(no study, no change)* | |
| *Darwiniothamnus* (Asteraceae) | 4 | 5(–7) | Andrus (2002), AT pers. obs. (CDS) |
| **Croton* (Euphorbiaceae) | 4 | 4 *(no study, no change)* | |
| **Peperomia* (Piperaceae) | 4 | 4 *(no study, no change)* | |
| **Aristida* (Poaceae) | 4 | 4 *(no study, no change)* | |
| **Lecocarpus* (Asteraceae) | 3 | 4 | AT pers. obs. (CDS) |

Table 5.6 (*cont.*)

| Genus | Endemic taxa in the radiation recognised by Porter (1979) | Current status of radiation | Reference to change |
|---|---|---|---|
| *Tiquilia* (Boraginaceae) | 3 | 4, or 3 +1 from an additional colonisation | Moore *et al.* (2006) |
| **Lippia* (Verbenaceae) | 3 | 4 | AT pers. obs. (CDS) |
| **Varronia* (Boraginaceae) | 3 (+ 1 from a separate colonisation) | 3–4 | Weeks *et al.* (2010) |
| **Verbena* (Verbenaceae) | 3 + 2 (from 2 colonisations) | 3 | AT pers. obs. (CDS) |
| *Jasminocereus* (Cactaceae) | 3 | 4 (including *Brachycereus*) | Lendel *et al.* (2006) |
| **Paspalum* (Poaceae) | 3 | 3 *(no study, no change)* | |
| **Salvia* (Lamiaceae) | 3 | 2 | Harley (1999) |
| *Galvezia* (Scrophulariaceae) | 2 | 3 | Tye & Jäger (2000) |
| **Amaranthus* (Amaranthaceae) | 2 + 2 (from 2 colonisations) | 2 (+1 from an additional colonisation) | Eliasson (1985) |
| **Lithophila* Amaranthaceae) | 2 | 2(–3) | Eliasson (1985) |
| **Delilia* (Asteraceae) | 2 | 2 *(no study, no change)* | |
| **Pectis* (Asteraceae) | 2 | 2 *(no study, no change)* | |
| **Tournefortia* (Boraginaceae) | 2 | 2 *(no study, no change)* | |
| **Linum* (Linaceae) | 2 | 2 *(no study, no change)* | |
| **Psidium* (Myrtaceae) | 2 | 2 *(no study, no change)* | |
| **Trichoneura* (Poaceae) | 2 | 2 *(no study, no change)* | |
| **Psychotria* (Rubiaceae) | 2 | 2 *(no study, no change)* | |
| *Solanum* (Solanaceae) | 2 | 2 | Nuez *et al.* (2004) |
| **Lantana* (Verbenaceae) | 2 | 1 | AT pers. obs. (CDS) |
| **Chrysanthellum* (Asteraceae) | 1 | 2 | Lawesson *et al.* (1987) |

The most complete survey of breeding systems in Galapagos plants so far (McMullen, 1987) found a high proportion of self-compatible species and no evidence for the evolution of outcrossing mechanisms. McMullen (1987) concluded that the proportions of dioecious and hermaphroditic species were similar to those in mainland Ecuador, but that the proportion of self-compatible species was relatively high. Thus, although selection for autogamy within the islands may have been less severe than at first supposed, due to a larger number of native pollinators than first assumed, selection for outcrossing may also be weak. However, recent studies have shown that, although many Galapagos endemic plants *may* reproduce sexually through selfing, they are in fact predominantly outcrossing and, even though fertilisation rate may sometimes be low with outcrossing, it greatly exceeds the rate through selfing (e.g. *Scalesia divisa*, Nielsen *et al.*, 2000; *Scalesia affinis*, Nielsen *et al.*, 2003a; *Scalesia cordata*, Philipp & Nielsen, 2010; a variety of Dry Zone species, Philipp *et al.* 2006; *Jasminocereus*, Jaramillo *et al.*, 2010). Further, some species appear to suffer inbreeding depression (e.g. Nielsen *et al.*, 2007), and outcrossing may be essential for establishing or maintaining variability in small populations (Nielsen *et al.* 2003b). Thus, the relative importance of selection pressures that favour autogamy and those that favour outcrossing remains unclear in the endemic flora as a whole.

## Conclusions

During the three decades since 1979, the numbers of articles covering aspects of the evolution of Galapagos endemic vascular plants were: 12 in 1980–1989; 12 in 1990–1999; 53 in 2000–2009. Fig. 5.5 appears to show three phases of study, with between zero and two publications appearing most years up to 1998, three or four in most years from 1999 to 2005, then five to fourteen per year from 2006 to 2009. The increase in publications since 1999 was in part stimulated by an active search by scientists of the Charles Darwin Research Station in Galapagos, during the period 1997–2007, for external collaborators with access to appropriate facilities to help investigate the flora. Since 1979, and especially in the past 10 years, there has been a revolution in our knowledge of Galapagos plant origins and phylogenies. Most phylogenetic studies during the 1980s were morphological, whereas most since 1990 were based on DNA, while studies on breeding systems also increased dramatically since 1999.

### *Level of study of different taxa, and objectives of the studies*

Some groups of Galapagos endemics, often those of commercial interest (e.g. tomatoes and cottons), as well as the iconic 'plant equivalent of Darwin's finches'

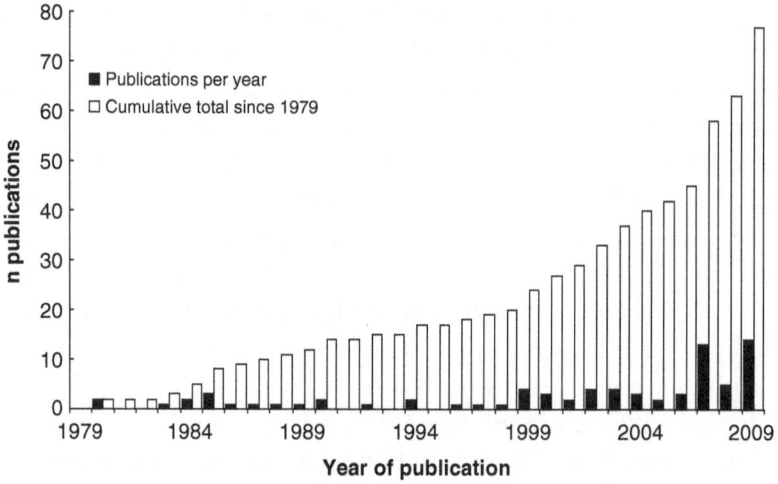


Figure 5.5 Numbers of evolutionary studies that included Galapagos endemic vascular plant taxa published per year since 1979. Studies in press by mid-2009 are included in the 2009 total, since although they were not all published by the date of going to press, and may not have been during 2009, the year had not then closed and it is likely that others not known to us at that point will have been published during 2009 and 2010.

genus *Scalesia* and the physically striking *Opuntia* have received disproportionate scientific interest (Table 5.7), while others remain comparatively poorly understood or have received no significant study during the past 30 years. Many of the studies that have included Galapagos endemics have not had their primary focus on the archipelago (especially phylogenetic investigations of whole genera or higher taxa), but these do contribute to our understanding of the development of the Galapagos flora, especially of its origins. However, as the publications on *Acnistus/Iochroma* and *Opuntia* show, getting just one student interested in a plant can generate a series of detailed studies (Table 5.7).

The footnotes to the evolutionary studies in Table 5.7 show that 60 of them had a primary focus on determining the phylogeny, nearest relatives or mainland origins of the taxa studied, characteristics which all emerge together from a single study. Fifteen studies investigated population genetics, while 18 were focused primarily or in large part on aspects of breeding system, including hybridisation.

### *How close are we to a comprehensive knowledge of the origins and evolution of the Galapagos flora? Directions for future research*

Information gathered since 1979 using a plethora of new techniques has in many cases confirmed conclusions drawn by Porter (1979), but in other cases has altered our perceptions of overall patterns, or of the origins and relationships of particular

Table 5.7 *Evolutionary studies of Galapagos endemic vascular plant groups since 1979.*

| Family | Genus | n endemic species | n endemic taxa | Evolutionary studies since 1979 | n studies/ genus | n studies/ sp | n studies/ taxon | Notes |
|---|---|---|---|---|---|---|---|---|
| Aizoaceae | *Sesuvium* | 1 | 1 | Eliasson, 1996[1] | 1 | 1.00 | 1.00 | plus 1 non-endemic |
| Amaranthaceae | *Alternanthera* | 6 | 12 | Sánchez-del Pino, 2007[1] | 1 | 0.17 | 0.08 | plus 4 non-endemics |
| | *Amaranthus* | 3 | 3 | Eliasson, 1985[1] | 1 | 0.33 | 0.33 | plus 1 non-endemic |
| | *Blutaparon* | 1 | 1 | Eliasson, 1985[1] | 1 | 1.00 | 1.00 | |
| | *Lithophila* | 2 | 3 | Eliasson, 1985[1] | 1 | 0.50 | 0.33 | |
| | *Pleuropetalum* | 1 | 1 | Eliasson, 1985[1] | 1 | 1.00 | 1.00 | |
| Asteraceae | *Darwiniothamnus/ Erigeron* | 3 | 6 | Lawesson & Adsersen, 1987[1]; McMullen & Viderman, 1994[3]; Andrus, 2002[1]; Nesom, 2008[1]; Andrus *et al.*, 2009; Tye, 2010[1] | 6 | 2.00 | 1.00 | |
| | *Encelia* | 1 | 1 | Fehlberg & Ranker, 2007[1] | 1 | 1.00 | 1.00 | |
| | *Lecocarpus* | 3 | 4 | Adsersen, 1980[1]; Rauscher, 2002[1]; Philipp *et al.*, 2004[3]; Brok & Adsersen, 2007[1] | 4 | 1.33 | 1.00 | |
| | *Macraea* | 1 | 1 | Eliasson, 1984[1]; Panero *et al.*, 1999[1] | 2 | 2.00 | 2.00 | |

Table 5.7 (*cont.*)

| Family | Genus | n endemic species | n endemic taxa | Evolutionary studies since 1979 | n studies/ genus | n studies/ sp | n studies/ taxon | Notes |
|---|---|---|---|---|---|---|---|---|
| | *Scalesia* | 15 | 21 | Schilling *et al.*, 1994[1]; Spring *et al.*, 1999[1]; Nielsen, 2004[2]; Nielsen *et al.*, 2000[3]; Nielsen *et al.*, 2003a[3]; Nielsen *et al.*, 2003b[2,3]; Nielsen *et al.*, 2007[3]; Blaschke & Sanders, 2009[1]; Lindhardt *et al.*, 2009[1,3]; Philipp & Nielsen, 2010[3] | 9 | 0.60 | 0.43 | |
| Boraginaceae | *Tiquilia* | 4 | 4 | Moore *et al.*, 2006[1] | 1 | 0.25 | 0.25 | |
| | *Tournefortia* | 2 | 2 | McMullen, 2007[3] | 1 | 0.50 | 0.50 | plus 1 non-endemic |
| | *Varronia* | 4 | 4 | Weeks *et al.*, 2010 | 1 | 0.25 | 0.25 | |
| Burseraceae | *Bursera* | 0 | 1 | Weeks & Tye, 2009[1,2,3] | 1 | — | 1.00 | plus 1 non-endemic |
| Cactaceae | *Brachycereus* | 1 | 1 | Lendel *et al.*, 2009[1] | 1 | 1.00 | 1.00 | |
| | *Jasminocereus* | 1 | 3 | Lendel *et al.*, 2009[1]; Jaramillo *et al.*, in press[3] | 2 | 2.00 | 0.67 | |
| | *Opuntia* | 6 | 14 | Browne *et al.*, 2003[1,2]; Helsen *et al.*, 2007[1,2]; Helsen *et al.*, 2009a[2]; Helsen *et al.*, 2009b[2]; Griffith & Porter, 2009[1] | 5 | 0.83 | 0.36 | |

| | | | | | | | | |
|---|---|---|---|---|---|---|---|---|
| Convolvulaceae | *Cuscuta* | 2 | 2 | Stefanović *et al.*, 2007[1] | 2 | 1.00 | 1.00 | |
| | *Ipomoea* | 3 | 3 | McMullen, 2009[3] | 1 | 0.33 | 0.33 | |
| Cucurbitaceae | *Sicyos/Sicyocaulis* | 2 | 2 | Sebastian *et al.*, in 2010[1] | 1 | 0.50 | 0.50 | |
| Euphorbiaceae | *Acalypha* | 4 | 4 | Seberg, 1984[1] | 1 | 0.25 | 0.25 | |
| | *Chamaesyce* | 7 | 8 | Huft & van der Werff, 1985[1] | 1 | 0.14 | 0.13 | |
| | *Croton* | 1 | 4 | Mauchamp, 1997[3] | 1 | 1.00 | 0.25 | |
| | *Euphorbia* | 1 | 1 | Steinmann & Porter, 2002[1] | 1 | 1.00 | 1.00 | |
| Fabaceae | *Phaseolus* | 1 | 1 | Delgado-Salinas *et al.*, 1999[1] | 1 | 1.00 | 1.00 | plus 3 introduced spp |
| Lamiaceae | *Salvia* | 2 | 2 | Harley, 1999 [1] | 1 | 0.50 | 0.50 | plus 1 non-endemic |
| Lycopodiaceae | *Huperzia* | 1 | 1 | Holub, 1985[1] | 1 | 1.00 | 1.00 | |
| Malvaceae | *Gossypium* | 2 | 2 | Wendel & Percival, 1990[1]; Wendel & Percy, 1990[1,2]; Small *et al.*, 1998[1]; Small & Wendel, 2000[1,2]; Alvarez *et al.*, 2005[1]; Alvarez *et al.*, 2006[2,3]; Guo *et al.*, 2007[1]; Lacape *et al.*, 2007[2] | 8 | 4.00 | 4.00 | |
| Mimosaceae | *Vachellia* | 1 | 1 | Seigler & Ebinger, 2005[1] | 1 | 1.00 | 1.00 | plus 2 non-endemics |
| Passifloraceae | *Passiflora* | 2 | 3 | Lawesson, 1988[1]; Porter-Utley unpubl.[1] | 2 | 1.00 | 0.67 | |
| Poaceae | *Urochloa* | 1 | 1 | Tye, 2008[1] | 1 | 1.00 | 1.00 | plus 1 introduced sp |
| Polygalaceae | *Polygala* | 3 | 5 | Forest *et al.*, 2007[1] | 1 | 0.33 | 0.20 | |

Table 5.7 (*cont.*)

| Family | Genus | n endemic species | n endemic taxa | Evolutionary studies since 1979 | n studies/ genus | n studies/ sp | n studies/ taxon | Notes |
|---|---|---|---|---|---|---|---|---|
| Scrophulariaceae | *Galvezia* | 1 | 3 | Elisens, 1989[1]; Elisens, 1992[1]; Tye & Jäger, 2000[1] | 1 | 1.00 | 0.33 | |
| Solanaceae | *Acnistus/Iochroma* | 1 | 1 | Smith & Baum, 2006[1]; Smith & Baum, 2007[3]; Smith *et al.*, 2008a[3]; Smith *et al.*, 2008b[3] | 4 | 4.00 | 4.00 | |
| | *Capsicum* | 1 | 1 | Walsh & Hoot, 2001[1] | 1 | 1.00 | 1.00 | |
| | *Lycium* | 1 | 1 | Bernardello, 1986[1]; Levin *et al.*, 2007[1] | 2 | 2.00 | 2.00 | |
| | *Nolana* | 1 | 1 | Dillon, *et al.*, 2007[1]; Tu *et al.*, 2008[1] | 2 | 2.00 | 2.00 | |
| | *Solanum* (Sect. *Lycopersicon*) | 1 | 2 | Rick, 1980[2]; Rick, 1983[2,3]; Peralta & Spooner, 2001[1]; Nesbitt & Tanksley, 2002[1]; Darwin *et al.*, 2003[1]; Nuez *et al.*, 2004[1,2] | 6 | 6.00 | 3.00 | plus 2 introduced spp |

[1] Studies with a focus on phylogeny, nearest relatives and mainland origins.
[2] Studies with a focus on population genetics.
[3] Studies with a focus on breeding systems or hybridisation.

endemic groups. However, despite this recent surge of interest in Galapagos plant evolution, much remains to be done before we can be fairly confident of the origins and relationships of a large part of the Galapagos endemic vascular flora. Most of the radiations remain unstudied by molecular techniques (Table 5.6) and significant questions remain unanswered about even those that have received some investigation, such as the detailed patterns of radiation among islands within the archipelago. Such questions remain as invitations to enthusiastic PhD students. Additionally, 75 endemics generated by phyletic evolution in 59 genera (some genera having colonised Galapagos more than once) remain unstudied by modern techniques. More extensive study of such single endemics, which account for far more (70%) of the colonisation events than do the radiations, will provide more data about the relative importance of source areas and the processes of dispersal to and from oceanic islands.

In summary, only 30 out of the 117 presumed colonisation events (26%) have been subjected to molecular study and, in many of these cases, the study has not fully addressed the main questions of origin and radiation. Considering that the Galapagos flora has attracted more such research than most archipelagos, this reveals the extent of the limitations to our knowledge of island floras worldwide. We cannot claim a comprehensive knowledge of the evolution of the Galapagos flora based on this sample, although we are able to generalise patterns more accurately than was possible 30 years ago.

Also needed are further investigations into the remaining 'doubtfully native' species, in order to refine our knowledge of the level of endemicity in the Galapagos flora.

A general gap in knowledge, despite the recent surge in climatic studies, concerns the part effects of climate change on plant evolution in the archipelago. We have no clear picture of changes beyond the part 20,000 years, and changes in vegetation zones even during this period are still open to debate.

A further general gap concerns the precision of knowledge of divergence times and colonisation dates of the endemic plants; molecular clock estimates are still few and imprecise. The new geological evidence shows that islands have been present in the Galapagos region for at least 14 million, and probably tens of millions of years, but as yet we have little molecular support for their early colonisation by plants. Filling this gap will require improvements in molecular dating techniques. However, studies in progress, and the current high level of interest in these questions, suggest that our knowledge of Galapagos plant evolution will continue to be expanded and refined in the coming years.

## Acknowledgements

We thank the following for helpful discussion and comments on parts or all of the manuscript: Emily Coffey, Cynthia Froyd, Ole Hamann, Jacqueline van Leeuwen and Marianne Lindhardt. We thank David Neill, Katherine Porter-Utley and Charlotte Taylor for discussion of particular taxa. Terry Gosliner, Peter Jørgensen, Tim Mottley and Peter Murgatroyd helped by providing literature or links to information sources. This is contribution number 184 of the Tropical Biology Program of Florida International University.

## References

Adsersen, A. & Adsersen, H. (1993). Cyanogenic plants in the Galapagos Islands: ecological and evolutionary aspects. *Oikos* 66: 511–520.

Adsersen, H. (1980). Revision of the Galapagos endemic genus *Lecocarpus* (Asteraceae). *Bot. Tidsskr.* 75: 63–76.

Adsersen, H. (1987). Galapagos rare plants refound. *Threatened Plants Newsl.* 18: 25–26.

Alvarez, I. & Wendel, J. F. (2006). Cryptic interspecific introgression and genetic differentiation within *Gossypium aridum* (Malvaceae) and its relatives. *Evolution* 60: 505–517.

Alvarez, I., Cronn, R., & Wendel, J. F. (2005). Phylogeny of the New World diploid cottons (*Gossypium* L., Malvaceae) based on sequences of three low-copy nuclear genes. *Plant Syst. Evol.* 252: 199–214.

Andrus, N. G. (2002). The origin, phylogenetics and natural history of *Darwiniothamnus* (Asteraceae: Astereae), an endemic shrub of the Galapagos Islands. MSc thesis: Florida International University, Miami.

Andrus, N. G., Tye, A., Nesom, G., Bogler, D., Lewis, C., Noyes, R. *et al.* (2009). Phylogenetics of *Darwiniothamnus* (Asteraceae: Astereae): molecular evidence for multiple origins in the endemic flora of the Galapagos Islands. *J. Biogeogr.* 36: 1055–1069.

Bailey, K. (1976). Potassium-Argon ages from the Galapagos Islands. *Science* 192: 465–467.

Baker, H. G. (1955). Self-compatibility and establishment after 'long-distance' dispersal. *Evolution* 9: 347–349.

Barrett, S. C. H. (1998). The reproductive biology and genetics of islands plants. In P. Grant (ed.), *Evolution on Islands*, pp. 18–34. Oxford, Oxford University Press.

Blaschke, J. D. & Sanders, R. W. (2009). Preliminary insights into the phylogeny and speciation of *Scalesia* (Asteraceae), Galapagos Islands. *J. Bot Res. Inst. Texas* 3: 177–191.

Bernardello, L. M. (1986). Revisión taxonómica de las especies sudamericanas de *Lycium* (Solanaceae). *Bol. Acad. Nac. Cienc. Córdoba* 57: 173–356.

Brok, C. S. & Adsersen, H. (2007). Morphological variation among populations of *Lecocarpus* (Asteraceae) on the Galapagos Islands. *Bot. J. Linn. Soc.* 154: 523–544.

Browne, R. A., Anderson, D. J., White, M. D. & Johnson, M. A. (2003). Evidence for low genetic divergence among Galápagos' *Opuntia* cactus species. *Notic. Galápagos* 62: 11–13.

Bush, M. B., Colinvaux, P. A., Steinitz-Kannan, M., Overpeck, J. T., Sachs, J., Cole, J. *et al.* (2010). Forty years of paleoecology in the Galapagos. *Galapagos Res.* 67: 55–61.

Carlquist, S. (1974). *Island Biology.* New York, Columbia University Press.

Christie, D. M., Duncan, R. A., McBirney, A. R., Richards, M. A., White, W. M., Harpp, K. S. *et al.* (1992). Drowned islands downstream from the Galapagos hotspot imply extended speciation times. *Nature* 355: 246–248.

Colinvaux, P. A. (1972). Climate and the Galapagos Islands. *Nature* 240: 17–20.

Colinvaux, P. A. (1984). The Galápagos climate: present and past. In R. Perry (ed.), *Key Environments: Galápagos*, pp. 55–69. Oxford, Pergamon.

Colinvaux, P. A. & Schofield, E. K. (1976a). Historical ecology in the Galapagos Islands: I. A Holocene pollen record from El Junco Lake, Isla San Cristobal. *J. Ecol.* 64: 989–1012.
Colinvaux, P. A. & Schofield, E. K. (1976b). Historical ecology in the Galapagos Islands: II. A Holocene spore record from El Junco Lake, Isla San Cristobal. *J. Ecol.* 64: 1013–1028.
Collins, L. (1996). Environmental changes in Caribbean shallow waters relative to the closing Tropical American Seaway. In J. B. C. Jackson, A. G. Coates & A. F. Budd (eds), *Evolution and Environment in Tropical America*, pp. 130–167. Chicago, University of Chicago Press.
Conroy, J. L., Overpeck, J. T., Cole, J. E., Shanahan, T. M. & Steinitz-Kannan, M. (2008). Holocene changes in eastern tropical Pacific climate inferred from a Galápagos lake sediment record. *Quat. Sci. Rev.* 27: 1166–1180.
D'Arcy, W. G. (1982a). *Jaltomata werfii* [*sic*] (Solanaceae), a new species from the Archipelago de Colon (Galapagos Islands). *Phytologia* 52: 9–10.
D'Arcy, W. G. (1982b). *Jaltomata werffii*: a correction. *Phytologia* 52: 230.
Darwin, S. C., Knapp, S. & Peralta, I. E. (2003). Taxonomy of tomatoes in the Galapagos Islands: native and introduced species of *Solanum* section *Lycopersicon* (Solanaceae). *Syst. Biodiv.* 1: 29–53.
DeJoode, D. R. & Wendel, J. F. (1992). Genetic diversity and origin of the Hawaiian Island cotton, *Gossypium tomentosum*. *Am. J. Bot.* 79: 1311–1319.
Delgado-Salinas, A., Turley, T., Richman, A. & Lavin, M. (1999). Phylogenetic analysis of the cultivated and wild species of *Phaseolus* (Fabaceae). *Syst. Bot.* 24: 438–460.
Dillon, M. O., Tu, T., Soejima, A., Yi, T., Nie, Z., Tye, A. *et al.* (2007). Phylogeny of *Nolana* (Nolaneae, Solanoideae, Solanaceae) as inferred from granule-bound starch synthase I (GBSSI) sequences. *Taxon* 56: 1000–1011.
Dodson, C. H. (1993). New orchid species and combinations from Ecuador. *Orquideología* 19: 77–110.
Duncan, R. A. & Hargreaves, R. B. (1984). Plate tectonic evolution of the Caribbean region in the mantle reference frame. *Mem. Geol. Soc. Am.* 162: 81–93.
Eliasson, U. (1984). Chromosome number of *Macraea laricifolia* Hooker fil. (Compositae) and its bearing on the taxonomic affinity of the genus. *Bot. J. Linn. Soc.* 88: 253–256.
Eliasson, U. (1985). Identity and taxonomic affinity of some members of the Amaranthaceae from the Galapagos Islands. *Bot. J. Linn. Soc.* 91: 415–433.
Eliasson, U. (1996). 38. Aizoaceae. *Flora of Ecuador* 55: 13–27.
Eliasson, U. H. (2004). The evolutionary patterns of the plant family Amaranthaceae on the Galapagos and Hawaiian Islands. *J. Torrey Bot. Soc.* 131: 105–109.
Elisens, W. J. (1989). Genetic variation and evolution of the Galapagos shrub snapdragon. *Natl. Geogr. Res.* 5: 98–110.
Elisens, W. J. (1992). Genetic divergence in *Galvezia* (Scrophulariaceae) – evolutionary and biogeographic relationships among South American and Galapagos species. *Am. J. Bot.* 79: 198–206.
Fairbanks, R. G. (1989). A 17,000-year glacio-eustatic sea level record: influence of glacial melting rates on the Younger Dryas event and deep-ocean circulation. *Nature* 342: 637–642.
Fehlberg, S. D. & Ranker, T. A. (2007). Phylogeny and biogeography of *Encelia* (Asteraceae) in the Sonoran and Peninsular Deserts based on multiple DNA sequences. *Syst. Bot.* 32: 692–699.

Forest, F., Chase, M. W., Persson, C., Crane, P. R. & Hawkins, J. A. (2007). The role of biotic and abiotic factors in evolution of ant dispersal in the milkwort family (Polygalaceae). *Evolution* 61: 1675–1694.

Garay, L. A. (1978). 225(1). Orchidaceae (Cypripedioideae, Orchidoideae, Neottioideae). *Flora of Ecuador* 9: 1–305.

Geist, D. (1996). On the emergence and submergence of the Galápagos islands. *Notic. Galápagos* 56: 5–9.

Grehan, J. (2001). Biogeography and evolution of the Galapagos: integration of the biological and geological evidence. *Biol. J. Linn. Soc.* 74: 267–287.

Griffith, M. P. & Porter, J. M. (2009). Phylogeny of Opuntioideae (Cactaceae). *Int. J. Plant Sci.* 170: 107–116.

Guo, W. Z., Sang, Z. Q., Zhou, B. L. & Zhang, T. Z. (2007). Genetic relationships of D-genome species based on two types of EST SSR markers derived from *G. arboreum* and *G. raimondii* in *Gossypium*. *Plant Sci.* 172: 808–814.

Hamann, O. (1979). Taxonomic and floristic notes from the Galapagos Islands. *Bot. Notiser* 132: 435–440.

Hamann, O. & Wium-Andersen, S. (1986). *Scalesia gordilloi* sp. nov. (Asteraceae) from the Galapagos Islands, Ecuador. *Nordic J. Bot.* 6: 35–38.

Harley, R. M. (1999). Lamiaceae. In P. M. Jørgensen & S. León-Yánez (eds.), *Catalogue of the Vascular Plants of Ecuador. Monogr. Syst. Bot. Missouri Bot. Gard.* 75.

Harpp, K. S., Wanless, V. D., Otto, R. H., Hoernle, K. & Reinhard, W. (2005). The Cocos and Carnegie aseismic ridges: a trace element record of long-term plume–spreading center interaction. *J. Petrol.* 46: 109–133.

Hartley, A. J. & Chong, G. (2002). Late Pliocene age for the Atacama Desert: implications for the desertification of western South America. *Geology* 30: 43–46.

Helsen, P., Verdyck, P., Tye, A., Desender, K., van Houtte, N. & van Dongen, S. (2007). Isolation and characterization of polymorphic microsatellite markers in Galapagos prickly pear (*Opuntia*) cactus species. *Molec. Ecol. Notes* 7: 454–456.

Helsen, P., Browne, R. A., Anderson, D. J., Verdyck, P. & van Dongen, S. (2009a). Galapagos' *Opuntia* (prickly pear) cacti: extensive morphological diversity, low genetic variability. *Biol. J. Linn. Soc.* 96: 451–461.

Helsen, P., Verdyck, P., Tye, A. & van Dongen, S. (2009b). Low levels of genetic differentiation between *Opuntia echios* varieties on Santa Cruz (Galápagos). *Plant Syst. Evol.* 279: 1–10.

Holub, J. (1985). Transfers of *Lycopodium* species to *Huperzia*: with a note on generic classification in Huperziaceae. *Folia Geobot. Phytotax.* 20: 67–80.

Hooker, J. D. (1847). An enumeration of the plants of the Galapagos Archipelago with descriptions of those which are new. *Trans. Linn. Soc.* 20: 163–233.

Huft, M. J. & van der Werff, H. (1985). Observations on *Chamaesyce* (Euphorbiaceae) in the Galapagos Islands. *Madroño* 32: 143–147.

Iturralde-Vinent, M. A. (2006). Meso-Cenozoic Caribbean paleogeography: implications for the historical biogeography of the region. *Int. Geol. Rev.* 48: 791–827.

Jansen, R. K. (1985). The Systematics of *Acmella* (Asteraceae-Heliantheae). *Syst. Bot. Monogr.* 8: 1–115.

Jaramillo, P. J., Trigo, M. M., Ramírez, E. & Mauchamp, A. (2010). Insect pollinators of *Jasminocereus thouarsii*, an endemic cactus of the Galapagos Islands. *Galapagos Res.* 67: 21–25.

Johnston, I. M. (1931). The flora of the Revillagigedo Islands. *Proc. Calif. Acad. Sci.* (4th series) 20 (2): 9–104.

Jørgensen, P. M. & León-Yánez, S. (eds.) (1999). Catalogue of the Vascular Plants of Ecuador. *Monogr. Syst. Bot. Missouri. Bot. Gard.* 75.

Kearns, D. & Wunderlin, R. (1999). Cucurbitaceae. In P. M. Jørgensen & S. León-Yánez (eds.), *Catalogue of the Vascular Plants of Ecuador. Monogr. Syst. Bot. Missouri Bot. Gard.* 75.

Kellogg, J. N. & Vega, V. (1995). Tectonic development of Panama, Costa Rica and the Colombian Andes: constraints from Global Positioning System geodetic studies and gravity. *Geol. Soc. Am. Spec. Pap.* 295: 75–90.

Lacape, J. M., Dessauw, D., Rajab, M., Noyer, J.-L. & Hau, B. (2007). Microsatellite diversity in tetraploid *Gossypium* germplasm: assembling a highly informative genotyping set of cotton SSRs. *Molec. Breed.* 19: 45–58.

Lawesson, J. E. (1987). Buddlejaceae: a new family for Galapagos. *Notic. Galápagos* 45: 28.

Lawesson, J. E. (1988). Contributions to the flora of the Galapagos Islands, Ecuador. *Phytologia* 65: 228–230.

Lawesson, J. E. & Adsersen, H. (1987). Notes on the endemic genus *Darwiniothamnus* (Asteraceae-Astereae) from the Galapagos Islands. *Opera Bot.* 92: 7–15.

Lawesson, J. E., Adsersen, H. & Bentley, P. (1987). An updated and annotated checklist of the vascular plants of the Galapagos islands. *Rep. Bot. Inst. Univ. Aarhus* 16.

Lea, D. W., Pak, D. K. & Spero, H. J. (2000). Climate impact of late Quaternary equatorial Pacific sea surface temperature variations. *Science* 289: 1719–1724.

Lea, D. W., Pak, D. K., Belanger, C. L., Spero, H. J., Hall, M. A. & Shackleton, N. J. (2006). Paleoclimate history of Galápagos surface waters over the last 135,000 yr. *Quat. Sci. Rev.* 25: 1152–1167.

Lendel, A., Eggli, U. & Nyffeler, R. (2006). Phylogenetic relationships in the tribe Trichocereeae (Cactaceae) inferred from cpDNA sequence data analysis. Abstract, American Society of Plant Taxonomists, Botanical Society of America Annual Meeting, Chico State University.

Levin, G. A. & Moran, R. (1989). The vascular flora of Isla Socorro, Mexico. *San Diego Soc. Nat. Hist. Mem.* 16: 1–71.

Levin, R. A., Bernardello, G., Venter, A. M., Shak, J. R. & Miller, J. S. (2007). Evolutionary relationships in tribe Lycieae (Solanaceae). *Acta Hort.* 745: 225–239.

Lindhardt, M. S., Philipp, M., Tye, A. & Nielsen, L. R. (2009). Molecular, morphological and experimental evidence for hybridisation between threatened species of the Galapagos endemic genus *Scalesia* (Asteraceae). *Int. J. Plant Sci.* 170: 1019–1030.

Loope, L. L. (1989). Climate change and island biological diversity. In P. M. Vitousek, L. L. Loope & H. Adsersen (eds.), *Islands: Biological Diversity and Ecosystem Function. Ecol. Stud.* 15: 123–132. Berlin, Springer.

Mauchamp, A. (1997). Monoecy in the dioecious *Croton scouleri*, endemic to Galápagos. *Notic. Galápagos* 58: 15–17.

McMullen, C. K. (1987). Breeding systems of selected Galapagos Islands angiosperms. *Am. J. Bot.* 74: 1694–1705.

McMullen, C. K. (1993). Flower-visiting insects of the Galapagos Islands. *Pan-Pac. Entomol.* 69: 95–106.

McMullen, C. K. (2007). Pollination biology of the Galápagos endemic, *Tournefortia rufo-sericea* (Boraginaceae). *Bot. J. Linn. Soc.* 153: 21–31.

McMullen, C. K. (2009). Pollination biology of a night-flowering Galápagos endemic, *Ipomoea habeliana* (Convolvulaceae). *Bot. J. Linn. Soc.* 160: 11–20.

McMullen, C. K. & Viderman, D. M. (1994). Comparative studies on the pollination biology of *Darwiniothamnus tenuifolius* (Asteraceae) and *Plumbago scandens* (Plumbaginaceae) on Pinta Island and Santa Cruz Island, Galapagos. *Phytologia* 75: 30–38.

Moore, M. J., Tye, A. & Jansen, R. K. (2006). Patterns of long-distance dispersal in *Tiquilia* subg. *Tiquilia* (Boraginaceae): implications for the origins of amphitropical disjuncts and Galapagos Islands endemics. *Am. J. Bot.* 93: 1163–1177.

Nesbitt, T. C. & Tanksley, S. D. (2002). Comparative sequencing in the genus *Lycopersicon* for the evolution of fruit size in the domestication of cultivated tomatoes. *Genetics* 162: 365–379.

Nesom, G. L. (2008). Classification of Subtribe Conyzinae (Asteraceae: Astereae). *Lundellia* 11: 8–38.

Nielsen, L. R. (2004). Molecular differentiation within and among island populations of the endemic plant *Scalesia affinis* (Asteraceae) from the Galapagos Islands. *Heredity* 93: 434–442.

Nielsen, L. R., Philipp, M., Adsersen, H. & Siegismund, H. R. (2000). Breeding system of *Scalesia divisa* Andersson, an endemic Asteraceae from the Galapagos Islands. *Norske Vidensk.-Akad. I. Matem.-Naturvidensk. Skr. Ny Ser.* 39: 127–138.

Nielsen, L. R., Siegismund, H. R. & Philipp, M. (2003a). Partial self-incompatibility in the polyploid endemic species *Scalesia affinis* (Asteraceae) from the Galapagos: remnants of a self-incompatibility system? *Bot. J. Linn. Soc.* 142 : 93–101.

Nielsen, L. R., Cowan, R. S., Siegismund, H. R., Adsersen, H., Philipp, M. & Fay, M. F. (2003b). Morphometric, AFLP and plastid microsatellite variation in populations of *Scalesia divisa* and *S. incisa* (Asteraceae) from the Galapagos Islands. *Bot. J. Linn. Soc.* 143: 243–254

Nielsen, L. R., Siegismund, H. R. & Hansen, T. (2007). Inbreeding depression in the partially self-incompatible endemic plant species *Scalesia affinis* (Asteraceae) from Galapagos islands. *Evol. Ecol.* 21: 1–12.

Nuez, F., Prohens, J. & Blanca, J. M. (2004). Relationships origin, and diversity of Galapagos tomatoes: implications for the conservation of natural populations. *Am. J. Bot.* 91: 86–99.

O'Connor, J. M., Stoffers, P., Wijbrans, J. R. & Worthington, T. J. (2007). Migration of widespread long-lived volcanism across the Galápagos Volcanic Province: evidence for a broad hotspot melting anomaly? *Earth Planet. Sci. Lett.* 263: 339–354.

Panero, J. L., Jansen, R. K. & Clevinger, J. A. (1999a). Phylogenetic relationships of subtribe Ecliptinae (Asteraceae: Heliantheae) based on chloroplast DNA restriction site data. *Am. J. Bot.* 86: 413–427.

Peralta, I. E. & Spooner, D. M. (2001). Granule-bound starch synthase (GBSSI) gene phylogeny of wild tomatoes (*Solanum* L. section *Lycopersicon* [Mill.] Wettst. subsection *Lycopersicon*). *Am. J. Bot.* 88: 1888–1902.

Philipp, M. & Nielsen, L. R. (2010). Reproductive ecology of *Scalesia cordata* (Asteraceae), an endangered species from the Galapagos Islands. *Bot. J. Linn. Soc.* 162: 496–503.

Philipp, M., Hansen, L. B., Adsersen, H. & Siegismund, H. R. (2004). Reproductive ecology of the endemic *Lecocarpus pinnatifidus* (Asteraceae) in an isolated population in the Galapagos Islands. *Bot. J. Linn. Soc.* 146: 171–180.

Philipp, M., Böcher, J., Siegismund, H. R. & Nielsen, L. R. (2006). Structure of a plant-pollinator network on a pahoehoe lava desert of the Galapagos Islands. *Ecography* 29: 531–540.

Porter, D. M. (1976). Geography and dispersal of Galapagos Islands vascular plants. *Nature* 264: 745–46.

Porter, D. M. (1979). Endemism and evolution in Galapagos Islands vascular plants. In D. Bramwell (ed.), *Plants and Islands*, pp. 225–256. London, Academic Press.

Porter, D. M. (1980). The vascular plants of Joseph Dalton Hooker's *An enumeration of the plants of the Galapagos Archipelago; with descriptions of those which are new. Bot. J. Linn. Soc.* 81: 79–134.

Porter, D. M. (1983). Vascular plants of the Galapagos: origins and dispersal. In R. I. Bowman, M. Berson & A. E. Leviton (eds.), *Patterns of Evolution in Galapagos Organisms*, pp. 33–96. San Francisco, American Association for the Advancement of Science.

Rauscher, J. T. (2002). Molecular phylogenetics of the *Espeletia* complex (Asteraceae): evidence from nrDNA ITS sequences on the closest relatives of an Andean adaptive radiation. *Am. J. Bot.* 89: 1074–1084.

Ravenna, P. (1979). New species of Iridaceae from Galapagos and Colombia. *Bot. Notiser* 132: 463–466.

Rick, C. M. (1966). Some plant-animal relations on the Galapagos Islands. In R. I. Bowman (ed.), *The Galapagos*, pp. 215–224. Berkeley, University of California Press.

Rick, C. M. (1980). Petroselinum (*Pts*), a new marker for chromosome 6. *Rep. Tomato Genet. Coop.* 30: 32.

Rick, C. M. (1983). Genetic variation and evolution of Galápagos tomatoes. In R. I. Bowman, M. Berson & A. E. Leviton (eds.), *Patterns of Evolution in Galápagos Organisms*, pp. 97–106. San Francisco, American Association for the Advancement of Science.

Rick, C. M. & Fobes, J. F. (1975). Allozymes of Galápagos tomatoes: polymorphism, geographic distribution, and affinities. *Evolution* 29: 443–457.

Rico-Arce, M. de L. (2007). *A Checklist and Synopsis of American Species of* Acacia *(Leguminosae: Mimosoideae).* London, Royal Botanic Gardens Kew.

Sánchez-del Pino, I. (2007). Phylogeny and floral evolution of the subfamily Gomphrenoideae (Amaranthaceae). PhD thesis: City University of New York.

Schilling, E. E., Panero, J. L. & Eliasson, U. H. (1994). Evidence from chloroplast DNA restriction site analysis on the relationships of *Scalesia* (Asteraceae, Heliantheae). *Am. J. Bot.* 81: 248–254.

Sebastian, P., Schaefer, H. & Renner, S. S. (2010). Darwin's Galapagos gourd: providing new insights 175 years after his visit. *J. Biogeogr.* 37: 975–978.

Seberg, O. (1984). Taxonomy and phylogeny of the genus *Acalypha* (Euphorbiaceae) in the Galapagos archipelago. *Nordic J. Bot.* 4: 159–190.

Seigler, D. S. & Ebinger, J. E. (2005). New combinations in the genus *Vachellia* (Fabaceae: Mimosoideae) from the New World. *Phytologia* 87: 139–178.

Seltzer, G. O., Rodbell, D. T., Baker, P. A., Fritz, S. C., Tapia, P. M., Rowe, H. D. *et al.* (2002). Early warming of tropical South America at the Last Glacial-interglacial transition. *Science* 296: 1685–1686.

Simbaña, W. & Tye, A. (2009). Reproductive biology and responses to threats and protection measures of the total population of a Critically Endangered Galápagos plant, *Linum cratericola* (Linaceae). *Bot. J. Linn. Soc.* 161: 89–102.

Small, R. L. & Wendel, J. F. (2000). Phylogeny, duplication, and intraspecific variation of Adh sequences in new world diploid cottons (*Gossypium* L., Malvaceae). *Molec. Phylogen. Evol.* 16: 73–84.

Small, R. L., Ryburn, J. A., Cronn, R. C., Seelanan, T. & Wendel, J. F. (1998). The tortoise and the hare: choosing between noncoding plastome and nuclear Adh sequences for phylogeny reconstruction in a recently diverged plant group. *Am. J. Bot.* 85: 1301–1315.

Smith, S. D. & Baum, D. A. (2006). Phylogenetics of the florally diverse Andean clade Iochrominae (Solanaceae). *Am. J. Bot.* 93: 1140–1153.

Smith, S. D. & Baum, D. A. (2007). Systematics of Iochrominae (Solanaceae): patterns in floral diversity and interspecific crossability. *Acta Hort.* 745: 241–254.

Smith, S. D., Ané, C. & Baum, D. A. (2008a). The role of pollinator shifts in the floral diversification of *Iochroma* (Solanaceae). *Evolution* 62: 793–806.

Smith, S. D., Hall, S. J., Izquierdo, P. R. & Baum, D. A. (2008b). Comparative pollination biology of sympatric and allopatric Andean *Iochroma* (Solanaceae). *Ann. Mo. Bot. Gard.* 95: 600–617.

Snell, H. M., Stone, P. A. & Snell, H. L. (1996). A summary of the geographical characteristics of the Galapagos Islands. *J. Biogeogr.* 23: 619–624.

Spring, O., Heil, N. & Eliasson, U. (1999). Chemosystematic studies on the genus *Scalesia* (Asteraceae). *Biochem. Syst. Ecol.* 27: 277–288.

Stefanović, S., Kuzmina, M. & Costea, M. (2007). Delimitation of major lineages within *Cuscuta* subgenus *Grammica* (Convolvulaceae) using plastid and nuclear DNA sequences. *Am. J. Bot.* 94: 568–589.

Steinmann, V. W. & Porter, J. M. (2002). Phylogenetic relationships in Euphorbieae (Euphorbiaceae) based on ITS and *ndh*F sequence data. *Ann. Missouri Bot. Gard.* 89: 453–490.

Trueman, M. & d'Ozouville, N. (2010). Characterizing the Galapagos terrestrial climate in the face of global climate change. *Galapagos Res.* 67: 26–37.

Tu, T. Y., Dillon, M. O., Sun, H. & Wen, J. (2008). Phylogeny of *Nolana* (Solanaceae) of the Atacama and Peruvian deserts inferred from sequences of four plastid markers and the nuclear LEAFY second intron. *Molec. Phylogen. Evol.* 49: 561–573.

Tye, A. (2006). Can we infer island introduction and naturalization rates from inventory data? Evidence from introduced plants in Galapagos. *Biol. Invas.* 8: 201–215.

Tye, A. (2008). The grasses *Urochloa multiculma* and *U. fusca* (Poaceae) in the Galapagos archipelago. *Syst. Biodiv.* 6: 293–303.

Tye, A. (2010). The Galápagos endemic *Darwiniothamnus alternifolius* (Asteraceae, Astereae) transferred to *Erigeron*. *Novon* 20: 111–112.

Tye, A. & Jäger, H. (2000). *Galvezia leucantha* subsp. *porphyrantha* (Scrophulariaceae), a new shrub snapdragon endemic to Santiago Island, Galapagos, Ecuador. *Novon* 10: 164–168.

van Leeuwen, J. F. N., Froyd, C. A., van der Knaap, W. O., Coffey, E. E., Tye, A. & Willis, K. J. (2008). Fossil pollen as a guide to conservation in the Galápagos. *Science* 322: 1206.

Walsh, B. M. & Hoot, S. B. (2001). Phylogenetic relationships of *Capsicum* (Solanaceae) using DNA sequences from two noncoding regions: the chloroplast *atpB-rbcL* spacer region and nuclear *waxy* introns. *Int. J. Plant Sci.* 162: 1409–1418.

Weeks, A., Baird, K. E. & McMullen C. K. (2010). Origin and evolution of endemic Galápagos *Varronia* species (Cordiaceae). *Molec. Phylog. Evol.* 57: 948–954.

Weeks, A. & Tye, A. (2009). Phylogeography of palo santo trees (*Bursera graveolens* and *Bursera malacophylla*; Burseraceae) in the Galapagos archipelago. *Bot. J. Linn. Soc.* 161: 396–410.

Wendel, J. F. & Percival, A. E. (1990). Molecular divergence in the Galapagos-Islands-Baja-California species pair, *Gossypium klotzschianum* and *G. davidsonii* (Malvaceae). *Plant Syst. Evol.* 171: 99–115.

Wendel, J. F. & Percy, R. G. (1990). Allozyme diversity and introgression in the Galapagos Islands endemic *Gossypium darwinii* and its relationship to continental *Gossypium barbadense*. *Biochem. Syst. Ecol.* 18: 517–528.

Werner, R., Hoernle, K., van den Bogaard, P., Ranero, C., von Huene, R. & Korich, D. (1999). Drowned 14-m.y.-old Galapagos Archipelago off the coast of Costa Rica; implications for tectonic and evolutionary models. *Geology* 27: 499–502.

Wiggins, I. L. & Porter, D. M. (eds.) (1971). *Flora of the Galapagos Islands*. Stanford, CA, Stanford University Press.

# 6

# The plants of the Caribbean islands: a review of the biogeography, diversity and conservation of a storm-battered biodiversity hotspot

MICHAEL MAUNDER, MELISSA ABDO, ROSALINA BERAZAIN, COLIN CLUBBE, FRANCISCO JIMÉNEZ, ÁNGELA LEIVA, EUGENIO SANTIAGO-VALENTÍN, BRETT JESTROW AND JAVIER FRANCISCO-ORTEGA

The Caribbean islands, despite centuries of large-scale habitat loss, still contain habitats of amazing beauty and, importantly, these habitats provide an integral foundation for the islands' sustainable development. These surviving habitats contain unique plant diversity found nowhere else on the planet and for some families, most notably the Asteraceae, globally unique lineages. Despite the Caribbean being one of the first contact zones between Western science and tropical biodiversity, much remains to be discovered and to be conserved.

Scattered between two large continental land masses and abutted by the Atlantic Ocean, the Caribbean islands comprise the Bahamas Archipelago, Greater and Lesser Antilles and the islands located off the northern coast of South America. The islands are separated from the adjacent continental land masses by relatively narrow straits, for instance 210 km between Cuba and the Yucatan. They are the most important insular system of the New World, and are recognised globally as a priority for conservation investment (Olsen & Dinerstein, 1998; Myers *et al.*, 2000; Eken *et al.*, 2004; Rodrigues *et al.*, 2004; Smith *et al.* 2004a, 2004; Kier *et al.*, 2009).

The cultural diversity of the Caribbean is quite astonishing – a vibrant cultural, political and linguistic mosaic of nations and peoples that reflects a turbulent colonial past. The islands include relatively affluent economies supported by agriculture, tourism and banking but also includes nations characterised by extreme poverty and vulnerability to natural disaster, for instance Haiti, the poorest country in the Western Hemisphere.

The Caribbean islands encompass a spectacular range of marine and terrestrial habitats that includes mangroves and seagrass pastures, dry coastal scrub with cactus formations, lowland woodlands, high altitude pine woodlands and cloud forest with tree ferns; the latter includes the extraordinarily wet cloud forests of Dominica with a recorded 10 000 mm of rain per year. Areces-Mallea *et al.* (1999) recognise

*The Biology of Island Floras*, eds. David Bramwell and Juli Caujapé-Castells. Published by Cambridge University Press. 

104 distinct vegetation types in the Caribbean; this ecological diversity contains a native flora characterised by a high level of endemism at both the species and genus levels. We record a total of 185 endemic genera restricted to the Caribbean islands, 182 as recorded by Francisco-Ortega *et al.* (2007, 2008) and the recently described *Garciadelia* (Euphorbiaceae), *Hispaniolanthus* (Capparaceae), and *Linnaeosicyos* (Cucurbitaceae) (Schaefer *et al.* 2008; Cornejo & Ildis, 2009; Jestrow *et al.* 2010).

We argue that the botanical diversity of the Caribbean islands is uniquely threatened; massive and ongoing landscape change has reduced most areas of natural habitat to degraded fragments. While the flora may have evolved to cope with episodic hurricane events it may not survive this massive and continuing loss of habitat. The Caribbean has a long history of botanical exploration. The early collectors, such as Plumier and Sloane, were amongst the first Western scientists to systematically document the diversity of the tropics. However, the Caribbean is still under-studied, with new genera and species continuing to be discovered and many important plant areas requiring exploration and protection. In parallel to new field discoveries, molecular phylogenetics is refining our understanding of the hotspot's botanical diversity (Santiago Valentin & Olmstead, 2004; Francisco-Ortega *et al.*, 2007; Santiago Valentin & Francisco-Ortega, 2008). The Caribbean islands will continue to be a crucible for understanding tropical botany and will be a vital testing ground for the application of both conservation biology and sustainable development (Maunder *et al.*, 2008).

## Origins of the Caribbean flora

Caribbean plant diversity has been forged by the region's complex geological history which has involved major eustatic/isostatic sea-level changes, climatic changes, meteor impact, volcanism and migration of lithospheric plate boundaries (Iturralde-Vinent & MacPhee, 1999; Pindell & Kennan, 2001; Hedges, 2001; Fritsch & McDowell, 2003, Kress & Horvitz, 2005). These great geological forces have created a mosaic of habitats profoundly influenced by soil type, altitude and rainfall with episodic, and occasionally catastrophic, hurricanes adding an extra ecological dynamic.

The Greater Antilles originated in the early Cretaceous (*c.*130 Ma) as a volcanic chain of sea mountains between North and South America in an area close to the present Panamanian Isthmus (Pindell *et al.*, 2005). These islands, the proto-Greater Antilles, moved northeastward until they collided first with the Yucatan Peninsula (*c.*84 Ma) and then with the Bahamas Platform in the early Eocene (*c.*56 Ma); by the middle Eocene (*c.*49 Ma) most of the Greater Antilles and part of the Lesser Antilles were above sea level as established pieces of land mass (Graham, 2003a, 2003b).

Based on this model, the middle Eocene (49 Ma) can be regarded as the starting point for the current biota of the Caribbean. Accordingly, the flora of the Caribbean islands is relatively young when compared with that of South and North America, or ancient island systems such as New Caledonia or the Seychelles. Support for middle Eocene origin of the Antillean flora comes from the palaeontological record with the oldest fossil deposits for the Greater Antilles from Cuba and Jamaica dating from this geological epoch (Graham *et al.*, 2000).

Some of the endemic genera seem to have originated before the middle Eocene. The monotypic Cuban endemic genus *Microcycas* A DC. is sister to the widespread neotropical cycad genus *Zamia* (Zamiaceae) (Bogler & Francisco-Ortega, 2004; Chaw *et al.*, 2005). Lower Eocene (*c.*55 Ma) fossils of *Zamia* (Hermsen *et al.*, 2006) suggest that the split between *Microcycas* and *Zamia* occurred long before the middle Eocene and that both genera might have originated in the mainland. This suggests that *Microcycas* reached Cuba some time after the middle Tertiary before becoming extinct elsewhere. A second hypothesis is that some of the proto-Greater Antilles biota escaped destruction during the major environmental changes that occurred in the islands prior to the middle Eocene. At the end of the Cretaceous the Yucatan area experienced a catastrophic asteroid impact that may have led to the virtual extinction of all the Caribbean biota (Hedges, 1996; Iturralde-Vinent, 2004–2005); others, however, consider it unlikely that the asteroid collision produced such a massive extinction (Crother & Guyer, 1996).

Molecular phylogenetic studies for vertebrates suggest a pre-middle Eocene origin for at least four endemic groups of tetrapods. Studies of the endemic Antillean mammal genus *Solenodon* Brandt, the frog *Eleutherodactylus* Duméril & Bibron, the lizard *Cricosaura* Gundlach & Peters and the worm-lizard *Amphisbaena* L. revealed that these taxa diverged long before the middle Eocene (Hedges, 1996; Hass *et al.*, 2001; Roca *et al.*, 2004). The presence of these pre-middle Eocene lineages may be explained by 'recent' dispersal from the mainland during the Tertiary of now extinct continental taxa (Hedges, 1996; Hass *et al.*, 2001), or these ancient lineages originated on the islands and survived the major environmental changes since the late Cretaceous (Duellman, 1999; Pregill & Crother, 1999).

The Caribbean has a unique geographical situation as it is located between two major continents, South and North America, two potential sources for the Caribbean biota. A large proportion of the Antillean biota are believed to have migrated through a late Cretaceous–early Palaeocene land connections between Meso and South America, the proto-Greater Antillean arc (Savage, 1982). Recent phylogenetic studies within the Piperaceae provide further support for the presence of early divergent lineages in the Caribbean Islands; *Verhuellia* Miq., a genus with three species endemic in Cuba and Hispaniola, is sister to the rest of the Piperaceae (Wanke *et al.*, 2007; Samain *et al.*, 2008).

A primary Gondwana origin has been suggested for the Caribbean biota through South America (Howard, 1973). However, other authors consider that many of the taxa of tropical North America, including those present in the Antilles, are linked to those that existed in Laurasia during the early Tertiary; for instance, Gentry (1982) stressed the importance of a Laurasian-derived origin for the neotropical flora. However, it appears that both Gondwana and Laurasia have been source areas for the Caribbean; data indicate that out of 12 913 seed-plant taxa native to the Caribbean islands 1495 (11%) belong to Laurasian plant families (Acevedo-Rodríguez & Strong, 2008). Historical biogeography research coupled with molecular phylogenetics suggest that at least two of the endemic Caribbean genera, *Pictetia* DC. and *Poitea* Vent., originally came from Tertiary North American groups (Lavin *et al.*, 2001b). In contrast, similar studies conducted by Santiago-Valentín and Olmstead (2003) and Olmstead *et al.* (2008) showed that four endemic genera of the subfamily Goetzeoideae have their closest continental relatives in South America and Madagascar. *Doerpfeldia* Urb. is another endemic genus that appears to have a Gondwana origin, as it is sister to the Brazilian *Ampeloziziphus* Ducke; these two genera are sister to *Bathiorhamnus* Capuron, a genus endemic to Madagascar.

Three genera (*Acidocroton* Griseb., *Domingoa* Schltr. and *Scolosanthus* Vahl.) are sister to genera restricted to South America and Central America (van den Berg *et al.*, 2000; Berry *et al.*, 2005; Motley *et al.*, 2005). Three of the endemic Caribbean genera (*Cubanola* Aiello, *Hebestigma* Urb. and *Phyllacanthus* Hook.f.) have their nearest relatives distributed both in North America and Central America (Lavin *et al.*, 2001b, 2003; Wojciechowski *et al.*, 2004; Motley *et al.*, 2005). Two genera (*Brya* P. Browne and *Doerpfeldia* Urb.) are sister to lineages from South America (Richardson *et al.*, 2000; Lavin *et al.*, 2001a; Wojciechowski *et al.*, 2004). Only one genus, *Thymopsis* Benth., is sister to a North American taxon, *Hymenopappus* L'Hér. (Baldwin *et al.*, 2002). Similarly, only one genus, *Rhodogeron* Griseb., is sister to the endemic genus *Sachsia* (Liu *et al.*, 2004; Anderberg *et al.*, 2005). The closest relatives for 11 genera (*Bonania* A. Rich., *Chascotheca* Urb., *Fuertesia* Urb., *Grimmeodendron* Urb., *Lasiocroton* Griseb., *Leptocereus* (A.Berger) Britton & Rose, *Leucocroton* Griseb., *Microcycas* (Miq.) A.DC., *Poitea* Vent., *Roigella* Borhidi & M. Fernandez Zeq and *Suberanthus* Borhidi & M. Fernandez Zeq) have widespread distributions in North, Central and South America (summarised from Francisco-Ortega *et al.*, 2007). *Pictetia* is sister to a lineage that has taxa both from the New World and Africa (Lavin *et al.*, 2001a, 2001b).

A number of the Caribbean endemic lineages have links to the Pacific. For instance, *Siemensia* Urb., a Cuban genus, is sister to a clade restricted to New Caledonia (Motley *et al.*, 2005). A similar biogeographical pattern has also been reported for the Hispaniolan genus *Selleophytum* Urb., which forms a clade with the Hispaniolan endemic *Narvalina domingensis*; this clade is sister to a group composed of two Polynesian genera, *Fitchia* Hook f. and *Oparanthus* Sherff. (Kimball & Crawford, 2004).

Several of the Caribbean genera are sister to other endemic genera. Francisco-Ortega *et al.* (2007, 2008) identified 10 lineages (33 genera) composed exclusively of endemic genera in four families. The most extraordinary example is provided by species from 10 of the endemic genera of the tribe Senecioneae; they form a distinct clade within this tribe (Pelser *et al.*, 2007; Francisco-Ortega *et al.*, 2008). Six of these 10 Antillean lineages collapse in polytomies with other lineages, or are sister to clades that have a widespread distribution. In contrast, a connection with the North American flora is provided by the Cuban lineage *Harnackia* Urb.–*Lescaillea* Griseb. which is sister to the Mexican genus *Boebrastrum* (Loockerman *et al.*, 2003). The *Isidorea* DC.–*Portlandia* P. Browne lineage is part of a clade that contains the unispecific genus *Osa* Aiello from Central America, and *Catasbea* L.; the latter is confined to the Bahamas, the Greater Antilles and South Florida (Motley *et al.*, 2005). The *Bellonia* L.–*Pheidonocarpa* L. E. Skog radiation falls in a clade that has four additional species endemic to the Greater Antilles (Roalson *et al.*, 2005).

Only one lineage of the basal angiosperms, the monocots, has endemic genera (23) in the Caribbean. There are no endemic genera in the 'Early-Diverging Eudicot' grade, and therefore the remaining endemic genera belong to the 'Core Eudicot' clade, although a recent phylogenetic study identifies an undescribed genus endemic in the Greater Antilles belonging to the Canellaceae (Magnoliids) (Salazar & Nixon, 2008). The endemic genera are distributed in the Asterids (103 genera), Rosids (45 genera), Caryophyllales (five genera) and Santalales (one genus, *Dendropemon* (Blume) Rchb.). The 'Core Eudicot' clade also has most of the plant biodiversity in the neotropics (Smith *et al.*, 2004b, 2004).

## Botanical diversity of the Caribbean islands

While the Caribbean's land areas are relatively small, a total land area of *c.*229, 550 km$^2$, the islands support a vascular flora of 205 native plant families, with 1447 genera and 10 948 species (Acevedo-Rodriguez & Strong, 2008). The *c.*7800 species of endemic vascular plants (Acevedo-Rodriguez & Strong, 2008) equates to *c.*11% of the total number of endemics restricted to islands worldwide (70 000) (Kreft *et al.*, 2008).

Work by Francisco-Ortega *et al.* (2007, 2008) and Jestrow *et al.* (2010) shows an amazingly high level of endemism at the genus level with a high proportion of monotypic genera, with a total of 185 endemic genera in 47 plant families; none of these endemic genera occurs on the coast of the Yucatan, Barbados, or on the island systems off the Venezuelan coast. This represents an extraordinary packing of endemic lineages into highly fragmented and increasingly threatened landscapes.

A large proportion of the 185 endemic genera are represented by a single species (96), with 107 of the 185 endemic taxa restricted to a single island, Cuba holding 65 endemic genera (Francisco-Ortega *et al.*, 2007). The five endemic genera with the highest number of species are *Wallenia* Sw. (40 species), *Calycogonium* DC. (36 species), *Dendropemon* Blume (36 species), *Leucocroton* Griseb. (28 species) and *Scolosanthus* Vahl. (27 species). The five families with the highest number of Caribbean island endemic genera are: Asteraceae (41), Rubiaceae (27), Orchidaceae (12), Euphorbiaceae (9) and Fabaceae (10). These five families are also among the seven most generically diverse (> 100 genera) in the New World (the others being the Melastomataceae and Poaceae). Therefore, at the family level, patterns of generic endemicity in the hotspot appear congruent with those of tropical regions of the New World (Smith *et al.*, 2004b). However, it is noteworthy that at least three families (Araceae, Cyperaceae and Lauraceae) that have a high number of endemic genera in the neotropics do not have any genera endemic to the Caribbean islands. Only one of the 25 families endemic to the neotropics, the Theophrastaceae, has two endemic Caribbean genera, *Neomezia* Votsch. and *Theophrasta* L. (Ståhl, 2004). The highly threatened Cuban cycad *Microcycas* is the only endemic gymnosperm genus in the Caribbean (Stevenson, 1992). The most species-rich non-endemic Caribbean genera are those with wide distributions (for instance, *Eugenia* L. with 218 endemic taxa, *Pilea* Lindl. with 205 endemic taxa and *Rondeletia* L. with 145 endemic taxa) (Acevedo-Rodriguez & Strong, 2008).

Recent molecular studies reveal new levels of generic endemism for the Caribbean radiations. For instance, with the large genera such as *Miconia* Ruiz & Pavón (Melastomataceae) or *Croton* L. (Euphorbiaceae), species in these groups are likely to be assigned to other genera once additional taxonomic/systematic studies are completed (Michelangeli *et al.*, 2008; van Ee *et al.*, 2008). A similar situation exists for the endemic species previously accommodated in the large genera *Eupatorium* L. and *Senecio* L. (Asteraceae). Many of the *Eupatorium* species are currently placed in five endemic genera (King & Robinson, 1987) whereas those in *Senecio* are in 10 endemic genera (Nordenstam, 1978, 2006a, 2006b; Nordenstam & Lundin, 2002).

Molecular data have also been important to reinstate the endemic status of at least one Antillean genus. *Selleophytum* Urb., a Haitian endemic (Urban, 1915), was considered by Sherff and Alexander (1955) to belong to *Coreopsis* L.; however, a recent molecular phylogeny by Kimball and Crawford (2004) demonstrated that it should be recognised as a distinct genus as originally suggested by Urban. In addition, phylogenetic reconstructions based on morphological traits have been useful for assigning a generic status for *Salcedoa*, one of the most recently described endemic genera (Jiménez-Rodríguez *et al.*, 2004).

One of the regionally important drivers for endemism in the Caribbean are the extensive areas of serpentine (nickel-rich) soil; this is a particular feature of Cuba (Borhidi, 1991). While serpentine soils account for only 7% of Cuba's land area, around 920 of Cuba's 6375 plant species are endemic to that habitat. A total of 28 genera are represented in Cuba only by serpentine obligate species; of these, 25 are endemic to Cuba and 18 are monospecific (Reeves *et al.*, 1999).

## Comparisons with other insular biodiversity hotspots

The Caribbean islands are remarkably rich in endemic taxa (Kier *et al.*, 2009) and specifically in endemic genera (Francisco-Ortega *et al.*, 2007); only the Malesian Islands (with at least 382 single-island endemic genera) and Madagascar (260 endemic genera) have more endemic genera than the Caribbean with 205 (Francisco-Ortega *et al.* 2007). The high proportion of unispecific endemic genera detected in the Caribbean (52%) is characteristic of other island groups with the notable exception of Hawaii (Francisco-Ortega *et al.*, 2007).

One of the major differences between the Caribbean islands and other tropical island systems concerns the distribution of endemic genera in the seed plant phylogeny (Soltis *et al.*, 2005). For instance, both Madagascar and the Malesian Islands have endemic genera in two of the major lineages of the Magnoliids: Magnoliales and Laurales. The Caribbean, on the other hand, does not have any endemic genus belonging to these two major lineages (Francisco-Ortega *et al.*, 2007). The Canellales have at least one genus, *Takhtajania* M.Baranova & J.-F.Leroy (Winteraceae), endemic to Madagascar (Soltis *et al.*, 2005). The genus *Pleodendron* Tiegh., a member of the Canellales long considered as an Antillean-endemic, has been discovered in Central America (Hammel & Zamora, 2005); this finding eliminates endemism of Caribbean island genera from the early branches of the angiosperms. However, a molecular phylogeny for the Canellaceae has established *Cinnamodendron* Endl. as polyphyletic, indicating that the five Greater Antillean taxa need to be assigned to a new endemic genus (Salazar & Nixon, 2008).

The 'Early-Diverging Eudicot' grade does not possess any Caribbean endemic genera; in contrast, two of the five lineages of this paraphyletic group have endemic genera in Madagascar or the Malesian Islands. Within the 'Core Eudicots', Madagascar and the Malesian Islands, as well as the Caribbean, have endemic genera within the Asterids, Rosids and Santalales. The Caryophyllales do not have any genus endemic to the Malesian Islands, although both the Caribbean and Madagascar have genera within this group. The Caribbean islands have a relatively recent geological history when compared with most of the large continental islands of the Pacific and Indian Ocean such as Madagascar, New Zealand, New Caledonia, New Guinea and the Fiji Islands (Nunn, 1994). Therefore, it is not surprising that the early diverging angiosperms do not have any genus endemic to the Caribbean.

## Case studies in diversity and conservation

### *The Caribbean daisies (Asteraceae)*

The Caribbean daisies, Asteraceae, are an overlooked and extraordinary flagship family for the Caribbean. The Caribbean has 36 major Asteraceae lineages that correspond with the current tribal classification of the family; 17 of the 36 lineages are represented with endemic taxa in the Caribbean, with 12 having both endemic genera and species. In addition the Caribbean is unique in having an endemic monotypic tribe. The Caribbean holds almost one-third of the world's insular endemic Asteraceae taxa (41 of the 136 genera, *c*.30%), while Madagascar and the Mascarenes hold 27 endemic genera (*c*.20%), Polynesia contains 22 genera (*c*.16%) and the combined island systems of Indonesia, Melanesia and the Philippines have only 9 endemic genera (6.6%) (Francisco-Ortega *et al*., 2008). Given the tiny land area of the Caribbean, this represents an extraordinary richness of endemic lineages found in a single biodiversity hotspot. A total of 32 of the 41 Caribbean endemic genera are single-island endemics, with Cuba holding 17, Hispaniola 11 and 4 endemic to Jamaica; of the 41 endemic genera a total of 26 are unispecific (Francisco-Ortega *et al*., 2008).

While there is no comprehensive IUCN Red List for the Caribbean flora, based on our own field experience and published sources we believe that the following 14 endemic Asteraceae genera are likely to be classified as Critically Endangered *sensu* IUCN: *Antillia* R. M. King & H. Rob.(1 species), *Ciceronia* Urb.(1 species), *Ekmaniopappus* Borhidi (2 species), *Feddea* Urb. (1 species), *Harnackia* Urb. (1 species), *Herodotia* Urb. & Ekman (1 species), *Herreranthus* B. Nord. (1 species), *Ignurbia* B. Nord. (2 species), *Lescaillea* Griseb. (1 species), *Rhodogeron* Griseb. (1 species), *Salcedoa* F. Jimenez & L. Katinas (1 species), *Shafera* Greenm. (1 species), *Tetraperone* Urb. (1 species) and *Thymposis* Benth. (2 species) (Francisco-Ortega *et al*., 2008; Maunder *et al*., 2008). A number of genera of conservation concern are edaphic endemics known only to occur on areas with serpentine soils in Cuba, for instance *Ciceronia*, *Feddea*, *Harnackia*, *Herreranthus*, *Koehneola* Urb., *Lescaillea*, *Oldfeltia* B. Nord. & Lundin, *Rhodogeron* and *Shafera*.

Two endemic genera, *Salcedoa* from the Dominican Republic and *Feddea* from Cuba, clearly illustrate the challenges of conserving endemic lineages The genus *Feddea* was first described by Urban (1925) based on materials collected by Erik Leonard Ekman in the 'Lomas de Cuaba' (Baracoa, Eastern Cuba) in 1915. *Feddea* is recorded from the serpentine soils of eastern Cuba, with 16 historic locations in Holguin and Guantanamo provinces; despite extensive fieldwork only one location with a total of five plants was found (Cariaga *et al.*, 2008; Francisco-Ortega *et al.* 2008). What is remarkable is that, following investigations by Cariaga *et al.*, (2008), *Feddea* has been placed in its own monotypic tribe, the only tribe in the Asteraceae that is endemic to an island.

*Salcedoa* is a monotypic genus from the Dominican Republic, known from a single locality in the Cordillera Septentrional (Jiménez-Rodriguez *et al.*, 2004). Only 15 plants were recorded from an area of degraded habitat subject to fragmentation from urban and agricultural development. The news of this discovery reached the national media, and currently *Salcedoa* has been adopted as the symbol for conservation in the province of Salcedo. There are plans to create a protected area where this genus occurs. Recent fieldwork conducted by botanists of the National Botanic Garden of Dominican Republic has located further populations of this species. The example of *Salcedoa* illustrates an urgent need to conduct biodiversity inventories in areas where there are still pockets of the original vegetation and to initiate on-site conservation action.

### *The Caribbean palms (Arecaceae)*

The Caribbean contains 121 endemic of palm and an increasing number of threatened taxa (Zona *et al.*, 2007; Ronçal *et al.*, 2008). The Caribbean has 21 genera of palms with three endemic genera, *Zombia* L.H. Bailey, *Hemithrinax* Hook. f. and *Calyptronoma* Griseb. (Roncal *et al.*, 2008). Several genera, such as *Pseudophoenix* H. Wendl. ex Sarg., *Roystonea* O. F. Cook, *Acoelorraphe* H. Wendl. and *Thrinax* L.f. ex Sw., are near-endemics to the Caribbean islands, also occurring in southern Florida, the Yucatan Peninsula, the Atlantic coast of Central America and the northern coast of South America. The remaining genera are shared with North, Central or South America.

There are 135 species of palms in the Caribbean. A few species, such as *Acrocomia aculeate* (Jacq.) Lodd. Ex Mart., *Desmoncus polyacanthos* Mart. and *Sabal palmetto* (Walter) Lodd. ex Schult & Schult f., are widespread, with ranges extending well outside the Caribbean islands; however, 121 species are strictly endemic to the Caribbean (Roncal *et al.*, 2008). As is typical for the Caribbean, species richness is concentrated in the Greater Antilles with Cuba holding 94 species, Hispaniola 31 species, Jamaica 11 species, Puerto Rico and the Virgin Islands 10 species, the Lesser Antilles 12 species, the Bahamas, including Turks and Caicos, 8 species and the Cayman Islands 2 species.

Phylogenetic studies (Roncal *et al.*, 2008) have identified five palm lineages unique to the Caribbean: the endemic genus *Calyptronoma* with three species; the Caribbean clade of *Bactris* Jacq. ex Scop., with three species; the Caribbean clade of *Gaussia* with three species, with *G. spirituana* Moya & Leiva, recognised as threatened; the endemic genus *Hemithrinax* (all four species threatened); and the endemic genus *Zombia* clade (with one threatened species).

To date only one Caribbean palm is recorded as extinct. *Roystonea stellata* León was endemic to Cuba, but repeated exploration for the palm has failed to locate it (Zona *et al.*, 2007). A number of endemic palm taxa are highly threatened. The category of Critically Endangered (CR) is assigned to eight taxa (plus three natural hybrids of *Copernicia* Mart. ex Endl.): *Attalea crassispatha* (Mart.) Burret; *Coccothrinax borhidiana* O. Muniz; *C. camagueyana; C. crinita* (Griseb. & H.Wendl. ex C.H. Wright) Becc. ssp. *crinati; C. victorini; Copernica brittonorum* León; *C. fallaensis* León; and *Pseudophoenix lediniana* Read (Zona *et al.*, 2007). Cuba, which has the greatest number of palm species, also has the greatest number of CR taxa (nine); Haiti is in second place, with two CR taxa, the endemic *Attalea crassispatha* and *Pseudophoenix lediniana* (Zona *et al.*, 2007).

The remaining taxa are point endemics known from very few populations with limited extent of occurrence. For example, *Coccothrinax borhidiana* is known only from a 1-km strip of coastal limestone scrub near Matanzas, Cuba. The endemic palms of the Caribbean, like those of the Mascarenes (Maunder *et al.*, 2002) have lost their commensal species. It is easy to imagine a pre-Colombian scene with the fruiting palms alive with flocks of the extinct endemic macaws (*Ara* spp.) while on the forest floor groups of the extinct giant hutias, and perhaps ground sloths, would have been feeding on fallen fruit.

### *The systematics and conservation of Goetzea (Solanaceae)*

We present the case of the Caribbean Goetzeoideae as an example of an endemic lineage in which both phylogenetic and conservation research has played a major part in developing conservation initiatives. The systematic placement of the four genera in the Goetzeaceae (*Coeloneurum* Radlk., *Goetzea*, *Espadaea* A. Rich. and *Henoonia* Griseb.) has been problematic and, after being assigned to different families, they were suggested as belonging to the Solanaceae (Wettstein, 1895). Several classifications (e.g. Takhtajan, 1997) have placed the group in their own family 'Goetzeaceae', this being proposed as the only plant family endemic to the Caribbean islands. The taxonomic impasse was resolved through molecular phylogenetic analyses (Olmstead *et al.*, 2008) that rejected the family Goetzeaceae but revealed new and unexpected relationships within the Solanaceae. The group is among the basal lineages of Solanaceae, forming the subfamily Goetzeoideae, a clade that also includes the Brazilian monotypic genus *Metternichia* J. G. Mikan, and the Madagascar monotypic *Tsoala* Bosser & D'Arcy, two other genera historically considered as being 'taxonomically obscure' (Olmstead *et al.*, 2008).

*Goetzea elegans*, the 'Matabuey', is endemic to Puerto Rico; historically (from the late nineteenth century to the 1930s) the species was reported from low-elevation mesic forests. However, for most of the twentieth century, the species was known from a single site in the karst region of northwestern Puerto Rico, with no reproductive trees being observed until a fruiting individual was discovered in the 1980s. After its formal designation as an Endangered Species in 1985, active field search led to the discovery of additional reproductive individuals in northwestern Puerto Rico and the discovery of new sites in eastern Puerto Rico (Fajardo and the nearby Vieques Island). The species, however, has never again been observed in several of the historical sites. Almost all of the known populations of *G. elegans* occur in ravines, in relictual patches of native forest that escaped clearing. Research has shown that it is bird pollinated, primarily by a common native species (*Coereba flaveola*); it is highly self-incompatible and requires outcrossing for successful seed set, a problem for a species with such fragmented populations. Further studies indicated that dispersal of seed and subsequent plant recruitment was not occurring efficiently in the wild. An additional challenge is predation of seeds and seedlings by introduced rats. Current genetic studies show that the extant populations exhibit considerable diversity among them, a fact that highlights the need to preserve all the populations in order to preserve important genotypes. Successful propagation from parental trees representing the whole distribution of the species – and therefore, its genetic makeup – is making possible the establishment of *ex situ*, self-reproducing populations in protected areas. This supplemental conservation measure, together with the protection of wild populations and habitats, active research/management programmes and a vigorous education effort could lead to this becoming a highly successful conservation case for Caribbean flora.

### *Conserving Jamaica's 'Cockpit country'*

Jamaica's 'Cockpit country' is one of the last wildernesses of the Caribbean. The terrain is extraordinary: a vast, rugged, hilly limestone landscape dominated by forested Cockpit karst. This landscape is characterised by steep conical hills and Cockpit depressions ranging from convex to nearly flat, rocky floors. The Cockpit country's flora is diverse, and each Cockpit hilltop is unique, often with notably different species associations and highly restricted plant taxa.

From an aerial vantage, the region has an 'egg-carton' shape; on the ground, it is readily apparent that nearly the entire region is riddled with rocky crevices, steep outcrops, solution holes and caves. This inhospitable and largely treacherous terrain has meant that most of the Cockpit's native forests and flora have, *de facto*, fared better than more accessible areas; yet the landscape is increasingly fragmented and degraded by small-scale clearing for agriculture, and by widespread tree harvesting carried out to gather 'yamsticks' (large slender trunks stuck in the ground to support yam vines in agricultural plots). The looming threat of bauxite mining compounds the endangered status of Cockpit country.

However, there exists compelling evidence in support of the feasibility and importance of the conservation of Jamaica's Cockpit country and its amazing plants. In addition to its amazing geology and botanical diversity, the Cockpit is home to 27 of Jamaica's 28 endemic birds and endemic bats, butterflies and herpetofauna. The Cockpit country forms a major part of the Cockpit Country-Martha Brae watershed that supplies fresh water to many parts of the island.

Work by Fairchild Tropical Botanic Garden and the Institute of Jamaica over the last decade has served to renew and emphasise botanical investigations and conservation of the Cockpit country. The team first set out to identify 'lost' species from the Cockpit country (plants known only from their type locality or those that have not been collected or seen in decades) and through extensive fieldwork have to date successfully rediscovered and studied 15 of the working list of 20 lost Cockpit country plants. Ongoing scientific explorations into botanical *terra incognita* of the Cockpit country (sites hitherto unexplored and undocumented by botanists) are leading to exciting and significant discoveries, such as several endemic plant species that are new to science.

The new checklist (Proctor & Abdo, unpublished data) reveals 907 plant taxa in the Cockpit; this equates to an astounding 9% of the Caribbean flora in a region that is a mere 0.2% of the total Caribbean land area. Based on this same work, scientists have learned that the Cockpit's fern flora is exceedingly diverse, with over 160 taxa occurring in the region. Cockpit country alone harbours over 35% of Jamaica's endemic flora, despite on a national level comprising only about 6% of Jamaica's entire land area. Finally, the study verifies that at least 67 endemic plant taxa occur only in Cockpit country and nowhere else on earth.

Today, the Cockpit country is under serious threat, and is in need of urgent protective action. The most significant threat to the region is that from bauxite mining; indeed, prospecting permits have been released, covering large swathes of the Cockpit country. Yet most Jamaicans express support for conservation of the Cockpit country; and a large stakeholder consortium, in addition to citizens and scientists, continue to speak out about the importance of strict conservation for Cockpit country.

## History of Caribbean botany

The Caribbean was one of the first contact zones between tropical biotas and Western science. The European discovery of the Caribbean in 1492 broke the biological seal on the area; subsequently botanical samples began to trickle out to

Europe where they became objects of curiosity and intense speculation (Crosby, 2003). The first Western naturalist to comment on this 'new world' was Gonzalo Fernandez de Oviedo y Valdes (1478–1557) who travelled in the Caribbean and Central America. Fernandez de Oviedo described maize and manihot and introduced avocado, papaya, pineapple and banana to European readers (Petersen, 2001). In the sixteenth century, Caribbean plants slowly started accumulating in European plant collections; for instance, the royal apothecary Hugh Morgan (*c.*1530–1613) cultivated Caribbean cacti (*Melocactus* Link & Otto) in London (Rowley, 1997). Philip II of Spain in 1554 ordered his officials in Hispaniola to send 'all the seeds and plants found' and in 1570 he commissioned a very early natural history expedition by requesting that Dr Francisco Hernandez collect samples of plants and medicines (Barrera-Osorio, 2006).

The early colonisation of the Caribbean was driven in large part by the quest for plants of commercial value, for instance Caribbean mahogany. The trade from the Caribbean in plant products defined the Western world's view of the Caribbean as a source of commercial plant products. While this trade generated enormous wealth for Europe (for instance, the sixteenth-century Augsburg banking family, the Fuggers, generated huge wealth through a trade monopoly on lignum vitae wood, Crosby, 2003), this trade also initiated a highly damaging pattern of island degradation where successive natural resources have been exhausted and abandoned in an unstable plantation based economy (Grove, 1995; Maunder *et al.*, 1998).

The colonial administration was influenced by eighteenth-century utopian environmentalists who promoted forest and watershed protection (Grove, 1995). One of the world's first protected areas was established on Tobago in 1762, the Main Ridge Reserve, and in 1765 the first botanical garden in the Caribbean was established on the island of St Vincent (Grove, 1995).

However, field botany came late to the Caribbean, virtually 200 years after the European discoveries of the 1490s. We propose that the history of field botany in the Caribbean starts with Charles Plumier (1646–1704), who made three important voyages, the first to Martinique and Haiti in 1689. The other pioneer who established the foundations for Caribbean botany was the British collector and intellectual, Sir Hans Sloane (1660–1753). As physician to the Governor of Jamaica, Sloane was resident in Jamaica between 1687 and 1688 when he collected some 800 new species of plant (Petersen, 2001). Sloane's and Plumier's publications provided important sources of generic and specific names for Linnaeus, Jacquin and Swartz. N. J. Jacquin (1727–1817) spent five years (1755–1759) collecting in the Caribbean and described many new Caribbean species in his beautifully illustrated publications. After Jacquin the Swedish botanist O. Swartz (1760–1818) advanced field botany, his *Florae India Occidentale* (1797–1806) being the first flora for the Caribbean.

We recognise Erik Leonard Ekman as the greatest Caribbean botanist. He started his career with the Swedish Museum of Natural History in Stockholm in 1908. Ekman landed in Cuba in 1917 and stayed for 10 years, except for a short visit to Haiti during 1917. He returned to Haiti in 1924 where he spent a period of intensive fieldwork lasting until 1928, when he moved to the Dominican Republic. Ekman died suddenly at the age of 47 on 15 January 1931, in the Dominican Republic. He described more than 2000 species new to science and collected around 36 000 species, amounting to more than 150 000 specimens including duplicates; indeed, between 1914 and 1931 Ekman collected approximately 100 000 specimens from Cuba and Hispaniola (Howard, 1996; Mejía *et al.*, 2001).

A study of the Caribbean's endemic genera allows an analysis of the region's collecting history. Approximately half of the endemic genera (87, 49%) were published in two time periods 1854–1878 (34 genera) and 1904–1928 (53 genera) (Francisco-Ortega *et al.*, 2007). These two periods represent 'golden ages' for botanical exploration in the Caribbean. The first period was marked by the effort of the North American plant explorer Charles Wright, who collected extensively in Cuba between 1856 and 1866. However, the second period (1904–1928) was the most active. During that time 54 genera were published, mostly as the result of the work by the German botanist Ignatz Urban, who between 1898 and 1928 published the nine-volume *Symbolae Antillanae.* Approximately half of the genera described by Urban were based on the collections made by Ekman.

Most of the endemic genera were published between the second half of the nineteenth and the first half of the twentieth centuries, long before systematics and phylogenetics merged to provide tools for plant classification. Thirty-eight of the endemic genera (21%) were published after 1950, and the 10 most recent were described between 2004 and 2006 (Francisco-Ortega *et al.*, 2007). The 1979–2003 period was the third most active for the number of published endemic genera (16 genera). However, most of the descriptions for these genera are new circumscriptions based on material collected prior to 1950; therefore, their basionyms were published much earlier. Only eight of 38 genera described between 1950 and 2006 are based on material collected after 1950. A cumulative analysis of the chronology of discovery of endemic genera in the Caribbean suggests that new genera created from taxonomic revisions are steadily increasing. New genera obtained from recent field explorations have decreased, and is possibly approaching saturation. These results show that botanical explorations conducted before 1950 may have already yielded much of the endemic diversity available at the generic level. Indeed, the worry is that a number of these historical collections are derived from lost habitats and taxa now extinct.

We believe that there are still large collecting gaps in the region. On the island of Hispaniola, for instance, Haiti remains poorly collected and documented (Zanoni

1989) with surviving upland mesic habitats such as Massif de la Hotte being clear and urgent priorities. Based on our own research experience, we know that many endemic species reported for Haiti are only known from old herbarium collections, several dating back to Ekman's exploration days. Likewise, the karst hills ('mogotes') and the dry forests of the Dominican Republic, especially areas of the Barahona Peninsula and the Llanura Costera, are in urgent need of collection (García, pers. comm.; Zanoni 1989). George Proctor (pers. comm.), estimates that less than one-quarter of the karst hills of Jamaica (especially the Cockpit Country) have been explored. There is much left to do in the Caribbean.

## Conservation of the Caribbean flora

The plant diversity of the Caribbean has been subject to centuries of overharvesting and habitat loss. As with so many islands, the Caribbean has both a high level of endemism and a growing human population (Cincotta *et al.*, 2000). There is a long history of extinction in the Caribbean: the region has suffered the extinction of several vertebrates, including six species of endemic macaws, four endemic species of birds, at least two species of reptiles, and unique guilds of endemic rodents, ground sloths, insectivores and an extinct monkey (*Xenothrix*) from Jamaica (Raffaele *et al.*, 1998; Woods, 1990; Pregill & Crother, 1999; Ottenwalder, 2001; Turvey *et al.*, 2007).

Available information on the conservation status of these plants is limited to the few taxa found in the *IUCN Red List of Threatened Species* (IUCN, 2006) and in publications on threatened plant species from Cuba and Jamaica (Borhidi & Muñiz, 1983; Kelly, 1988; Leiva, 1989; Rankin-Rodríguez & Areces-Berazaín, 2003; Berazaín-Iturralde *et al.*, 2005). The 2008 IUCN Red List recognises 163 Cuban species as globally threatened, with 209 similarly listed form Jamaica. Introductory reviews on conservation aspects of Antillean plants are also scattered in several introductory papers on biodiversity of the Caribbean islands (e.g. Howard, 1977; Adams, 1997; Smith *et al.*, 2004b). At least 17 of the unispecific genera are highly threatened (Francisco-Ortega *et al.*, 2008) (e.g. *Asciadium* Griseb., *Doerpfeldia* Urb., *Euchorium* Ekman & Radlk., *Euleria* Urb., *Harnackia* Urb., *Kodalyodendron* Borhidi & Acuña, *Lepturidium* Hitchc. & Ekman, *Microcycas*, *Nodocarpaea* A.Gray, *Salcedoa*, *Phyllacanthus* Hook. f., *Rhodogeron*, *Solonia* Urb., *Stahlia*, *Synapsis*, *Tetrasiphon* and *Tetraperone* Urb.). In addition, several of the genera with four or more species have at least one threatened species each (e.g. *Gyrotaenia* Griseb., *Leptocereus* (A. Berger) Britton & Rose, *Leucocroton*, *Lasiocroton*, *Phialanthus* Griseb. and *Wallenia*). From this study it is clear that more comprehensive examinations of current distribution patterns and the conservation biology of the plant genera endemic to the Caribbean islands are immediate priorities for the hotspot.

Many Caribbean endemic species have very restricted areas of occurrence, and DNA phylogenies are giving new perspectives to conservation biology strategies for the region. For instance, *Rhodogeron* and *Feddea* (Asteraceae) are confined to a few places with serpentine soils in Central and Eastern Cuba, respectively (Liu *et al.*, 2004; Cariaga *et al.*, 2008). These two genera represent unique evolutionary lineages with a high priority for conservation because of their rarity and their phylogenetic patterns (Francisco-Ortega *et al.*, 2008). A study of threatened plants in the Dominican Republic (Peguero & Jimenez, 2008) provides a useful overview, indicating that many threatened taxa have one or two populations only (for example, the cacti *Melocactus prearrupticola* Areces, *Pereskia marcanoi* Areces and *P. quisqueyana* Liogier survive in single populations with less than 100 individuals).

The Caribbean grasses are poorly studied, with nine endemic genera; this family ranks sixth among those with a high number of endemic genera. Poaceae represents a clear example of the many research and conservation gaps that need to be filled in the Caribbean. For example, the status of the monotypic *Triscenia* Griseb. is of great concern as this genus has not been found in the wild in the last 45 years; historically the species was considered common in serpentine areas of eastern Cuba (Catasús-Guerra, 1997, 2002).

Other endemic species have not been found since their original description. Some taxa are rediscovered after extensive periods as 'lost' species; for instance, *Solanum orthacanthum* O. E. Schulz was rediscovered in 1988, 65 years after its description in 1913 (Peguero & Jimenez, 2008). Similarly *Tapura haitiensis* Urban & Ekman was lost to science between its description in 1926 and re-collection in 1982 (Prance, 1984). In their recent phylogenetic study of the Nyctaginaceae, Douglas and Manos (2007) report that despite intensive fieldwork they were unable to rediscover the monotypic Cuban genus *Caribea* Alain.

It is estimated that at least 23 of the flowering plants endemic to Cuba are currently extinct (Berazaín-Iturralde *et al.*, 2005 ; Zona *et al.*, 2007). Clearly, some of the taxa mentioned in our review have not been found in the wild for a long period of time. We believe that additional fieldwork targeting this kind of species is an immediate priority for the region.

The Caribbean hotspot is ranked as the fifth 'hottest' in terms of five major biodiversity factors (Myers *et al.*, 2000; Brooks *et al.*, 2002). Glam *et al.* (2010) identified Cuba as part of a global assessment of projected habitat loss and one of the countries most in need of conservation support. The region has been subject to massive levels of forest loss. While the FAO (2005) records relatively high forest cover for a number of countries (e.g. Cuba 22.2%, Jamaica 31.4% and Dominican Republic 28.4%), the surviving primary forest areas (and the richest refugia for biodiversity) will be significantly smaller, with some countries possessing virtually no primary vegetation, e.g. Haiti (Blair Hedges, 2006). Some vegetation types have been entirely lost; for instance, Greater Antillean lowland rainforest may only survive as fragments in Guantanamo Province, Cuba (Blair Hedges, 2006). While the forest loss has been catastrophic there are tremendously valuable habitat areas surviving. The forest giants within the Caribbean are Cuba, Puerto Rico, Jamaica and the Dominican Republic. They contain the largest tracts of forest left in the region and have the potential for retaining intact ecosystems into the future. Cuba, with 2 348 000 hectares of forest land, has *c.*22% of its land area forested. However, a large proportion of this is planted forestry land.

After the European colonisation the original biota was subject to large-scale clearance for plantations and the first wave of invasives impacted the native biodiversity. Tthese included species with massive impacts on native biodiversity – commensals such as the rat and the pig and introductions such as the mongoose and invasive plant species including *Cinchona* on Jamaica and exotic grasses such as *Imperata* establishing new fire prone savannah communities. Associated with the plantation economy was a boom-and-bust cycle of natural resource utilisation that successively exhausted different resources. Today the timber reserves of the Caribbean have been largely exhausted and tourism, the exploitation of landscapes and culture, has become a dominant land use.

Primary forest habitats are a clear priority for conservation, but the rich dry woodlands and scrub communities that contain many endemics should not be ignored. While conservation in biodiversity hotspots is often driven by endemism (in some cases a relatively recent phenomenon), conservation planning should also take into account the evolutionary history of the area (see Secrest *et al.*, 2007). The Caribbean contains a number of habitats of immense evolutionary value that hold large samples of the region's plant diversity; these evolutionary crucibles are clear conservation priorities. The serpentine vegetation of Cuba is globally unique and represents the product of millions of years of natural selection. The young serpentine soils of Cuba, while accounting for 36% of the serpentine habitat, only hold 14% of the serpentine endemics, while in contrast the older sites (available for plant colonisation for 10–30 Ma) contain 81% of the Cuban serpentine endemics (Reeves *et al.*, 1999). The Caribbean contains relictual areas of seasonally dry tropical forest (SDTF), one of the region's most threatened habitats. SDTF exists as fragments through the neotropics, Caribbean, the Horn of Africa and southwestern Africa. In the Caribbean they are remarkably rich in cacti, euphorbs and, in particular, legumes and appear to have played an important role in the diversification of Caribbean legumes (Lavin and Beyra Matos, 2008). Found through Cuba, Hispaniola, Jamaica and Puerto Rico are spectacular karst uplands or mogotes; these limestone hills are rich in endemic vegetation and often represent the last wild habitats in an increasingly agricultural landscape. The conservation of these karst landscapes is essential both for retaining endemic biodiversity and also for maintaining hydrological systems. It is worth noting that the Massif de la Selle is the major water source for the Haitian capital Port au Prince, while the Massif de la Hotte provides water for the cities of Les Cayes, Port Salut and Tiburon.

While retaining primary habitats will be one goal, this needs to be supplemented by adopting restoration ecology as a tool for conservation. Caribbean forests can be regenerated, either by actively managed restoration or through the spontaneous regrowth of forest that occurs after changes of land use (Rivera *et al.*, 2000; Lugo & Helmer, 2004). These new forests are not as rich or complex as the originals but they should be regarded as ecological investments building biodiversity, soils, watersheds and carbon reserves over time. Restoration and conservation corridors are now being built into regional land use planning, for instance the Spinal Forest initiative of the Environmental Foundation of Jamaican (EFJ). Some islands have a proven success in restoring and growing forest resources, most notably Puerto Rico; others, such as Haiti, mired in extreme poverty and political unrest, will suffer the continued loss of forest and the irreplaceable biodiversity they hold.

In 2006 the Organization of Eastern Caribbean States (OECS) updated the 'St George's Declaration of Principles for Environmental Sustainability', stating that the OECS is 'firmly convinced that the wellbeing of humanity depends upon preserving a healthy environment with all its ecological systems, a rich variety of plants and animals, fertile soils, pure water and clean air'. The conservation of Caribbean plant diversity should be recognised as a key contributor to sustainable development; the forests of the Caribbean are key elements in creating liveable landscapes, protecting communities from catastrophic floods and landslides, protecting watersheds and furnishing communities with timber, medicinals and potentially ecotourism income.

While there are many challenges facing plant conservationists in the Caribbean (see Maunder *et al.*, 2008), including the urgent need to build capacity for botany and conservation and to develop tools for species recovery and management of invasives, we argue that the greatest challenge will be the retention of extensive areas of wild habitats. These habitats, economically important as slope stabilisers and watersheds, will be the refuges for the Caribbean's biodiversity. If large enough, and connected through habitat types and altitude, they will offer migration opportunities as climate change (Malcolm *et al.*, 2006) is added to the long list of challenges facing the Caribbean's biodiversity.

## Acknowledgements

We dedicate this paper to two extraordinary Caribbean botanists, George Proctor and Milciades Mejia. Both gentlemen have selflessly committed their careers to Caribbean botany, the mentoring of young botanists and the conservation of Caribbean plants and habitats. This is contribution number 185 of the Tropical Biology Program of Florida International University.

## References

Acevedo-Rodríguez, P. & Strong, M. T. (2008). Floristic richness and affinities in the West Indies. *Bot. Rev.* 74 (1): 5–36.

Adams, C. D. (1997). Regional overview: Caribbean Islands. In S. D. Davis, V. H. Heywood, O. Herrera-MacBryde, J. Villa-Lobos & A. C. Hamilton (eds.), *Centres of Plant Diversity: A Guide and Strategy for their Conservation. Vol. 3. The Americas*, pp. 233–258. Cambridge, WWF & IUCN, IUCN Publications Unit.

Anderberg, A. A., Eldenäs, P., Bayer, R. J. & Englund, M. (2005). Evolutionary relationships in the Asteraceae tribe Inuleae (inc. Plucheeae) evidenced by DNA sequences of ndhF; with notes on the systematic positions of some aberrant genera. *Organisms Divers. Evol.* 5: 135–146.

Areces-Mallea, A., Weakley, A. S., Li, X., Sayre, R. G., Parrish, J. D., Tipton, C. V. *et al.* (1999). *A Guide to Caribbean Vegetation Types: Preliminary Classification System and Descriptions.* Washington, DC, Nature Conservancy.

Baldwin, B. G., Wesa, B. L. & Panero, J. L. (2002). Nuclear rDNA evidence for major lineages of Helenioid Heliantheae (Compositae). *Syst. Bot.* 27: 161–198.

Barrera-Osorio, A. (2006). *Experiencing Nature: The Spanish American Empire and the Early Scientific Revolution.* Austin, University of Texas Press.

Berazaín-Iturralde, R., Areces-Berazaín, F., Lazcano-Lara, J. C. & González-Torres, L. R. (2005). *Lista roja de la flora vascular cubana. Documentos 4.* Gijón, Spain, Jardín Botánico Atlántico de Gijón.

Berry, P. E., Hipp, A. L., Wurdack, K. J., van Ee, B. & Riina, R. (2005). Molecular phylogenetics of the giant genus *Croton* and tribe Crotoneae (Euphorbiaceae *sensu stricto*) using ITS and TRNL-TRNF DNA sequence data. *Am. J. Bot.* 92: 1520–1534.

Blair Hedges, S. (2006). An overview of the evolution and conservation of West Indian amphibians and reptiles. *Applied Herpetology* 3: 281–292.

Bogler, D. J. & Francisco-Ortega, J. (2004). Molecular systematic studies in cycads: evidence from trnL intron and ITS2 rDNA sequences. *Bot. Rev.* 70: 260–273.

Borhidi, A. (1991). *Phytogeography and Vegetation Ecology of Cuba.* Budapest, Hungary, Akadémiai Kiadó.

Borhidi, A. & Muñiz, O. (1983). *Catálogo De Plantas Cubanas Amenazadas O Extinguidas.* Havana, Cuba, Editorial Academia.

Brooks, T. M., Mittermeier, R. A., Mittermeier, C. G., da Fonseca, G. A. B., Rylands, A. B., Konstant, W. R. *et al.* (2002). Habitat loss and extinction in the hotspots of biodiversity. *Conserv. Biol.* 16: 909–923.

Cariaga, K. A., Pruski, J. F., Oviedo, R., Anderberg, A. A., Lewis, C. E. & Francisco-Ortega, J. (2008). Phylogeny and systematic position of *Feddea* (Asteraceae: Feddeeae): a taxonomically enigmatic and critically endangered genus endemic to Cuba. *Syst. Bot* 33: 193–202.

Catasús-Guerra, L. (1997). Las gramíneas (Poaceae) de Cuba, I. *Fontqueria* 46: 1–259.

Catasús-Guerra, L. (2002). Las gramíneas (Poaceae) de Cuba, II. *Cavanillesia* Alt. 3: 1–163.

Chaw, S.-M., Walters T. W., Chang, C.-C., Hu, S-H. & Chen, S-H. (2005). A phylogeny of cycads (Cycadales) inferred from chloroplast mstK gene, trnK intron, and nuclear rDNA ITS region. *Molec. Phylogenet. Evol.* 37: 214–234.

Cincotta, R. P., Wisnewski, J. & Engelman, R. (2000). Human population in the biodiversity hotspots. *Nature* 404: 990–992.

Cornejo, X. & Ildis, H. H. (2009). *Hispaniolanthus*. A new genus of Capparaceae endemic to Hispaniola. *Harvard Pap. Bot.* 14: 9–14.

Crosby, A. W. (2005). *The Columbian Exchange: biological and cultural consequences of 1492*. Praeger, Conneticut.

Crother, B. I. & Guyer, C. (1996). Caribbean historical biogeography: was the dispersal-vicariance debate eliminated by an extraterrestrial bolide? *Herpetologica* 52: 440–465.

Douglas, N. A. & Manos, P. S. (2007). Molecular phylogeny of Nyctaginaceae: taxonomy, biogeography, and characters associated with a radiation of xerophytic genera in North America. *Am. J. Bot.* 94: 856–872.

Duellman, W. E. (1999). The West Indies and Middle America: contrasting origins and diversity. In B. I. CROWTHER (ed.), *Caribbean Amphibians and Reptiles*, pp. 357–369. Academic Press, San Diego, CA.

Eken, G., Bennun L., Brooks, T.M., Darwall, W., Fishpool, L. C. D., Foster, M. *et al.* (2004). Key biodiversity areas as site conservation targets. *BioScience* 54: 1110–1118.

FAO (2005). *Global Forest Resources Assessment 2005: Progress Towards Sustainable Forest Management*. Rome, United Nations Food and Agriculture Organization.

Francisco-Ortega, J., Santiago-Valentín, E., Acevedo-Rodríguez, P., Lewis, C. E., Pipoly, J., III, Meerow, A.W. *et al.* (2007). Seed plant genera endemic to the Caribbean Island biodiversity hotspot: a review and a molecular phylogenetic perspective. *Bot. Rev.* 73: 183–234.

Francisco-Ortega, J., Ventosa, I, Oviedo, R., Jimenez, F., Herrera, P., Maunder, M. *et al.* (2008). Caribbean Island Asteraceae: systematics, molecules, and conservation on a biodiversity hotspot. *Bot. Rev.* 74: 112–131.

Fritsch, P. W. & McDowell, T. D. (2003). Biogeography and phylogeny of Caribbean plants: introduction. *Syst. Bot.* 28: 376–377.

Gentry, A. H. (1982). Neotropical floristic diversity: phytogeographical connections between Central and South America, Pleistocene climatic fluctuations, or an accident of the Andean orogeny? *Ann. Mo. Bot. Gard.* 69: 557–593.

Glam, X., Bradshaw, C. J. A., Tan, H. T. W. & Sochi, N. S. (2010). Future habitat loss and the conservation of plant biodiversity. *Biological Conservation*. 143: 1594–1602.

Graham, A. (2003a). Historical phytogeography of the Greater Antilles. *Brittonia* 55: 357–383.

Graham, A. (2003b). Geohistorical models and Cenozoic paleoenvironments of the Caribbean region. *Syst. Bot.* 28: 378–386.

Graham, A., Cozadd, D, Areces-Mallea, A. & Frederiksen, N. O. (2000). Studies in Neotropical paleobotany. XIV. A palynoflora from the Middle Eocene Saramaguacán formation of Cuba. *Am. J. Bot.* 87: 1526–1539.

Grove, R. H. (1995). *Green Imperialism: Colonial Expansion. Tropical Island Edens and the Origins of Environmentalism, 1600–1860*. Cambridge, Cambridge University Press.

Hammel, B. E. & Zamora, N. A. (2005). *Pleodendron costaricense* (Canellaceae), a new species for Costa Rica. *Lankesteriana* 5: 211–218.

Hass, C. A., Maxson, L. R. & Hedges, S. B. (2001). Relationships and divergence times of West Indian amphibians and reptiles: insights from albumin immunology. In C. A. Woods & F. E. Sergile (eds.), *Biogeography of the West Indies: Patterns and Perspectives*, 2nd edition, pp. 157–174. Boca Raton, FL, CRC Press.

Hedges, S. B. (1996). Historical biogeography of West Indian vertebrates. *Ann. Rev. Ecol. Syst.* 27: 163–196.

Hedges, S. B. (2001). Biogeography of the West Indies: an overview. In C. A. Woods & F. E. Sergile (eds.), *Biogeography of the West Indies: Patterns and Perspectives*, 2nd edition, pp. 15–33. Boca Raton, FL, CRC Press.

Hermsen, E. J., Taylor, T. N., Taylor, E. L. & Stevenson, D. W. (2006). Cataphylls of the middle Triassic cycad Antarcticycas schopfii and new insights into cycad evolution. *Am. J. Bot.* 93: 724–738.

Howard, R. A. (1973). The vegetation of the Antilles. In A. Graham (ed.), *Vegetation and Vegetational History of Northern Latin America*, pp. 1–38. Amsterdam, Elsevier.

Howard, R. A. (1977). Conservation and the endangered species of plants in the Caribbean Islands. In G. T. Prance & T. S. Elías (eds.), *Extinction is Forever*, pp. 105–114. New York, The New York Botanical Garden.

Howard, R. A. (1996). Ignatz Urban and the 'Symbolae Antillanae'. *Flora Greater Antilles Newsletter* 10. www.nybg.org/bsci/fga/Newsletter/FGANno10.htm

Iturralde-Vinent, M. A. (2004–2005). La paleogeografía del Caribe y sus implicaciones para la biogeografía histórica. *Revista Jard. Bot. Nac. Univ. Habana* 25–26: 49–78.

Iturralde-Vinent, M. A. & Macphee, R. D. E. (1999). Paleogeography of the Caribbean region: implications. *Bull. Am. Mus. Nat. Hist.* 238: 1–95.

IUCN (2001). IUCN red list categories and criteria: Version 3.1. IUCN Species Survival Commission. Gland, Switzerland and Cambridge, UK, IUCN. www.redlist.org/info/categories_criteria2001.

IUCN (2006). 2006 IUCN red list of threatened species. www.redlist.org

Jestrow, B., Jiménez, F. & Francisco-Ortega, J. (2010). Generic Limitations in the Antillean *Adelieae* (Euphorbiaceae) with the description of the Hispaniolan endemic genus Garciadelia. *Taxon* 59: 1801–1814.

Jiménez-Rodriguez, F., Katinas, L., Telleria, M. C. & Crisci, J. V. (2004). *Salcedoa* gen. nov. a biogeographic enigma in the Caribbean Mutisieae (Asteraceae). *Syst. Bot.* 29: 987–1002.

Kelly, D. L. (1988). The threatened flowering plants of Jamaica. *Biol. Conserv.* 46: 201–216.

Kier, G., Kreft, H., Lee, T. M., Jetz, W., Ibisch, P. L., Nowicki, C. *et al.* (2009). A global assessment of endemism and species richness across island and mainland regions. *Proc. Natl. Acad. Sci. USA* 106: 9322–9327.

King, R. M. & Robinson, H. (1987). The genera of the Eupatorieae (Asteraceae). *Monogr. Syst. Bot. Mo. Bot. Gard.* 22: 1–581.

Kimball, R. T. & Crawford, D. J. (2004). Phylogeny of Coreopsidae (Asteraceae) using ITS sequences suggests lability in reproductive characters. *Molec. Phylogenet. Evol.* 33: 127–139.

Kreft, H., Jetz, W., Mutke, J., Kier, G. & Barthlott, W. (2008). Global diversity of island floras from a macroecological approach. *Ecol. Lett.* 11: 116–127.

Kress, W. J. & Horvitz, C. (2005). Habitat alteration in the Caribbean: natural and human induced. In G. A. Krupnick & W. J. Kress (eds.), *Plant Conservation: A Natural History Approach*, pp. 147–151. Chicago, University of Chicago Press.

Lavin, M. & Beyra Matos, A. (2008). The impact of ecology and biogeography on legume diversity, endemism, and phylogeny in the Caribbean region: A new direction in historical biogeography. *Bot. Rev.* 74 (1): 178–196.

Lavin, M., Wojciechowski, M. F., Richman, A., Rotella, J., Sanderson, M. J. & Beyra-Matos, A. (2001a). Identifying Tertiary radiations of Fabaceae in the Greater Antilles: alternatives to cladistics vicariance analysis. *Int. J. Plant Sci.* 162: S53–S76.

Lavin, M., Pennington, R. T., Klitgaard, B. B., Sprent, J. I., Cavalcante-de-Lima, H. & Gasson, P. E. (2001b). The dalbergioid legumes (Fabaceae): delimitation of a pantropical monophyletic clade. *Am. J. Bot.* 88: 503–533.

Lavin, M., Wojciechowski, M. F., Gasson, P. E., Hughes, C. & Wheeler, E. (2003). Phylogeny of Robinioid legumes (Fabaceaea) revisited: *Coursetia* and *Gliricidia* recircumscribed, and a biogeographical appraisal of the Caribbean endemics. *Syst. Bot.* 28: 387–409.

Leiva, A. (1989). Notes on the history of botanical exploration and conservation of nature in Cuba. In D. G. Campbell & H. D. Hammond (eds.), *Floristic Inventory of Tropical Countries: The Status of Plant Systematics, Collections, and Vegetation, Plus Recommendations for the Future*, pp. 327–335. New York, The New York Botanical Garden.

Liu, H., Trusty, J., Oviedo, R., Anderberg, A. & Francisco-Ortega, J. (2004). Molecular phylogenetics of the Caribbean genera *Rhodogeron* and *Sachsia* (Asteraceae). *Int. J. Plant Sci.* 165: 209–217.

Loockerman, D. J., Turner, B. L. & Jansen, R. K. (2003). Phylogenetic relationships within the Tageteae (Asteraceae) based on nuclear ribosomal ITS and chloroplast ndhF gene sequences. *Syst. Bot.* 28: 191–207.

Lugo, A. E. & Helmer, E. (2004). Emerging forests on abandoned land: Puerto Rico's new forests. *Forest Ecol. Manag.* 190 (2–3): 145–161.

Malcolm, J. R., Liu, C. R., Neilson, R. P., Hansen, L., & Hannah, L. (2006). Global warming and extinctions of endemic species from biodiversity hotspots. *Cons. Biol.* 20: 538–548.

Maunder, M., Culham, A. & Hankamer, C. (1998). Picking up the pieces: botanical conservation on degraded oceanic islands. In P. L. Fiedler & P. M. Kareiva (eds.), *Conservation Biology*, 2nd edition, pp. 317–344. New York, Chapman & Hall.

Maunder, M., Page, W., Mauremootoo, J., Payendee, R., Mungroo, Y., Maljkovic, A. *et al.* (2002). The decline and conservation management of the threatened endemic palms of the Mascarene Islands. *Oryx* 36: 56–65.

Maunder, M., Leiva, A., Santiago-Valentin, E., Stevenson, D.W., Acevedo-Rodriguez, P., Meerow, A. *et al.* (2008). Plant conservation in the Caribbean biodiversity hotspot. *Bot. Rev.* 74 (1): 197–207.

Mejía, M., García, R. & Rodríguez, S. (2001). Datos biográficos de Erik Leonard Ekman. *Moscosoa* 12: 1–3.

Michelangeli, F. A., Judd, W. S., Penneys, D. S., Skean, J. D., Bequer-Granados, E. R., Goldenberg, R. *et al.* (2008). Multiple events of dispersal and radiation in the tribe Miconieae (Melastomataceae) in the Caribbean. *Bot. Rev.* 74 (1): 53–77.

Motley, T. J., Wurdack, K. J. & Delprete, P. G. (2005). Molecular systematics of the Catesbaeeae-Chiococceae complex (Rubiaceae): flower and fruit evolution and biogeographic implications. *Am. J. Bot.* 92: 316–329.

Myers, N., Mittermeier, R. A., Mittermeier, C. G., da Fonseca, G. A. B. & Kent, J. (2000). Biodiversity hotspots for conservation priorities. *Nature* 403: 853–858.

Nordenstam, B. (1978). Taxonomic studies in the tribe Senecioneae (Compositae). *Opera Bot.* 44: 1–83.

Nordenstam, B. (2006a). New genera and combinations in the Senecioneae of the Greater Antilles. *Compositae Newslett.* 44: 50–73.

Nordenstam, B. (2006b). *Ignurbia*, a new genus of the Asteraceae-Senecioneae from Hispaniola. *Willdenowia* 36: 463–468.

Nordenstam, B. & Lundin, R. (2002). *Oldfeltia*, a new genus of the Compositae-Senecioneae from Cuba. *Compositae Newslett.* 38: 65–70.

Nunn, P. D. (1994). *Oceanic Islands.* Oxford, Blackwell.

Olmstead, R. G., Bohs, L., Migid, H. A., Santiago-Valentin, E., Garcia, V. F. & Collier, S. M. (2008). A molecular phylogeny of the Solanaceae. *Taxon* 57 (4): 1159–1181.

Olsen, D. & Dinerstein, E. (1998). The Global 200: a representation approach to conserving the earth's most biologically valuable ecoregions. *Conserv. Biol.* 12: 502–515.

Ottenwalder, J. A. (2001). Systematics and biogeography of the West Indian genus *Solenodon*. In C. A. Woods & F. E. Sergile (eds.), *Biogeography of the West Indies: Patterns and Perspectives*, 2nd edition, pp. 253–329. Boca Raton, FL, CRC Press.

Peguero, B. & Jimenez, F. (2008). Inventario preliminary de plantas endemicas locales en peligro de extinction en la Republica Dominicana. *Moscosoa* 16: 84–94.

Pelser, P. B., Nordenstam, B., Kadereit, J. W. & Watson, L. E. (2007). An ITS phylogeny of tribe Senecioneae (Asteraceae) and a new delimitation of Senecio L. *Taxon* 56: 1077–1104.

Petersen, R. H. (2001). *New World Botany: Columbus to Darwin.* Ruggel, Liechtenstein, Gantner Verlag.

Pindell, J. J. & Kennan, L. (2001). Kinematic evolution of the Gulf of Mexico and Caribbean. In R. H. Fillon, N. C. Rosen, P. Weimer, *et al.* (eds.), *Petroleum Systems of Deep Water Basins: Global and Gulf of Mexico Experience. Proceedings Gulf Coast Section, Society for Sedimentary Geology (SEPM), 21st Annual Bob F. Perkins Research Conference*, 2–5 December, pp. 193–220. Houston, TX, Society for Sedimentary Geology (SEPM).

Pindell, J., Kennan, L., Maresch, M. V., Stanek, K.-P., Draper, G. & Higgs, R. (2005). Plate-kinematics and crustal dynamics of circum-Caribbean arc-continent interactions and tectonic controls on basin development. In H. G. Avé Lallemant & V. B. Sissons (eds.), Proto-Caribbean margins, in Caribbean–South American plate interactions, Venezuela. *Special Pap. Geol. Soc. Am.* 394: 7–52.

Prance, G. T. (1984). The rediscovery of *Tapura haitiensis* (Dichapetalaceae). *Brittonia* 36 (1): 15–17.

Pregill, G. K. & Crother, B. I. (1999). Ecological and historical biogeography of the Caribbean. In B. I. Crother (ed.), *Caribbean Amphibians and Reptiles*, pp. 335–356. San Diego, CA, Academic Press.

Raffaele, H., Wiley, J., Garrido, O., Keith, A. & Raffaele, J. (1998). *A Guide to the Birds of the West Indies.* Princeton, NJ, Princeton University Press.

Rankin-Rodríguez, R. & Areces-Berazain, F. (2003). Contribución a la actualización taxonómica y localización geográfica de especies amanazadas y endémicas en Cuba I. *Revista Jard. Bot. Nac. Univ. Habana* 24: 81–128.

Reeves, R. D., Baker, A. J. M., Borhidi, A. & Berazain, R. (1999). Nickel hyperaccumulation in the serpentine flora of Cuba. *Ann. Bot.* 83: 29–38.

Richardson, J. E., Fay, M. F., Cronk, Q. C. B., Bowman, D. & Chase, M. W. (2000). A phylogenetic analysis of Rhamnaceae using rbcL and trnL-F plastid DNA sequences. *Am. J. Bot.* 87: 1309–1324.

Rivera, L. W., Zimmerman, J. K., & Aide, T. M. (2000). Forest recovery in abandoned agricultural lands in a karst region of the Dominican Republic. *Plant Ecol.* 1248: 115–125.

Roalson, E. H., Boggan., J. K., Skog, L. E. & Zimmer, E. A. (2005). Untangling Gloxiniae (Gesneriaceae). I. Phylogenetic patterns and generic boundaries inferred from nuclear, chloroplast, and morphological cladistics datasets. *Taxon* 54: 389–410.

Roca, A. G., Bar-Gal, K., Elzirik, E., Helgen, K. M., María, R., Springer, M. S. *et al.* (2004). Mesozoic origin for the West Indian insectivores. *Nature* 429: 649–651.

Rodrigues, A. S. L., Akcakaya, H. R., Andelman, S. J., Bakarr, M. I., Boitani, L., Brooks, T. M. *et al.* (2004). Global gap analysis: priority regions for expanding the global protected-area network. *BioScience* 54: 1092–1100.

Roncal, J., Zona, S. & Lewis, C. E. (2008). Molecular phylogenetic studies of Caribbean palms (Arecaceae) and their relationships to biogeography and conservation. *Bot. Rev.* 74 (1): 78–102.

Rowley, G. D. (1997). *A History of Succulent Plants.* Mill Valley, CA, Strawberry Press.

Salazar, J. & Nixon, K. (2008). New discoveries in the Canellaceae in the Antilles: How phylogeny can support taxonomy. *Bot. Rev.* 74 (1): 103–111.

Samain, M-S., Mathieu, G., Wanke, S., Neinhuis, C. & Goetghebeur, P. (2008). *Verhuellia* revisited – unravelling its intricate taxonomic history and a new subfamilial classification of *Piperaceae. Taxon* 57: 583–587.

Santiago-Valentín, E. & Francisco-Ortega, J. (2008). Plant evolution and biodiversity in the Caribbean Islands-prspectives from molecular markers. *Bot. Rev.* 74 (1): 1–4.

Santiago-Valentín, E. & Olmstead, R. G. (2003). Phylogenetics of the Antillean Goetzeoideae (Solanaceae) and their relationships within the Solanaceae based on chloroplast and ITS DNA sequence data. *Syst. Bot.* 28: 452–460.

Santiago-Valentín, E. & Olmstead, R. G. (2004). Historical biogeography of Caribbean plants: introduction to current knowledge and possibilities from a phylogenetic perspective. *Taxon* 53: 299–319.

Savage, J. M. (1982). The enigma of the Central American herpetofauna: dispersals or vicariance? *Ann. Mo. Bot. Gard.* 69: 464–547.

Schaefer, H., Kocyan, A. & Renner, S. S. (2008) *Linnaeosicyos* (Cucurbitaceae): A new genus for *Trichosanthes amara*, the Caribbean sister species of all Sicyeae. *Syst. Bot.* 33: 349–355.

Secrest, W., Brooks, T. M., da Fonseca, G. A., Konstant, W. R., Mittermeier, R. A., Purvis, A. *et* al. (2007). Hotspots and the conservation of evolutionary history. *Proc. Natl. Acad. Sci. USA* 99 (4): 2067–2071.

Sherff, E. E. & Alexander, E. J. (1955). Compositae-Heliantheae-Coreopsidinae. *N. Am. Fl.* II 2: 1–149.

Smith, M. L., Hedges, S. B., Buck, W., Hemphill, A., Inchaustegui, S., Ivie, M. A. *et al.* (2004a). Caribbean Islands. In R. A. Mittermeier, R. R. Gil, M. Hoffman, *et al.* (eds.), *Hotspots Revisited: Earth's Biologically Richest and Most Threatened Terrestrial Ecoregions*, pp. 112–118. CEMEX, Mexico DF.

Smith, N., Mori, S. A., Henderson, A., Stevenson, D. W. & Heald, S. V. (eds.) (2004b). *Flowering Plants of the Neotropics*. Princeton, NJ, Princeton University Press.

Soltis, D. E., Soltis, P. S., Endress, P. K. & Chase, M. W. (2005). *Phylogeny and Evolution of Angiosperms*. Sunderland, MA, Sinauer Associates.

Ståhl, B. (2004). *Theophrastaceae*. In K. Kubitzki (ed.), *The Families and Genera of Vascular Plants. Vol. VI. Flowering Plants. Dicotyledons. Celastrales, Oxalidales, Rosales, Cornales, Ericales*, pp. 472–478. Berlin, Springer.

Stevenson, D. W. (1992). A formal classification of the extant cycads. *Brittonia* 44: 220–223.

Takhtajan, A. (1997). *Diversity and Classification of Flowering Plants*. New York, Columbia University Press.

Turvey, S. T., Oliver, J. R., Narganes Storde, Y. M. & Rye, P. (2007). Late Holocene extinction of Puerto Rican native land mammals. *Biol. Lett.* 3 (2): 193–196.

Urban, I. (1915). Sertum antillanum II. *Repert. Spec. Nov. Regni Veg.* 13: 465–484.

van Ee, B. W., Berry, P. E., Riina, R., & Gutierrez Amaro, J. E. (2008). Molecular phylogenetics and biogeography of the Caribbean centred *Croton* subgenus Moacroton (Euphorbiaceae s.s.). *Bot. Rev.* 74 (1): 132–165.

van den Berg, C., Higgins, W. E., Dressler, R. L., Whitten, W. M., Soto-Arenas, M. A., Culham, A. *et al.* (2000). A phylogenetic analysis of Laeliinae (Orchidaceae) based on sequence data from Internal Transcribed Spacers (ITS) of nuclear ribosomal DNA. *Lindleyana* 15: 96–114.

von Wettstein, R. (1895). Solanaceae. In A. Engler & K. Prantl (eds.), *Die Naturlichen Pflanzenfamilien* 4 (3b): 4–38. Leipzig, Engelmann.

Wanke, S., Jaramillo, M. A., Borsch, T., Samain, M. S., Quandt, D. & Neinhuis, C. (2007). Evolution of Piperales – matK gene and trnK intron sequence data reveal lineage specific resolution contrast. *Molec. Phylogenet. Evol.* 42: 477–497.

Wojciechowski, M. F., Lavin, M. & Sanderson, M. J. (2004). A phylogeny of legumes (Leguminosae) based on analysis of the plastid matK gene resolves many well-supported subclades within the family. *Am. J. Bot.* 91: 1846–1862.

Woods, C. A. (1990). The fossil and recent land mammals of the West Indies: and analysis of the origin, evolution, and, extinction of an insular fauna. *Atti Convegni Lincei* 85: 641–680.

Zanoni, T. A. (1989). Hispaniola. In D. G. Campbell & H. D. Hammond (eds.), *Floristic Inventory of Tropical Countries: The Status of Plant Systematics, Collections, and Vegetation, Plus Recommendations for the Future*, pp. 337–346. New York, The New York Botanical Garden.

Zona, S., Verdecia, R., Leiva-Sánchez, A., Lewis, C. E. & Maunder, M. (2007). Conservation status of West Indian palms (Arecaceae). *Oryx* 41: 300–305.

# 7

# The biogeography of Madagascar palms

JOHN DRANSFIELD AND MIJORO RAKOTOARINIVO

Madagascar has an extraordinary palm flora, diverse in species and evolutionary lines, a diversity that poses interesting evolutionary questions. The palms in Madagascar are in stark contrast to two other families of largely tropical monocotyledons, Zingiberaceae and Araceae, where the Madagascar representation is low in diversity despite being high in endemism. The contrast between the palm flora of Madagascar and that of the whole of continental Africa is remarkable, the former with over three times the number of species as the latter. With the discovery of yet more new palms in recent fieldwork the contrast is set to become even more striking. The phylogenetic relationships of Madagascar palms are complex, with groups variously linked to related taxa in differing parts of the tropical world, while the relationships of the most diverse group of Madagascar palms, subtribe Dysidinae, remain obscure.

There has been intense interest in the Madagascar palm flora over the last 25 years and yet further new taxa continue to be discovered, suggesting that we are still some way from a complete cataloguing of the flora. To put our discussion into context, some description of the history of palm exploration on the island seems necessary.

## History of palm exploration in Madagascar

The first herbarium specimens of palms from Madagascar collected by the early French naturalists Commerson and du Petit Thouars reached Europe in the late eighteenth and early nineteenth century. These were collections of the genus *Dypsis* and were duly described by the German palm botanist, von Martius (1823–50). Gradually more and more collections were made, by naturalists and missionaries,

*The Biology of Island Floras*, eds. David Bramwell and Juli Caujapé-Castells. Published by Cambridge University Press. 

and described piecemeal by botanists such as Wendland, Baker and Baillon. By the early 1900s the Italian palm botanist Odoardo Beccari began a grand synthesis, describing genera and species from Madagascar, based largely on material accumulating in the Paris Herbarium, and in the herbarium at the Royal Botanic Gardens, Kew, resulting in the large folio monograph, *Palme del Madagascar* (Beccari, 1912–14). This highly detailed monograph was, however, very soon out of date, the continuing botanical exploration of the island largely by French naturalists enriching the Paris Herbarium. Perrier de la Bâthie, with substantial field experience in Madagascar, described many of the novelties together with Henri Jumelle, culminating in 1945 in the publication of the fascicle of the *Flore de Madagascar et des Comores* devoted to palms (Jumelle & Perrier de la Bâthie, 1945). Perhaps surprisingly, there remained many good collections in the Paris Herbarium, not annotated by Perrier and Jumelle, including material of undescribed taxa that were never considered in the production of the flora account.

The most significant post-Second World War collectors of palm material were Humbert and Guillaumet, the former penetrating remote parts of the island previously unexplored, and returning with astonishing material, including the first collections of the genus *Marojejya*, which he duly described (Humbert, 1955a). The promised description of further novelties never occurred and large quantities of Humbert collections remained unnamed in the Paris Herbarium. Harold E. Moore, working towards a complete survey of the world's palm genera, visited Madagascar in the early 1970s, looking for material of all the endemic genera, but with only partial success. The mid to late 1970s saw an increasingly unwelcoming political situation in Madagascar, preventing foreign scientists conducting fieldwork. By the mid 1980s Madagascar was once again open to foreign researchers and intensive fieldwork began on the palms based at Royal Botanic Gardens, Kew. On the first major expedition in 1986 all the endemic genera were rediscovered together with material of two new genera and many new species. This productive period of palm fieldwork culminated in 1995 in the publication of a full taxonomic monograph of Madagascar palms (Dransfield & Beentje, 1995b). In it over 70 of the *c.*170 species treated were described as new. Since then new taxa have continued to be discovered and described and reports of over 20 more are in press or preparation. It would seem that any visit to an area previously not sampled will yield new taxa. For example, a visit by one of us (MR) to Andilamena in 2006 produced material of a new species of *Ravenea*, remarkable for its small size, dichotomously branching stems and plumose leaflets, characters not otherwise associated with the genus (Rakotoarinivo, 2008a), and a second climbing species of *Dypsis* (the first species, *D. scandens*, was itself not discovered until 1994).

Most of the new species discovered in recent years belong to existing genera but in the last 20 years four new genera have been described, one as late as 2008 and representing an evolutionary line new to the island. *Tahina spectabilis*, a massive hapaxanthic palm, belongs to Chuniophoeniceae, a tribe previously unknown in Madagascar (Dransfield *et al.*, 2008a).

It is difficult to predict how many more taxa will be discovered. There remain significant areas of the island (e.g. the Tsaratanana massif) that are still under-represented for palms in the herbarium, but even forest fragments in the east coast lowlands are likely to yield further novelties.

At the same time as the scientific exploration of the palm flora, commercial and amateur palm seed collectors have made numerous collections of palm seed to feed the almost obsessive interest in the palms of Madagascar among amateur growers. Indeed, the first intimations of two newly described genera, *Voanioala* and *Lemurophoenix*, came from Mardy Darian, an amateur grower who visited the island through the difficult days of the 1970s and early 1980s. Vast quantities of seed have been harvested and distributed over the last two decades. Sometimes the seed has been correctly identified, but all too often the seed is exported identified with local names, impossible to match with confidence to scientific names. Some of the earlier introductions have reached maturity and attempts can then be made to identify them, taking into account the fact that palms in good cultivation may look very different from how the same species appears in the wild. Among the recent introductions have been distinct new taxa and four species have been described from cultivation, initially with no idea whence they came (Dransfield & Marcus, 2002; Hodel & Marcus 2004; Hodel *et al.* 2005, 2009). Two of these, *Dypsis carlsmithii* and *D. robusta*, have since been found in the wild.

## Composition of the Madagascar palm flora

What is most striking about the palm flora of Madagascar is the very high level of endemism at both species and generic levels (see Table 7.1). Seven species, widespread in Madagascar, are not endemic. While the coconut (*Cocos nucifera*) and date palm (*Phoenix dactylifera*) are clearly introduced by humans, always occurring in association with cultivation, the status of the remaining non-endemic species is less clear. *Raphia farinifera*, regarded as almost emblematic of Malagasy rural life (with a generic name based on a Malagasy word, *rofia*) is almost certainly an early introduction by humans (see below). *Borassus aethiopum* is widespread in tropical Africa. In Madagascar it is restricted to the northwest where it was previously known as *B. sambiranensis* (see Bayton, 2007) and thought to be endemic. As with *Raphia farinifera* the palm does not occur in natural vegetation

Table 7.1 *Palm genera in Madagascar and their endemicity.*

| Genus (total no. of species worldwide in parentheses) | No. of species in Madagascar (endemic in parentheses) | Extra-Madagascar distribution |
|---|---|---|
| *Beccariophoenix* (2) | 2 (2) | — |
| *Bismarckia* (1) | 1 (1) | — |
| *Borassus* (6) | 2 (1) | Widespread in tropical Africa, southern and southeast Asia to New Guinea |
| *Cocos* (1) | 1 | Pantropic in cultivation |
| *Dypsis* (165) | 161 (161) | Comores, Pemba |
| *Elaeis* (2) | 1 | Widespread in tropical Africa, Central America and northern South America |
| *Hyphaene* (8) | 1 | Africa, Arabia to India and Sri Lanka |
| *Lemurophoenix* (1) | 1 (1) | — |
| *Marojejya* (2) | 2 (2) | — |
| *Masoala* (2) | 2 (2) | — |
| *Orania* (25) | 3 (3) | South Thailand through Malesia to New Guinea |
| *Phoenix* (14) | 2 | Canaries, Cape Verde Islands, Mediterranean, Africa, Arabia, through Indian subcontinent to east Asia and western Malesia |
| *Raphia* (20) | 1 | Tropical Africa, central and eastern South America |
| *Ravenea* (20) | 8 (8) | Comores |
| *Satranala* (1) | 1 (1) | — |
| *Tahina* (1) | 1 (1) | — |
| *Voanioala* (1) | 1 (1) | — |

but is found near habitation. *Hyphaene coriacea* is very widespread in savanna and coastal vegetation in the west of the island and it seems likely that it is indigenous. It is possible, also, that *Phoenix reclinata* is native, as it is unlikely that the palm would have been introduced for any use. Similarly wild-type African oil palm, *Elaeis guineensis*, could well have reached Madagascar without the aid of humans, and so may be indigenous.

The remaining more than 166 species of palm in Madagascar are endemic. As mentioned above, approximately 20 further species have been recognised and are currently being prepared for publication, making a total of 188 species (Rakotoarinivo & Dransfield, 2010).

## Endemicity at the generic level

At the generic level, eight out of 17 genera are endemic. Two of the non-endemic genera, *Dypsis* and *Ravenea*, are endemic to the islands of the Western Indian Ocean; *Ravenea* is restricted to Madagascar and the Comores and *Dypsis* to Madagascar, the Comores and Pemba off the coast of Tanzania. The non-endemic genus with the most unusual distribution is *Orania*, with three species in Madagascar and the remaining 22 species in Malesia and southern Thailand.

## Subfamilial representation

Four of the five currently recognised palm subfamilies are present in Madagascar (Dransfield *et al.*, 2008b); only the monotypic Nypoideae is absent and there is no fossil evidence to suggest that it was ever present on the island. With over 170 species, Madagascar has a rich palm flora, including two substantial radiations, yet the flora is as remarkable for what is not there as for the diversity that has evolved.

### *Calamoideae*

Calamoideae, a diverse and very widely distributed subfamily that includes the rattan palms, so characteristic of the equatorial forest of West Africa, Asia and Malesia, is represented by a single species of the genus *Raphia*. *Raphia*, with about 20 species, is abundant in humid tropical Africa and with one species (*R. taedigera*) in the New World. *Raphia farinifera*, a very familiar palm in Madagascar, is also present in East Africa, where it is variable in its habit and fruit shape in contrast to its behaviour in Madagascar where fruit shape is more or less uniform and the habit is always solitary. *Raphia farinifera* is a palm very closely associated with human activity – it is perhaps the most utilised palm of Madagascar and is frequently planted in valley bottoms when forest is cleared for rice cultivation. It does not occur in primary forest and, despite its immense ethnobotanical significance to the Malagasy, it seems highly likely that the palm was introduced into Madagascar when humans first colonised over 1500 years ago (Dransfield & Beentje, 1995b). The complete lack of calamoid diversity in Madagascar is in marked contrast to the situation in West Africa and the Asian and Malesian tropics.

### *Coryphoideae*

Subfamily Coryphoideae comprises eight tribes, distributed widely throughout the tropics and subtropics. They include some of the most drought- and cold-tolerant palm species. Representatives of three tribes are present in Madagascar.

Tribe Phoeniceae is represented by two non-endemic taxa of the 14 species in the only genus, *Phoenix*. *P. dactylifera* is undoubtedly introduced and naturalised around towns and villages in the southwest, while *P. reclinata*, quite possibly native, is found scattered around the island in drier areas.

Tribe Chuniophoeniceae, very well supported with molecular data, comprises four genera – monotypic *Nannorrhops* in Arabia and neighbouring southwest Asia, monotypic *Kerriodoxa* in southern Thailand, *Chuniophoenix* with three species in northern Vietnam, Hainan and southern China and monotypic *Tahina* in northwestern Madagascar, an astonishing disjunct distribution without obvious explanation (Dransfield *et al*., 2008a, 2008b).

### *Tribe Borasseae*

Borasseae is divided into two subtribes, each with four genera worldwide. Hyphaeninae consists of *Hyphaene*, largely restricted to Africa and the drier lands of coastal Indian Ocean as far as Sri Lanka, with one species, *H. coriacea*, also in Madagascar, *Medemia*, restricted to Sudan and Egypt, and *Bismarckia* and *Satranala* endemic to Madagascar. Lataniinae comprises *Lodoicea* in the Seychelles, *Latania* in the Mascarenes, *Borassodendron* in the Malay Peninsula and Borneo and *Borassus*, present in Africa, Madagascar, the Indian subcontinent and Southeast Asia through Malesia to New Guinea.

### *Subtribe Hyphaeninae*

The single species of *Hyphaene* to occur on Madagascar, *H. coriacea*, occurs scattered through the drier parts of the island where it is sometimes intensively harvested for fibre, thatch and wine and may possibly receive some rudimentary management. Outside Madagascar it is very much a plant of the coastal regions of East Africa from the Horn to Natal. Variation is extensive and many names have been published, but now included in synonymy (Dransfield, 1986). In Madagascar it was long known as *H. shatan*. Could it have been introduced by humans? Unlike *Raphia farinifera*, there seems not to be active planting of the species and it can be found far from human habitation. On balance it would seem most likely that it is actually native.

*Bismarckia*, with the single species *B. nobilis*, is a feature of savannas predominantly in the west of the island from sea level up to about 600 m elevation, occurring where groundwater is near the surface. Two forms occur, one with glaucous foliage, the other with dull green foliage – these forms can occur in uniform stands or intermingled. To the north of Mahajanga in remote areas near the coast, huge stands of *Bismarckia* occur, stretching for tens of kilometres as far as the eye can see. Little is known about its biology and dispersal despite its being one of the most popular and widespread ornamental palms in the drier tropics and subtropics.

The endemic *Satranala* with the single species, *S. decussilvae*, was not discovered until 1991, described and named in 1995 (Dransfield & Beentje, 1995a). It is a rare fan palm of the middle storey of humid rainforest along the east coast of Madagascar from the Masoala Peninsula (on both east and west coasts) southwards to Pointe à Larré, near Soanierana Ivongo. In its scattered localities, it often occurs in colonies of many individuals. As with all members of Borasseae, it is dioecious. The fruit has an unusual ridged endocarp that lacks an apical germination pore, present in other members of the tribe; uniquely in the palms, this splits in two vertically, on germination. Little is known of its natural history.

## *Ceroxyloideae*

Subfamily Ceroxyloideae with three tribes worldwide is represented in Madagascar by tribe Ceroxyleae with the single genus *Ravenea*. Twenty species are recognised in *Ravenea* and two more have just been described (Rakoroarinivo & Dransfield, 2011). All but two are restricted to Madagascar; *R. hildebrandtii* and *R. moorei* are endemic to the Comores. *Ravenea* together with *Ceroxylon* (Andes in South America), *Juania* (Juan Fernandez off the coast of Chile) and *Oraniopsis* (Queensland) make up tribe Ceroxyleae (Ceroxyloideae), distributed across the Southern Hemisphere (Dransfield *et al.*, 2008b). For long thought to be a classic example of Gondwanan vicariance, this interpretation of the origin and distribution of the group has recently been called into question by Trénel *et al.* who convincingly argue for an origin of the clade that resulted in Ceroxyleae long after the breakup of Gondwana and suggest that long-distance dispersal is the more likely explanation of the astonishing distribution (Trénel *et al.*, 2007).

Within a rather constrained morphology, *Ravenea* has radiated into most habitats (Beentje, 1994; Dransfield & Beentje, 1995a). *Ravenea nana* and *R. glauca* are palms of high elevation in the mountains. *R. sambiranensis*, in one its forms, is a palm of white sand forest along the coast, while *R. musicalis* is an aquatic that starts its life as a completely submerged plant with flaccid leaflets that offer little resistance to water currents, eventually growing into a squat bottle-palm with robust aerial stems to *c.*8 m tall, growing in up to 1 m of water. *Ravenea xerophila* is the most xerophytic of all Madagascar palms, occurring in the spiny forest of the southwest. *Ravenea louvelii* and *R. dransfieldii* are squat litter-trapping palms of the ever-wet rainforests of the eastern escarpment. The diminutive *R. delicatula*, only recently described, is unique in the presence of fascicled rather than regularly arranged leaflets and dichotomously branched stems. It is a palm of slightly drier montane forest to the west of the eastern escarpment.

## *Arecoideae*

### *Tribe Oranieae*

There is one genus in *Oranieae*, containing about 25 species distributed from southern Thailand, the Malay Peninsula, Borneo, Sumatra, the Philippines through the Moluccas to New Guinea, where there is greatest diversity of species. Three species occur spatially widely separated in Madagascar. Previously included in *Sindroa* and *Halmoorea*, the Madagascar species are clearly congeneric with the Malesian taxa (Dransfield *et al.*, 2008b). These are generally tall tree palms (a few of the New Guinea species are short tree palms of the understorey); two of the Madagascar species, *O. trispatha* and *O. ravaka*, have distichous, while the third species, *O. longisquama*, has spiral phyllotaxy. (In New Guinea, both spiral and distichous species are known.) In Madagascar the species are confined to the eastern rainforest occurring from Farafangan a northwards to the Masoala Peninsula. As in Southeast Asia, the Malagasy believe that *Orania* species are poisonous, so they are not felled for palm cabbage. The relatively large fruit may, at least in *O. trispatha*, be dispersed by water, as adult palms are often abundant along watercourses.

### *Tribe Cocoseae*

Besides the cultivated and semi-naturalised coconut and the African oil palm which may possibly occur naturally in Madagascar, tribe Cocoseae is represented in Madagascar by two endemic members of subtribe Attaleinae, *Voanioala* and *Beccariophoenix.* Cocoseae is predominantly New World in distribution; of the 18 genera, only four genera occur in the Old World – *Elaeis* (Elaeidinae) in Africa and the Americas, *Jubaeopsis* with a single species in South Africa and the two endemic Madagascar genera (all three members of Attaleinae). *Cocos* itself is of unknown origin; although its wide infraspecific diversity probably evolved in the Western Pacific or the Eastern Indian Ocean, its closest relatives are New World palms.

### *Voanioala*

*Voanioala*, with its single species *V. gerardii*, is restricted to rainforest in northeastern Madagascar and is one of the rarest palms in Madagascar. Properly confirmed records are all from around the Bay of Antongil, but there are rumours of its occurrence near to Toamasina, much further south on the east coast. The palm is a sub-canopy tree palm with a strongly stepped trunk. The fruit with its heavily sclerified endocarp with irregular intrusions into the endosperm, is about 10 cm long; fruit lie beneath the few parent trees known, seemingly undispersed, and it may well be that the natural disperser of this large and heavy fruit is no longer extant.

This species has the highest chromosome number of any known monocot ($2n$ = 550–606 ± 3) (Johnson, 1989), the only other cocosoid with a polyploidy number being South African *Jubaeopsis caffra* with $2n$ = 160–200. *Voanioala* occupies a position sister to a clade comprising New World members of Attaleinae and *Cocos* itself (Dransfield *et al.*, 2008b).

### *Beccariophoenix*

*Beccariophoenix*, until recently thought to be monotypic and on the verge of extinction, is now known to consist of at least two species, *B. madagascariensis* with at least three widely separated populations, and *B. alfredii*, highly restricted on the high plateau southwest of Antsirabe, where it is known from one very dense population of well over 1000 individuals (Rakotoarinivo *et al.*, 2007). A small and highly threatened population of *Beccariophoenix* in the coastal lowlands near Ampasimanolotra fits neither of the two species and may represent a third taxon. All taxa of *Beccariophoenix* are robust tree palms, remarkable for the massive torpedo-like peduncular bract that is circumscissile at the tip of the peduncle, the only cocosoid with such a circumscissile bract (Dransfield *et al.*, 2008b). Felling of trees for the edible growing point or cabbage has been one of the major reasons for the decline of the species near habitation. Indeed, when first described by Perrier de la Bâthie in 1915, it was even then thought to be on the verge of extinction. *Beccariophoenix* occupies a position sister to all other members of Cocoseae. Together with *Voanioala* they represent remarkable relics of cocosoid evolution in Madagascar.

### *Tribe Areceae*

Of the 11 recognised subtribes and 10 genera as yet unplaced in subtribes, only one group within tribe Areceae is present in Madagascar – Dypsidinae – but this subtribe accounts for the major bulk of palm species diversity on the island.

### *Subtribe Dypsidinae*

This is the quintessentially Madagascar palm grouping. Currently, Dypsidinae is a rather poorly understood clade with four genera recognised – *Marojejya*, *Masoala*, *Lemurophoenix* and *Dypsis*. There is some support from molecular data for the monophyly of the subtribe (Dransfield *et al.*, 2008b), but in terms of morphological characters, the group is so varied that no synapomorphies are distinguishable and the only feature that unites them is that they are clearly members of tribe Areceae, restricted to Madagascar, the Comores and Pemba off the coast of Tanzania. Relationships within the subtribe also remain unresolved, mostly due to as yet inadequate sampling; in some studies *Lemurophoenix* is embedded within *Dypsis*, for example (Lewis & Doyle, 2002), yet the genus has a suite of morphological characters completely at variance with *Dypsis* (e.g. multistaminy as opposed to three or six stamens, a corky-warted fruit as opposed to a smooth fruit, apical embryo, as opposed to basal embryo and endocarp with a basal heart-shaped button, lacking in *Dypsis*). The study sampling the largest number of species of *Dypsis* resolves *Lemurophoenix* as sister to the rest of the Dypsidinae (Baker *et al.*, 2011).

*Marojejya*, with its two species *M. insignis* and *M. darianii*, are robust squat palms of forest undergrowth, with stems rarely exceeding 6 m tall, and massive leaves, the bases of which trap leaf-litter from surrounding dicot trees. *Marojejya insignis* occurs from the Marojejy Massif in the far northeast of the island and in scattered localities, mostly at elevations of 300–1000 m but occasionally at sea level, all the way south along the eastern escarpment to the Fort Dauphin area. It is remarkable that this strikingly distinct palm remained unknown until Humbert discovered it on Marojejy in 1949 (Humbert, 1955a). Until the late 1980s it was only known from the Marojejy and a locality in east Masoala. *Marojejya darianii*, now known from a few localities from Masoala and Sahavary south towards Toamasina, is distinctive in having massive only partially or completely undivided leaves with striking basal auricles that not only trap litter, but also trap rainwater. Adventitious roots from the stem surface grow into the resulting water and debris-filled tank and presumably absorb water and nutrients directly. Roots from neighbouring dicot trees also grow through the leaf litter on the ground and into the juvenile palms, following the edges of the leaf bases, resulting eventually as leaf sheaths fall in zigzag roots apparently hanging out of the crown of the palm. In both species of *Marojejya*, inflorescences are mostly unisexual, but both sexes occur on the same individual. The inflorescences are compact with congested branches that scarcely emerge from the subtending leaf bases. They bear a remarkable resemblance to the form of inflorescence found in completely unrelated *Elaeis*.

Being a genus of squat, litter-trapping palms, *Masoala* bears a strong resemblance to *Marojejya*, but here the inflorescences are bisexual with staminate and pistillate flowers borne within the same inflorescence. The inflorescences are erect, with long peduncles that carry the flowers out from the leaf bases. *Masoala madagascariensis* is a palm of the Masoala Peninsula, Marojejy, and a few scattered localities southwards towards Toamasina. *Masoala kona*, in contrast, occurs much further south in forest on the eastern escarpment between Ifanadiana and Vondrozo. Both species tend to occur in scattered populations of few individuals and are thus particularly vulnerable.

*Lemurophoenix*, with a single species, *L. halleuxii*, is restricted to a very few localities around the Bay of Antongil in the northeast of the island. It is a massive tall single-stemmed palm that eventually reaches the forest canopy. At all localities populations are endangered by illegal felling of the trees, even within the boundaries of the Masoala National Park, and outside reserves are further threatened by forest clearance for shifting cultivation. Fruit exported from Madagascar about a decade ago to an amateur grower in Australia seems to suggest the presence of a second species; the fruits are ellipsoidal rather than spherical and the pericarp is smooth rather than corky-warted, yet the embryo is apical and the endosperm shallowly ruminate as is *L. halleuxii*. Unfortunately the source of the fruits has not been divulged so we have been unable to investigate further.

*Dypsis*, with 165 species described is the second largest genus of all palms (after *Calamus*). In the past species now included in *Dypsis* have been described in 13 additional genera. The morphological distinctions between these genera were considered by Dransfield and Beentje (1995b) to be completely unreliable. Many intermediates between genera existed, and some taxa had even been described, sometimes by the same author, in different genera; without complete material it was often impossible to say to which of the 14 genera a specimen might belong and, even in the case of *Dypsis sensu stricto*, with three rather than six stamens, the position of the stamens – whether antesepalous or antepetalous – suggested that the three-staminate condition probably arose at least twice. In the absence of a rigorous phylogenetic analysis and with the practical need to provide a useful working taxonomy, based often on incomplete material, the decision was taken to lump all genera into *Dypsis*. The background for this decision is discussed in detail in Dransfield and Beentje (1995b). After an initial period in which growers expressed dissatisfaction with the broad genus *Dypsis*, the current taxonomy is now widely followed. It at least has the merit of allowing the generic placement of new taxa. Further phylogenetic work, particularly using molecular data not available at the time of the 1995 revision, will be crucial in reassessing the relationships within this protean genus and it may well be necessary to resurrect some of the genera sunk into synonymy in 1995. Dransfield and Beentje (1995b) recognised 18 informal groupings in the genus, based on combinations of habit, inflorescence structure, floral morphology and fruit.

In size, the species range from massive tree palms that reach the forest canopy or emerge from it, to some of the smallest palms in the entire family – *D. tenuissima* has stems not much more than 3 mm in diameter and is rarely more than 30 cm tall. Perhaps surprisingly there are very few acaulescent species. Aerial branching is not infrequent and there are some squat, litter-trapping palms in which the marcescent leaf bases allow the accumulation of detritus that slowly rots within the crown, presumably enhancing the nutrient supply to the palm. Leaves vary from massive pinnate leaves with over 100 pinnae on each side of the rachis, to entire bifid leaves and there is even one species with a completely entire leaf (*D. integra*). Leaflet tips are usually acute or acuminate, but rarely praemorse. In the climbing species terminal leaflets are widely spaced, reflexed and reduced in size, giving the leaf apex and its leaflets a grapnel function. With this range of form, *Dypsis* encompasses most of the possible recorded leaf morphology seen in reduplicate pinnate-leaved palms. Inflorescences can be interfoliar or infrafoliar, and spicate to branched to four orders. Flowers are unisexual and borne in triads of a central pistillate and two lateral staminate flowers. Staminate flowers have six or three stamens; where there are three stamens these can be opposite the sepals or opposite the petals, suggesting that the three-staminate condition has perhaps arisen at least twice by reduction from six stamens. There is even one rare instance (some populations of *Dypsis lantzeana* around the Bay of Antongil) where stamen number is reduced to just one, the only instance in the whole family (Dransfield & Beentje, 1995b). Fruits vary in size from *c*.5 to 30 mm in diameter but stigmatic remains are always basal, and endosperm can be homogeneous or ruminate, sometimes minutely, sometimes deeply so, but the embryo is always basal.

*Dypsis* has radiated into most of the major terrestrial habitats of Madagascar, apart from the spiny forests of the southwest (although one species, *Dypsis decaryi*, occurs on the fringes of this habitat). The genus occurs high in the mountains of Tsaratanana and Marojejy, at sea level, in humid tropical forests to dry sclerophyll forest, there is at least one rheophyte (*D. crinita*) and one true aquatic (*D. aquatilis*), and two species of climber (*D. scandens* and *D. andilamenensis*, Rakotoarinivo & Dransfield, 2010), the only climbing members of tribe Areceae.

## Main factors governing palm distribution within Madagascar

The distribution pattern of palms in Madagascar depends on various elements such as the topography, the nature of the geological substrate and, most importantly, the climatic type at the regional or local scale. These parameters allowed Humbert (1955b) to define two major provinces for the phytogeography of Madagascar – the Eastern and the Western regions. Humbert justified this classification on the basis that each has a distinct flora, and this is strongly borne out by the biogeography of palms in Madagascar. Of the indigenous palm species currently known, 90% inhabit and are endemics to the Eastern region of the island.

The Eastern region has in fact a humid climatic type and its primary vegetation is characterised mainly by the various ranges of humid forest, from lowland to mountain forest. As almost everywhere else in the world, palms are prolific in these vegetation types (Dransfield *et al.*, 2008b). The climate of this part of Madagascar is defined by the eastern trade wind (Alizé) that has an influence throughout the year in this region. This wind is responsible for the high precipitation and moisture rate and explains thus the spatial variation of palm diversity as well as the different centres of endemism. For the Eastern region, Humbert (1955b) distinguished four domains – the Eastern Lowlands (0–800 m), the Central Highlands (800–1800 m, including the eastern and western slopes), the high mountains (> 1800 m) and the Sambirano.

The Eastern Lowlands have undeniably the highest diversity rate for palms in Madagascar as 124 species have been recorded in the lowland rainforest (Rakotoarinivo, 2008b). From Daraina to Taolagnaro, the lowland rainforest is consistently characterised by the presence of some or all of the following species – *Dypsis fibrosa, D. nodifera, D. pinnatifrons, Ravenea robustior* and *R. sambiranensis*. Recent biogeographic analysis resulted in the conclusion that the palms flora reaches a peak of abundance at around 300–400 m elevation in the primary humid forest (Rakotoarinivo, 2008b). The average number of palm species in one humid forest is between 10 and 15 but the richest areas may have more than 30 species in the same scale. Areas having high richness are characterised by the presence of a large number of locally or regionally endemic species, while the palm flora is poor when there are only widespread species in the area. Floristically, the most important palm sites in Madagascar are areas bordering the Bay of Antongil in the northeast of the island. The eastern part of the Makira chain (Ambinanitelo areas), Antanambe forest (Mananara Avaratra) and the western coast of Masoala Peninsula (Antalavia/Tampolo) have respectively in their forests 43, 41 and 37 species of palms. Research on palms of Madagascar produced the conclusion that the diversity rate becomes higher with high humidity, permanent warm temperature and less seasonality as well as with various geological substrates and topographic features (Dransfield & Beentje, 1995b; Rakotoarinivo, 2008b). This explains the striking difference of species richness even within the rainforest along its latitudinal gradient; there are many species of palms in the northeast of Madagascar compared with the southeast. This can be attributed to the fact that the area between Sambava and Toamasina has a bioclimatic perhumid type characterised by the highest precipitation rate in Madagascar (annual precipitation 3000–3500 mm) and the absence of a dry season. The zone southwards from Toamasina is less humid (annual precipitation 1500–2500 mm) and with two dry months (Donque, 1972). Paucity of palms in the southeast may be also be explained by the rarity of primary forest in this area at the present time. This may be the result of human intervention as well as Quaternary palaeoclimatic variations. As in almost all parts of Madagascar, this zone had some periods of vegetation contraction during that period and it is supposed that some species have experienced local extinction and only areas sheltered from bad weather were protected (Straka, 1996). In the southeast, only a few areas seem to have been able to escape periodic droughts. These areas considered as refuges are now characterised by high local endemism but also vicariance of some species characteristic of the north-eastern rainforest. Ifanadiana, Farafangana and Vondrozo provide examples where restricted species have been recorded (*Dypsis ifanadianae, D. interrupta* and *D. laevis*, respectively) but also where species such as *Dypsis integra, D. simianensis, Orania longisquama, O. trispatha, Ravenea julietiae* and *R. lakatra*, more frequent in the northern part of the rainforest, are found. On the other hand, centres of endemism are frequent in the northeast and, consequently, the endemism rate is particularly high. Each area between Ampasimanolotra and Daraina seems to have its own local endemic as well as regional endemic species; the rate becomes higher towards the Bay of Antongil. This richness may be explained by the fact that this part of Madagascar has been less affected by Quaternary climatic variation (Straka, 1996) and it has preserved its rich flora through time.

The Central Highlands (800–1800 m) have great climatic variation but this area is globally characterised by its subhumid bioclimatic type (Cornet, 1974). The mean annual temperature and the precipitation rate decrease compared to the coastal area but the humidity is locally very high due to the occasional or permanent occurrence of cloud cover, drizzle or fog. Humbert (1955b) distinguished the eastern slopes of the Highlands from its western slopes. For palms, the areas are also different. In the eastern escarpment of the Highlands, the humid mountain forest contains 64 species (Rakotoarinivo, 2008b) where evident variation in flora as well as species richness is noted. As in the lowlands, areas in the northeast are much richer. The Andasibe area, Zahamena, Ambatovaky and the lower part of Marojejy mountain have the most diverse palm flora in this zone. Numerous local endemic palms are found in addition to the widespread species of the mountain forest such as *Dypsis baronii, D. catatiana, D. heterophylla, D. procumbens* and *Ravenea madagascariensis*. The western slopes have a very different and more or less poor palm flora. Primary vegetation in this zone is sparse and also the climate is relatively dry with its 7-month dry season. Rain is brought seasonally by the monsoon from the equator while the current of the southeastern trade wind has lost its humidity after having crossed the ridge top of the Highland plateau. Of the widespread species on the eastern slopes of the Central Highlands, only *Dypsis baronii* and *Ravenea madagascariensis* are present on the western slopes where *D. onilahensis* is most frequently found. In total, there are only 14 species of palms recorded throughout the western slopes; the distribution pattern is characterised by restricted ranges. For instance, *Dypsis ambositrae* has only been recorded in the Ambositra and Itremo areas, *Beccariophoenix alfredii* in the west of Antsirabe and *Ravenea glauca* from Andringitra and Isalo.

The high mountains have a characteristic palm flora despite paucity in numbers. Only 23 species (Rakotoarinivo, 2008b) have been recorded from areas above 1800 m elevation but almost all of them are endemic to one massif or to one area. The climatic type is classified as 'mountain' (Cornet, 1974) with its high humidity and low mean temperature. Primary vegetation types are sclerophyllous forest and mountain thicket. Humbert (1955b) emphasised five sectors in this zone, according to the distinctiveness of their flora. These are Tsaratanana, Marojejy, Ankaratra, Andringitra and Andohahela; the biogeography of palms in this zone also reflects this division. Tsatananana and Marojejy have the greatest diversity of palms. As relatively adjacent, the palm flora around the summits is more or less similar and characterised by the vicariance of species such as *Dypsis acuminum, D. andrianatonga, D. baronii* and *D. oreophila*. However, rare species endemic to each massif also exist; *Dypsis ankaizinensis, D. montana* and *D. tsaratananensis* are endemic to Tsaratanana and *D. heteromorpha* and *D. pumila* to Marojejy. Palms are less abundant in the southern mountains; only *Ravenea glauca* has so far been recorded from the summit area of Andringitra while no palms have been recorded yet for the summits of Andohahela and Ankaratra.

The Sambirano domain is located in northwest Madagascar. Despite its western position, this zone is characterised by high humidity due to the simultaneous influence of the eastern trade wind and the equatorial monsoon current, coupled with high temperatures throughout the year (Humbert, 1955b). These conditions explain the relatively high diversity of the palm flora in this region. In a recent census 23 species were recorded (Rakotoarinivo, 2008b): 10 of them are endemic to Sambirano (e.g. *Dypsis andrianatonga, D. ankaizinensis, D. occidentalis*), nine are shared with the Eastern Lowland rainforest (e.g. *Dypsis lastelliana, D. canaliculata, Orania longisquama* and *Ravenea sambiranensis*), three with the adjacent western domain (*Dypsis madagascariensis, Bismarckia nobilis* and *Hyphaene coriacea*) and one is shared with the mainland of Africa: *Borassus aethiopum*.

As mentioned earlier, the Western region is poor in palms compared with the Eastern part of Madagascar and the palm flora is totally different. The climate is in fact characterised by constantly high temperatures (mean ± 26°C; Cornet, 1974) throughout the year and evident drought for all major parts of the year since the rain is brought only by the seasonal monsoon, from the equator. The precipitation rate decreases the further south one goes; this allowed Humbert (1955b) to divide this region into two distinct phytogeographic domains: the western domain and the southern domain.

The western domain, from Antsiranana southwards to just north of Toliara, has various types of habitat (e.g. plain, canyon, limestone, sands) as well as vegetation types (dry forest, subhumid forest, riverine forest, savanna and mangrove) but the most striking aspect of palm distribution in this region is the abundance and sometimes dominance of just three species: *Bismarckia nobilis, Dypsis madagascariensis* and *Hyphaene coriacea*. However, 14 species of palms have been recorded for the western domain of Madagascar and nearly all of them are restricted to around the Sambirano domain, where rainfall varies from 1200 to 1500 mm per year. Good examples are *Dypsis leptocheilos, D. rivularis* and *Tahina spectabilis*.

Further south, the palm flora is also characterised by the presence of *Borassus madagascariensis* on the plain between Maevatanana and Mahabo, and then by *Ravenea rivularis* around Sakaraha and Analavelona plateau.

Palms are usually absent in the southern domain of Madagascar except in certain areas such as Antanimora, Ampanihy and the western slopes of the Anosy chain. Only two species have been recorded: *Dypsis decaryi* and *Ravenea xerophila*. The paucity of palms in this region is correlated with the extreme drought in this area – about 10–11 dry months and a rainfall of only about 300 mm per year (Cornet, 1974).

All of these distribution patterns of palms demonstrate that the species diversity in one area can vary greatly depending on the regional climatic type. The higher the humidity, the higher the number of palms species that can be recorded. Quite simply lowland areas are richer than highlands, the eastern part is more favourable than the central or the western part and finally there is evidence of a latitudinal gradient – there are more species in the north compared to the south. The combination of these parameters explains the wonderful diversity of palms in the northeast of Madagascar, between Ampasimanolotra and Marojejy, where the most important environmental variables for palms coincide.

The great diversity coupled with high rates of local endemicity have serious consequences for conservation.

## A note about spines

All endemic species of palm in Madagascar are unarmed, apart from *Borassus madagascariensis*. The only spiny palms are the non-endemic or introduced taxa *Hypahaene coriacea*, *Borassus aethiopum*, *Phoenix dactylifera* and *P. reclinata*, *Elaeis guineensis* and *Raphia farinifera*. This is in marked contrast to the palms of Africa where almost all palms are armed in some way, apart from the genera *Podococcus*, *Sclerosperma* and *Jubaeopsis*. It is thought that the heavy armature found in many African palms, belonging to different clades, is correlated with the abundance of large herbivores such as ungulates and elephants. Perhaps the absence of such armature in Madagascar palms indicates that there has been no selection pressure for spines to have evolved and that herbivory was less significant in Madagascar than in Africa.

## Conclusions

The palm flora of Madagascar is diverse, with representatives of many evolutionary lines, showing various relationships with African, Indian Ocean island and coastal, mainland Southeast Asian, Malesian and Southern Hemisphere palm floras. The distribution of two evolutionary lines (Ceroxyleae and Cocoseae) have in the past been explained in terms of the splitting of Gondwana and subsequent vicariance. However, recent evolutionary studies suggest that the timing of divergence of these two lines occurred far too late to have been influenced by the rafting of Gondwanic fragments and can more easily be explained in terms of long-distance dispersal. The extraordinary disjunctions in the genus *Orania*, in Madagascar and Malesia, and in the Chuniophoeniceae, with *Tahina* in Madagascar, *Nannorrhops* in Arabia and western Asia, *Kerriodoxa* in Thailand and *Chuniophoenix* in Hainan and Vietnam, are paradoxical and very difficult to explain. The palms of Madagascar remain incompletely known, of great interest and severely threatened.

## References

Baker, W. J., Norup, M. V., Clarkson, J. J., Couvreur, T. L. P., Dowe, J. L., Lewis, C. E. *et al.* (2011). Phylogenetic relationships among arecoid palms (Arecaceae: Arecoideae). *Annals of Botany* doi: 10. 1093/aob/mcr020.

Bayton, R. (2007). A revision of *Borassus* L. (Arecaceae). *Kew Bull.* 62: 561–586.

Beccari, O. (1912–14). *Palme del Madagascar.* Firenze, Italy, Istituto Micrografica Italiano, 59 pp.

Beentje, H. J. (1994). A monograph of *Ravenea* (Palmae: Ceroxyloideae). *Kew Bull.* 49: 623–671.

Cornet, A. (1974). Essai de cartographie bioclimatique à Madagascar. Note Explicative n°55. Paris, Editions ORSTOM, 41 pp.

Donque, G. (1972). The Climatology of Madagascar. In R. Battistini & G. Richard-Vindard G. (eds.), *Biogeography and Ecology of Madagascar*, pp. 87–144. The Hague, Junk.

Dransfield, J. (1986). Palmae. *Flora of Tropical East Africa.* Rotterdam, A. A. Balkema.

Dransfield, J. & Beentje, H. J. (1995a). *Satranala* (Coryphoideae: Borasseae: Hyphaenina), a new palm genus from Madagascar. *Kew Bull.* 50: 85–92.

Dransfield, J. & Beentje, H. J. (1995b). *The Palms of Madagascar.* Royal Botanic Gardens Kew and The International Palms Society. 475 pp.

Dransfield, J. & Marcus, J. (2002). *Dypsis* 'stumpy'. *Palms* 46: 47–51.

Dransfield, J., Rakotoarinivo, M, Baker, W. J., Bayton, R. P., Fisher, J. B., Horn, J. W. *et al.* (2008a). A new Coryphoid palm genus from Madagascar. *Bot. J. Linn. Soc.* 156: 79–91.

Dransfield, J., Uhl, N. W., Asmussen, C. B., Baker, W. J., Harley, M. M. & Lewis, C. E. (2008b). *Genera Palmarum, The Evolution and Classification of Palms.* Royal Botanic Gardens, Kew, International Palm Society and Bailey Hortorium. Kew Publishing, 732 pp.

Hodel, D. R. & Marcus, J. (2004). The White Powder *Dypsis*: a new species from cultivation. *Palms* 48: 90–93.

Hodel, D. R., Marcus, J. & Dransfield, J. (2005). *Dypsis robusta*, a large new palm from cultivation. *Palms* 49: 128–130.

Hodel, D. R., Marcus, J. & Dransfield, J. (2009). *Dypsis plumosa*, the Madagascar Queen Palm. *Palms* 53 (4): 161–165.

Humbert, H. (1955a). Une merveille de la nature à Madagascar. Première exploiration du massif du Marojejy et de ses satellites. *Mém. Inst. Scient. Madagascar* B 6: 1–210.

Humbert, H. (1955b). Les territoires phytogéographiques de Madagascar. In Colloques internationaux du C.N.R.S., 59: Les divisions écologiques du Monde. Moyen d'expression, nomenclature, cartographie. Paris, juin-juillet 1954. *Année biologique*, 3e série, 31: 439–448.

Johnson, M. A. T. (1989). An unusually high chromosome number in *Voanioala gerardii* (Palmae: Arecoideae: Cocoeae: Butiinae) from Madagascar. *Kew Bull.* 44: 207–210.

Jumelle, H. & Perrier de la Bâthie, H. (1945). Palmiers. In H. Humbert (ed.), *Flore de Madagascar et des Comores*, 30 edition. Imprimerie officielle, Antananarivo.

Lewis, C. E. & Doyle, J. J. (2002). A phylogenetic analysis of tribe Areceae (Arecaceae) using two low-copy nuclear genes. *Plant Syst. Evol.* 236: 1–17.

Rakotoarinivo, M. (2008a). A remarkable *Ravenea* from the montane forest of Andilamena, Madagascar. *Palms* 52: 11–17.

Rakotoarinivo, M. (2008b). Analayse de la distribution et de la conservation des Palmiers (Arecaceae) de Madagascar par l'utilisation du système d'information géographique. Université d'Antananarivo, These de Doctorat, 200 pp.

Rakotoarinivo, M. & Dransfield, J. (2010). New species of *Dypsis* and *Ravenea* (Arecaceae) from Madagascar. *Kew Bull.* 65: 279–303.

Rakotoarinivo, M., Ranarivelo, T. & Dransfield, J. (2007). A new species of *Beccariophoenix* from the high plateau of Madagascar. *Palms* 51: 63–75.

Straka, H. (1996). Histoire de la végétation de Madagascar oriental dans les derniers 100 millenaires. In W. R. Lourenco (ed.), *Biogeographie de Madagascar*, pp. 37–47.

Trénel, P., Gustafsson, M. H. G., Baker, W. J., Asmussen-Lange, C. B., Dransfield, J. & Borchsenius, F. (2007). Mid-Tertiary dispersal, not Gondwanan vicariance explains distribution patterns in the wax palm subfamily (Ceroxyloideae: Arecaceae). *Molec. Phylogen. Evol.* 5: 272–288.

von Martius, C. F. P. (1823–50). *Historia Naturalis Palmarum*. 3 vols. Munich.

# 8

# Evolution and biogeography of the flora of the Socotra archipelago (Yemen)

LISA M. BANFIELD, KAY VAN DAMME AND ANTHONY G. MILLER

The Socotra archipelago is an ancient continental fragment situated in the Western Indian Ocean that has been isolated for at least 18 million years (Ma). Although often described as the 'Galapagos of the Indian Ocean', compared to the Galapagos and other island systems, this Arabian Sea archipelago is relatively poorly studied and often overlooked in works on island biology. Nevertheless, the Socotra archipelago is characterised by relatively high floristic diversity and endemism with local adaptations and radiations, making it a good example of a system suitable for the study of island biogeography. We discuss origins, affinities, disjunctions and adaptations in the flora and explore underlying environmental and evolutionary processes that may have helped the Socotra archipelago to its present uniqueness.

## Location

The Socotra archipelago (also spelt 'Soqotra') consists of four islands situated in the southeast of the Gulf of Aden; Socotra, Abd al Kuri and 'The Brothers', Samha and Darsa, with several additional rocky outcrops. They are located in the Indian Ocean between 12° 06′ and 12° 42′ N and 52° 03′ and 54° 32′ E. The largest island, Socotra, lies *c.*380 km south of the Arabian Peninsula and *c.*225 km east of Cape Guardafui, Somalia (Fig. 8.1).

## Landscapes and topography

Socotra is the largest and highest island of the archipelago with a surface area of approximately 3600 km$^2$ and reaching an altitude of *c.*1550 m. Its landscape is dominated by the granite Haggeher mountains (Figs 8.1 and 8.2a) and limestone plateaus and escarpments, floristically the most diverse areas of the archipelago.

*The Biology of Island Floras*, eds. David Bramwell and Juli Caujapé-Castells. Published by Cambridge University Press. 

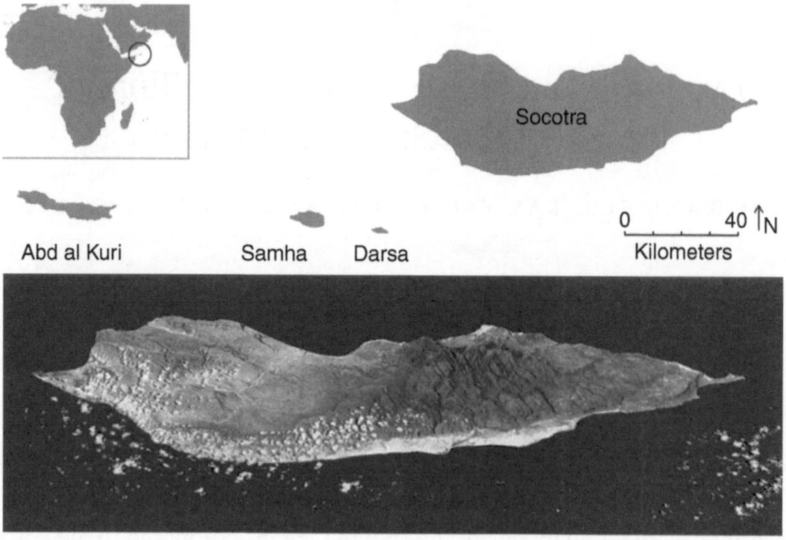


Figure 8.1 (above) Location of the Socotra archipelago in the Western Indian Ocean and its four main islands; (below) topography of Socotra island. Image below Landsat 7 ETM draped over USGS DEM, courtesy Kay Van Damme.

Dense woodland occurs at the cooler and wetter higher altitudes and valleys sheltered from monsoon winds (Fig. 8.2a). Exposed rock and steep cliffs provide further habitats for additional species. Permanent streams and intermittent watercourses ('wadis') flow from the upland peaks to the surrounding lowlands. The limestone plateaus on Socotra average 700–800 m in altitude but reach 1000 m. These plateaus descend gradually toward the interior central basin or fall steeply to the coastal plains or the sea (Fig. 8.2c). The plains reach an average 8 km in width and consist of alluvial soils, compacted gravels, rocks, white sands and elevated fossil reefs.

Abd al Kuri the second largest island, lies closer to Somalia (*c.*100 km) than to Socotra itself (*c.*145 km). Its surface is *c.*133 $km^2$ and the island is dominated by a limestone plateau reaching *c.*740 m. Samha is situated *c.*50 km southeast of Socotra and is 41 $km^2$. It is largely desertic with no permanent surface water. Like Abd al Kuri, it is dominated by an imposing limestone plateau, here reaching 680 m in altitude. Darsa is the smallest and lowest island in the archipelago, only 17 $km^2$ in surface area and 350 m in altitude. It is also desertic and lacks permanent water. All the islands are inhabited apart from Darsa, although Samha and Abd al Kuri are largely deserted during the winter (NE Asian) monsoon.

The limestone plateaus display typical karstic features, with areas of strongly eroded, bare rock and deep fissures or sinks, within which soils accumulate and support plant life (Fig. 8.2b). On all of the islands sea-facing escarpments, cliffs and higher elevations receive considerably larger amounts of moisture through

Figure 8.2 The varied topography and vegetation of Socotra island. (a) The granite pinnacles of the Haggeher Mountains (asl ca. 1550 m), with dense woodland forming part of the montane mosaic vegetation, including *Pteridium aquilinum* (Dennstaedtiaceae), *Euclea divinorum* (Ebenaceae), and endemics *Draceana cinnabari* (Ruscaceae), *Indigofera socotrana* (Fabaceae), *Coelocarpum haggierensis* (Verbenaceae) and *Hypericum* species (Clusiaceae). Photo by Lisa Banfield. (b) Karstic features and succulent shrubland of Sirahon limestone plateau (asl 450 m) dominated by *Jatropha unicosta* (Euphorbiaceae) and *Adenium obesum* subsp. *sokotramum* (A pocynaceae). Photo by Steve Scott, RBGE. (c) Southern coastal plain with *Croton socotranus* (Leguminosae) shrubland with a view to the southern escarpment of the limestone plateau of Qatariyah (asl 30 m). Photo by Anthony Miller. (d) The arid central Zahr Basin (asl 60 m) with succulent shrubland including the endemics *Euphorbia schimperi* (Euphorbiaceae) and *Dendrosicyos socotranus* (Cucurbitaceae). Photo by Lisa Banfield.

orographic rainfall and sea fogs. These constitute important plant habitats and support many endemic species. These habitats are considered as wet refugia within the region, playing an important role in the evolution of the Socotran flora (Miller & Morris, 2004; discussed further below).

Figure 8.2 (*cont.*)

The vegetation of Socotra has been classified broadly, in order of increasing altitude, as coastal, croton shrubland, succulent shrubland, semi-evergreen woodland, open and woody-based herb communities, submontane shrubland and a montane mosaic of shrubland, woodland and grassland (Miller & Morris, 2004: 48–49).

## Socotran floristics in perspective

The Socotra archipelago currently counts 835 vascular plant species, of which 308 (37%) are endemic (figures from Kilian & Hein, 2006, added to those of Miller & Morris, 2004). These species represent 114 families and 433 genera, of which 15 are endemic. The ferns count 30 species and there is a single gymnosperm (*Ephedra foliata*). The family Poaceae is the most diverse (90 species), followed by Fabaceae (70 species) and then Asteraceae, Euphorbiaceae, Apocynaceae (including Asclepiadoideae), Acanthaceae and Rubiaceae with over 20 species each. Together these families total more than 40% of the flora (Miller & Morris,

2004). The archipelago counts 74 bryophytes: one hornwort, 30 liverworts and 43 mosses, six of which are endemic (Kürschner, 2003), and 130 algae and seagrasses, of which phytogeography has been studied (Schils & Coppejans, 2003). Several names of endemic plants on Socotra refer to explorers of the previous century who contributed to the botanical knowledge of the archipelago, such as I. B. Balfour, G. Schweinfurth and J. R. Wellsted.

Local diversity and percentage of endemism in vascular plants compares well with some other continental islands. When surface area of these islands is taken into account, the archipelago has a greater number of endemics per $km^2$ than all the islands considered, except for the Seychelles, New Caledonia and Hawaii. It equals diversity per $km^2$ of the Canary Islands and Jamaica (Table 8.1).

Socotra, being the largest island with a greater diversity of habitats, is the most diverse island. The current count for Samha and Abd al Kuri islands is 122 species each, with Samha having eight endemic species and Abd al Kuri 12. Only 12 species have been recorded from Darsa. However, the smaller islands are only accessible by boat, and sea conditions prevent explorations until a time of year when the islands are already dry. They are therefore explored for less than Socotra island itself.

The uniqueness of the Socotra archipelago is also reflected in the fauna. Although several groups are virtually absent (flightless mammals, amphibians and primary freshwater fish), most groups show high levels of endemism. For example, 95% of the more than 100 species and 75% of the genera of terrestrial molluscs are endemic, with major radiations in three families (Neubert, 2006; 2009). Other well-studied examples are the reptiles, with *c.*90% endemic species (Rössler & Wranik, 2004) and crustacean isopods, with 73% (Taiti & Ferrara, 2004). Many groups remain poorly studied, but in most groups relict taxa and peculiar affinities were noted (e.g. Cheung & DeVantier, 2006).

These high levels of biodiversity, as well as a unique culture and history of sustainable land use, have led the islands to be designated a UNESCO World Heritage Site, UNESCO Man and Biosphere Reserve, WWF Global 200 Ecoregion and Plantlife International Centre of Plant Diversity. They form part of Conservation International's Horn of Africa Biodiversity Hotspot, and the *Dracaena (cinnabari)* (Ruscaceae) woodland is globally unique.

## The abiotic environment: geology, climate and refugia

As shown above, the Socotra archipelago has high diversity and endemism of both flora and fauna. The main reasons for this are an ancient geological history with a long period of isolation, diverse geomorphology and topography, a complex climatic history, a current climate that varies spatially and altitudinally, and the

Table 8.1 *Levels of diversity, endemism, and endemics per km² of vascular plant species of selected continental and oceanic islands or archipelagos. Those underlined are continental, those not are oceanic.*

| Island or archipelago | Endemics/ km² | Number of species | Number of endemics | % endemism | Area (km²) |
|---|---|---|---|---|---|
| Seychelles[1] | 0.56 | 1 640 | 250 | 15.2 | 444 |
| New Caledonia[1] | 0.14 | 3 094 | 2 480 | 80.2 | 16 648 |
| Hawaii[1] | 0.09 | 1 180 | 906 | 89.8 | 10 434 |
| Socotra archipelago[2] | 0.08 | 835 | 308 | 37.0 | 3 791 |
| Canary Islands[1] | 0.08 | 1 300 | 570 | 44.3 | 7 275 |
| Jamaica[1] | 0.08 | 3 308 | 906 | 37.2 | 11 189 |
| Mascarenes[1] | 0.07 | 878 | 329 | 37.5 | 4 507 |
| Galápagos[1] | 0.03 | 529 | 211 | 39.9 | 7 845 |
| Corsica[3] | 0.03 | 2 150 | 240 | 11.0 | 8 680 |
| Cuba[1] | 0.03 | 6 514 | 3 229 | 49.6 | 105 805 |
| Azores[1] | 0.03 | 197 | 67 | 34.0 | 2 335 |
| Cyprus[4] | 0.02 | 1 620 | 162 | 10.0 | 9 251 |
| Cape Verde[1] | 0.02 | 224 | 66 | 29.4 | 4 033 |
| Trinidad & Tobago[5] | 0.01 | 2 500 | 59 | 2.4 | 5 126 |
| Madagascar[1] | 0.01–0.008 | 8 000–10 000 | 5 000–8 000 | 68.4 | 587 713 |
| Sicily[3] | 0.009 | 2 402 | 255 | 9.4 | 25 662 |
| New Guinea[6] | 0.007 | 14 000–20 000 | 6 460 | 46.6 | 865 000 |
| New Zealand[1] | 0.007 | 2 371 | 1 942 | 81.9 | 268 021 |
| Borneo[6] | 0.005 | 45 500–25 000 | 4 089 | 9–16 | 748 168 |
| Java[6] | 0.004 | 4 598–5 800 | 565 | 9–12 | 138 793 |
| Sulawesi[6] | 0.004 | 5 000–6 050 | 743 | 12–15 | 180 680 |
| Tasmania[7] | 0.004 | 1 600 | 320 | 20.0 | 68 000 |
| Sumatra[6] | 0.002 | 8 000–11 200 | 1 278 | 11–16 | 443 065 |
| Crete[3] | 0.001 | 1 710 | 200 | 11.1 | 8 336 |

[1] Whittaker & Fernández-Palacios (2007: 66); [2] Miller & Morris (2004: 8); Kilian & Hein (2006); [3] Blondel & Aronson (1999: 60); [4] Thompson (2005: 36); [5] van den Eynden *et al.* (1998); [6] Roos *et al.* (2004); [7] Brown *et al.* (1994). All area figures taken from UNEP Islands Directory (1998) for consistency.

apparent persistence of wet and dry refugia habitats. Combined, these factors have produced a variety of ecological niches reflected in the high species diversity and percentage endemism.

### *Geology and geological history*

The Socotra archipelago is a continental fragment closest to the Horn of Africa. The islands sit on a granite micro-plate, the Socotra Platform, which was once part of eastern Gondwanaland. It consists of Precambrian granite dated at *c.*800 Ma (see Beydoun & Bichan, 1970; Fleitmann *et al.*, 2004). This basement was formed as part of the Afro-Arabian Plate, largely through volcanic activity, and was situated adjacent to the south of the Dhofar region of Oman (Beydoun & Bichan, 1970; Samuel *et al.*, 1997).

Major tectonical instability such as the breakup of Gondwanaland in the Late Mesozoic caused longitudinal stresses in the surrounding plates. Mesozoic continental rifts were formed, similar to the present-day African Rift, which subsequently became the seas separating the Arabian and the African plates. Sea floor spreading in one of these rifts, the Gulf of Aden, coincided with the separation of the Socotra Platform, moving it away from the Arabian Plate. Oceanic spreading in the eastern Gulf of Aden has been dated at around 18 Ma (Leroy *et al.*, 2004; Cheung & DeVantier, 2006 and references therein; Van Damme, 2009), marking the beginning of the archipelago's isolation.

During Pleistocene glaciations, lower sea levels caused the Socotra Platform to become exposed to the extent that Socotra, Samha and Darsa were connected by an extended coastal plain. This created a single land mass termed 'Greater Socotra' (Van Damme, 2006). Abd al Kuri, on the other hand, has been continuously isolated since the split from the Arabian Plate due to a deep sea trench between itself and Samha.

The archipelago has a diverse geology, consisting of a granite core base which crops out in eroded areas and at the highest peaks, the Haggeher mountains of Socotra. The base is overlain with marine limestone deposits resulting from major marine transgressions that pre-date the split of the archipelago from the mainland. Karstic Palaeocene-Eocene plateaus make up the largest surface area, reaching up to 700 m in thickness and deposited on Cretaceous limestone or directly on the granite base (Beydoun & Bichan, 1970). Lowland areas consist mainly of Quaternary deposits of both alluvial and marine origins (Beydoun & Bichan, 1970; Fleitmann *et al.*, 2004).

## *Climate: past and present*

The climate of the archipelago varies greatly, both temporally and spatially. The northeast and southwest Asian monsoons are the most pronounced influences, but the islands also experience occasional tropical cyclones and effects of the Indian Ocean Dipole (IOD) and the El Niño Southern Oscillation (ENSO). The northeast Asian (winter) monsoon brings the lowest temperatures of 16°C at high altitudes (Culek *et al.*, 2006) and the greatest amount of precipitation. Although no figures are currently available, maximum annual rainfall possibly reaches 1000 mm in the Haggeher mountains (Miller & Morris, 2004: 7). Fogs, mists and cloud bring an additional source of moisture, with relative humidity reaching 100% or averaging 65–89% (Culek *et al.*, 2006) depending on location and time of year. The southwest Asian (summer) monsoon brings little moisture but strong, desiccating winds averaging over 50 km $h^{-1}$ (14 m $s^{-1}$) (Morris, 2002) and gusting to 144 km $h^{-1}$ (40 m $s^{-1}$). The hottest temperature recorded is 43.45°C (Culek *et al.*, 2006). In general, the higher altitudes are coolest and can experience substantial rainfall, cloud cover, and mists and fogs, with the lowland areas and coastal plains being far more arid.

Climate reconstructions for Socotra have been made by measuring the deposited layers within speleothems, the size of which is related to the amount of rainfall and hence to monsoon strengths (Burns *et al.*, 2003; Fleitmann *et al.*, 2004; De Geest, 2006). Preliminary analysis for the last 55 000 years indicates that the climate has been very unstable and changeable. More than one abrupt change has been detected, including a rapid transition (over *c.*25 years) between dry and wet phases occurring 48 000 years ago (Fleitmann *et al.*, 2004). An independent reconstruction found similar evidence of abrupt changes over the last 6000 years but also detected differing local patterns in geographically separated speleothems (e.g. De Geest, 2006), suggesting that changes have not been spatially uniform (De Geest, pers. comm.). Data indicates that the archipelago has generally become drier, as with surrounding areas, but that there have been significantly wetter periods in the past (De Geest, pers. comm.; Scholte & De Geest, 2010).

Other evidence also points to recent aridification of the climate. Socotri people have reported a decrease in the duration of cloud cover, and disused field systems in some areas suggest that rainfall has been more abundant in those areas in the past (Miller & Morris, 2004: 43). *Dracaena cinnabari*, while actively regenerating in higher, wetter areas, is present only as relict populations in lower, drier areas (Attorre *et al.*, 2007; Miller & Morris, 2004) (Fig. 8.3). The general absence of drought-adapted plant species at higher altitudes of the Haggeher mountains, Socotra, would suggest that these peaks have remained wet enough to sustain non-xeromorphs.

### *Wet and dry refugia*

Wet and dry refugia on the Socotra archipelago have been described previously (Miller & Morris, 2004: 10). These refugia are likely to have been extremely dynamic in the past, with vegetation retreating or expanding in response to climate changes. Wet refugia occur on the higher peaks which receive more rainfall, and cliffs, escarpments and some plateaus that receive moisture from mists and fogs from the ocean (Figs 8.2a and 8.2b). These are the most diverse areas of the island. The Haggeher mountains of Socotra is the largest and most diverse of the

Figure 8.3 (a) The globally unique *Dracaena cinnabari* (Ruscaceae) woodland on limestone plateau of Firmihin, with the peaks of the Haggeher Mountains in the back, Socotra island. Here, regeneration of the woodland still occurs (Photo by Steve Scott, RBGE); (b) Limestone plateau at Homhil, Socotra island. Only mature individuals of *Dracaena cinnabari* remain, thought to be linked with aridification of the archipelago. Photo by Kay Van Damme.

wet refugia, supporting an estimated 200 species, of which 116 are endemic to the archipelago and 55 are endemic to that refugium (Miller & Morris, 2000). Approximately 200 of the total 835 vascular plant species are confined to the wet refugia, including 42% (130) of the endemic species (Banfield, 2006). Species confined to wet refugia are twice as likely to be endemic as those also found in the drier areas, and cliffs are particularly important for 22% of those endemics (Banfield, 2006). Due to the small area of many of the cliffs and escarpments and their vulnerability to aridification, 78 of the 130 endemics confined to the wet refugia have been assigned an IUCN threatened status category (examples in Table 8.2).

Table 8.2 *Examples of threatened endemic species which are confined to the wet refugia of the Socotra archipelago (after Banfield, 2006).*

| Species | Threatened status |
|---|---|
| *Aloe squarrosa* (Aloaceae) | Vulnerable |
| *Begonia samhaensis* (Begoniaceae) | Endangered |
| *Boswellia dioscorides, B. bullata and B. popoviana* (Burseraceae) | Vulnerable |
| *Dirachma socotrana* (Dirachmaceae) | Vulnerable |
| *Dracaena cinnabari* (Ruscaceae) | Vulnerable |
| *Duvaliandra dioscorides* (Asclepiadoideae; Apocynaceae) | Critically endangered |
| *Erythroseris amabilis* (Asteraceae) | Endangered |
| *Exacum socotranum* (Gentianaceae) | Endangered |
| *Nesocrambe socotrana* (Brassicaceae) | Endangered |
| *Pelargonium insularis* (Geraniaceae) | Critically endangered |
| *Pulicaria aromatica* (Asteraceae) | Endangered |

Dry refugia are found in lowland and foothill regions around the coast, dry limestone plateaus and the large, inland drainage basin (Figs 8.2c and 8.2d). Although not as diverse in endemics, xeromorphic adaptations of many remarkable endemic species (discussed below) show that the dry refugia have also played an important role in evolutionary processes.

There are several possible reasons why the wet refugia support a greater diversity and higher number of endemics. The dry refugia generally form continuous belts across the low and mid altitudes, with no obvious barriers to dispersal or gene flow. In contrast the wet refugia occur in separate locations throughout the archipelago, providing isolated habitats where speciation can occur. Also, wet refugia have a greater geological and landscape heterogeneity, creating a diversity of ecological niches and vegetation types.

The only population genetics study carried out to date on plant taxa of the archipelago examined the two endemic species of *Begonia* (Begoniaceae), both confined to wet refugia. *B. samhaensis* is restricted to a bare-rock habitat at the highest altitudes of Samha. The total range is extremely small at just $< 150 \times 50$ m (Hughes *et al.*, 2003). *B. socotrana* is restricted to cliffs and cracks between boulders on Socotra at high altitudes. The analysis showed genetic/geographic clustering in *B. socotrana* and evidence for significant isolation-by-distance (Hughes *et al.*, 2003). Although this is likely to be influenced by past and current dispersability, Hughes *et al.* (2003) suggested it could also be the result of genetic drift following bottlenecks, and that these bottlenecks were caused by retraction of populations during previous dry periods.

## Adaptations of the flora to aridity

The strong influence of an arid climate can be observed in the xeromorphic adaptations of many plant species. Xeromorphs are found throughout the archipelago, but are absent from wetter peaks of the Haggeher mountains of Socotra. An early classification of adaptations to aridity was published by Kearney and Shantz (1911), who divided plants into four categories – drought evaders, escapers, endurers and resistors.

Many of the small herbs that grow throughout the archipelago can be considered drought evaders. These are annual plants that germinate, grow, flower and set seed quickly following rain, hence evading dry periods. They are under-recorded on the archipelago due to the sporadic nature of rainfall and the plants' short life span. Examples include *Campanula balfourii* (Campanulaceae) and other endemic species listed in Table 8.3.

Drought escapers are perennial species that use underground storage organs such as tubers and bulbs to store nutrients. This allows the plant to lose its above-ground biomass and become dormant during dry periods. Again, there are many examples across the Socotra archipelago (Table 8.3), including the two endemic *Begonia* species. *Ledebouria grandifolia* (Hyacinthaceae) also has a bulb, but this plant will keep its leaves throughout the dry season if the rain has been plentiful.

The archipelago also has several examples of drought endurers (Table 8.3). These are generally shrubby perennials with deep root systems and adaptations to prevent water loss such as small, waxy leaves or surface indumentum. Throughout the coastal plains and foothills there are many drought endurers. They contribute a significant amount of above-ground biomass, especially during the dry season. Some of these species show convergence in vegetative appearance. For example, *Placopoda virgata* (Rubiaceae), *Ballochia amoena* (Acanthaceae) and *Cryptolepis intricata* (Asclepiadoideae, Apocynaceae) are superficially very similar. All have a compact shrubby habit and small, shiny, waxy leaves.

The most notable and famous plant species of the archipelago can be regarded as drought resistors: succulent plants that store water within their tissues to allow extended photosynthesis into the dry season. The most outstanding examples, perhaps, are the pachycauls or bottle trees, whose succulence gives them a swollen, bloated appearance. These include *Adenium obesum* subsp. *sokotranum* (Apocynaceae), which is widespread and locally common on Socotra island. This endemic subspecies is far larger than the African and Arabian subspecies, reaching 5 m in height and several metres in girth. As well as storing water, it has a shiny coating that reflects solar radiation (Fig. 8.4a).

Another outstanding example is *Dendrosicyos socotrana*, the only arborescent species of the Cucurbitaceae family (Fig. 8.4b). This endemic species, absent on

Table 8.3 *Examples of endemic plants species with adaptations to aridity, classified in the four categories of Kearney and Shantz (1911).*

| Drought evaders | Drought escapers | Drought endurers | Drought resistors |
|---|---|---|---|
| *Campanula balfourii* (Campanulaceae) | *Begonia socotrana* and *B. samhaensis* (Begoniaceae) | *Ballochia amoena* (Acanthaceae) | *Adenium obesum* subsp. *sokotranum* (Apocynaceae) |
| *Erucastrum rostratum* (Brassicaceae) | *Dipcadi balfourii*, *D. guichardii* and *D. kuriensis* (Hyacinthaceae) | *Croton socotranus* (Euphorbiaceae) | *Aloe jawiyon*, *A. perryi*, and *A. squarrosa* (Aloaceae) |
| *Helichrysum gracilipes* (Asteraceae) | *Drimia porphyrostachys* (Hyacinthaceae) | *Cryptolepis intricata* (Asclepiadoideae; Apocynaceae) | *Dendrosicyos socotrana* (Cucurbitaceae) |
| *Melhania milleri* (Sterculiaceae) | *Dorstenia socotrana* (Moraceae) | *Farsetia socotrana* (Brassicaceae) | *Dorstenia gigas* (Moraceae) |
| *Nanorrhinum kuriensis* (Scrophulariaceae) | *Habenaria socotrana* (Orchidaceae) | *Jatropha unicostata* (Euphorbiaceae) | *Duvaliandra dioscoridis* (Asclepiadoideae) |
| *Polycarpaea balfourii* (Caryophyllaceae) | *Ledebouria grandifolia* and *L. insularis* (Hyacinthaceae) | *Justicia rigida* (Acanthaceae) | *Echidnopsis* species (Asclepiadoideae) |
| | | *Lachnocapsa spathulata* (Brassicaceae) *Placopoda virgata* (Rubiaceae) | *Euphorbia spiralis*, *E. schimperi* and *E. hamaderoensis* (Euphorbiaceae) |
| | | *Pulicaria stephanocarpa* (Asteraceae) | *Socotrella dolichocnema* (Asclepiadoideae; Apocynaceae) |

Figure 8.4 Drought-resisting bottle trees of the Socotra archipelago. (a) *Adenium obesum* subsp. *sokotrana*, larger than its mainland relatives. Photo by Kay Van Damme. (b) *Dendrosicyos socotranus*, the only arborescent species of the Cucurbitaceae family. Photo by Steve Scott, RBGE. (c) The cliff-dwelling *Dorstenia gigas*. Photo by Patrick Home Robertson, RBGE.

Abd al Kuri but present elsewhere in the archipelago, can grow over 6 m, making it one of the largest trees on the islands. This species has abundant parenchyma cells with undulating walls, allowing them to expand in order to hold more water during rainy periods (Olsen, 2003). Dead leaves persist on the branches, providing shade which reduces transpiration from living leaves (Olsen, 2003).

*Dorstenia gigas* (Moraceae), also endemic, has a similar appearance to *Adenium obesum* but is much smaller (< 2.5 m) and is restricted to rock faces and cliffs (Fig. 8.4c). *D. gigas* is far larger than any of the 169 species of its genus and is often dubbed the 'giant *Dorstenia*'. In fact, the three bottle trees mentioned here could be considered examples of insular gigantism, although examples are more commonly given from the animal kingdom (e.g. Whittaker & Fernández-Palacios, 2007: 186–194; for Socotran fauna see Cheung & DeVantier, 2006).

The Socotra archipelago has a diversity of smaller succulent plant species that are also drought resistors, for example, three endemic *Aloe* (Aloaceae) species (Table 8.3). There are 11 succulent species of the Asclepiadoideae (Apocynaceae), including five endemic species of *Echidnopsis*, and the monospecific, endemic genera *Socotrella* and *Duvaliandra*. There are also several succulent *Euphorbias* (Table 8.3).

## *Dracoid growth forms*

Dracoid growth forms occur in plant species that have a single stem or trunk with tips that successively ramify at the ends. The ramification creates almost even-sized

Figure 8.5 Dracoid growth forms on Socotra, a proposed adaptation to arid environments (Beyhl, 1996). (a) *Dracaena cinnabari* (Ruscaceae). Photo by Patrick Home Robertson, RBGE. (b) *Euphorbia arbuscula* (Euphorbiaceae), with a ring of green understorey plants beneath; the dense canopy reduces solar radiation and evaporation, increasing water availability for the tree and allowing other plants to exploit the greater amount of soil moisture. Photo by Lisa Banfield.

branches that spread to create an umbrella-shaped crown (Fig. 8.5). According to Beyhl (1996), dracoid growth forms occur in 15 unrelated species. They are found in regions governed by a similar climate: the Canary Islands, Cape Verde Islands, Yemen, Oman and countries around the Mediterranean. Interestingly, these regions contain so-called fragments of a 'Tethyan flora' (see below). Three plant species of the Socotra archipelago have a dracoid growth form: *D. cinnabari*, *Euphorbia arbuscula* (Euphorbiaceae) and *Euryops arabicus* (Asteraceae) (Beyhl, 1996).

Research from other areas has shown a selective pressure on crown shape, suggesting its importance in maximising probabilities for reproduction and survival (e.g. Kuppers, 1989; Yamada & Suzuki, 1996). Beyhl (1996) suggests that dracoid growth forms have evolved in adaptation to climates that are arid but which have atmospheric moisture available, as in many areas in the Socotra archipelago. The widely spread branches and tightly packed leaves (or photosynthetic branches in the case of *E. arbuscula*) may increase the surface area available to cause condensation of water from surrounding fogs and mists. The dense canopies also create a cooling shade which reduces solar radiation and evaporation in the area around the trunk or stem, benefiting both the tree and other plant species growing beneath it (Fig. 8.5b).

Distributions of these plants fit with the hypothesis that a dracoid growth form is advantageous in arid environments, but with high levels of mist. *Euryops arabicus* is found in high-mist areas on the limestone plateaus and Haggeher mountains from 500 m to 1350 m (Miller & Morris, 2004). *Dracaena cinnabari* is found at similar altitudes, although current regeneration is limited to the highest, wettest peaks. *Euphorbia arbuscula*, on the other hand, is found at lower altitudes in drier habitats where atmospheric moisture is minimal. Therefore, the main benefit of a dracoid growth form for this species may be reduced evaporation of soil moisture around the base of the trunk, rather than increased water uptake from the atmosphere.

### *Longevity*

In response to stresses such as cycles of climate change, and particularly to periods of aridity, species may evolve a strategy of longevity: long individual life spans with energy and resources allocated to persistence rather than to reproduction. On the Socotra archipelago, this strategy can be observed in endemic species such as *Dendrosicyos socotrana*, *Commiphora ornifolia* (Burseraceae), *Sterculia africana* var. *socotrana* (Sterculiaceae), *Lannea transulta* (Anacardiaceae) and, possibly, *Euphorbia socotrana* (Euphorbiaceae). All are found at lower, drier altitudes, are long-lived, slow-growing tree species mainly found in low densities, and are not often observed flowering or fruiting. A risk is attached to this life strategy; if drought conditions are extended the species may fail to reproduce. As the island gets drier, such species may be at increased risk of extinction.

## Evolutionary processes

### *Radiations*

Spectacular radiations are a typical phenomenon for islands, and these are particularly interesting when they can be linked to a founder effect. An ancestor becomes isolated due to a vicariant or single colonisation event, after which it can no longer share a gene pool with the original population. Following a bottleneck, the single ancestral lineage differentiates into several daughter lineages and the original form does not survive. A causal relationship with the environment may or may not apply (adaptive or non-adaptive radiation; see Whittaker & Fernández-Palacios, 2007: 217–238).

Several radiations have been suggested in plant genera of the Socotra archipelago; genera with relatively numerous species which are all endemic. Of a total of 30 species of *Echidnopsis* (Asclepiadoideae-Apocynaceae), five are endemic to Socotra island. Molecular analysis of the genus found that the four Socotran endemics sampled were monophyletic and have evolved from a single colonising ancestor (Thiv & Meve, 2007). *Echidnopsis* species on Socotra are small, succulent plants that are confined to the limestone plateaus (Miller & Morris, 2004). These species do not have overlapping areas of occurrence, suggesting an adaptive radiation, but they also occupy very similar niches, suggesting a non-adaptive radiation. However, there may be more inconspicuous variations in the micro-habitat of each species, such as pollinator relationships, that may have caused adaptive radiation. Detailed ecological studies required to answer such questions are yet to be carried out.

Seven of the 24 species of *Boswellia* are endemic to the Socotra archipelago. As with *Echidnopsis*, these frankincense trees grow throughout the island but their ranges rarely overlap (only *B. dioscorides* and *B. ameero* in Firmihin, Socotra). Each species is confined to either cliffs and rocks or ground soil. Those found on the cliffs develop swollen 'holdfasts' to support their growth from the rocks. Only one *Boswellia* species found outside the archipelago grows on cliffs. Species found on the ground have exclusively pinnate leaves, whereas those on the cliffs also have simple leaves. These observations could indicate that *Boswellia* trees have radiated to fill cliff niche habitats on Socotra, possibly to escape more arid habitats, and that there has also been selective pressure on leaf shape. However, the taxonomy of the Socotran *Boswellia* remains problematic; individual trees have been observed with intermediate leaf shapes (possible hybridisation) or even two distinct leaf shapes on a single leaf whorl on one tree. There are therefore a number of interesting features and peculiarities about the way in which *Boswellia* grows on Socotra: the unusual cliff habitat, the association between leaf shape and habitat, the presence of both simple and pinnate leaves on single plants, and possible evolution or extinction through hybridisation.

Examples of other possible radiations include *Pulicaria* (Asteraceae), of which all seven species on the archipelago are endemic, and *Hypericum* (Clusiaceae) with five species, all endemic. In the fauna, the terrestrial molluscs, a group closely associated with the vegetation, are a major example of radiation (e.g. Neubert, 2005).

### *Anagenesis*

Anagenesis is an evolutionary process where a single species evolves due to genetic drift or directional selection, leading to the development of a single species rather than

Table 8.4 *Suggested examples of anagenesis (top half of table) in the Socotra archipelago where families are represented by a single endemic species only, and (lower half of table) families represented by a single non-endemic species.*

| Family | Endemic species |
|---|---|
| Amaryllidaceae | *Crinum balfourii* |
| Colchicaceae | *Iphigenia socotrana* |
| Dioscoreaceae | *Dioscorea lanata* |
| Dirachmaceae | *Dirachma socotrana* |
| Ebenaceae | *Euclea divinorum* |
| Goodeniaceae | *Scaevola socotraensis* |
| Malpighiaceae | *Acridocarpus socotranus* |
| Meliaceae | *Turraea socotrana* |
| Ruscaceae | *Dracaena cinnabari* |
| Thymelaeaceae | *Gnidia socotrana* |
| | **Non-endemic species** |
| Avicenniaceae | *Avicennia marina* |
| Buxaceae | *Buxanthus pedicellatus* |
| Ephedraceae | *Ephedra foliata* |
| Erythroxylaceae | *Erythroxylon socotranum* |
| Flacourtiaceae | *Dovyalis abyssinica* |
| Globulariaceae | *Poskea socotrana* |
| Myrsinaceae | *Myrsine africana* |
| Oxalidaceae | *Oxalis corniculata* |
| Pittisporaceae | *Pittosporum viridiflorum* |
| Salvadoraceae | *Salvadora persica* |

a radiation. Again, the ancestral form becomes extinct. There are many examples of possible anagenesis on the Socotra archipelago in families and genera represented by only a single endemic species. For example, the family Goodeniaceae has only one species, *Scaevola socotraensis*, an endangered endemic limited to the rare habitat of the permanent springs of Socotra. Other examples are listed in Table 8.4 below.

Traditionally, these examples of suggested anagenesis have been considered indications of a vicariant origin of the flora. Many of these species are confined to the woodland flora found in the wet refugia of the Haggeher mountains on Socotra. As discussed above, these areas have potentially been more stable climatically, and in the absence of available niches speciation may have been constrained to anagenesis due to directional selection or genetic drift rather than radiation. However, it is also possible that other species of these genera were present on the archipelago previously, but have become extinct due to aridification, marine transgressions or other factors.

Although anagenesis may have occurred in the examples given, there are also many cases where families are represented on the archipelago by a single non-endemic species, including seven fern families and examples listed in Table 8.4. These species may share a gene pool with populations outside the archipelago, preventing genetic drift and, depending on differences in the ecosystems of the separate populations, directional selection.

### *Anacladogenesis*

Anacladogenesis is unlike anagenesis in that the progenitor of a group of species remains unchanged, but radiation occurs in isolated populations (Whittaker & Fernández-Palacios, 2007: 208). There are several possible examples of this on the archipelago. *Polycarpaea* (Caryophyllaceae) has eight species in total, of which seven are endemic. *Helichrysum* (Asteraceae) has 13 species, of which 12 are endemic. Other large genera include *Hibiscus* (Malvaceae: 15 species, 11 endemic) and *Heliotropium* (Boraginaceae: 17 species, 11 endemic). Some of these genera have more than one non-endemic species in the archipelago, indicating that multiple colonisation or vicariant events may have occurred and that gene flow with the wider population continues. However, phylogenetic analysis is required to determine whether a single ancestor or multiple ancestors have radiated, leading to the endemic species.

## Biogeography of the flora

### *Floristic classifications and affinities*

The flora of the Socotra archipelago has traditionally been linked with areas of northeast tropical Africa and Arabia. White (1983) placed the archipelago in the Somalia-Masai Regional Centre of Endemism, along with southeast Sudan, northeast Uganda, some parts of Kenya, Ethiopia, Tanzania, and an uncertain area of southern Arabia. Within the centre of endemism, White also described a Somalo-Arabian Domain, classifying the Socotra archipelago with Arabia and Somalia particularly. Genera particularly diverse in the Somalia-Musai Regional Centre of Endemism include *Boswellia* (Burseraceae), *Acacia* (Fabaceae), *Commiphora* (Burseraceae), *Cadaba* (Capparaceae), *Euphorbia* (Euphorbiaceae), *Indigofera* (Fabaceae) and *Maerua* (Capparaceae). Takhtajan (1986) categorised the archipelago in a similar way, within an Eritreo-Arabian floristic subregion, but placed it in a separate Socotra Province. Studying the bryophyte flora, Kürschner (2000) found that more than 50% of the bryophytes have a tropical origin, with a particular link to Africa.

Discussion to date has focused on the origin and affinities of endemic taxa and their disjunctions (see below). However, analysis of the distribution of non-endemics is another way of uncovering affinities. The distribution of all non-endemic vascular plant species, based on Miller and Morris (2004), is shown in Table 8.5. Country level statistics are likely to be underestimates.

Table 8.5 *Distribution of non-endemic plant species of the Socotra archipelago, based on data from Miller and Morris (2004).*

| Area/climatic region | Number of species | % | Number of unique species* |
|---|---|---|---|
| **Tropical regions** | 380 | 80.0 | |
| Pantropical | 32 | 6.7 | |
| Old World tropics | 348 | 73.3 | 65 |
| Tropical Africa | 279 | 58.7 | 30 |
| Tropical northeast Africa | 129 | 27.2 | 23 |
| Somalia | 62 | 13.0 | 17 |
| Ethiopia | 26 | 5.3 | 0 |
| Kenya | 15 | 3.2 | 0 |
| Eritrea | 4 | 0.8 | 0 |
| Djibouti | 4 | 0.8 | 0 |
| Madagascar | 24 | 5.1 | 0 |
| Tropical Asia (including trop. Arabia) | 195 | 40.1 | 16 |
| Tropical Asia (excluding trop. Arabia) | 153 | 32.2 | 0 |
| Indian subcontinent | 100 | 21.0 | 0 |
| Pakistan | 63 | 13.5 | 0 |
| Southeast Asia | 15 | 3.2 | 0 |
| Tropical Arabia/Arabian Peninsula | 61 | 12.8 | 16 |
| Yemen | 35 | 7.4 | 2 |
| Oman | 9 | 2.0 | 0 |
| Saudi Arabia | 18 | 3.8 | 0 |
| (Middle East) | (214) | (45.1) | 15 |
| New World tropics | 64 | 13.5 | 0 |
| **Temperate regions** | 176 | 37.0 | 0 |
| Temperate Asia | 124 | 26.1 | 0 |
| America | 9 | 1.9 | 0 |
| Europe | 3 | 0.6 | 0 |

*Species found on the Socotra archipelago and the corresponding area/climatic region and not elsewhere.

The analysis confirms a close affinity to tropical areas, particularly the Old World tropics and tropical Africa. In total, 80% of taxa are found in tropical regions and 57.5% in tropical Africa. This correlates well with Kürschner's analysis of bryophyte origin; 23 species are shared only with tropical northeast Africa, and 17 only with Somalia. There are no species unique to tropical Asia and Socotra, whereas 16 are found only in tropical Arabia (some parts of Yemen, Oman and Saudi Arabia) and Socotra.

Perhaps surprising when considering the past classifications of the flora is the high percentage of species shared with tropical Asia (32.2%) excluding tropical Arabia. Of the non-endemics, 21% are found on the Indian subcontinent, also surprising, and 13.5% are found in Pakistan. This exceeds the number shared with the closer areas of Somalia (13%) and Yemen (7.4%). This is contrary to the affinities considered previously, and possibly indicates a high level of dispersal to and from the Indian subcontinent or via other areas. Another surprising result is the high percentage of non-endemics that are also found in temperate regions (37%).

## *Disjunctions of endemic taxa*

Study of the possible origins of the endemic taxa of Socotra has a long tradition, going back as early as Balfour (1888), and more recently by several authors independently (e.g. Mies, 1998; Kürschner, 2000; Miller & Morris, 2004). Traditionally these studies have considered the interesting disjunctions of endemic taxa.

### *Tethyan disjunctions*

Tethyan disjunctions, including Macaronesian-Socotran disjunctions, have been given strong attention in the literature (e.g. Balfour, 1888; Mies, 1998). These disjunctions are thought to be relicts of a flora that was continuous in Laurasia bordering the Tethys Ocean, occurring in the Mesozoic to the late Tertiary. Areas around this ocean had similar, subtropical climates and evidence suggests they shared a similar flora before increasing aridity caused it to die out across many of the mainland areas (e.g. Rodríguez-Sánchez & Arroyo, 2008 and references therein). The vegetation survived in disjunct areas where the climate was still suitable. Taxa with a Tethyan disjunction are therefore thought to be vicariant.

*Hemicrambe* and *Nesocrambe* (Brassicaceae, Brassicae tribe) are thought to have a Tethyan origin. *Hemicrambe* is a genus of two species, one endemic to Socotra (*H. fruticosa*) and one endemic to Morocco (*H. fruticulosa*). *Nesocrambe* has a single species endemic to Socotra. These are both confined to very limited wet refugia habitats. *Hemicrambe* has traditionally been considered basal within the Brassicaceae family due to seemingly vestigial seed characteristics (Gomez-Campo, 1977). Both genera have a perennial woody habit, a characteristic among the Brassicae tribe mainly confined to other taxa with Tethyan distributions (Miller *et al.*, 2002). As these taxa retracted to wetter areas during aridification, rapid radiation occurred in their place, producing the majority of the Brassicae tribe we see today: annuals and biennials able to withstand the arid climate (Hedge, 1976). Recent molecular analysis by Franzke *et al.* (2009) produced a divergence age of the sister families Brassicaceae and Cleomaceae of 19 Ma (Lower Miocene) and suggested that their last common ancestor was adapted to a warm, humid (sub)tropical climate.

*Campylanthus* (Scrophulariaceae) is a genus of 15 species, with two in Macaronesia, 12 in northeastern Africa and Arabia (including one in Socotra) and one in Pakistan (Hjertson, 2003). It is a genus of uncertain position within its family and is taxonomically isolated, possibly indicating a great divergence age. The isolation of the Macaronesian species is thought to have occurred following the retreat of the

Table 8.6 *Examples of disjunctions shown in genera found on the Socotra archipelago (after Miller & Morris, 2004: 9–10).*

| Tethyan disjunctions | African dry-corridor disjunctions | Boreotropical/ New World disjunctions |
|---|---|---|
| *Campylanthus* (Scrophulariaceae) | *Camptoloma* (Scrophulariaceae) | *Chapmannia* (Fabaceae) |
| *Dracaena* (Ruscaceae) | *Chorisochora* (Acanthaceae) | *Erythroxylon* (Erythroxylaceae) |
| *Hemicrambe* (Brassicaceae) | *Thamnosma* (Rutaceae) | *Indigofera* (Fabaceae) |
| *Nesocrambe* (Brassicaceae) | *Wellstedia* (Boraginaceae) | *Zygocarpum* (Fabaceae) |

Tethyan flora, but the African and Arabian species may have been more continuous until the Pleistocene (Hjertson, 2003) (see below). Phylogenetic analysis is currently being carried out on *Campylanthus* at the State Museum of Natural History, Stuttgart, Germany.

*Dracaena*, and in particular the arborescent species, is also suggested to be a Tethyan element. Species are found on Socotra, Morocco, Macaronesia, the Arabian Peninsula and northeast tropical Africa, and fossils have been recorded from southern France (e.g. Marrero *et al.*, 1998).

### *African dry-corridor disjunctions*

Some genera display a disjunct distribution between arid areas of Africa. These are thought to be remnants of previously continuous populations that were spread along a dry corridor during the period of aridification during the Pleistocene. This corridor is thought to have previously joined the vegetation of the northern part of White's Somalia-Masai centre of endemism and the southern part of the Zambesian centre of endemism, as well as to the Karoo-Namib region. Many disjunctions have been attributed to this dry corridor and its possible influence has long been discussed (e.g. Verdcourt, 1969; White, 1983). Taxa thought to have an African dry-corridor disjunction are therefore found in semi-desert areas in southwest Africa, Somalia, Ethiopia, Socotra and mainland Arabia, and the disjunctions are attributed to vicariance.

*Wellstedia* (Boraginaceae) has six species distributed in Socotra (the endemic *W. socotrana*), Somalia, Ethiopia, South Africa and Namibia. Preliminary analysis found *W. socotrana* to be vicariant sister to the Somalian species *W. somalensis* (Thulin & Johansson, 1996). The genus has unclear relationships within Boraginaceae as the fruits are uniquely capsular, so it is generally allocated a distinct subfamily (Wellstedioideae) (e.g. Retief & van Wyk, 2008; Thulin & Johansson, 1996). In the past it has been placed in an entirely separate family, Wellstediaceae (e.g. Novak, 1943) (see Table 8.6 for further examples).

### *Boreotropical/New World disjunctions*

Many plant taxa have disjunct distributions between North America and Africa, with a predominance of characters adapted to seasonally dry or xeric vegetation (Lavin *et al.*, 2000). These taxa are thought to have originated from a boreotropical dry woodland vegetation that covered large areas of the Eurasian and North American land masses (Laurasia) in Cretaceous and Tertiary times. This formed a single vegetation zone with limited barriers to dispersal. The continental land masses are thought to have been joined by land bridges (for example, Lavin *et al.*, 2000 and references therein). As with the Tethys flora, extinctions and retraction of the vegetation was caused by climate changes as well as continental migration northwards and the severance of land connections.

*Chapmannia*, *Zygocarpum* and *Indigofera* (all Fabaceae, Papillionoideae) have distributions that suggest a boreotropical origin. The genus *Chapmannia* has seven species: four in Socotra, one in Somalia and two in America. Both genera are closely related to others that have Tertiary fossil records (Lavin *et al.*, 2000). Phylogenetic analysis supported a vicariant origin of the present-day distributions of *Chapmannia* and *Zygocarpum*, rather than dispersal between Africa and America (Lavin *et al.*, 2000). In contrast, the boreotropical disjunction of *Indigofera* has been explained by dispersal, as recent analysis of the Indigofereae tribe indicated an oldest age of only 16 Ma (Schrire *et al.*, 2009).

### *Intra-island disjunctions*

Interesting disjunctions can also be observed within the archipelago, and in particular between the isolated and separate wet refugia, with some endemic species present in all refugia and others far more restricted. Investigations so far have been limited to comparison of dispersal capabilities of restricted and non-restricted species, measured by analysing seed characteristics (Banfield, 2006). Some patterns were found but they were not statistically significant. For example, seeds of non-restricted species were either anemochorous or endozoochorous, whereas seeds of restricted species were also exozoochorous, barochorous, or had explosive fruits (Banfield, 2006).

Possible vicariant species pairs occur in endemic species found on separate islands within the archipelago. For example, *Dipcadi kuriensis* (Hyacinthaceae) (Abd al Kuri) and *D. balfourii* (Socotra), and *Ledebouria samhaensis* (Hyacinthaceae) (Samha) and *L. grandifolia* (Socotra). Further investigation of limiting ecological factors or population genetic studies are required to further understand the disjunctions within the archipelago.

### *Faunal disjunctions*

Interesting disjunctions are also found in fauna taxa. For example, examples of important radiations and endemism are found in the invertebrates, such as the

Figure 8.6 Enigmatic taxa of the Socotra archipelago. (a) *Dirachma socotrana* (Dirachmaceae), one of only two species of the genus *Dirachma*, with unclear affinities due to a unique combination of characters, including flowers with eight parts. Photo by Lisa Banfield. (b) *Punica protopunica* (Punicaceae/Lythraceae), the only known relative of the pomegranate. Photos, habit: Steve Scott, RBGE; fruit and flower: Patrick Home Robertson, RBGE.

terrestrial molluscs and isopods (e.g. Taiti & Ferrara, 2004; Neubert, 2005), both of significant biogeographical value.

## *Enigmatic taxa*

Some plant taxa of Socotra have no clear affinities or easily recognisable relationships, as the previously mentioned *Dendrosicyos socotrana*, and the genera *Campylanthus* and *Wellstedia*. Perhaps the most striking example of an enigmatic taxon, however, is the genus *Dirachma* (Dirachmaceae), with one species endemic to Socotra (*D. socotrana*) (Fig. 8.6a) and one endemic to Somalia (*D. somalensis*). These species have a unique combination of characters including flowers with eight parts (octohomomerous), free petals positioned opposite the stamens, deeply lobed ovaries with one ovule per chamber and nectarines at the petal bases (Baas *et al.*, 2001). The family has been described as 'a living fossil'. Relationships have been suggested with Geraniaceae, Tiliaceae, Sterculiaceae, Samydaceae and Greyiaceae. Molecular and morphological analysis found two differing relationships, but placed the family in the Rhamnales along with Rhamnaceae, Elaeagnaceae and Barbeyaceae (Thulin *et al.*, 1997).

*Punica protopunica* (Fig. 8.6b) is also an enigmatic taxon, being the only known relative of the widely cultivated pomegranate *P. granatum*, which is found naturally from the Himalayas to the Balkans (Miller & Morris, 2004). Relationships are unclear. Cladistic analysis has placed *Punica* within the Lythraceae family as, along with several other genera, the family becomes monophyletic following its inclusion (Johnson & Briggs, 1984). This has been supported by Graham *et al.* (2005), who also carried out molecular and morphological analysis and suggested a Cretaceous Laurasian, rather than Gondwanan, origin. However, there are no clear synapomorphies for the expanded Lythraceae family, and others have preferred to place *Punica* within its own family, Punicaceae (e.g. Miller & Morris, 2004).

These enigmatic species are often considered ancient relicts. However, uncertain taxonomic relationships make them difficult to analyse using molecular techniques, as their closest living relatives are uncertain and divergence is likely to be large and difficult to date. However, these are perhaps the species that are most likely to show a vicariant origin due to their uncertain affinities possibly reflecting a greater period of isolation.

### *Vicariance vs dispersal*

As with other islands, dispersal and vicariance hypotheses have been formulated to explain the origin of the Socotran flora, and the examples given above have been discussed in these terms. Traditionally, due to the described disjunct distributions, the proposed occurrences of anagenesis (indicating a possible lack of available niches to enable adaptive radiation) and enigmatic taxa, a vicariant origin of the flora of the Socotra archipelago has been emphasised.

To build on and test the vicariance vs dispersal hypotheses, Socotran endemics are increasingly included in molecular analyses. Some have supported the vicariant hypothesis, as with *Chapmannia* and *Zygocarpum* (Lavin *et al.*, 2000) discussed above. In addition, divergence of the enigmatic *Dendrosicyos socotrana* has been estimated using molecular clocks at around 22 (30–14) Ma (Schaefer *et al.*, 2009), which coincides with the estimated separation of the archipelago.

Other dated molecular phylogenies suggest a younger age for some groups that could only have reached the island after its separation from the mainland. Dispersal was suggested, for example, for *Aerva* (Amaranthaceae) (Thiv *et al.*, 2006) and *Echidnopsis* (Asclepiadoideae; Apocynaceae) (Thiv & Meve, 2007), and is perhaps most likely for many of the groups with closest African and Arabian affinities. The high proportion of non-endemics shared with the Indian subcontinent shown above may also indicate a dispersal origin for endemic taxa with Indian affinities. A number of families with a single, non-endemic species present on the archipelago, and examples of anacladogenesis, also indicate possible dispersals and continued gene flow with surrounding populations.

For other taxa, a clear vicariance/dispersal origin cannot be confirmed. For example, the genus *Exacum*, which has endemic representatives on Socotra, was originally thought to be a group with a typical Gondwanan distribution (e.g. Klackenberg, 1985), but more recent evidence shows that its distribution may only be partly 'vicariance-like' (Yuan *et al.*, 2005).

Despite these emerging studies, hypotheses of vicariance and dispersal are not mutually exclusive (Miller & Morris, 2004) and there is no general, overall hypothesis for the Socotran flora. Information is still relatively scattered and a discussion of new data versus older hypotheses is yet to start.

## Threats to the flora

The flora of the Socotra archipelago remains relatively untouched compared to floras of many other islands. Although there are a few alien species, none have become invasive and those posing the greatest threat have been eradicated or are being managed. However, increasing interaction with the mainland is making the arrival of invasive species more likely. Large-scale infrastructure developments, disproportionate to local needs and dismissive of environmental impacts, are damaging important habitats, as seen in an ongoing road-building programme. Traditional land management, practised since the arrival of humans several thousand years ago, has shaped the flora as we see it today. These practices are becoming less adhered to as Socotra is increasingly influenced by the outside world. Serious overgrazing followed by soil erosion is found in some areas. Although the flora has survived abrupt changes between a wet and dry climate, intensified and continued arid periods would threaten many of the endemic species for which the archipelago is famous. This is perhaps already apparent in the lack of regeneration of flagship species, such as *Dracaena cinnabari*. For an in-depth discussion of conservation issues on Socotra island see Van Damme & Banfield (2011).

## Conclusion

The Socotra archipelago has a diverse and distinctive flora with many notable species. It has a greater number of endemic species per $km^2$ than several islands known for having high diversity. This is likely related to a diverse geomorphology and variable climatic conditions providing a diversity of ecological niches, combined with a long period of isolation. Wet refugia of the archipelago support a disproportionate number of species and endemics, although adaptations to aridity are found in many notable endemics.

The archipelago has traditionally been linked floristically with Africa, particularly northeast tropical Africa, and Arabia. Analysis of the distributions of non-endemics supports this, but also highlights a link with the Indian subcontinent. Some taxa in the flora and fauna show interesting disjunctions with significant biogeographical value.

Discussions regarding plant biogeography generally focus on the 'dispersal vs vicariance' debate. It is clear from the above that both scenarios are likely to have played an important part in forming the Socotran vegetation we see today. Molecular phylogenies published to date have shown support for both the vicariance and dispersal hypotheses.

In comparison to many other islands, research of biogeography on Socotra island is relatively recent, and intensive discussions or reviews are only now beginning to take shape.

Despite the flora being relatively intact now, conservation efforts will be needed to preserve the flora and fauna of this unique archipelago.

## References

Attorre, F., Francesconne, F., Taleb, N., Scholte, P., Saed, A., Alfo, M. et al. (2007). Will dragonblood survive the next period of climate change? Current and future potential distribution of *Dracaena cinnabari* (Socotra, Yemen). *Biol. Conserv.* 138: 430–439.

Baas, P., Jansen, S. & Smits, E. (2001). Vegetative anatomy and affinities of *Dirachma socotrana* (Dirachmaceae). *Syst. Bot.* 26 (2): 231–241.

Balfour, I. B. (1888). Botany of Socotra. *Trans. R. Soc. Edinb.* 31: i–lxxv & 1–446.

Banfield, L. M. (2006). The Biodiversity and Conservation of the Wet Refugia of the Socotra Archipelago. MSc thesis: Royal Botanic Garden Edinburgh.

Beydoun, Z. R. & Bichan, H. R. (1970). The geology of Socotra Island, Gulf of Aden. *Q. J. Geol. Soc. Lond.* 125: 413–446.

Beyhl, F. E. (1996). The Enta tree (*Euphorbia arbuscula* Balf. Fil.), a succulent tree of dracoid habitus from the island of Socotra. In H. J. Dumont (ed.), *Soqotra: Proceedings of the First International Symposium on Soqotra Island, Aden 1996: Present and Future.* United Nations Development Programme, Conservation and Sustainable Use of Biodiversity of Soqotra Archipelago, Technical Series, Volume 1. New York, pp. 115–123.

Blondel, J. & Aronson, J. (1999). *Biology and Wildlife of the Mediterranean Region.* Oxford, Oxford University Press.

Brown, M. J., Jarman, S. J. & Kantvilas, G. (1994). Conservation and reservation of non-vascular plants in Tasmania, with special reference to lichens. *Biodivers. Conserv.* 3: 263–278.

Burns, S. J., Fleitmann, D., Matter, A., Kramers, J. & Al-Subbary, A. A. (2003). Indian Ocean climate and an absolute chronology over Dansgaard/Oeschger Events 9 to 13. *Science* 301: 1365–1367.

Cheung, C. & Devantier, L. (2006). *Socotra: A Natural History of the Islands and their People* (K. Van Damme, ed.), Hong Kong, Odyssey Books and Guides, Airphoto International Ltd, Hong Kong.

Culek, M., Kral, K., Habrova, H., Adolt, R., Pavlis, J. & Madera, P. (2006). Socotra's annual weather pattern. In C. Cheung & L. DeVantier. *Socotra: A Natural History of the Islands and their People* (K. Van Damme, ed.), pp. 42–46. Hong Kong, Odyssey Books and Guides, Airphoto International Ltd, Hong Kong.

De Geest, P. (2006). Speleothems and palaeoclimate. In C. Cheung & L. DeVantier. *Socotra: A Natural History of the Islands and their People* (K. Van Damme, ed.), pp. 36–37. Hong Kong, Odyssey Books and Guides, Airphoto International Ltd, Hong Kong.

Fleitmann, D., Matter, A., Burns, S. J., al-Subbary, A. K. & al-Aowah, M. A. (2004). Geology and Quaternary climate history of Socotra. *Fauna Arabia* 20: 27–43.

Franzke, A., German, D., al-Shebaz, I. A. & Mummenhoff, K. (2009). Arabidopsis family ties: molecular phylogeny and age estimates in Brassicaceae. *Taxon.* 58: 425–437.

Gomez-Campo, C. (1977). Studies on *Cruciferae*: III. *Hemicrambe townsendii*. nom. nov. An example of geographic disjunction. *An. Instit. Bot. A. J. Cavanilles* 34 (1): 151–155.

Graham, S. A., Hall, J., Sytsma, K. & Shi, S. H. (2005). Phylogenetic analysis of the Lythraceae based on four gene regions and morphology. *Int. J. Plant Sci.* 166 (6): 995–1017.

Hedge, I. C. (1976). A systematic and geographic survey of the Old World *Cruciferae*. In J. G. Vaughan, A. J. Macleod & B. M. G. Jones (eds.), *The Biology and Chemistry of the Cruciferae*, pp. 47–68. London, Academic Press.

Hjertson, M. (2003). Revision of the disjunct genus *Campylanthus* (Scrophulariaceae). *Edinb. J. Bot.* 60 (2): 131–174.

Hughes, M., Hollingsworth, P. M. & Miller, A. G. (2003). Population genetic structure in the endemic Begonia of the Socotra Archipelago. *Biol. Conserv.* 113: 277–284.

Johnson, L. A. S. & Briggs, B. G. (1984). Myrtales and Myrtaceae: a phylogenetic analysis. *Ann. Mo. Bot. Gard.* 71: 700–756.

Kearney, T. H. & Shantz, H. L. (1911). The water economy of dry land crops. In *US Department of Agriculture Yearbook*, pp. 351–361.

Kilian, N. & Hein, P. (2006). New and noteworthy records for the vascular plant flora of Socotra Island, Yemen. *Englera*. 28: 57–77.

Klackenberg, J. (1985). The genus *Exacum* (Gentianaceae). *Opera Botanica a Societate Botanica Lundensi* (Lund, Copenhagen) 84: 1–144.

Kuppers, M. (1989). Ecological significance of above-ground architectural patterns in woody-plants: a question of cost-benefit relationships. *Trends Ecol. Evol.* 4 (12): 375–379.

Kürschner, H. (2000). Bryophyte flora of the Arabian Peninsula and Socotra. *Bryophytum Bibliotheca*. 55: 1–131.

Kürschner, H. (2003). Nineteen new records to the bryophyte flora of Socotra Island. Additions to the bryophyte flora of the Arabian Peninsula and Socotra 5. *Willdenowia* 33: 445–458.

Lavin, M., Thulin, M., Labat, J-N. & Pennington, R. T. (2000). Africa, the odd man out: molecular biogeography of Dalbergioid legumes (*Fabaceae*) suggests otherwise. *Syst. Bot.* 25 (3): 449–467.

Leroy, S., Gentss, P., Fournier, M., d'Acremont, E., Patriat, P., Beslier, M. O. *et al.* (2004). From rifting to spreading in the eastern Gulf of Aden: a geophysical survey of a young oceanic basin from margin to margin. *Terra Nova*. 16 (4): 185–192.

Marrero, A., Almeida, R. S. & González-Martín, M. (1998). A new species of the wild dragon tree, *Dracaena* (Dracaenaceae) from Gran Canaria and its taxonomic and biogeographic implications. *Bot. J. Linn. Soc.* 128 (3): 291–314.

Mies, B. (1998). The phytogeography of Soqotra: evidence for disjunctive taxa, especially with Macaronesia. In H. J. Dumont (ed.), *Soqotra: Proceedings of the First International Symposium on Soqotra Island, Aden 1996: Present and Future.* United Nations Development Programme, Conservation and Sustainable Use of Biodiversity of Soqotra Archipelago, Technical Series, Volume 1. New York, pp. 83–105.

Miller, A. G. & Morris, M. (2000). *Final Report on the Biodiversity and Resource Use Survey: Target Areas.* Report for the UNDP/GEF project for the Conservation and Sustainable Use of the Biodiversity of Socotra Archipelago – YEM/96/G32.

Miller, A. G. & Morris, M. (2004). *The Ethnoflora of the Soqotra Archipelago.* Edinburgh: Royal Botanic Garden Edinburgh.

Miller, A. G., Atkinson, R., Alkhulaidi, A. W. & Taleb, N. (2002). *Nesocrambe*, a new genus of *Cruciferae (Brassiceae)* from Soqotra, Yemen. *Willdenowia*. 32: 61–67.

Morris, M. (2002). *Manual of Traditional Land Management Practices in the Socotra Archipelago.* Report for the UNDP/GEF project for the Conservation and Sustainable Use of the Biodiversity of Socotra Archipelago – YEM/96/G32.

Neubert, E. (2005). The continental malacofauna of Arabia and adjacent areas. IV. Revision of the family Cerastidae in the Soqotran Archipelago. I. The genus *Passamaella* L. Pfeiffer, 1877, with description of a new genus and species (Gastropoda: Pulmonata: Cerastidae). *Arch. Molluskenkunde* 134: 1–21.

Neubert, E. (2006). Land snails of the Socotra Archipelago. In C. Cheung & L. DeVantier. *Socotra: A Natural History of the Islands and their People* (K. Van Damme, ed.), pp. 125–127. Hong Kong, Odyssey Books and Guides, Airphoto International Ltd, Hong Kong.

Neubert, E. (2009). The continental malacofauna of Arabia and adjacent areas. VI. Pomatiidae of Arabia, Socotra and Northeast Africa, with descriptions of new genera and species (Gastropoda: Caenogastropoda: Littorinoidea). *Fauna Arabia* 24: 47–127.

Novack, F. A. (1943). Cited in: Retief, E. & van Wyk, A. E. (2008). The genus *Wellstedia* (Boraginaceae: Wellstedioideae) in southern Africa. *Bothalia* 38 (1): 57–63.

Olsen, M. E. (2003). Stem and leaf anatomy of the arborescent Cucurbitaceae *Dendrosicyos socotrana* with comments on the evolution of pachycauls from lianas. *Plant Syst. Evol.* 239: 199–214.

Raven, P. H., Evert, R. F. & Eichhorn, S. E. (1992). *Biology of Plants*, 5th edition. New York, Worth Publishers.

Retief, E. & van Wyk, A. E. (2008). The genus *Wellstedia* (Boraginaceae: Wellstedioideae) in southern Africa. *Bothalia* 38 (1): 57–63.

Rodrigues-Sanchez, F. & Arroyo, J. (2008). Reconstructing the demise of Tethyan plants: climate-driven range dynamics of *Laurus* since the Pliocene. *Global Ecol. Biogeogr.* 17 (6): 685–695.

Roos, M. C., Kessler, P. J. A., Gradstein, S. R. & Baas, P. (2004). Species diversity and endemism of five major Malesian islands: diversity-area relationships. *J. Biogeogr.* 31: 1893–1908.

Rössler, H. & Wranik, W. (2004). A key and annotated checklist to the reptiles of the Socotra archipelago. *Fauna Arabia* 20: 505–534.

Samuel, M. A., Harbury, N., Bott, R. & Thabett, A. M. (1997). Field observations from the Socotran platform: their interpretation and correlation to Southern Oman. *Mar. Pet. Geol.* 14 (6): 661–673.

Schaefer, H. M., Heibl, C. & Renner, S. (2009). Gourds afloat: a dated phylogeny reveals an Asian origin of the gourd family (Cucurbitaceae) and numerous oversea dispersal events. *Proc. R. Soc. B.* 276: 843–851.

Schils, T. & Copperjans, E. (2003). Phytogeography of upwelling areas in the Arabian Sea. *J. Biogeogr.* 30 (9): 1339–1356.

Scholte, P. & De Geest, P. (2010). The climate of Socotra Island (Yemen): a first-time assessment of the timing of the monsoon wind reversal and its influence on precipitation and vegetation patterns. *J. Arid Environ.* 74: 1507–1515.

Schrire, B. D., Lavin, M., Barker, N. P. & Forest, F. (2009). Phylogeny of the tribe Indigofereae (Legumnosae-Papilionoideae): geographically structures more in succulent-rich and temperate settings than in rich grass-rich environments. *Am. J. Bot.* 96 (4): 816–852.

Taiti, S. & Ferrara, F. (2004). The terrestrial Isopoda (Crustacea: Oniscidae) of the Socotra Archipelago. *Fauna Arabia* 20: 211–236.

Takhtajan, A. (1986). *Floristic Regions of the World.* Berkeley, CA, University of California Press.

Thiv, M. & Meve, U. (2007). A phylogenetic study of Echidnopsis Hook. f. (Apocynaceae-Asclepiadoideae): Taxonomic implications and the colonization of the Socotran Archipelago. *Plant Syst. Evol.* 265: 71–86.

Thiv, M., Thulin, M., Kilian, N. & Linder, H. P. (2006). Eritreo-Arabian affinities of the Socotran flora as revealed from the molecular phylogeny of Aerva (Amaranthaceae). *Syst. Bot.* 31 (3): 560–570.

Thompson, J. D. (2005). *Plant Evolution in the Mediterranean.* Oxford, Oxford University Press.

Thulin, M. & Johansson, A. N. B. (1996). Taxonomy and biogeography of the anomalous genus *Wellstedia.* In L. J. G. van der Maesen, X. M. van der Burgt & J. M. van Medenbach de Rooy (eds.), *Biodiversity of African Plants,* Proceedings of XIVth Congress of the Association pour l'É tude taxonomique de la Flore d'Afrique Tropicale (AETFAT), pp. 73–86. Wageningen, Netherlands, Kluwer.

Thulin, M., Bemer, B., Richardson, J., Niklasson, J., Fay, M. F. & Chase, M. (1997). Family relationships of the enigmatic rosid genera *Barbeya* and *Dirachma* from the Horn of Africa region. *Plant Syst. Evol.* 213: 109–119.

UNEP Islands Directory (1998). *Island Index.* http://islands.unep.ch/index.htm (accessed 2 June 2009).

Van Damme, K. (2006). Island size and glacial maxima. In C. Cheung & L. DeVantier. *Socotra: A Natural History of the Islands and their People* (K. Van Damme, ed.), pp. 20–21. Hong Kong, Odyssey Books and Guides, Airphoto International Ltd, Hong Kong.

Van Damme, K. (2009). Socotra Archipelago. In R. Gillespie & D. Clague (eds.), *Encyclopedia of Islands,* pp. 846–851. Berkeley, CA, University of California Press.

Van Damme, K. & Banfield, L. M. (2011). Past and present human impacts on the biodiversity of Socotra island (Yemen): implications for future conservation. *Zoology in the Middle East.* 3: 31–88.

van den Eynden, V., Oatham, M. P. & Johnson, W. (1998). How free access internet resources benefit biodiversity and conservation research: Trinidad and Tobago's endemic plants and their conservation status. *Oryx* 42: 400–407.

Verdcourt, B. (1969). The arid corridor between the north-east and south-west areas of Africa. *Palaeoecol. Africa* 4: 140–144.

White, F. (1983). Vegetation Map of Africa. UNESCO. *Nat. Resour. Res.* 20: 1–356.

Whittaker, R. J. & Fernández-Palacios, J. M. (2007). *Island Biogeography: Ecology, Evolution and Conservation,* 2nd edition. Oxford, Oxford University Press.

Yamada, T. & Suzuki, E. (1996). Ontogenic change in leaf shape and crown form of a tropical tree, *Scaphium macropodum* (Sterculiaceae) in Borneo. *J. Plant Res.* 109: 211–217.

Yuan, Y.-M., Wohlhauser, S., Möller, M., Klackenberg, J., Callmander, M. W. & Küpher, P. K. (2005). Phylogeny and biogeography of *Exacum* (Gentianaceae): a disjunctive distribution in the Indian Ocean Basin resulted from long distance dispersal and extensive radiation. *Syst. Bot.* 54 (1): 21–34.

# 9

# Biogeography and conservation of the flora of New Caledonia

DAVID BRAMWELL

## Geography

Straddling the Tropic of Capricorn, New Caledonia is situated in the South Pacific Ocean between latitudes 18° 00′ and 23° 50′ S, and longitudes 154° 45′ and 176° 20′ E. It is an archipelago made up of the islands of Grande Terre, Iles Belep, d'Entrecasteaux Récifs, Île des Pins, the Loyalty Islands and several other small islands such as the Chesterfield Islands and Walpole Island (Fig. 9.1). New Caledonia is an isolated archipelago situated some 1500 km east of Australia with the nearest land further to the east at the New Hebrides Islands and Fiji. New Caledonia is one of the extant land masses of an ancient continent known as Zealandia and now almost completely submerged. The island was formerly part of Australasia but separated from it about 60–80 million years ago (Ma) and drifted to its present position by about 50 Ma. It is, therefore, one of the most important and isolated surviving fragments of Gondwanaland. In the process, much or all of the area of New Caledonia appears to have been submerged for up to 20 Ma. During this period, oceanic mantle was deposited over the original schistic rock, eventually forming a thick layer of ultrabasic substrates including peridotite, ferricrete, serpentinite and laterite. These were once widespread over the islands and still cover about 30% of the land surface.

The island of Grande Terre with an area of over 16 000 $km^2$ occupies nearly 90% of the land surface of the group and is by far the largest island. It is a long island with a North–South axis and a central mountain range with five peaks that exceed 1500 m in elevation.

## Climate and vegetation

As with many oceanic islands, the original vegetation of New Caledonia has been severely damaged both by extensive agricultural development and by mining

*The Biology of Island Floras*, eds. David Bramwell and Juli Caujapé-Castells. Published by Cambridge University Press. 

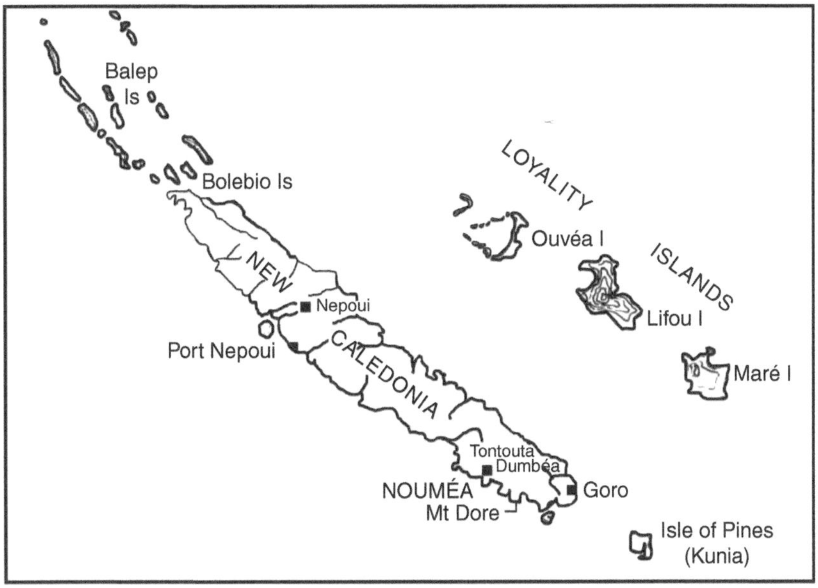


Figure 9.1 The New Caledonia archipelago. The main island is also known as Grande Terre.

activities, in this case principally nickel mining. The climate of the islands is predominantly tropical, and the highly seasonal rainfall originates from easterly trade winds. Annual rainfall averages about 1500 mm on the Loyalty Islands, 2000 mm in the lower areas of eastern Grande Terre and 2000–4000 mm in the montane region of Grande Terre. The western part of Grande Terre falls within the rain shadow of the central mountains, and rainfall is lower, averaging only about 1200 mm annually. The year has two seasons: April through November are drier, cooler months with a temperature range of 17–28°C, and December to March are warmer and much wetter with temperatures up to 33°C. During this season the effect of the southeast trade winds moderates temperatures though occasional typhoons may affect the islands.

The diverse climatic and geological conditions and the mountainous relief of the main island have given rise to a series of vegetation types each with an exceptionally rich diversity of endemic species. The principal primary types are described in Morat *et al.* (1984).

### *Humid evergreen forest*

The humid evergreen rainforests of New Caledonia once covered over 70% of the land area but now occupy only about 22%, mainly in refuges in the higher rainfall areas of the east and in the central mountains of Grande Terre where some

3000 km$^2$ still survive. Both the Loyalty Islands and Ile de Pins also still retain areas of rainforest. Evergreen rainforest is a dense tropical forest with a closed canopy of trees growing up to 20 m in height with taller emergent trees sometimes occurring. The humid evergreen forest is rich with over 2000 species and a high level of endemism of about 82.5%, but has suffered since pre-European times from agricultural clearance and forest fires and more recently from mining. The current distribution of many of the forest endemics is now fragmented and severely reduced, and numerous species face competition from invasive aliens. Like the Australian rainforests and the temperate forests of New Zealand, conifers of the Southern Hemisphere families Araucariaceae and Podocarpaceae are mixed with angiosperm trees, including the Southern Hemisphere genus *Nothofagus* and trees and shrubs of families Myrtaceae and Proteaceae.

The humid evergreen forest is subdivided into two basic types, the lowland rainforest found on Grand Terre and the Loyalty Islands and the mountain forest of Grande Terre. In the lowland forests gymnosperms predominate with *Agathis, Araucaria, Falcatifolium* and *Dacrydium* species. The angiosperm flora includes species of *Montrouziera, Calophyllum (C. neocaledonicum), Hernandia, Kermadecia* and *Macadamia* and stands of *Callistemon* and *Nothofagus* sometimes occur. The main mountain forest elements include species of *Araucaria, Agathis, Podocarpus, Libocedrus, Metrosideros, Weinmannia, Dacrydium* and *Nothofagus*. In all, over 100 plant families are found in the humid forests, the most predominant, according to Morat *et al.* (1984), being Rubiaceae, Araliaceae, Myrtaceae, Sapotaceae, Proteaceae, Leguminosae, Araucariaceae, Podocarpaceae and Guttiferae. The most prominent floristic elements of the New Caledonian rainforests are derived from the so-called Antarctic flora of the ancient southern Gondwana continent.

## *Maquis*

Maquis vegetation is characterised by low, sclerophyllous, evergreen, heath-like plants with stunted *Araucaria* trees and is found from sea level to the mountain regions on Grande Terre and other nearby islands. The maquis forests on Grande Terre are unique. Their scrublike structure resembles that of Mediterranean climate woodlands. However, the maquis of New Caledonia, rather than a climatic climax ecosystem, is a response to the special conditions of the ultrabasic substrate which formed in the deep ocean and is rich in metals, including nickel, magnesium, chromium and manganese, which are toxic to many plants (Jaffré, 1976). A rather impoverished maquis, however, also occurs on schistose acid soils, usually at low elevations. Bradford and Jaffré (2004), on the other hand, report the presence of high-altitude acidic soil maquis with a considerable degree of local microendemism. Usually regarded as a fire-maintained transition form, maquis generally has a high level of species endemism (89%) and richness (nearly 1150 species). Important families in the maquis include Myrtaceae, Dilleniaceae, Proteaceae, Epacridaceae, Casuarinaceae and Cunoniaceae and, according to Morat *et al.* (1995), many of the endemic species that are restricted to the maquis have spectacular flowers (*Hibbertia, Xanthostemon* and *Bikkia*, for example). These authors also point out that, though the montane maquis is structurally similar to the lowland formation, the floristic composition is rather different. Maquis on schist is, however, dominated by species of *Melaleuca* and *Codia.* The higher mountain maquis still occupies a major part of its original area, about 100 $km^2$, while lowland maquis is now the most widespread natural vegetation type in the islands. Lowland maquis once occupied only about 5% of the islands' area, but has spread, mainly due to fire disturbance, to occupy about 23% of New Caledonia (4500 $km^2$).

## *Sclerophyll forest*

The dry tropical forests of New Caledonia once extended throughout the lowlands of the western coast where they occur below an altitude of about 300 m, usually on sedimentary rocks. The sclerophyllous forests are unique in that they represent some of the world's richest dry forests with very high numbers of endemic species. Over 400 species of plants have been recorded, 59 of them unique to this habitat, and overall endemism within the dry forests (some 223 species) is about 58% (Bouchet *et al.* 1995). Unfortunately, the sclerophyll forests are now reduced to to a tiny proportion of their original distribution. In its optimum state, the sclerophyll forest probably covered all the lowland area of the west coast, about 4500 $km^2$. Today, pristine stands of sclerophyll forest occur in less than 100 $km^2$. These represent only about 2% (10 000 ha) of the original extension. In most places the sclerophyll forest is very degraded. In its climax condition, the sclerophyll forest is a dense woodland with a canopy of between 10 and 15 m, with trees of *Terminalia, Cupaniopsis* and *Diospyros.* It has a rather dense shrubby understorey formed of *Codiaeum, Gardenia, Pittosporum* and *Eugenia* among others and where lianas are common with over 50 reported species. The sclerophyll forest differs notably from the rainforest by the absence of gymnosperms, palms, *Pandanus* and Winteraceae, and none of the five angiosperm families endemic to New Caledonia has representatives in this formation. In its most degraded form, the sclerophyll forest can form open communities of secondary savanna vegetation of grassland and *Melaleuca quinquenervia* shrubs and grasses and sedges, or a closed bushy thicket of the native *Acacia spirorbis* or the introduced *Leucaena lezicocephala* (coffee bush). In the most degraded habitats the native flora competes with invasive exotic species such as *Leucaena lezicocephala*, *Psidium guajuva* (guava) and *Lantana camara* (lantana).

### *Mangrove*

New Caledonian mangrove vegetation covers some 200 $km^2$ or just over 1% of the total land area. The principal mangrove swamps are on the west coast of Grande Terre where they extend over 80% of the coastline in estuarine, lagoon and coastal habitats. They are, however, rather less common on the east side of the island where they are concentrated in estuarine habitats. Mangrove is rare on the smaller islands. Approximately 25 species occur in the mangrove forests but none of them is endemic, the nearest being *Rhizophora lamarckii*, a species of probable hybrid origin found on the east coast of Grande Terre and on the northeast coast of Australia.

### *Littoral and marshland vegetation*

The strand vegetation of New Caledonia consists mainly of Indo-Pacific species found throughout the region and of halophyte communities dominated by *Salicornia* and *Sueda* species. At the extreme southern end of the main ultrabasic mountain chain there is an area of small lakes and streams known as the Plaine des Lacs. Here the wetland species are all New Caledonian endemics (Morat *et al.*, 1995) such as *Chorizandra cymbaria, Pancheria communis, Cloezia aquarum* and the local endemic gymnosperms, *Dacrydium guillauminii* and *Decussocarpus minor.*

## Flora

Estimates of the size of the native flora of New Caledonia vary between 3200 and 3400 species with possibly up to a further 10% yet to be described (Morat, 1993; Morat *et al.*, 1995). Of these some 79% (almost 2650) are endemic species, making New Caledonia the third highest ranked island group for endemism after Hawaii (89%) and New Zealand (82%). There are 863 native genera, of which 110 are endemic. Of 191 families of plants present, five are considered to be endemic to New Caledonia (Amborellaceae, Paracryphiaceae, Strasburgeriaceae, Oncothecaceae and Phellinaceae). Modern familes such as Asteraceae, Lamiaceae, Melastomataceae and Graminae are under-represented in the flora whereas families of Gondwanan origin such as Proteaceae, Cunoniaceae and Myrtaceae are well represented in terms of both abundance and species diversity. Many taxa are considered to be relicts of the flora of the Gondwana land

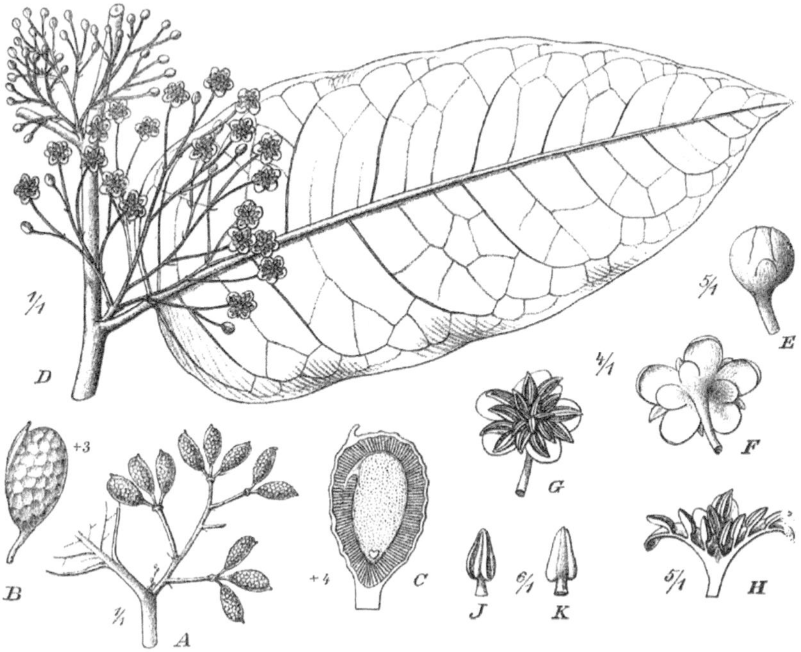


Fig. 1. *Amborella trichopoda* Baill. *A* Infl. fructigera. *B* Fructus. *C* Id. longit. sectus. *D* Infl. ♂. *E* Alabastrum ♂. *F*, *G* Flos ♂. *H* Id. longit. sectus. *J*, *K* Stamina. — *A*—*C* reiter. ex Monogr. Fig. 3, ceter. icon. origin.

Figure 9.2 *Amborella trichopoda*, sister group to all the angiosperm families. From A. Engler (1911), *Das Pflanzenreich* IV, 101, fig. 1.

mass and occurred there during the Cretaceous. One of the most striking New Caledonian examples is the gymnosperms with 43 endemic species in 15 genera, making New Caledonia the tropical region with the highest level of species richness of this group in the world today including 13 of the 19 known species of *Araucaria*. *Parasitaxus usta*, the only known parasitic gymnosperm, is endemic to New Caledonia, where it is parasitic on another member of the Podocarpaceae, *Falcatifolium taxoides*.

Undoubtedly, the flora is adapted to a great diversity of habitats and mineral contents in the soils; this, along with geographic isolation, has contributed to the high levels of endemism (Morat, 1993). The endemic families Strasburgeriaceae and Oncothecaceae, for example, occur only in forests on the ultrabasic substrates. The most species-rich families of angiosperms are Myrtaceae, Cunoniaceae, Pandanaceae and Palmae, a family with 16 endemic genera.

Perhaps the most remarkable endemic family is the Amborellaceae. The family consists of only a single species, *Amborella trichopoda* (Fig. 9.2), currently accepted as the most basal lineage in the angiosperms clade (Soltis & Soltis, 2004) and considered to be putatively the most ancient angiosperm in the world.

## Floristic affinities

The floristic affinities of the native flora of New Caledonia present few surprises. According to Morat *et al.* (1986), affinities are by far strongest with Australia (27.3%), New Guinea (17.1%) and Malesia (13.4%). Relationships with other Pacific islands floras are weaker (6.3%). This is consistent with the hypothesis that the New Caledonian flora has been formed from two principal groups: first, the ancient Gondwanan element dating from the splitting of New Caledonia from Australia; and, second, a long-distance dispersal element arriving more recently from Malesia and New Guinea. Recent support for the ancient element hypothesis comes from the discovery of three New Caledonian endemic genera, *Amphorogyne*, *Paracryphia* and *Phelline*, as leaf cuticle fossils in the early Miocene of New Zealand (Pole, 2010), suggesting that they were once more widely distributed. This gives an indication that at least part of the extant flora of New Caledonia is the result of contraction of a Gondwana flora, or possibly migration, from more southerly ancient land masses.

## Biogeography and evolution

Though the view that endemic flora of New Caledonia is a relict one dating from the breakup of the Gondwana continent was held until recently, this classical view is no longer valid. Geological evidence indicates that the Norfolk Island–New Caledonian ridge was completely submerged during the Palaeocene until the Late Eocene (Pelletier, 2006) when New Caledonia re-emerged following the obduction of oceanic mantle over the original Cretaceous base. This obduction gave rise to the diversity of ultrabasic substrates now found on the island. The Gondwana relict flora, therefore, cannot apparently have reached New Caledonia prior to the mid or late Eocene period and cannot be more than 40 Ma. This hypothesis is confirmed by recent molecular studies on the origin and evolution of diversity in New Caledonia. There are, therefore, three potential explanations for the presence of Gondwana relicts in New Caledonia: (1) incomplete submergence providing mountain top refuges; (2) the existence of now-submerged land masses with a Gondwana flora persisting into the Tertiary Period; and (3) long-distance dispersal.

There seems to be no geological evidence for the first hypothesis of unsubmerged refuges on New Caledonia itself. In the second case, Ladiges and Cantrill (2007), in a review of current knowledge of the tectonic rifting of eastern Gondwana in the context of possible land continuity or 'stepping stones' between Australia and New Caledonia, suggest that the geological history of the region, though complex, indicates evidence of land persisting into the Cenozoic, Palaeocene and Eocene. This possibly allowed old biota to persist and could explain the presence of 'relictual' groups in New Caledonia. The relatively harmonious nature of the flora and the diversity of higher-level taxa, including groups with very limited dispersal capacity, is perhaps not consistent with an exclusively long-distance dispersal origin in post-Eocene times from far-off areas such as Australia. The nearby areas ('stepping stones') hypothesised to have remained above sea level may have served to maintain a nearby sample of the original Gondwanan flora.

The third possiblity, long-distance dispersal, can apply to both the old and the more recent elements. Pole (1994) suggests that all the present biota of New Caledonia arrived from neighbouring sources. His hypothesis implies, in the case of relict endemic groups, that the original source populations became extinct in their centres of origin after arriving at New Caledonia (Ladiges & Cantrill, 2007). Recent molecular phylogenetic studies have focused on the origin and evolution of biodiversity in New Caledonia. Most of these confirm a post-Eocene arrival to New Caledonia and suggest that floristic richness can be explained by adaptation to the great diversity of soil types and habitats rather than long periods of isolation. New Caledonian diversity appears to have resulted from post-Eocene adaptive diversification with more recent small-scale speciation processes. The island's entire endemic flora would, therefore, be the result of total recolonisation and diversification since re-emergence in the Eocene. This explanation of diversity is consistent with dating from molecular phylogenetic studies. Recent phylogenetic studies of groups in a growing number of plant families indicate a complex mix of local radiation and diversification in basal lineages that appear to have reached New Caledonia long ago and in derived groups that probably arrived more recently.

Molecular phylogenetic surveys indicate that in many cases the New Caledonian species are nested within larger clades of taxa from Australia, New Zealand or New Guinea, implying recent dispersal event (Grandicolas *et al.*, 2008). Where New Caledonian clades are found to be sister groups to taxa from other areas, then deciding whether dispersal or vicariance hypotheses are valid requires dating the original dichotomy. The clades that have been studied so far, however, do not show very clear-cut results in this respect. Evidence for nested relationships of New Caledonian taxa within groups from other regions comes from *Nothofagus* (Swenson *et al.*, 2001; Cook & Crisp, 2005), Sapotaceae (Bartish *et al.*, 2005) and *Meryta* (Tronchet *et al.*, 2005). Bartish *et al.* (2005) date the oldest New Caledonian species of Sapotaceae to 32.4 Ma using a molecular clock, placing their origin to the Oligocene–Miocene transition. This does not, however, necessarily date the original arrival to New Caledonia, as molecular studies cannot take into account earlier local extinctions.

In other groups of plants molecular studies suggest a single origin for New Caledonian species. In the case of *Araucaria*, for example, Setoguchi *et al.* (1998) using *rbcL* sequences demonstrate that all 13 New Caledonian species are sister to a single species, *Araucaria heterophylla* from Norfolk Island. They also show that the New Caledonian *Araucaria* species form a monophyletic group with very low differentiation in *rbcL* sequences among them, indicating rapid and possibly recent (3.7 Ma) adaptive radiation into new edaphic conditions. *Santalum austrocaledonicum*, a sandalwood endemic to New Caledonia differentiated from Vanuatu species, also separated only 1–1.5 Ma (Harbaugh & Baldwin, 2007).

Thus, there is no strong, unambiguous evidence for very old local Gondwanan radiations. Rather, many old Gondwanan groups are represented in New Caledonia by species of quite recent origin (Murienne *et al.*, 2005). As a whole, phylogenetic patterns and especially the dating are consistent with the geological framework presented above. Phylogenetic relicts remain puzzlingly enigmatic, and their presence cannot be explained in a simple way. Barker *et al.* (2007), based on molecular studies, place the divergence of Proteaceae from New Caledonia and Australia in the Eocene between 43 and 25 Ma. Despite such findings this does not suggest a Gondwanan vicariance event but rather a dispersal event after the re-emergence of New Caledonia. These authors do, however, propose a Gondwana relict hypothesis for other members of the family.

## Threats to the endemic flora

With over 3300 plant species, New Caledonia is classified as one of the world's biodiversity hotspots (Myers *et al.*, 2000). According to Jaffré *et al.* (1998) some 25% of these are at risk and at least five species are already extinct.

To qualify for hotspot classification a region has to have lost over 70% of its original vegetation and today, on New Caledonia, highly degraded grassland and niaouli (*Melaleuca quinquenervia*; an extremely invasive species) savanna communities occupy more than 6000 km$^2$, or 32% of the area. These are highly disturbed anthropogenic communities maintained by repeated fire and grazing by cattle and introduced deer.

Agriculture also tends to occupy land formerly covered by sclerophyllous forest, which is now only found in very small patches of usually less than 40 ha. A recent survey (Jaffré *et al.*, 1998) suggests that New Caledonia's sclerophyll forest is the most threatened tropical dry forest in the world. However, because it is almost exclusively restricted to ultrabasic substrates, which are essentially useless for agriculture because of the high concentrations of toxic metals, the maquis is much less vulnerable. Logging is, however, a considerable threat to the humid forests though the most valuable timber species are already gone, but even limited logging threatens the habitats of species with highly restricted ranges.

According to Jaffré *et al.* (1998), however, the biggest single threat to the flora is currently fire, often set by the human population to clear land, and they state that 'conservation of biodiversity outside and inside reserves would be greatly enhanced by a policy of fire control, including, e.g., establishment of firebreaks and plantation of appropriate tree species'. The maquis vegetation is the most vulnerable to widespread fires, especially during the dry season in areas with low precipitation, resulting in varying degrees of degradation to the ecosystem.

New Caledonia is rich in mineral deposits and nickel mining produces almost all the country's foreign exchange. In fact New Caledonia is responsible for producing almost half the world's nickel. Mining is a major threat to the island ecosystems, as open-cast methods have led to the clearing of extensive areas of forest, especially the most luxuriant ones on the higher peaks (Morat *et al.*, 1995). Mining has also taken its toll on the maquis at many sites, resulting in degradation or total destruction.

Open-cast mining produces bare slopes and extensive waste heaps and leads to water pollution and can be devastating for local ecosystems. Even though much stricter controls on mining and spoil deposits have been introduced, the threat of new mining activities still exists, especially as the world demand for nickel rises, with consequent high prices making its extraction increasingly viable.

A further major threat to the New Caledonian endemic flora comes from the intentional or accidental introduction of invasive alien species. Recent estimates of the size of the flora indicate that, in addition to the aproximately 3400 native species, there are about 1400 introduced species of vascular plants. Many of these are now naturalised and integrated into the New Caledonian landscape but are generally unable to colonise the ultrabasic substrates and, therefore, do not pose a threat to remaining natural vegetation in such habitats. Alien plants do, however, invade disturbed habitats and secondary vegetation, especially in the domain of the lowland sclerophyllous forest. The introduction and naturalisation of grazing animals such as the Java deer (*Cervus timorensis*) has also had an important negative effect on the regeneration of some native vegetation types (Morat *et al.*, 1995).

## Conservation

Jaffré *et al.* (1998), in their review of the state of conservation on New Caledonia, provide a list of threatened species applying the IUCN red list categories to the endemic flora. They evaluate almost 500 species and place 58 species or 11.6% of endemic plant species in the critically endangered category, and a total of 392 are vulnerable, endangered or critical with another five species already extinct. The highest proportion comes from the high mountain maquis and many are known only from a single locality.

Though New Caledonia has a system of protected areas including 13 Special Botanical Reserves, about 450 threatened species are not included within any conservation reserve (Jaffré *et al.*, 1998). In all, about 9% of the main island is protected but both the sclerophyllous forests and the low and mid-altitude maquis are very under-represented and hardly protected at all, though this latter vegetation type has the highest percentage of threatened plants.

A number of New Caledonian threatened species, including many gymnosperms, are in cultivation (Farjon & Page, 1999) with *ex situ* conservation programmes. An international group of specialists led by the Royal Botanic Garden, Edinburgh, is currently engaged in conservation research for New Caledonian *Araucaria* species including establishing *ex situ* conservation collections.

As Jaffré *et al.* (1998) conclude, however, the only way to conserve the unique flora and vegetation of New Caledonia and their unique genetic diversity is by the establishment of an adequate system of *in situ* reserves. The designation of New Caledonia as a biodiversity hotspot was due to the unique tropical forests on the islands but, in fact, the humid tropical forest ecosystem is the least threatened at the present time. The maquis and sclerophyllous forests are now considered to be the most vulnerable. As so many of New Caledonia's native plant species are endemic and have very restricted distributions, often only in one or a few localities, the effects of habitat destruction on biodiversity in general are correspondingly large. Estimates of the number of extinct species by Bouchet *et al.* (1995) and the destruction of whole ecosystems (Jaffré *et al.*, 1993) imply that many known species have become extinct, probably over 50; it seems that many more have been lost even before they have been described. If the loss of species is to be halted and unique biodiversity is to survive, there is no doubt that the existing protected area network must be improved to include a wider range of vegetation types, and sites of high species concentration should be identified and included. Mining and agricultural activities should be highly restricted in these areas. Effective management of protected areas and education to increase conservation awareness among the population of New Caledonia, particularly with respect to the terrestrial environment, is also required. Because New Caledonia is technically part of France, there are also international conservation funding problems to be resolved as funding is not forthcoming from the major international agencies, and conservation depends on occasional aid from NGOs such as WWF and Birdlife International. The development of a comprehensive local policy for the conservation of this unique biodiversity hotspot is, therefore, essential for its future wellbeing and should involve local communities who may well benefit from ecotourism and enhanced scientific interest in New Caledonia in the future.

## References

Barker, N. P., Weston, P. H., Rutschmann, F. & Sauquet, H. (2007). Molecular dating of the 'Gondwanan' plant family Proteaceae is only partially congruent with the timing of the break-up of Gondwana. *J. Biogeogr.* 34: 2012–2027.

Bartish, I. V., Swenson, U. V., Munzinger, J. & Anderberg, A. A. (2005). Phylogenetic relationships among New Caledonian Sapotaceae (Ericales): molecular evidence for generic polyphyly and repeated dispersal. *Am. J. Bot.* 92: 667–673.

Bouchet, P. H., Jaffré, T. & Veillon, J. M. (1995). Plant extinction in New Caledonia: protection of sclerophyll forests urgently needed. *Biodivers. Conserv.* 4: 415–428.

Bradford, J. & Jaffré, T. (2004). Plant species microendemism and conservation of montane maquis in New Caledonia: two new species of *Pancheria* (Cunoniaceae) from Roche Ouaïème. *Biodivers. Conserv.* 13: 2253–2273.

Cook, L. G & Crisp, M. D. (2005). Not so ancient: the extant crown group of *Nothofagus* represents a post-Gondwanan radiation. *Proc. R. Soc. B.* 272: 2535–2544.

Engler, A. (1911). *Das Pflanzenreich* IV, 101.

Farjon, A. & Page, C. N. (1999). *Conifers: Status Survey and Conservation Action Plan.* Gland, Switzerland, IUCN.

Grandicolas, P., Murienne, J., Robillard, T., Desutter-Grandicolas, L., Jourdan, H., Guilbert, E. *et al.* (2008). New Caledonia: a very old Darwinian island? *Philos. Trans. R. Soc. Lond. B Biol. Sci.* 27; 363 (1508): 3309–3317.

Harbaugh, D. T. & Baldwin, B. G. (2007). Phylogeny and biogeography of the sandalwoods (*Santalum*, Santalaceae): repeated dispersals throughout the Pacific. *Am. J. Bot.* 94: 1028–1040.

Jaffré, T. (1976). Composition chimique et conditions de l'alimentation minérale sur roches ultrabasiquesen Nouvelle-Calédonie. *Cah. ORSTOM, sér. Biologie* 11: 53–63.

Jaffré, T., Morat, P. H. & Veillon, J.-M. (1993). Étude floristique et phytogéographique de la forêt sclérophylle de Nouvelle-Calédonie. *Bull. Mus. Natl. Hist. Nat. B, Adansonia* 15: 107–146.

Jaffré, T., Bouchet, P. & Veillon, J. M. (1998). Threatened plants of New Caledonia: Is the system of protected areas adequate? *Biodivers. Conserv.* 7: 109–135.

Ladiges, P. Y. & Cantrill, D. (2007). New Caledonia–Australian connections: biogeographic patterns and geology. *Aust. Syst. Bot.* 20: 383–389.

Morat, P. (1993). Our knowledge of the flora of New Caledonia: endemism and diversity in relation to vegetation types and substrates. In 'The Terrestrial Biota of New Caledonia', *Biodivers. Lett.* 1: 72–81.

Morat, P. H., Veillon, J. M. & Mackee, H. S. (1984). Floristic relationships of New Caledonian rain forest phanerograms. In F. J. Radovsky, P. H. Raven & S. H. Sohmer (eds.), *Biogeography of the Tropical Pacific*, pp. 71–128. O'ahu, Hawaii, Association of Systematics Collections & Bernice P. Bishop Museum.

Morat, P., Jaffré, T., Veillon, J.-M. & Mackee, H. S. (1986). Affinités floristiques et considérations sur l'origine des maquis miniers de la Nouvelle-Calédonie. *Bull. Mus. Natl. Hist. Nat. B, Adansonia* 8: 133–182.

Morat, P., Jaffré, T. & Veillon, J. M. (1995). Neocaledonian region: Grand Terre, New Caledonia, France. In S. D. Davis, V. H. Heywood & A. C. Hamilton (eds.), *Centers of Plant Diversity: A Guide and Strategy for their Conservation.* Volume 2, Asia, Australasia and the Pacific. The World Wildlife Fund and The World Conservation Union. Cambridge, UK, IUCN Publications Unit.

Murienne, J., Grandcolas, P., Piulachs, M. D., Bellés, X., d'Haese, C., Legendre, F. *et al.* (2005). Evolution on a shaky piece of Gondwana: is local endemism recent in New Caledonia? *Cladistics* 21: 2–7.

Myers, N., Mittermeier, R. A., Mittermeier, C. G., da Fonseca, G. A. B. & Kent, J. (2000). Biodiversity hotspots for conservation priorities. *Nature* 403: 853–858.

Pelletier, B. (2006). Geology of the New Caledonia region and its implications for the study of the New Caledonian biodiversity. In C. Payri & B. Richer de Forges (eds.), *Compendium of marines species from New Caledonia, Forum BIOdiversité des Ecosystèmes Coralliens, Nouméa, Nouvelle-Calédonie.* Documents Scientifiques et Techniques IRD, II 7, pp. 17–30. Nouméa, France, Institut de Recherche pour le Développement.

Pole, M. (1994). The New Zealand flora – entirely long-distance dispersal? *J. Biogeogr.* 21: 625–635.

Pole, M. (2010). Was New Zealand a primary source for the New Caledonian flora? *Alcheringa* 34: 61–74.

Setoguchi, H., Asakawa Osawa, T., Pintaud, J. C., Jaffré, T. & Veillon, J. M. (1998). Phylogenetic relationships within Araucariaceae based on *rbcL* gene sequences. *Am. J. Bot.* 92: 316–329.

Soltis, D. E. & Soltis, P. S. (2004). *Amborella* not a 'basal angiosperm'? Not so fast. *Am. J. Bot.* 91: 997–1001.

Swenson, U., Backlund, A., McLoughlin, S. & Hill, R. S. (2001). *Nothofagus* biogeography revisited with special emphasis on the enigmatic distribution of subgenus *Brassospora* in New Caledonia. *Cladistics* 17: 28–47.

Tronchet, F., Plunkett, G. M., Jeremie, J. & Lowry, P. P. (2005). Monophyly and major clades of *Meryta* (Araliaceae) *Syst. Bot.* 30: 657–670.

# 10

# Phytogeography and relationships of the Pitcairn Islands flora

STEVE WALDREN AND NAOMI KINGSTON

## Geography

The Pitcairn Islands are a group of four islands that lie 23.9–24.7°S and 124.7–130.7°W in the south-central Pacific Ocean, roughly half way between Australia and South America. The Pitcairn group is at the southwestern extremity of the main groups of Polynesian islands, with very few islands to the east. The nearest islands are the Gambier group to the west, the southeastern extremity of French Polynesia, some 450 km from Oeno, the westernmost of the Pitcairn group. The easternmost of the Pitcairn group, Ducie, is some 1000 km from Easter Island further to the east (Fig. 10.1).

Oeno and Ducie are atolls, their highest land surface being only a few metres above sea level. Pitcairn Island is a 'high' volcanic island, while Henderson is a raised atoll or makatea island – an atoll which was elevated above sea level by the eruption of the Pitcairn volcano. All are oceanic islands, and have resulted from volcanic activity on the Pacific Plate; none of the islands in this part of the Pacific has ever had connections to, or formed a part of, continental land masses.

In this chapter we discuss the vascular plant diversity of the Pitcairn Islands, and summarise our understanding of the biogeographical affinities and the evolutionary origins of the flora. To understand this fully and set the context, we need to take a brief diversion into the geological origins of these islands, their climate and patterns of human occupation.

## Geology

The development of atolls from volcanic islands was first described by Darwin (1842). When volcanic eruptions from marine plates form islands, they are

*The Biology of Island Floras*, eds. David Bramwell and Juli Caujapé-Castells. Published by Cambridge University Press. 

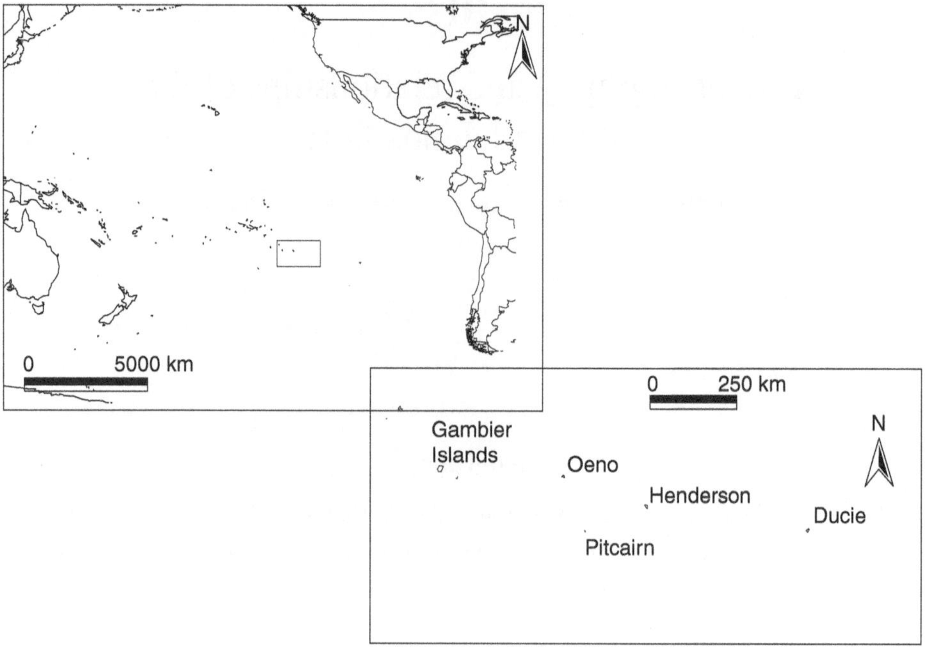


Figure 10.1 The location of the Pitcairn Island group. The top map shows the islands' location in terms of the Pacific; the central map shows the Pitcairn group of islands and some nearby island groups.

immediately subjected to aerial erosion, and also subsidence due to the placement of a large mass of rock on the oceanic lithosphere so the island gradually decreases in height with time. If the island is in warm water (above ~ 20°C monthly mean), coral growth may occur in the shallow waters around the island. If the upward growth of coral matches the rate of subsidence of the island, a fringing reef develops, and ultimately the process leads to the development of an atoll, where the volcanic island core is completely covered by reef limestones.

Many of the Pacific island groups show an obvious linear arrangement, perhaps most obvious in the isolated Hawaiian group, where the islands run in a WNW direction, with a continuation (as the Emperor Seamounts) in a NNW delineation. The simple explanation for this is the movement of the Pacific Plate to the WNW over a hotspot of volcanic activity which periodically results in island formation. Thus the younger of the Hawaiian islands are at the ESE end of the chain, and these are the highest volcanic islands in the chain. To the WNW the islands diminish in height, and in this direction the high islands are gradually replaced by atolls, and ultimately by submerged seamounts (guyots). The kink in the Hawaiian–Emperor chain is due to a realignment of the direction of seafloor spreading about 42 Ma (Mayes *et al.*, 1990).

The situation giving rise to the Pacific islands is likely to be far more complex, with various fracture zones, and regions of different plate thickness causing a complexity of island histories; some 'single' islands actually represent a number of submerged volcanoes in close proximity (e.g. Ducie). For a detailed discussion in relation to the Pitcairn group, see Spencer (1995), and for further details of the Pitcairn hotspot see Stoffers *et al.* (1990) and Hekinian *et al.* (2003). For our purposes, it is sufficient to note that the islands Oeno, Henderson and Ducie (also Crough seamount, to the east of Ducie) lie in what appears to be a single lineation. Estimated ages for the islands are 16 Ma for Oeno, 13 Ma for Henderson, 8 Ma for Ducie and 4 Ma for Crough. Pitcairn itself lies in a lineation resulting from a different hotspot which links (running WNW) with the Gambier group, Mururoa Atoll and the Duke of Gloucester Islands in the Tuamotu group. Pitcairn has been attributed to at least four periods of volcanic activity, with the main phase probably being from 0.95 to 0.76 Ma.

Thus Pitcairn is far younger than the other islands in the group, and the placement of this volume of material has resulted in lithospheric flexure. A near-volcanic moat of deeper water likely surrounds Pitcairn, but further out the lithosphere is bulged or arched upwards, and this process has pushed the atoll of Henderson some 30 m above sea level to form the existing makatea island.

This complex history of the islands has several consequences for the floras of these islands. First, the group represents a wide range of island types, from a relatively young, high volcanic island, to atolls barely above sea level, to a raised atoll, the interior of which is protected by 30 m high cliffs. This results in varied ecological conditions among the Pitcairn Islands, and this has influenced the species diversity and vegetation development among the islands. The second point to note relates to the development sequence from high islands to atolls and raised atolls – sea-level change, particularly during the Pleistocene, may have resulted in some present-day atolls being raised at some points in the past, and at other times completely submerged. In addition, some of the numerous submerged seamounts may have been emergent islands during lower Pleistocene sea levels. Crough seamount, previously mentioned, is 1000 m below sea level and hence would not have been emergent (Spencer, 1995), but the seamount named Adams on the Gambier–Pitcairn delineation reaches to within about 50 m of the surface and shows evidence of recent volcanic activity (Stoffers *et al.*, 1990; Binard *et al.*, 1992). The emergence of such seamounts in the vicinity of the Pitcairn group may have meant that they were less isolated at some stages of the Quaternary than they are at present (Fig. 10.2; Kingston *et al.*, 2003).

## Climate

The climate of the Pitcairn group is subtropical. Mean annual rainfall is about 1716 mm on Pitcairn, but with considerable annual variation. Mean temperatures

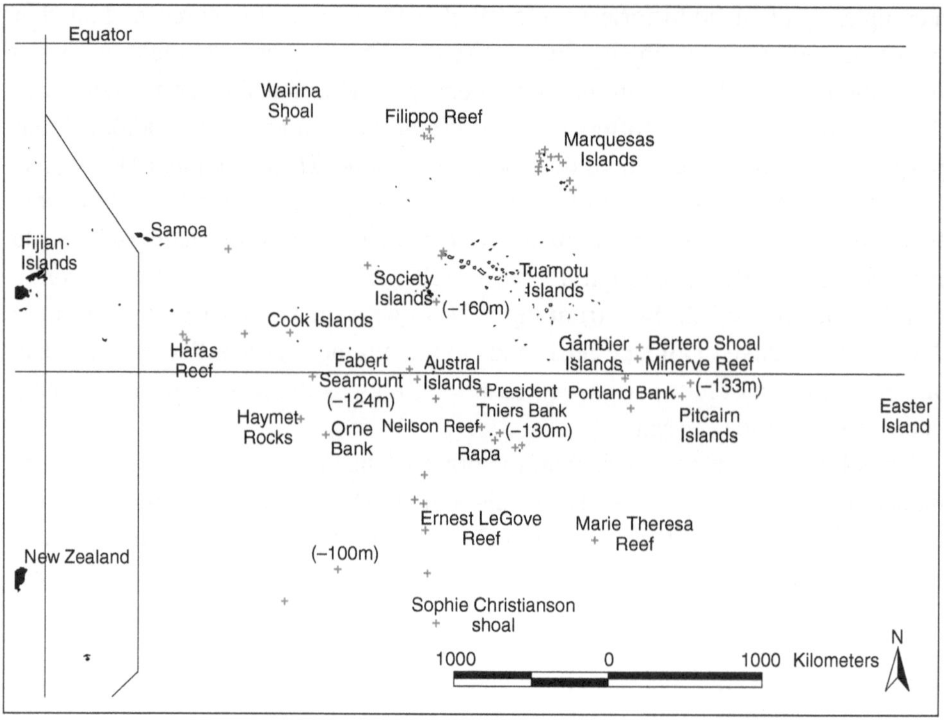


Figure 10.2 Locations of seamounts and reefs that may have emergent islands during periods of lowered sea levels. Potential islands are marked with crosses. Figure is derived from charts 4607 and 4061 of the Admiralty series.

range in summer from 17 to 28°C, and in winter from 13 to 23°C, with winter being wetter and windier (Spencer, 1995). As for the rest of this region of the South Pacific, the trade winds blow from the southeast, and Pitcairn is affected by complex climatic patterns which affect the whole of the Pacific region.

Sea currents in the South Pacific circle in an anti-clockwise direction, running west along the equator, south past New Zealand, east in the very south and north along the west coast of South America (Whittaker & Fernandez-Palacios, 2006). Pitcairn lies in the centre of this area and so receives warm currents predominately from the north.

Pitcairn, like other oceanic high islands, has a leeward and a windward side, with the leeward side being generally drier. This effect is caused by the southeast trade winds which move towards the intertropical convergence zone (ITCZ; moving seasonally, but almost coincident with the equator). As the trade winds blow over the ocean to the ITCZ, they pick up moisture from its surface. This atmospheric moisture is carried to the intervening islands, where it is precipitated as oreographic rainfall on the windward side (Mueller-Dombois & Fosberg, 1998). The higher the island the more rainfall is received, and the larger the discrepancy between high and low altitude rainfall. This explains the considerable annual variation in rainfall, and the year-round supply of rainfall found on Pitcairn. The other islands in the group likely have a lower and spatially more evenly distributed rainfall.

The rainfall pattern is interrupted during the cyclone season when heavy rains fall on both the leeward and the windward sides of the high islands. Cyclones form in the subequatorial tropics, when ocean temperatures are very warm, and move towards the subtropics in an eastward direction. As such they have pathways across the Pacific, along which the islands are the worst affected. The Western Pacific region is the worst affected by cyclones, but paths also lie across the Cook, Austral and Society islands. The Pitcairn group are out of the line of cyclones, but are occasionally hit by them (Gothesson, 1997; Pitcairn islanders, pers. comm.).

The climate of the Pitcairn group has not been static since its formation. It is possible to tentatively reconstruct the climatic history from the literature, although only from information for nearby island groups, as no quantitative studies have been carried out in the Pitcairn group. Glacial times were generally windier and stormier than interglacial times, and there have been 10 interglacial cycles in the last 1 million years, or since the formation of Pitcairn Island (Spencer, 1989). As glaciation also alters sea levels, climatic changes correspond strongly to sea level changes.

Of particular importance to our discussions below, the sea levels during the Pleistocene glaciations (between 1.77 Ma and 10 000 BP) probably had a maximum drop of 130–150 m below current sea levels (Menard, 1964; Spencer, 1989; Whittaker & Fernandez-Palacios, 2006). As a result of the lowered sea levels, islands in the Pacific such as Pitcairn would have been much higher and thus wetter on the windward sides during the Pleistocene (Nunn, 1999). However, as there is evidence that the tradewind system was weakened during these glaciations, the islands' leeward sides might have been much drier (Nunn, 1999). During the last interglacial (128 000–111 000 BP), sea levels were 2–6 m higher than during the current interglacial (Nunn, 1999), and thus many Pacific atolls, including Oeno and Ducie, may have been submerged. At the height of the last glaciation *c.*18 000 BP sea levels were between 75 and 110 m lower than present-day levels (Clark *et al.*, 1978; Bard *et al.*, 1996). The Bølling-Allerød warm period (*c.*13 000–11 000 BP) and subsequent rapid melting caused a sea-level rise and resultant cooling event, known as the Younger Dryas climatic reversal (Bard *et al.*, 1996; Nunn, 1999; Whittaker & Fernandez-Palacios, 2006). Other climatic fluctuations have subsequently occurred in the Holocene but, as we shall see, the lowering of sea levels around the last glacial maximum may have had a major effect on the colonisation of these islands by vascular plants.

## Human occupation

The islands were first colonised by Polynesians around the tenth century, and Pitcairn itself subsequently became a major stone manufacturing site serving islands of southeastern Polynesia, particularly the Gambier group. Henderson and to a lesser extent Oeno were also occupied by Polynesians; the settlement on Henderson was extensive, based around the beaches and surrounding cliffs in the north, northwest and northeast of the island. There is evidence that the population on Henderson, at first in regular contact with Pitcairn, subsequently became isolated and perhaps persisted for longer than the population in Pitcairn. The best evidence for this is a sudden replacement in Henderson cave profiles from tools manufactured from Tautama stone obtained from Pitcairn to tools made from one of the few local sources of material on Henderson, the shells of the giant clam, *Tridacna*. For further details of the Polynesian occupation and its subsequent demise see Irwin (1992) and Weisler (1995), and for a popular account see Diamond (2005).

The islands were probably abandoned some time around 1600, but were soon after settled by the *Bounty* mutineers and their Tahitian wives led by Fletcher Christian in 1790. Much has been written of the saga of the *Bounty* mutiny and the mutineers' descendants, varying from accounts that appear to be factually accurate to those which are clearly flights of fanciful imaginations. For the purposes of our own story, it is clear that both the mutineers and their descendants, and the Polynesians before them, cleared forested land for agriculture and also introduced many crop, ornamental and weed species.

Henderson was 'discovered' by the Portugese sailor de Quieros in 1606, and subsequently by the British East India Company's Captain Henderson in 1819 who named the island after himself. Though Henderson is uninhabited, the Pitcairn islanders regularly visit it to harvest miro (*Thespesia populnea*) and toa (*Cordia subcordata*) which are highly regarded for carving curios. Other visitors are sporadic and in some cases eccentric, for example, the castaway Robert Tomarchin and his chimpanzee, or the attempted purchase of the island by businessman Smiley Ratcliffe. Of greater interest was the temporary visit by some of the whalers from the wreck of the *Essex* (which formed part of the basis for Melville's *Moby Dick*), and the subsequent deprivations of those who sailed on for the Juan Fernandez islands (Nickerson, 1984). These visitors to Henderson have had moderately limited impact; both Polynesians and Pitcairners have damaged some of the coastal communities and introduced a few weedy species (Waldren *et al.*, 1999).

The Pitcairn islanders visit Oeno for holidays and to harvest fish from the lagoon; there is some impact on the vegetation around the vicinity of their camps, and several species have been deliberately introduced. Due to its remoteness, Ducie is seldom visited by Pitcairners, though cruise ships operating mainly between Tahiti and Easter Island regularly call at Henderson, Pitcairn and Ducie (but not Oeno).

The Pitcairn Islands are now British Overseas Territories, with the British High Commissioner to New Zealand being the Island Governor.

## Floristics and vegetation

The history of floristic recording has been summarised by Gothesson (1997) and Kingston (2001). Major collections from the islands are by Tait in 1912 (Henderson), the Whitney expedition of 1922 (mostly Pitcairn), by St John and Fosberg on the Mangarevan expedition of 1934 (Henderson, Oeno, Pitcairn), Twyford in 1955 (Pitcairn), Lintott in 1957 (Pitcairn), Florence, Waldren and Chepstow-Lusty in 1991 (Henderson, Pitcairn and Oeno) and Waldren and Kingston in 1997 (Pitcairn, a few collections from Henderson). Other important collections include Cuming in 1827 (Pitcairn), Ross in the 1930s (Pitcairn) and Paulay and Spencer in 1989 (Henderson and Pitcairn). Information on these collections is published in St John and Philipson (1960, 1962), St John (1987), Florence *et al.* (1995), Kingston (2001) and Kingston and Waldren (2003, 2005). In summary, Oeno and Pitcairn have been well surveyed, Henderson is much less completely covered due to its dense forest and difficult makatea terrain; given that we know that several taxa on Henderson appear to be geographically restricted (including *Asplenium polyodon*, *Hernandia ovigera* ssp. *Stokesii* and *Meryta brachypoda*), a more thorough survey of Henderson's plateau forest may well reveal additional local species.

The flora of the Pitcairn group is generally depauperate, reflecting the geographic isolation of the group. Ducie has two reliably recorded vascular plant species, both widespread Indo-Pacific coastal species: the shrubs or small trees *Argusia argentea* and *Pemphis acidula*. Both taxa occur on the other three islands in the Pitcairn group, though both are rare on Pitcairn (Waldren *et al.*, 1995a; Kingston & Waldren, 2005). The Whitney expedition of 1922 reported other vascular plant taxa, but identifications were uncertain and no voucher specimens were made. *Pemphis* had not been seen on Ducie prior to 1991, and hence there is evidence of some species turnover. The extreme remoteness and uniformity of the habitat likely contributes to the very low vascular plant diversity of Ducie.

The vegetation of Ducie consists of taller shrubs on the seaward side, with smaller and often more closely packed plants on the lagoon side, though there are also numerous sparsely vegetated and completely open areas; all of these features are clearly visible on the Google Earth image of Ducie (accessed 29 July 2009). Though the flora and vegetation of Ducie is of relatively little botanical interest, the vegetation forms important nesting areas for seabird colonies of international significance (Brooke, 1995a, 1995b).

Oeno also has a depauperate vascular flora (Table 10.1), though with a total of 18 native taxa recorded, it is considerably more diverse than Ducie, probably reflecting

Table 10.1 *Alpha[1], beta[2] and gamma[3] diversity of native species recorded from 10 × 10 m relevés on Pitcairn, Henderson and Oeno islands.*

| Statistic | Pitcairn | Henderson | Oeno |
|---|---|---|---|
| Max alpha | 14 | 23 | 7 |
| Min alpha | 1 | 2 | 2 |
| Mean alpha | 6.43 | 9.56 | 4.57 |
| Gamma | 63 | 61 | 16 |
| Beta | 9.79 | 6.38 | 3.50 |
| No. of relevés | 83 | 220 | 54 |
| Total native taxa | 81 | 63 | 17 |
| Endemic taxa | 10 | 8 | 1 |

[1] Alpha diversity is the number of species in each sample (relevé).
[2] Beta diversity is calculated as gamma diversity divided by mean alpha diversity.
[3] Gamma diversity is the total number of species sampled in relevés.

less isolation and a larger area of the main island. At the southern end of Oeno, a forest of tall (6 m) *Pisonia grandis* and *Argusia argentea* has developed, and large specimens of *Asplenium australasicum* are restricted to this forest (Waldren *et al.*, 1995b). Even so, most of the native species of Oeno are widespread Indo-Pacific coastal species, and the vegetation is largely dominated by open scrub of shrubby *A. argentea* (to 2 m), though stands of taller *A. argentea* (to 5 m) occur around the coast in addition to communities dominated by other shrubs (*Suriana maritima, Hedyotis romanzoffiensis*) and the grass *Lepturus repens*. However, there is one taxon of particular note, the endemic *Bidens hendersonensis* var. *oenoensis*, known from only one collection, now considered extremely rare and not found when Florence, Waldren and Chepstow-Lusty visited the island in 1991.

The uplifted atoll of Henderson has much greater vascular plant diversity than the atolls, with 63 native taxa recorded, eight of which are endemic (Table 10.1). The 30 m cliffs of Henderson doubtless limit the storm overwash that is likely to be regular across the low-lying atolls. However, during a storm in 1991 SW observed coral rubble that was overturned and waves were breaking over the 25 m cliffs at the southwest point of Henderson. This part of the island lacks a fringing reef, and wind-blown spray had killed leaves of several species for up to 1 km inland. Much of the interior of Henderson is covered with dense low forest, largely dominated by *Pisonia grandis*, but with several other tree species present (Waldren *et al.*, 1995b). Many of the forest trees are partially decumbent, and together with several vines, this makes progress through the forest difficult. The forested interior clearly exerts some influence on the microclimate, and this in turn appears to influence the distribution of taxa. The epiphytic fern *Pyrrosia serpens* is common in the forested interior, but gradually becomes scarcer closer to the plateau margins and is absent from the beach swale forests of the northern part of the island. Several bryophytes and lichens show similar distributions, which may be limited by lower humidity and increased salt spray closer to the plateau margin.

Less than 30% of Pitcairn Island is covered by native forest, and these areas are limited to remote valleys. Higher-altitude native *Metrosideros collina* forests are characterised by a high epiphytic and epilithic cover, commonly *Peperomia* spp., *Vittaria elongata*, *Trichomanes* spp. and bryophytes. Further down the slopes other native tree species such as *Homalium taypau*, *Cyathea medullaris*, *Cerbera manghas*, *Coprosma benefica*, *Xylosma suaveolens* and *Glochidion* spp. become more frequent, with a lush ground flora dominated by pteridophytes. The endemic *Homalium taypau* forest may have been managed for timber in the past and so forests that are pure stands of *Homalium taypau* may not represent natural vegetation. At lower altitudes forests are typified by the trees *Pandanus tectorius* and *Hibiscus tiliaceus*, with locally *Glochidion* spp., *Pisonia austro-orientalis*, *Celtis pacifica*, *Hernandia sonora* and *Cyclophyllum barbatum*. Large areas are covered with non-native scrub vegetation, and by almost monospecific *Syzygium jambos* plantations. Weedy scrub is also very common across the island, covering approximately 15% of the island area, caused by clearance, goat trampling and invasion by scrubby non-native species such as *Lantana camara*, *Psidium guava*, *Conyza bonariensis* and the grasses *Paspalum conjugatum* and *Sorghum sudanense*. Fernlands, dominated by *Dicranopteris linearis*, also cover large areas, including both eroding areas and ridge tops. Coastal vegetation comprises rock and cliff communities with limited strand vegetation. Rocky habitats include taxa such as *Peperomia* spp., *Dianella intermedia*, *Asplenium shuttleworthianum*, *Cocculus ferrandianus* and *Bidens mathewsii*, while scrubby habitats tend to be dominated by non-native species such as *Sonchus oleraceus* and *Eleusine indica*. The drier and settled north coast is more heavily invaded and disturbed, with *Pandanus tectorius* and *Cocos nucifera* forming a tall coastal forest with virtually no ground flora, except for the large invasive species *Crinum asiaticum*. The major environmental factor affecting the composition of the vegetation communities is altitude, although anthropogenic factors have also had a significant impact, especially due to clearance and introduced species (Kingston & Waldren, 2003).

## Vascular plant diversity

The alpha, beta and gamma diversity measures given in Table 10.1 reflect the floristic and ecological diversity of the three island types. The atoll Oeno has low total species (gamma) diversity, and the maximum number of native species recorded in a 10 × 10 m relevé was only 7. Beta diversity is relatively low at 3.5, reflecting the general uniformity of the vegetation. Gamma diversity is much higher on Henderson, though a far higher number of relevés were taken. Henderson had the highest alpha diversity in any relevé, with 23 taxa being recorded from two relevés in the forested interior and another in the north beach forest. However, the more exposed coastal communities of Henderson had a low diversity similar to the atolls; one relevé sampled on the East Beach had only two species present (*Pemphis acidula* and *Heliotropium anomalum*). Beta diversity was moderate at Henderson (Table 10.1), the value of 6.38 being intermediate between Pitcairn and Oeno. This may reflect the generally higher diversity in each relevé at Henderson (mean gamma diversity was highest here) and the widespread nature of most species, resulting in a certain uniformity to at least the forest relevés. Thus although there may be some ecological differentiation on Henderson (see *Pyrrosia* example above), this is not very marked. Patchiness of soil development and nutrient availability may account for some of the variation in forest communities (Waldren *et al.*, 1995b; S. Waldren & L. Scally, unpublished data).

The high island of Pitcairn has a higher number of recorded vascular plant taxa than the other islands, and this is reflected in the slightly higher gamma diversity (Table 10.1). These diversity indices are based on considerably fewer relevés on Pitcairn compared to Henderson. It is interesting to note that on both Henderson and Oeno the relevés sampled all but one of the native taxa present. However for Pitcairn, there are several native species that were not sampled in the 83 relevés. Most of these were very scarce species, often restricted to inaccessible locations; two were considered extinct during fieldwork in 1991 and 1997, but were subsequently rediscovered (*Abutilon pitcairnense* and *Myrsine* aff. *niauensis*; see below). Mean and maximum alpha diversity were lower at Pitcairn than at Henderson, while beta diversity was higher on Pitcairn. This suggests greater turnover of species among relevés, reflecting a less-uniform vegetation for Pitcairn. Several distinct forest and coastal communities occur on Pitcairn, compared to the more uniform communities of Henderson – we believe that these diversity values reflect not only the diversity of species, but also of vegetation communities and habitats. The high island of Pitcairn has higher habitat diversity than the other islands, resulting in higher beta and gamma diversities.

Also of interest on Pitcairn are the two relevés from which a single native species was recorded, and both species are remarkable. *Ophioglossum reticulatum* was the only native species recorded from a relevé below Long Ridge, while *Hibiscus australensis* was recorded as dominant in a relevé beneath Tibi's Cave. These relevés serve as a reminder that the flora of Pitcairn contains numerous introduced species, many of which are fully naturalised and some are invasive; these relevés starkly reveal the threats to native species from invasive non-natives.

## Biogeography and evolution of the flora

The Pitcairn group has the highest floristic association with island groups that have all island types (atoll, makatea and high), such as the Australs and Societies, even though these islands are not necessarily the closest in terms of distance (Table 10.2). The Pitcairn flora is essentially a subset of the larger flora of these island groups. Data analyses of the Pitcairn flora have shown it to be most closely associated with that of the Austral group of islands (Kingston *et al.*, 2003). This is supported by the presence of several species originally thought to be endemic to the Australs and Rapa being found in the Pitcairn group (e.g. *Hibiscus australense*, *Peperomia rapensis* and *Senecio stokesii*).

The colonisation pathways of the flora are closely linked to the habitats in which the taxa occur, with strand species arriving through hydrochory and taxa occurring further inland arriving through zoochory and anemochory. Therefore, it is not simply dispersal of propagules to islands which limits their occurrence, but also the availability of a vacant niche in which they can establish a viable population before events that might lead to their extinction are experienced.

The endemic taxa are listed in Table 10.3; one further species (*Glochidion pitcairnense*) is endemic to the Pitcairn group, being restricted to Henderson and Pitcairn. A small number of taxa are endemic to a region that includes the Gambier Islands and Pitcairn (*Pilea sancti-johannis*, *Pisonia austro-orientalis*); in these cases the populations on Pitcairn are likely to represent a major part of the global population, given the highly degraded vegetation of the Gambier group (Diamond, 2005). Interestingly, Oeno, Pitcairn and Henderson each have an endemic *Bidens* taxon. These are woody species of the section *Campylotheca*, restricted to Pacific Islands. *Bidens matthewsii* of Pitcairn is a low shrub which reaches about 75 cm, most common on the south side of the island. *Bidens hendersonensis* var. *hendersonensis* is a tall spindly plant fairly common in the plateau forest of Henderson, where it reaches a height of 3–4 m; *B. hendersonensis* var. *oenonensis* is a poorly known and extremely rare endemic to Oeno. Another woody-based composite from Henderson, *Senecio stokesii*, shows similar growth forms to the *Bidens* species. In the plateau forests, the growth form is tall and spindly like *B. hendersonensis*, reaching 2–3 m, with thin lanceolate leaves; however *S. stokesii* also grows in open makatea vegetation, where it assumes the habit of a compact dwarf shrub more like *B. mathewsii*, to 0.5 m in height, with much smaller internodes and shorter, more succulent leaves and few florets per capitulum (Table 10.4). It is not known whether these pronounced morphological differences are genetically determined, but they may illustrate an early phase in the evolution of sister taxa; we do not consider they merit taxonomic recognition at present.

Notwithstanding the intraspecific morphological variation just described in *Senecio stokesii*, the Pitcairn and Henderson *Bidens* species do not seem to be

Table 10.2 *Vascular plant species in common among the Pitcairn Islands of Pitcairn, Henderson and Oeno, considered both separately and together as a group, and some nearby Polynesian island groups. Island groups are ranked in order of increasing distance from Pitcairn.*

| | Pitcairn | Henderson | Oeno | Pitcairn grp. | Gambiers | Tuamotus | Rapa | Marquesas | Australs | Societies | Cooks |
|---|---|---|---|---|---|---|---|---|---|---|---|
| Pitcairn | **** | | | | | | | | | | |
| Henderson | 33 | **** | | | | | | | | | |
| Oeno | 10 | 17 | **** | | | | | | | | |
| Gambiers | 46 | 36 | 16 | 57 | **** | | | | | | |
| Tuamotus | 33 | 42 | 16 | 50 | 43 | **** | | | | | |
| Rapa | 47 | 31 | 8 | 58 | 30 | 25 | **** | | | | |
| Marquesas | 42 | 35 | 8 | 56 | 46 | 32 | 27 | **** | | | |
| Australs | 60 | 52 | 16 | 83 | 64 | 64 | — | 82 | **** | | |
| Societies | 59 | 50 | 16 | 82 | — | 77 | 92 | 140 | 146 | **** | |
| Cooks | 58 | 44 | 16 | 76 | — | — | — | — | — | — | **** |
| **Total species** | **81** | **65** | **18** | **114** | **71** | **95** | **188** | **278** | **220** | **575** | **293** |

Table 10.3 *Vascular plant taxa endemic to the Pitcairn Islands.*[1,2]

| Taxon | Island | Nearest relative |
|---|---|---|
| *Nesoluma st-johnianum* | Henderson | Rapa; Austral/Society Islands |
| *Bidens hendersonensis* var. *hendersonensis* | Henderson | Oeno; Pitcairn; Rapa; Gambier/Tuamotu/Austral/Society/Marquesas Islands |
| *Santalum insulare* var. *hendersonensis* | Henderson | Marquesas (Fosberg & Sachet, 1985) |
| *Myrsine hosakae* | Henderson | Society Islands |
| *Alyxia fosbergii* | Henderson | Pitcairn; Rapa; Gambier/Tuamotu/Austral/Society/Marquesas Islands |
| *Geniostoma hendersonense* | Henderson | Rapa; Gambier/Marquesas Islands |
| *Peperomia rhomboidea* | Henderson | Society/Austral Islands |
| *Ixora fragrans* | Henderson | Tuamotu/Austral/Society/Marquesas Islands |
| *Bidens mathewsii* | Pitcairn | Henderson/Oeno; Rapa; Gambier/Tuamotu/Austral/Society/Marquesas Islands |
| *Glochidion comitum* | Pitcairn | Pitcairn; Rapa; Gambier/Tuamotu/Austral/Society/Marquesas Islands |
| *Homalium taypau* | Pitcairn | Austral/Tuamotu Islands |
| *Abutilon pitcairnense* | Pitcairn | Gambier Islands; Tuamotu/Marquesas Islands |
| *Myrsine aff. niauensis* | Pitcairn | Henderson; Rapa; Austral/Society Islands |
| *Peperomia pitcairnensis* | Pitcairn | Austral Islands |
| *Peperomia sp. nov. A* | Pitcairn | Henderson; Austral/Society Islands |
| *Peperomia sp. nov. B* | Pitcairn | Society Islands |
| *Coprosma benefica* | Pitcairn | Rapa; Austral/Society Islands |
| *Angiopteris chauliodonta* | Pitcairn | Rapa; Austral/Society Islands |
| *Ctenitis cumingii* | Pitcairn | Rapa; Austral/Society Islands |
| *Bidens hendersonensis* var. *oenonesis* | Oeno | Henderson; Pitcairn; Rapa; Gambier/Tuamotu/Austral/Society/Marquesas Islands |

[1] Ignores *Xylosma suaveolens* ssp. *haroldii*, *Bidens hendersonensis* var. *subspathulata* (= *B. hendersonensis* var. *hendersonensis*), *Athyrium pitcairnense* (= *Diplazium harpeodes*).

[2] *Geniostoma hendersonense* retained as per IPNI even though it is considered part of *G. rupestre* by Conn (1980); *Alyxia fosbergii* retained, though considered part of a very variable *A. scandens* by D. Middleton (pers. comm.).

Table 10.4 *Morphological variation among three contrasting populations of* Senecio stokesii *on Henderson Island. Values given are means (in mm) of three measurements made for each of eight individuals per population. Differences among populations are significant for all variables (ANOVA, P < 0.001).*

| Population | Petiole length | Internode length | Leaf thickness | Leaf breadth | Leaf length |
|---|---|---|---|---|---|
| North-South Trail | 27.8 | 43.0 | 0.463 | 35.2 | 88.5 |
| North West Point | 15.9 | 13.9 | 0.671 | 17.3 | 49.7 |
| South West Point | 13.6 | 7.0 | 1.113 | 20.7 | 50.2 |

very closely related. Apart from the differences in morphology, there are very pronounced differences in the aroma of the flowers, which may indicate differences in pollinators. It is possible that the *Bidens* species represent colonisation of the Pitcairn group on two occasions (the first to Henderson/Oeno with a later separate radiation to Pitcairn), with subsequent isolation responsible for the evolution of the two endemic varieties of *B. hendersonensis.* Sherff (1937) described two endemic varieties from Henderson (var. *hendersonensis* and var. *subspathulata*) but we find it impossible to separate these in the field as a range of intermediates occur.

The only other genera with more than one endemic taxon in the group are *Glochidion* and *Peperomia. Glochidion pitcairnense* is endemic to Pitcairn and Henderson, and relatively widespread on each. Florence (1997a) described *G. comitum* from a collection we made in 1991, a major difference from *G. pitcairnese* being the pubescent ovary. However, we found numerous examples of *Glochidion* on Pitcairn with pubescent unripe fruit in 1997, but it appears that hairs may develop after fruit ripening; Florence did not consider any of our 1997 collections to be *G. comitum.* As a result *G. comitum* remains very poorly known, and may possibly represent an extreme form of *G. pitcairnense.* There are at present insufficient data to determine evolutionary relationships in the Pitcairn *Glochidion.*

*Peperomia* is the only genus of angiosperms in the Pitcairn group with more than three species in the group, with some six taxa present. Bradley (2002) has examined the phylogenetic relationships of the southeastern Polynesian *Peperomia* in detail, with widespread island and/or achipelago endemism in the region. The following can be deduced from sequence data based on the ITS nuclear region, and the *trn*L-F plastid region. The widespread *Peperomia blanda* (Pitcairn) forms a single clade with no sister species in southeastern Polynesia and hence shows no speciation in southeastern Polynesia. This may reflect widespread and possibly relatively frequent dispersal, coupled with broad ecological tolerance. Where it occurs in Polynesia, *P. blanda* is often the commonest species and occurs in a variety of habitats (terrestrial, epilithic, epiphytic; low to high altitude, etc.). Its absence from Henderson is remarkable. The endemic *P. rhomboidea* (= *P. hendersonensis*, see Florence, 1997b) of Henderson is closely related to an undescribed species discovered in a very limited area of Tautama on Pitcairn in 1997. The Pitcairn species has remarkably stout inflorescence axes and is extremely succulent; ecologically, it is restricted to wet seepages on cliffs in the south of the island. By contrast, *P. rhomboidea* is widespread and common across Henderson, growing among coral rubble, in rock crevices and (more rarely) epiphytically. Both form part of a clade containing the widespread *P. pallida* (which is surprisingly absent from these islands). *Peperomia rapensis* is also recorded from Pitcairn; this taxon is characterised by micropubescence on the leaf surfaces and stems (Florence, 1997b). We have interpreted similar individuals which are completely glabrous as being *P. pitcairnensis*, though this determination is based only on an examination of an image of the type specimen. These glabrous and micropubescent forms of *P. rapensis* (*sensu lato*) are phylogetically linked with a clade containing *P. australana* from the Austral Islands. Finally, in 1997 we also obtained material of a *Peperomia* from Tautama that we took to be a hybrid involving *P. blanda*. Molecular evidence suggests that it is in fact phylogenetically linked with a clade containing taxa that are otherwise endemic to the Society Islands.

In summary, the *Peperomia* of the Pitcairn group are represented by four separate clades, and hence there have probably been at least four distinct colonisation events: (1) *P. blanda* (Pitcairn); (2) *P. pallida* group – *P. rhomboidalis* (Henderson) and *P.* undescribed sp. A (Pitcairn); (3) *P. australana* group – *P. rapensis* and *P. pitcairnensis* (Pitcairn); and (4) Society group – *P.* undescribed sp. B (Pitcairn).

So despite *Peperomia* being the most species-rich genus in these islands, there remains very little evidence for radiation. As with other endemics, most of the local evolution of *Peperomia* species can be explained by isolation by distance and/or pronounced founder events.

Stuessy *et al.* (1998) examined isolating mechanisms in the endemic flora of the Juan Fernandez group, where they found that geographic isolation is the main isolating mechanism driving speciation in the archipelago. However, they also documented ecological and reproductive isolation as drivers of speciation among closely related taxa in the islands. Geographic isolation explains almost all of the speciation within the Pitcairn Islands; while ecological and reproductive isolation cannot be ruled out for the allopatric taxa mentioned above, geographic isolation is probably more likely. The only taxa for which geographic isolation cannot explain the mode of speciation are the two Pitcairn *Glochidion*, and the *Peperomia rapensis – pitcairnensis* pair, also on Pitcairn. In both cases there are taxonomic uncertainties; for both there is no clear reproductive or ecological isolation among the taxa. A better example of ecological isolation might be found in the contrasting ecotypes of *Senecio stokesii*.

The only other genera for which more than two taxa occur in these islands are: (1) *Phymatosorus*: *P. scolopendria* (Pitcairn, Henderson, Oeno), *P. commutatum* and *P. powellii* (both Pitcairn); (2) *Nephrolepis*: *N. hirsutula* (Pitcairn, Henderson), *N. biserrata* (Pitcairn, Henderson?) and *N. cordifolia* (Pitcairn); (3) *Korthalsella*: *K. platycaula* and *K. rubescens* (both Henderson); (4) *Ophioglossum*: *O. reticulatum* and *O. nudicaule* (both Pitcairn); and (5) *Ipomoea*: *I. macrantha* (Pitcairn, Henderson) and *I. littoralis* (Pitcairn). As none of these taxa are endemic to the islands, they all represent separate colonisations by individual taxa. They are also typically widespread and often pantropic taxa, so might be expected to occur on the islands.

## Taxonomic problems

The taxonomic problems surrounding *Peperomia pitcairnensis/rapensis* and *Glochidion comitum/pitcairnense* have been alluded to above. Other taxonomic problems are mentioned in this section.

*Ctenitis cumingii* was described by Holttum (1985) and differs from *C. rapensis* (endemic to Rapa) and the more widespread *C. sciaphila* by relatively subtle microscopic characters. It seems that Holttum's species concept in delimiting these taxa was based on the likely complete isolation of these taxa (J. Camus, pers. comm. to SW). Even if we accept this species concept, it is clear that these taxa represent very closely related sister taxa.

The genus *Angiopteris* has been the subject of much taxonomic debate, some authors recognising a single, widespread and variable taxon (*A. evecta*), others numerous taxa within *Angiopteris* and a small number of sister genera. One taxonomic difficulty with these giant ferns is the lack of complete character information from voucher material – the fronds may exceed 4 m in length, and several important characters also exist in the fleshy 'stipules' that occur at the stipe base but these stipules are rarely preserved as vouchers. The distinction of *Angiopteris chauliodonta*, a taxon endemic to Pitcairn, remains uncertain; some authors refer the Pitcairn material to *A. longifolia*.

Florence *et al.* (1995) referred the Pitcairn *Cyathea* material to *C. medullaris*. Others have commented that the Pitcairn material does not seem to correspond to this taxon (D. Given, pers. comm.), and perhaps it might be better to retain *C. cumingii*, a taxon restricted to Pitcairn and Rapa. D. Given (pers. comm.) described *C. medullaris* as showing a continuum of variation moving across the Pacific islands. As with *Angiopteris*, one problem with *Cyathea* is the size of the plants and possible difficulty in retaining all taxonomically informative characters in herbarium material. The Polynesian *Cyathea* and *Angiopteris* would repay a molecular phylogenetic and phylogeographic study; such a study would, however, require extensive sampling over a large area which would prove a costly though highly enviable undertaking. We noted considerable morphological variation in the Pitcairn *Cyathea* in the field; interestingly, the Pitcairn islanders consider there are two types of this 'man fern', though we were unable to partition the variation into consistent entities. Again, a molecular population genetic study would be interesting.

Other taxonomic problems indicate the close relationship between material from the Pitcairn group and from elsewhere in Polynesia. For example, D. Middleton (pers. comm.) considers *Alyxia fosbergii* (endemic to Henderson) to be a part of the variable *A. scandens. A. scandens* is very rare on Pitcairn, though the Pitcairn material is morphologically very distinct from that of Henderson. Similarly Conn (1980) has suggested that the supposed morphological distinction of *Geniostoma hendersonense* is not consistent, and that the Henderson material is part of the widespread *G. rupestre.*

Most of these taxonomic uncertainties point again to the close biogeographical link between the flora of the Pitcairn group and the flora of eastern Polynesia, in particular that of the Gambier, Austral and Society islands. The close relationship between the endemics of the Pitcairn group and other archipelagoes of southeastern Polynesia is perhaps best illustrated by *Pandanus.* St John has described endemic *Pandanus* species from Henderson, Oeno and Pitcairn (see St John & Philipson, 1960, 1962; St John 1987); we now consider all of these to be forms of the variable *P. tectorius* which, although probably native in the Pitcairn, is also likely to have been introduced to all islands by the former Polynesian occupants, producing a considerable amount of local variation.

## Species turnover

Since 1985 there have been several new records of vascular plant species from the Pitcairn group. Some of these are likely to have been overlooked or simply not found in the past: *Trichomanes tahitense* (first record 1997) is moderately widespread on Pitcairn in suitable habitat, but is cryptic and easily overlooked; *Merremia turpethum* occurs in the very remote southern part of Henderson, and *Achyranthes aspera* occurs in the dense plateau forest of Henderson. Given the limited amount of botanical survey on Henderson, the presence of these taxa is likely to have gone undetected. However, we have evidence for other taxa being recent colonists. In 1997 we discovered *Lycopodiella cernua* along a roadside that Florence, Waldren and Chepstow-Lusty had closely investigated in 1991. Discussions with the Pitcairn islanders showed that this species was first noticed by them in an area of highly eroded soils know as 'The Hollow' in about 1995; the islanders were also convinced that Florence *et al.* were responsible for its arrival on Pitcairn! By 1997 it had colonised numerous parts of the island at moderate altitude, typically occurring on eroded banks, often alongside tracks. *Dicranopteris linearis* provides another example of a taxon that has spread considerably on Pitcairn. St John (1987) recorded the plant as rare in 1934, but by the 1990s this species had become common in disturbed and eroding slopes. Much of this erosion is likely caused by large numbers of feral goats, a major conservation problem on Pitcairn.

Also in 1997 we discovered *Haloragis* cf. *prostrata* scattered and local in coastal communities in the south of the island. By 2003 it was frequent in some areas (e.g. Ailehow), suggesting recent colonisation and spread. For one taxon we have good evidence of its increase on Pitcairn. *Lastreopsis pacifica* was first collected on Pitcairn in 1991 by Florence, Waldren and Chepstow-Lusty from a well-botanised site at Brown's Water. The material was at first mistakenly referred to *Ctenitis cumingii*, and a small number of individuals were reported (Florence *et al.*, 1995; Waldren *et al.*, 1995a), but subsequent investigations by J. Camus and S. Waldren showed that the Pitcairn material was close to *L. pacifica*. We revisited and surveyed the site in 1997 and found more plants, and found another single individual at Pulau. S. Waldren and N. Smyth revisited these sites in 2003 and subsequent monitoring showed that both the Brown's Water and Pulau populations had increased further (Smyth, 2008; Fig. 10.3). This seems to be a clear indication of a relatively recent colonisation and subsequent spread on Pitcairn.

*Ipomoea littoralis* was also first recorded from Pitcairn in 1997. A single plant of *Canavallia rosea* was discovered by Florence, Waldren and Chepstow-Lusty in 1991 near the southern cliffs of Henderson. Both of these taxa are widespread strand species, and occur frequently in the nearby Gambier group. These records may represent casual occurrences; only time will tell if they develop into permanent populations.

In contrast to these examples of what appear to be relatively recent, and in some cases perhaps transient, colonists, there are some examples of species that have declined considerably and a few that have become locally extinct. A grass and a small herb were recorded from Ducie by the Whitney expedition of 1992, though no vouchers exist; no such taxa have been found there since. Such turnover in a very isolated atoll prone to storm damage is perhaps to be expected. Several taxa appear to be extremely rare on Henderson, including *Pittosporum arborescens*, *Mertya brachypoda* and *Allophyllus rhomboidalis*. However, due to the difficulty

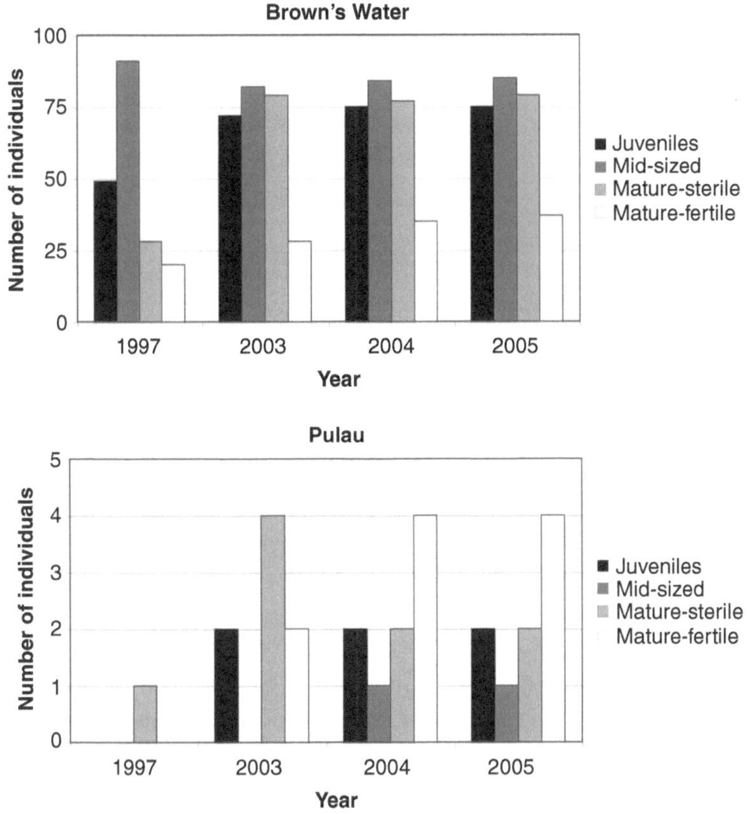


Figure 10.3 Changes in numbers of different sizes of *Lastreopsis pacifica* from the two known populations on Pitcairn at Brown's Water and Pulau. First recorded in 1991 at Brown's Water, the species seems to be increasing steadily in both populations.

in accessing much of Henderson, it is uncertain whether these represent declining taxa or whether they have always been very rare.

For Pitcairn, a small number of taxa have not been refound; these include *Ipomoea pes-caprae, Pemphis acidula* and *Capparis cordifolia.* All are widespread coastal species and their apparent decline may be due to natural disturbance and turnover; a similar example occurs in the form of the single individual of *Argusia argentea* known from Pitcairn. Apart from *I. pes-caprae,* all of these taxa are common on Henderson. Of greater concern is the extreme rarity of some endemic Pitcairn taxa, perhaps best illustrated by the case of *Abutilon pitcairnense.* There have been few collections made of this taxon, and we failed to find the species in 1991 and, despite extensive searching, in 1997. However, Carol Warren, an islander, discovered a single individual in about 2000 and had the foresight to propagate it from cuttings. This was fortunate as the single wild individual was destroyed by a landslide in 2004. *A. pitcairnense* was probably always rare on the island, though when vegetative it could be easily overlooked among *Hibiscus tileaceus*. Several other taxa occur in precariously low numbers; *Myrsine* aff. *niauensis*, collected once in 1957, was only rediscovered in 2003 and is currently known from a single individual, while only 11 individuals of the endemic *Coprosma benefica* are known. Stochastic loss threatens all of these critically rare taxa – we witnessed the largest example of *C. benefica* being blown over by a severe storm in 1997 and, though showing shoots from the base, this individual and one other had died by 2003. For further discussion of the threats to the flora and possible conservation actions, see Kingston and Waldren (2003, 2005).

## Invasive species

Finally, we should mention the non-native elements of the flora of these islands. Pitcairn, with its long history of Polynesian occupation and trade, and subsequent colonisation by the *Bounty* mutineers, has a large number of established non-native species. Most of these were important food or ceremonial plants for the Polynesians, or horticultural and timber species introduced by the descendents of the mutineers, and weedy species introduced by both. Problem invasive taxa on Pitcairn include *Syzygium jambos*, *Lantana camara*, *Canna indica*, *Paspalum conjugatum*, *Crinum asiaticum* (in coastal communities), *Sorghum sudanense*, *Psidium guajava* and several *Passiflora* species; several other taxa are also currently present in low number, but are highly invasive elsewhere in the Pacific and may represent future invasive problems for Pitcairn (see Kingston, 2001). There is a well-established and extensive herbaceous weed flora. One particular problem has occurred with the transfer of potted plants of *Thespesia populnea* (used by the islanders for carving curios) from Pitcairn to Oeno, as B. Bell (1997, pers. comm.) reported several of the herbaceous weed flora of Pitcairn occurring on Oeno. It is not known whether these weeds have survived on Oeno, but given the importance of Oeno for ground-nesting seabirds, the potential spread of non-native invasive plants there is of serious concern.

By contrast, the World Heritage site of Henderson has a remarkably low occurrence of non-native species. *Setaria verticillata* was discovered on North Beach in 1991 and promptly dug up and destroyed on a camp fire. A *Passiflora* species was also discovered on the North Beach, but was not refound in 1997. There is evidence that Operation Raleigh personnel misguidedly planted coconuts at the ends of trails that they cut inland from the North Beach. Several Polynesian introductions occur on Henderson, including *Cordyline fruticosa* (this often occurs on the clifftop above habitation caves on North Beach), *Thespesia populnea* and *Cordia subcordata*; the latter two species appear to be well-integrated with native vegetation in beach-back forests (Waldren *et al.*, 1995b, 1999), and may possibly be native. However, by comparison with other Pacific islands we have visited, we are struck by the comparative lack of non-native species on Henderson, and this surely adds to its ecological importance. However we must add a cautionary note – in 1997 we discovered a single plant of *Hedyotis romanzoffiensis*, previously unknown on the island. The individual occurs exactly in the position that the senior author had dried seeds in 1991, seed collections that included *H. romanzoffiensis* from Oeno. This demonstrates not only the susceptibility of these islands to invasion, but also that despite their best intentions, careless scientists can embarrassingly stimulate the invasion of a near-pristine ecosystem. For the record, we took a tissue sample of the individual and left the plant otherwise intact; a future study in colonisation genetics for some fortunate visitor.

## Conclusions

The flora of the Pitcairn group is clearly a subset of that occurring in the Polynesian archipelagos to the west: the Gambier, Society and Austral Islands (including Rapa Iti). The flora of these islands is in turn derived largely from that of the Malesian region; there is no evidence to support Brown's (1935) contention that the flora of these islands shows New World origins. There are no floristic links with islands further to the east, except for widespread species which are also common on islands to the west. Florence *et al.* (1995) have previously pointed out that many Polynesian genera reach their eastern limit in the Pitcairn group, further confirming biogeographic links to the west. Endemic species occur on Pitcairn and Henderson, with an endemic variety on the atoll Oeno. In all cases, these endemic species are closely related to, and in some cases possibly conspecific with, taxa occurring further west.

The flora of the atolls Ducie and Oeno consists of widespread Indo-Pacific coastal species that are mainly hydrochorous (Kingston *et al.*, 2003), and this is also true of much of the beachfront communities of Henderson. The prevalence of a relatively unspecialised flora of widespread coastal species on atolls has been commented on by others (van Balgooy, 1960; Sachet, 1967), including by Renvoize (1979) in his discussion of the Indian Ocean island floras, in the first edition of this book. Vascular plant diversity is low to very low on the atolls, though the presence of the highly threatened *Bidens hendersonensis* var. *oenoensis* on Oeno is of note.

The 30 m high plateau of Henderson clearly limits overwash by seawater and facilitates the development of a non-coastal community on the plateau. Species richness (gamma diversity) is much higher than that of the atolls, though somewhat lower than Pitcairn. However mean diversity among samples (mean alpha diversity) was higher for Henderson than Pitcairn, while turnover among samples (beta diversity) was lower on Henderson. This probably reflects the relatively uniform environment over the plateau of Henderson, with perhaps only subtle differences in habitat. *Pisonia grandis* is almost ubiquitous throughout the plateau and beach swale forests, as are *Pandanus tectorius*, *Phymatosorus scolopendria*, *Nephrolepis hirsutula* and the endemic *Peperomia rhomboidalis*. Subtle differences in wind-blown salt spray and humidity may facilitate the development of slightly different communities further inland, and the very deeply pinnacled makatea to the north-west of the island contains a distinct suite of species, but again, differences in vegetation communities are subtle (Waldren *et al.*, 1995b).

The diversity of habitats is considerably greater on Pitcairn than Henderson, and there is pronounced climatic and edaphic variation depending on altitude and exposure. These factors doubtless enable both a great diversity of species (gamma diversity) and of vegetation communities on Pitcairn compared to Henderson (Waldren *et al.*, 1995b; Kingston & Waldren, 2003). The upper zones of Pitcairn receive considerable orographic rainfall, and the resulting high humidities enable the development of a relatively rich epiphytic flora compared to Henderson. The southern slopes of Pitcairn are far more exposed than the northern, and we have witnessed spindrift in storms being blown to a altitude of over 200 m. These southern slopes and cliffs have a different flora and vegetation than the northern slopes (Kingston & Waldren, 2003). The flora and vegetation of Pitcairn therefore shows a higher degree of ecological specialisation, probably driven by the effects of altitude on climate and soil formation. Several plant species and many vegetation types are geographically and ecologically restricted on Pitcairn, and this is reflected in the higher turnover of species among samples (beta diversity).

Stoddart (1992) suggests that island type should be taken into account when discussing biogeographic schemes of island biodiversity, but we contend that habitat availability is likely to be the most important factor affecting species distribution (Kingston *et al.*, 2003), as acknowledged by MacArthur and Wilson (1967). Clearly, the geological history and stage of development of oceanic volcanic islands will play a major part in the determination of habitat diversity, but other factors (edaphic, climatic) may also be important. What seems to be required is better ecological understanding of Pacific insular floras.

Of the plant taxa that have colonised these islands, few show any evidence of radiation following colonisation. The endemic taxa are nearly all closely related to taxa from the Gambier, Society and Austral islands and Rapa Iti, reflecting the general biogeographic patterns mentioned above. This may imply a relatively young flora that has not had sufficient time for divergence, or that there is some, albeit infrequent, gene flow between the island populations through repeated colonisation events.

Few genera appear to have colonised these islands more than once; most that have done so are widespread Pteridophytes (*Nephrolepis*, *Phymatosorus*, *Ophioglossum*) typically with small wind-blown propagules. The most interesting genus is perhaps *Peperomia* with six taxa present, where at least four different taxa have colonised and subsequently speciated. Two examples where there may be some evidence of ongoing evolutionary change can be found in *Senecio stokesii* of Henderson and *Lastreopsis pacifica* of Pitcairn. *S. stokesii* shows considerable morphological variation in different populations but, as this variation appears to be correlated with environmental conditions, notably the degree of exposure, we cannot rule out extreme phenotypic flexibility as being the cause of this variation. A better understanding of the population genetic differentiation is clearly required here. For *Lastreopsis*, the Pitcairn material appears to differ slightly from that of the Society Islands in having an extremely small indusium which becomes submerged as the sporangia develop, and in being noticeably more glandular. This may represent differentiation caused by founder events, but again, genetic studies of population variation are required. For the moment, we do not consider that either the Pitcairn *Lastreopsis* or the variation among *Senecio stokesii* populations merits taxonomic recognition. Instead, we believe that they may offer examples of how evolutionary change and speciation may have occurred in other taxa on these islands.

Though small, the flora of the Pitcairn group is of particular interest, through its position at the eastern extremity of southeastern Polynesia. The threats to the flora, particularly on Pitcairn, are well known, and include habitat loss and degradation, invasive species spread and critically small population size (Kingston & Waldren, 2005). In addition, there are considerable threats and pressures faced by the small local human population, who are largely marginalised in an extremely remote location. Their ability to make a positive impact in the conservation of these islands is considerable, as the *Abutilon pitcairnense* case mentioned above clearly shows. Imaginative holistic solutions are required to sustain and develop the islanders' livelihoods while also furthering protection of their unique environment.

## Acknowledgements

We are grateful to the people of Pitcairn for their generous hospitality, cheery good nature and botanical expertise during several visits to the island. We are grateful to the Pitcairn Island Governors and to the Island Council for permission to visit and undertake scientific studies, and we are grateful to all at the Pitcairn Islands Administration for facilitating these visits, and for their support and encouragement.

## References

Bard, E., Hamelin, B., Arnold, M., Montaggioni, L., Cabioch, G., Faure, G. & Rougerie, F. (1996). Deglacial sea-level record from Tahiti corals and the timing of global meltwater discharge. *Nature* 382: 241–244.

Binard, N., Hekinian, R. & Stoffers, P. (1992). Morphostructural study and type of volcanism of submarine volcanos over the Pitcairn hot-spot in the South-Pacific. *Tectonophysics* 206: 245–264.

Bradley, U. (2002). *Biogeography and speciation of the genus* Peperomia *Ruiz and Pavon in Eastern Polynesia.* Unpublished PhD thesis: Trinity College, Dublin.

Brooke, M. de L. (1995a). The modern avifauna of the Pitcairn Islands. *Biol. J. Linn. Soc.* 56: 199–212.

Brooke, M. de L. (1995b). The breeding biology of the gadfly petrels *Pterodroma* spp. of the Pitcairn Islands: characteristics, population sizes and controls. *Biol. J. Linn. Soc.* 56: 213–231.

Brown, F. B. H. (1935). Flora of Southeastern Polynesia: III Dicotyledons. *Bull. Bernice P. Bishop Mus.* 130: 1–386.

Clark, J. A., Farrell, W. E. & Peltier, W. R. (1978). Global changes in postglacial sea level: a numerical calculation. *Quat. Res.* 9: 265–287.

Conn, B. J. (1980). A taxonomic revision of *Geniostoma* subg. *Geniostoma* (Loganiaceae). *Blumea* 26: 245–364.

Darwin, C. R. (1842). *The Structure and Distribution of Coral Reefs.* London, Smith Elder.

Diamond, J. (2005). *Collapse – How Societies Choose to Fail or Survive.* London, Penguin.

Florence, J. (1997a). New species of *Alyxia* Banks ex. R. Brown (Apocynaceae) and *Glochidion* J. R. & J. G. Forster (Euphorbiaceae) from the Pitcairn Islands (South East Pacific). *Novon* 7: 27–31.

Florence, J. (1997b). *Flore de la Polynésie français, volume 1.* Paris, ORSTOM editions.

Florence, J., Waldren, S. & Chepstow-Lusty, A. J. (1995). The flora of the Pitcairn Islands: a review. *Biol. J. Linn. Soc.* 56: 79–119.

Fosberg, F. R. & Sachet, M-H. (1985). *Santalum* in Eastern Polynesia. *Candollea* 40: 459–470.

Gothesson, L.-A. (1997). *Plants of the Pitcairn Islands including local names and uses.* Sydney, University of New South Wales.

Hekinian, R., Cheminee, J. L., Dubois, J., Stoffers, P., Scott, S., Guivel, C. *et al.* (2003). The Pitcairn hotspot in the South Pacific: distribution and composition of submarine volcanic sequences. *J. Volcanol. Geoth. Res.* 121: 219–245.

Holttum, R. E. (1985). Studies in the fern genera allied to *Tectaria* Cav. IV. The genus *Ctenitis* in Asia, *Malesia* and the Western Pacific. *Blumea* 31: 1–38.

Irwin, G. (1992). *The Prehistoric Exploration and Colonisation of the Pacific.* Cambridge, Cambridge University Press.

Kingston, N. (2001). *The Flora and Vegetation of Pitcairn Island – its Phytogeography and Conservation.* Unpublished PhD thesis: University of Dublin, Trinity College.

Kingston, N. & Waldren, S. (2003). The plant communities and environmental gradients of Pitcairn Island: the significance of invasive species and the need for conservation management. *Ann. Bot.* 92: 31–40.

Kingston, N. & Waldren, S. (2005). A conservation appraisal of the rare and endemic vascular plants of Pitcairn Island. *Biodivers. Conserv.* 14: 781–800.

Kingston, N., Waldren, S. & Bradley, U. (2003). The phytogeographical affinities of the Pitcairn Islands – a model for south-eastern Polynesia? *J. Biogeogr.* 30: 1311–1328.

MacArthur, R. H. & Wilson, E. O. (1967). *The Theory of Island Biogeography.* Monographs in Population Biology. Princeton, NJ, Princeton University Press.

Mayes, C. L., Lawyer, L. A. & Sandwell, D. T. (1990). Tectonic history and new isochron chart of the South Pacific. *J. Geophys. Res.* 95: 8543–8567.

Menard, H. W. (1964). *Marine Geology of the Pacific.* New York, McGraw-Hill.

Mueller-Dombois, D. & Fosberg, F. R. (1998). *Vegetation of the Tropical Pacific Islands.* New York, Springer-Verlag.

Nickerson, T. (1984). *The Loss of the Ship 'Essex' Sunk by a Whale and the Ordeal of the Crew in Open Boats.* Nantucket Historical Society.

Nunn, P. D. (1999). *Environmental Change in the Pacific Basin.* Chichester, John Wiley & Sons.

Renvoize, S. A. (1979). The origins of Indian Ocean floras. In D. Bramwell (ed.), *Plants and Islands*, pp. 107–129. London, Academic Press.

Sachet, M.-H. (1967). Coral islands as ecological laboratories. *Micronesica* 3: 45–49.

Sherff, E. E. (1937). Some Compositae of southeastern Polynesia (*Bidens*, *Coreopsis*, *Cosmos*, and *Oparanthus*). *Occ. Pap. Bernice P Bishop Mus.* Series 12, 19: 1–19.

St John, H. (1987). An account of the flora of Pitcairn island with new *Pandanus* species. *Pac. Plant Stud.* 46. Privately published, Honolulu.

St John, H. & Philipson, W. R. (1960). List of the flora of Oeno Atoll, Tuamotu archipelago, south-central Pacific Ocean. *Trans. R. Soc. New Zeal.* 88: 401–403.

St John, H. & Philipson, W. R. (1962). An account of the flora of Henderson Island, South Pacific Ocean. *Trans. R. Soc. New Zeal.* 1: 175–194.

Smyth, N. (2008). Syzygium jambos *(L). Alston (Myrtaceae) control, and conservation restoration and monitoring, of the threatened native flora of Pitcairn Island, South Central Pacific.* Unpublished PhD thesis: University of Dublin, Trinity College.

Spencer, T. (1989). Tectonic and environmental histories in the Pitcairn Group, Paleogene to present: reconstructions and speculations. *Atoll Res. Bull.* 322: 1–41.

Spencer, T. (1995). The Pitcairn Islands, South Pacific Ocean: plate tectonic and climatic conditions. *Biol. J. Linn. Soc.* 56: 13–42.

Stoddart, D. R. (1992). Biogeography of the tropical Pacific. *Pac. Sci.* 46: 276–293.

Stoffers, P., Hekinian, R., Ackermand, D., Binard, N., Botz, R., Devey, C W. *et al.* (1990). Active Pitcairn hot spot found. *Mar. Geol.* 95: 51–55.

Stuessy, T. F., Crawford, D. J., Marticorena, C. & Silva O. M. (1998). Isolating mechanisms and modes of speciation in endemic angiosperms of the Juan Fernandez Islands. In T. F. Stuessy & M. Ono (eds.), *Evolution and Speciation of Island Plants*, pp. 79–96. Cambridge, Cambridge University Press.

van Balgooy, M. M. J. (1960). Preliminary plant-geographical analysis of the Pacific. *Blumea* 10: 385–430.

Waldren, S., Florence, J. & Chepstow-Lusty, A. J. (1995a). Rare and endemic vascular plants of the Pitcairn Islands, south-central Pacific ocean: a conservation appraisal. *Biol. Conserv.* 74: 83–98.

Waldren, S., Florence, J. & Chepstow-Lusty, A.J. (1995b). A comparison of the vegetation communities from the islands of the Pitcairn group. *Biol. J. Linn. Soc.* 56: 121–144.

Waldren, S., Weisler, M. I., Hather, J. G. & Morrow, D. (1999). The non-native vascular plants of Henderson Island, South-Central Pacific Ocean. *Atoll Res. Bull.* 463.

Weisler, M. I. (1995). Henderson Island prehistory: colonisation and extinction on a remote Polynesian island. *Biol. J. Linn. Soc.* 56: 377–404.

Whittaker, R. J. & Fernandez-Palacios, J. M. (2006). *Island Biogeography: Ecology, Evolution and Conservation*, 2nd edition. Oxford, Oxford University Press.

# 11

# Chromosomes and evolution in New Zealand endemic angiosperms and gymnosperms

BRIAN G. MURRAY AND PETER J. DE LANGE

The New Zealand flora, like that of many other oceanic islands, is unique in several ways. It has a high level of endemicity (de Lange *et al.*, 2006) but this is largely at the species level, with approximately 82% (1944/2357) of taxa being endemic. However, only 65 of the 446 indigenous genera are regarded as endemic and there was only one endemic family, the Ixerbaceae, represented by the monogeneric *Ixerba brexioides* (Fig. 11.1a.b), a small woody tree of northern New Zealand forests, but the family has now been merged with the Strasburgiaceae (APG, 2009). The vascular flora is also remarkably small for an archipelago the size of New Zealand (263 830 km$^2$), in comparison, the much smaller Fijian archipelago (18 235 km$^2$) has an indigenous spermatophyte flora of 1318 taxa in 137 families and 490 genera (Smith, 1996) and, while complete statistics are not available, the even smaller Hawaiian archipelago (16 626 km$^2$) has an indigenous flowering plant flora of *c.*956 species in 87 families and 216 genera (Wagner *et al.*, 1990). So what makes New Zealand's flora so unusual? Perhaps the answer lies not so much in the size of the flora and its perceived diversity but rather in the problem of species and generic delimitation. Many New Zealand genera, most notably those that are species rich, either have poorly defined generic boundaries, e.g. *Celmisia*, *Raoulia* (Allan, 1961; Ward, 1997; Cross *et al.*, 2002; Smissen *et al.*, 2004), or hybridise readily, especially in situations of prolonged disturbance, e.g. *Chionochloa*, *Coprosma*, *Epilobium* and *Kunzea* (Oliver, 1935; Connor, 1967; Raven & Raven, 1976; Wichman *et al.*, 2002; de Lange *et al.*, 2005). Hybridism is a feature of the New Zealand flora and it occurs at both the species and genus level to varying degrees but at sufficient frequency to ensure that the existence of hybrids has caused difficulties with traditional taxonomic attempts to classify the flora (Morgan-Richards *et al.*, 2009). This recognition of hybridism in the New Zealand flora is not a recent discovery; indeed, early New Zealand botanical writings indicate an

*The Biology of Island Floras*, eds. David Bramwell and Juli Caujapé-Castells. Published by Cambridge University Press. 

Figure 11.1 (a,b) The small forest tree tawari (*Ixerba brexioides*) is the sole species of New Zealand's formerly endemic flowering plant family the Ixerbaceae. (a) An adult flowering tree in its typical forest habitat. (b) Close-up of the flowers, which are produced in profusion between September and December. (c) Kauri (*Agathis australis*), opinions vary as to whether this spectacular conifer is a truly ancient vicariant element of the New Zealand flora that survived the Oligocene marine transgression or a more recent Quaternary arrival.

awareness of hybrids as early as the mid 1860s and it was certainly postulated as both a driver for speciation and a feature of the indigenous flora by pioneering botanists such as Leonard Cockayne and Harry Allan (Cockayne, 1923, 1929; Allan, 1924; Cockayne & Allan, 1934). Although the prevalence of widespread hybridism has been questioned (Allan, 1961; Hair, 1966; Connor, 1985), its significance has come in for re-evaluation (Morgan-Richards *et al.*, 2009), especially with the advent of DNA-based methods when many examples have been confirmed (Molloy & Dawson, 1998; Lockhart *et al.*, 2001; Wichman *et al.*, 2002; Gardner *et al.*, 2004; Murray *et al.*, 2004; de Lange *et al.*, 2008).

Phylogenetic studies using DNA sequences have also prompted a re-evaluation of the 'age' of the New Zealand flora. It has long been accepted that the New Zealand flora is mostly ancient in origin, derived from a Gondwanan flora that had survived the tectonic upheavals that ended that land mass, through rafting off on what is now known as the Zealandia sub-continent (Fleming, 1949). This idea has been questioned by more recent studies (Trewick *et al.*, 2006; Campbell & Hutching, 2007) and it would now appear that very few (if any) of New Zealand indigenous spermatophytes have a truly ancient origin (Pole, 1994; Brownsey, 2001; McGlone *et al.*, 2001; Winkworth *et al.*, 2002; Knapp *et al.*, 2005; Trewick *et al.*, 2006) though the case for invertebrates needs further exploration (Gibbs, 2007). Even recent claims that the endemic conifer, kauri (*Agathis australis*) (Fig. 11.1c), survived the Oligocene marine transgression (Knapp *et al.*, 2007; Lee *et al.*, 2007) remain controversial (Pole, 2008). It now seems that most if not all of New Zealand's indigenous flora was derived over the last 26 Ma and, in some cases, as recently as the last 1–5 Ma, from Australia, the wider Pacific basin or further afield, leading many biogeographers to regard the current New Zealand archipelago as truly oceanic.

Another aspect of the New Zealand flora that may explain its small size, abundance of species-poor genera, and poor genus and species boundaries is the fact that the archipelago is tectonically active (Fig. 11.2a,b). Currently the sub-continent is in the throes of the Kaikoura Orogeny, a major phase of tectonic uplift that was initiated during the Miocene and continues to the present day. As a result, much of what has been interpreted as a low-lying, highly fragmented, if not fully submerged, land mass towards the end of the Oligocene (Campbell & Hutching, 2007; Gibbs, 2007) has been rapidly uplifted to form the main islands recognised as the New Zealand of today. During the Pliocene, the orogeny appears to have increased in pace, resulting in the creation of new mountain habitats that had previously not existed since the onset of the Rangitata Orogeny during the early Cretaceous. These events, coupled with associated and increasing volcanic activity stemming from a steepening and widening of the Pacific Plate subduction zone, and frequent glacials during the Pleistocene, have all contributed to a dynamic environment ideally providing novel habitats for new species but paradoxically also leading to sudden species destruction. Under these conditions a robust, resilient though remarkably plastic flora appears to have evolved: a flora that comprises species capable of hybridising with close and distant relatives to varying degrees.

Chromosome numbers are now known for approximately 82% of angiosperms and 100% of gymnosperms in the native vascular flora (de Lange *et al.*, 2006). The majority of the information comes from the series of 'Contributions to a chromosome atlas of the New Zealand flora', the first of which was published over 50 years ago in the *New Zealand Journal of Science* by Hair and Beuzenberg (1958a). The

Figure 11.2 (a) New Zealand is now a complex archipelago. Apart from the three main islands, New Zealand has thousands of smaller islands, each often supporting its own unique flora and fauna. Some of these islands have recently been reconnected with the main islands of New Zealand, such as those seen here at Cape Maria Van Diemen, the northernmost extremity of the North Island of New Zealand. (b) The onset of the Kaikoura Orogeny during the Miocene led to the formation of the main axial ranges that run north to south across the two main islands of New Zealand. These ranges created a diversity of new habitats for plants to colonise. It is in these mountains that most of New Zealand's floral diversity is now found.

series is ongoing, now published by the *New Zealand Journal of Botany*, and the latest number, 39, deals with the chromosomes of the Orchidaceae (Dawson *et al.*, 2007). As a consequence of this long-term interest, the New Zealand flora is among the best-studied floras from a chromosome number perspective in the world. This allows conclusions to be drawn on patterns of chromosomal evolution from a firm factual foundation.

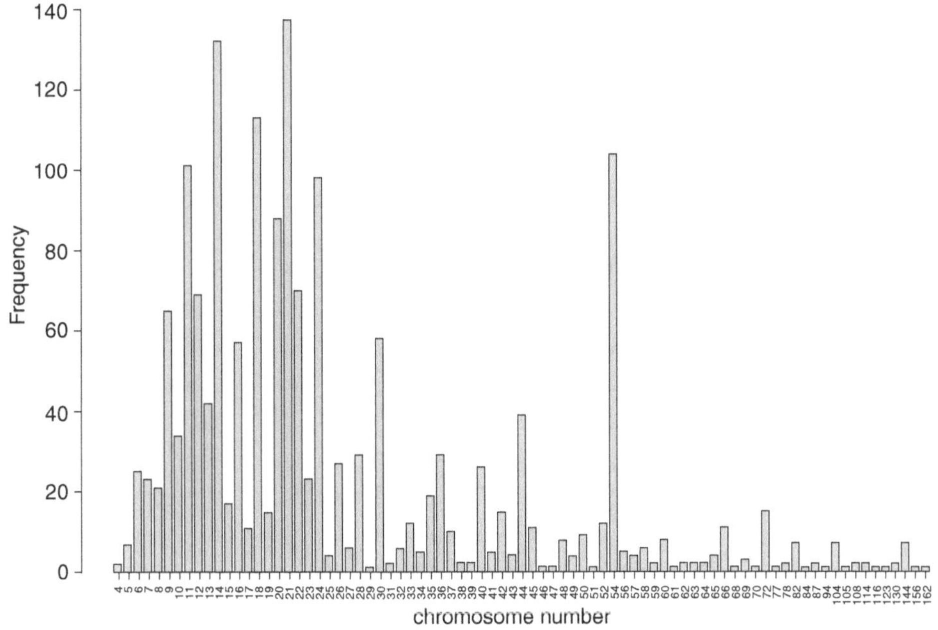


Figure 11.3 The distribution of haploid ($n$) chromosome numbers in the endemic and indigenous angiosperms of New Zealand.

Among the angiosperms, haploid (gametic, $n$) chromosome numbers range from 4 to 162 (Fig. 11.3). The great majority of species have high haploid chromosome numbers and there is an obvious disparity between even and odd numbers that becomes very apparent when $n$ exceeds 14. Otto and Whitton (2000) have suggested that this excess of even over odd haploid numbers of chromosomes can be used as an indicator of polyploidy, as chromosome doubling will automatically generate even numbers. Many workers have also suggested that haploid numbers > 10–14 are indicative of a polyploid ancestry (Stebbins, 1971; Goldblatt, 1980) so when these two observations are taken together it would appear that polyploidy has been a key process in the evolution of New Zealand plants. At this stage it is necessary to differentiate between neopolyploids, polyploids that are of relatively recent origin that are very similar and closely related to extant diploids (or lower polyploids) and palaeopolyploids which are more ancient in origin and their polyploid history is obscured by the absence of obviously diploid relatives (Ehrendorfer, 1980; Bennett, 2004; Guerra, 2008). Both of these types of polyploid are common in New Zealand angiosperms. Among the neopolyploids are many examples of very high multiples of the basic chromosome number; these include *Senecio* (10-ploid), *Ranunculus* (18-ploid) and *Poa* (38-ploid).

The neopolyploids can be analysed in several ways. Of considerable interest is whether these polyploids have arisen in the place where they are found, that is autochthonously, or whether the successful colonisers were polyploid when they arrived. Evidence for widespread autochthonous polyploidy can be obtained from the 82 genera where chromosome numbers are known for five or more species, five species being considered a reasonable number of species sampled. In 40 of the 82 genera, all the species have the same chromosome number but in the remaining 42 genera there are species within them that differ in ploidy. In some cases, such as *Hebe* (Wagstaff & Garnock-Jones, 1998; Wagstaff *et al.*, 2002), there is also phylogenetic evidence of a relatively recent, post-Pliocene arrival in New Zealand by long-distance dispersal of a diploid ancestral species. These examples of widespread autochthonous polyploidy provide a striking contrast to the situation discussed by Stuessey and Crawford (1998) who describe 'chromosomal stasis' in the floras of a number of other Pacific island groups. It is possible that the tectonically active, geologically and climatically diverse New Zealand archipelago, which is spread over some 2600 km (29° 14′ to 52° 37′ S) with a west–east extent of 1800 km (166° 28′ E to 176° 10′ W), has made available a wider range of new habitats that have been successfully exploited by these neopolyploids.

Further evidence for autochthonous polyploidy is seen among the examples of intraspecific variation in chromosome number in the New Zealand angiosperm flora. Although it is uncommon, only *c.*3% of the species show variable chromosome numbers and in almost all cases (see *Crassula* below as an exception; de Lange *et al.*, (2008) there are only two ploidy races, as in *Hebe* (Murray *et al.*, 1989; Dawson & Beuzenberg, 2000).

It is also possible that the occurrence of neopolyploidy in New Zealand may be related to the lowest diploid chromosome number found in a genus. In other words, the incidence of neopolyploidy may be greater where diploid chromosome number is low; further increase in chromosome number being less likely when large numbers of chromosomes are already present in the cell. A comparison of chromosome numbers in the genera that are exclusively composed of diploid species and those that contain both diploid and neopolyploid species does not suggest any obvious relationship (Fig. 11.4). In both groups, those with and those without neopolyploids, the distribution of genera with different chromosome numbers is not obviously different and in the genera that contain both diploid and neopolyploid species there is a very wide range of diploid numbers. Thus, chromosome number does not appear to be a defining factor that determines the presence or absence of polyploids in a genus.

An analysis of the incidence of neopolyploidy in the different families of New Zealand angiosperms shows that at this level there are 82 genera where chromosome numbers are known for five or more species distributed among 37 families. These families can be divided into two groups, one where all genera contain species that have the same chromosome number (diploid or polyploid) and a second

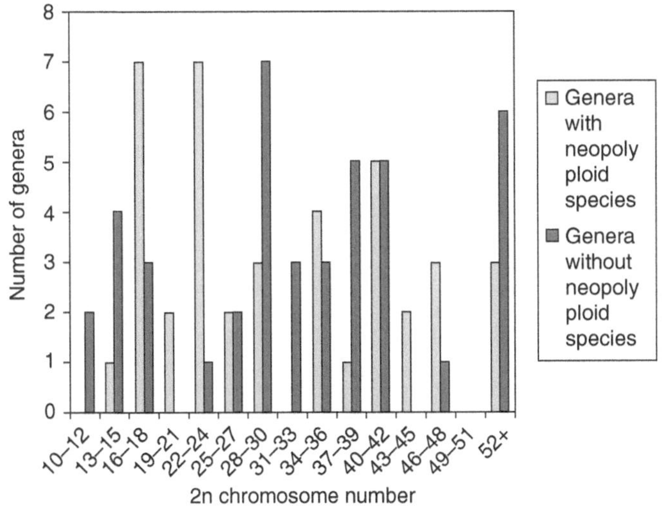


Figure 11.4 The distribution of chromosome numbers in genera that do or do not contain neopolyploid species.

where polyploidy is seen in some species within genera. Thirteen of the families contain only genera with no autochthonous polyploids, 11 contain genera, some of which contain species with different ploidy levels and others have only a single ploidy level, and in the remaining 13 families all the genera contain a mixture of diploid and polyploid species (Table 11.1). Thus, there seems to be little evidence for a familial bias in the distribution of polyploidy. Stebbins (1971) summarised data that showed a low percentage of polyploids among woody as compared to herbaceous plants. In New Zealand the situation is not so clear-cut and a survey of woody genera with and without polyploids shows that the distribution is more even: 18 genera have no polyploids and 13 have polyploids (Table 11.1).

We have also used the data from the 'Contributions' series along with a small number of other publications (Frankel & Hair, 1937; Murray *et al.*, 2004) to try and determine whether the autochthonous polyploids are autoploids or alloploids. Autoploids are defined as the results of chromosome doubling within a species and are characterised by the formation of chromosome multiples at meiotic metaphase I (MI). Alloploids, on the other hand, are hybrid polyploids derived from chromosome doubling following hybridisation between species that are genomically differentiated. These are characterised by bivalent formation at meiotic MI. Meiotic MI is illustrated for 201 species in 77 genera in the 'Contributions' series plus Frankel and Hair (1937) and Murray *et al.* (2004) and in only 10 species – six in *Hebe*, one in *Myoporum*, one in *Myosotis* and two in *Pomaderris* (4.98% of the total) – and these four genera (5.19% of the total) is there any evidence of multivalent formation. On the face of it this is highly suggestive of an allopolyploid origin

Table 11.1 *The familial distribution of genera that do or do not contain a polyploid series.*

| Family polyploid series | Genera without a polyploid series | Genera with a polyploid series |
|---|---|---|
| Alseuosmiaceae | *Alseuosmia** | |
| Apiaceae | *Aciphylla, Anisotome, Gingidia* | *Schizeilema* |
| Apocynaceae | *Parsonsia** | |
| Araliaceae | *Pseudopanax** | *Hydrocotyle* |
| Asphodelaceae | *Bulbinella* | |
| Asteliaceae | | *Astelia* |
| Asteraceae | *Abrotanella, Anaphaloides, Brachyglottis*, Craspedia, Euchiton, Lagenifera* | *Celmisia*, Leptinella, Olearia*, Raoulia*, Senecio* |
| Boraginaceae | | *Myosotis* |
| Brassicaceae | *Pachycladon* | |
| Campanulaceae | | *Wahlenbergia* |
| Coriariaceae | | *Coriaria** |
| Crassulaceae | | *Crassula* |
| Ericaceae | *Dracophyllum*, Gaultheria** | *Leucopogon** |
| Fabaceae | *Sophora** | *Carmichaelia** |
| Gentianaceae | *Gentianella* | |
| Geraniaceae | | *Geranium* |
| Gunneraceae | *Gunnera* | |
| Haloragaceae | | *Myriophyllum* |
| Iridaceae | | *Libertia* |
| Lobeliaceae | | *Lobelia* |
| Malvaceae | *Hoheria** | |
| Myrtaceae | *Metrosideros** | |
| Myrsinaceae | *Myrsine** | |
| Onagraceae | *Epilobium* | |
| Orchidaceae | *Nematoceras, Pterostylis* | *Thelymitra* |
| Phrymaceae | | *Mazus* |
| Pittosporaceae | *Pittosporum** | |
| Plantaginaceae | *Chionohebe*, Heliohebe*, Leonohebe*, Ourisia* | *Callitriche, Hebe*, Parahebe*, Plantago* |
| Poaceae | *Chionochloa, Cortaderia, Deschampsia* | *Agrostis, Deyeuxia, Elymus, Lachnagrostis, Poa, Puccinellia, Festuca, Hierochloe, Rytidosperma, Trisetum* |
| Polygonaceae | *Muehlenbeckia** | |
| Portulaceae | *Montia* | |
| Ranunculaceae | *Clematis** | *Ranunculus* |
| Rhamnaceae | | *Pomaderris** |
| Rosaceae | *Rubus** | *Acaena*, Geum* |
| Rubiaceae | *Nertera* | *Coprosma** |
| Thymelaeaceae | *Kelleria** | *Pimelea** |
| Violaceae | | *Melicytus** |

* Woody species.

for the majority of the New Zealand polyploid angiosperms; however, some caution is needed with this interpretation.

As indicated above, many of the polyploid genera appear to be recent arrivals in New Zealand of only one or a few propagules with consequent low levels of genetic differentiation between species, so it is problematic to suggest that there has been rapid genomic evolution that would prevent chromosome synapsis in newly formed hybrids that subsequently underwent polyploidy. Indeed, many studies suggest that the morphological diversity of the species-rich genera has been accompanied by only small genotypic changes (Winkworth *et al.*, 1999). The available evidence of chromosome differentiation, though limited to studies on a small number of genera, suggests that there has not been extensive genomic re-patterning. Regular meiotic pairing was observed in artificial hybrids of *Ranunculus* (Rendle & Murray, 1989) and fertile interspecific hybrids, indicative of genomic homogeneity, are common in many, though by no means all, interspecific hybrids in *Hebe* (Frankel & Hair, 1937). Other examples of autopolyploidy supported by molecular phylogenetic evidence are found in *Leucogenes* and *Lobelia* (which at that time were treated as *Pratia*) (Murray *et al.*, 2004; Knox *et al.*, 2008; Smissen & Breitwieser, 2008). Recently, Paun *et al.* (2009) have used genetic divergence, as a surrogate for chromosomal differentiation, to show that genomic differentiation between the progenitor species is an important driver of whether hybrids give rise to diploid (homoploid) or polyploid hybrid species.

With diploids in genera such as *Hebe* and *Lobelia* (Wagstaff & Garnock-Jones, 1998; Wagstaff *et al.*, 2002; Murray *et al.*, 2004) showing generally low levels of genetic divergence, the high frequency of polyploids in the New Zealand flora would appear unlikely to be derived from hybridisation events (i.e. alloploids). If this is the case then some explanation for the prevalence of diploid-like meiosis in the polyploids is required. Chromosome pairing in autoploids can be modified by the activity of meiotic pairing control genes such as *Ph*, which promotes exclusive bivalent formation in wheat (Riley & Chapman, 1958) and other genera (Murray *et al.*, 1984) despite widespread similarity between component genomes. Chiasma frequency is also an important factor in multivalent formation as a chiasma frequency greater than 0.5 per homologous chromosome is necessary for multivalent formation. A low chiasma frequency will result in bivalent formation in autoploids, as shown by Hauber (1986) in *Haplopappus spinulosus*. None of the MI cells illustrated in the 'Contributions', nor the clearly autoploid race of *Lobelia angulata* (Murray *et al.*, 2004), show evidence in their chromosome pairing of a restricted number of chiasmata; all show many ring bivalents with at least two chiasmata (i.e. $\geq 1$ chiasma per homologous chromosome, Fig. 11.5a), a prerequisite for multivalent formation. Thus an explanation for the widespread occurrence of bivalent formation in New Zealand endemic autopolyploids remains to be elucidated.

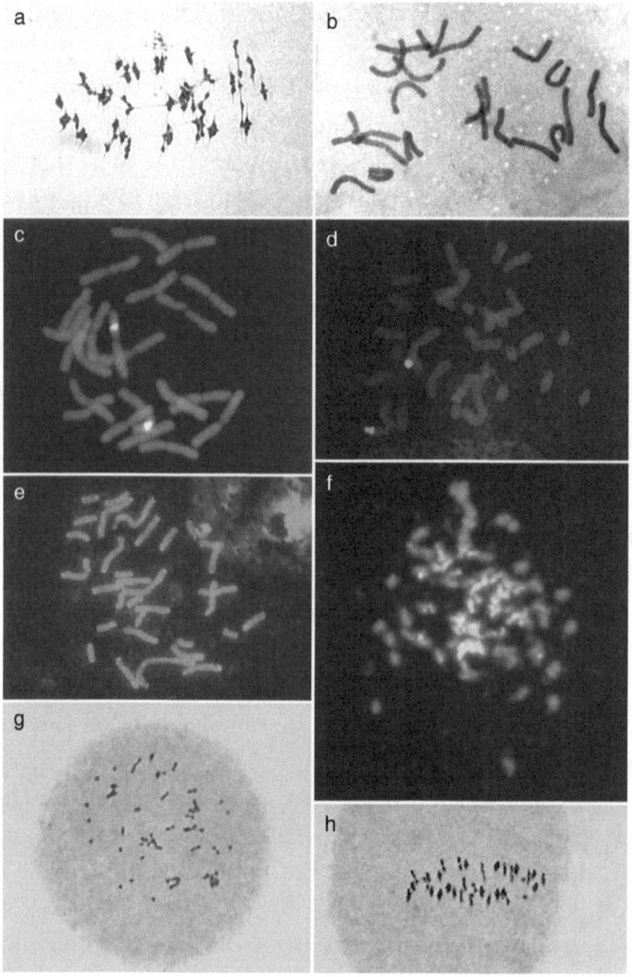


Figure 11.5 (a) Meiotic metaphase I in *Lobelia angulata* showing 70 bivalents, most of which have two chiasmata. (b) The somatic chromosomes of *Libocedrus plumosus* $2n = 22$ solid stained with orcein. (c) The somatic chromosomes of *Libocedrus bidwillii* $2n = 22$ stained with CMA showing a single fluorescent band located around the nucleolar organiser region. (d) The somatic chromosomes of *Podocarpus totara* $2n = 34$ showing the location of the 45S rDNA on one pair of chromosomes. (e) The somatic chromosomes of *Podocarpus totara* $2n = 34$ showing the telomeric and dispersed location of the *Arabidopsis*-like telomere sequence, TTTAGGG. (f) The application of GISH with *Lobelia angulata* DNA to the somatic chromosomes of a putative $13x$ hybrid between *L. angulata* and *L. perpusilla*. Positive hybridisation sites are dark as the *L. angulata* DNA was labelled with Texas Red. (g, h) Meiotic metaphase I in *Crassula ruamahanga*, (g) with $2n = 78$ showing a mixture of bivalents and univalents and (h) with $2n = 64$ and regular bivalent formation. (a) and (f) from Murray *et al.* (2004) by permission of Taylor & Francis; (d) and (e) from Murray *et al.* (2002) by permission of Oxford University Press; (g) and (h) from Lange *et al.* (2008) by permission of Oxford University Press.

In contrast to the widespread incidence of polyploidy, supernumerary, fragment or B chromosomes that are relative common in other floras have seldom been reported in New Zealand endemic angiosperms. This rarity may be real or possibly a consequence of only small numbers of individuals, sometimes only one, examined for each species. Supernumerary chromosomes have only been reported in six genera – *Celmisia*, *Dichelachne*, *Festuca*, *Hebe*, *Pomaderris* and *Ranunculus* – but no detailed studies have been made to ascertain the true nature of the additional chromosomes. In some species, such as *Ranunculus ranceorum*, several types of additional chromosomes were observed in a single population; some were large and indistinguishable from the normal complement but at least three smaller types that had no obvious homology to the normal complement were reported (de Lange & Murray, 2008).

More recently, a combination of population-based studies with the more extensive sampling of species and the application of molecular cytogenetic and phylogenetic techniques has made significant contributions to a better understanding of chromosome evolution in a number of endemic groups.

The 21 New Zealand endemic gymnosperms belong to five families: the Araucariaceae (1 species), Cupressaceae (2 species), the Prumnopityaceae (2 species) Phyllocladaceae (3 species) and Podocarpaceae (13 species). Chromosome numbers were published for all the species by Hair and Beuzenberg (1958a) at a time when the Podocarpaceae included the species now placed in the Prumnopityaceae and Phyllocladaceae. These Podocarpaceae species were interesting in that they showed a wide range of chromosome numbers, from $2n = 18$ to $2n = 38$, unlike genera of Northern Hemisphere conifers such as *Pinus* and *Picea* where all the species have the same chromosome number. Hair and Beuzenberg (1958b) followed up their account of chromosome numbers with a more detailed commentary describing a proposed scheme of chromosome evolution by Robertsonian change, the fusion/fission of chromosomes at or near their centromeres, to account for the variation in chromosome number in the species of Podocarpaceae. Later taxonomic studies have reassigned a number of the *Podocarpus* species to other genera such as *Dacrycarpus* and *Prumnopitys* (de Laubenfels, 1969) and some of the *Dacrydium* species to *Halocarpus*, *Lepidothamnus* and *Manoao* (Quinn, 1982; Molloy, 1995). *Phyllocladus* is now considered to be sufficiently distinct to be recognised at the family level (Phyllocladaceae) (Page, 1990). Davies *et al.* (1997) took this proposed Robertsonian relationship between the chromosomes of the species as their starting point for a study that applied differential chromosome banding techniques using DNA base-specific fluorochromes and silver nitrate staining to identify regions of constitutive heterochromatin and nucleolar organiser regions (NORs), respectively, to try and clarify the evolutionary relationships of the chromosomes of the Podocarpaceae. They also measured genome size using flow cytometry of propidium iodide stained nuclei and found that, as expected, there was a significant correlation between genome size and total chromosome length. Despite their relatively large size (Fig. 11.5b), the chromosomes of these species showed few bands and these were only observed after staining with the GC-specific fluorochrome Chromomycin A3 (CMA) (Fig. 11.5c). No significant DAPI-positive bands were observed, though occasionally small centromeric spots were seen in a few species. All the species showed a pair of CMA bands that were associated with either a secondary constriction or a centromere, but the bands were shown to be on chromosomes of very variable size and in different positions along the chromosome arms. Success with silver staining was variable but when positive bands were observed they were always associated with the single pair of CMA bands. The small number of bands meant that banding was not particularly useful for following possible evolutionary patterns in these plants but, together with the large differences in genome size that were observed, suggested that the proposed Robertsonian relationship(s) between the related species with different chromosome numbers were unlikely to be valid. Chromosome breakage/fusion would not be expected to contribute to the large differences in genome size that were observed.

A subsequent study by Murray *et al.* (2002) concentrated on *Podocarpus* species and used fluorescence *in situ* hybridisation (FISH) to map the location of a range of repetitive DNA sequences to the chromosomes of the New Zealand species and *P. lawrencei*, an Australian one. The location of the 45S (Fig. 11.5d) and 5S ribosomal RNA genes (rDNA) was unusual as they co-localised to a single pair of regions on the chromosomes of the New Zealand *Podocarpus* species, and the hybridisation sites were located in the same region as the CMA-positive bands (that were also often positive with silver nitrate). The hybridisation sites were on different chromosomes in the New Zealand species and two additional minor 45S sites were seen in the Australian *P. lawrencei*. This latter species was the only one that showed DAPI-positive bands after FISH, indicating blocks of AT-rich heterochromatin. It appears that in this group of *Podocarpus* species the number of nucleolar organiser regions is low, just on one (in the New Zealand species) or two (in the Australian species) pairs of chromosomes, and this provides a striking contrast to results from other gymnosperms such as *Pinus*, *Picea* and *Larix* where up to 12 pairs of sites have been reported (Hizume *et al.*, 1992; Lubaretz *et al.*, 1996; Jacobs *et al.*, 2000). FISH was also used to determine whether the telomere sequences of the New Zealand *Podocarpus* are comprised of the *Arabidopsis*-type telomere sequences $(TTTAGGG)_n$. The results were positive (Fig. 11.5e) but, in addition to signals located at the chromosome ends, positive signals for these sequences were found at many small interstitial sites distributed along the length of the chromosome arms of all the species. Interstitial hybridisation sites are found in other gymnosperms, but in these they are less numerous and often larger than those seen in *Podocarpus* (Schmidt *et al.*, 2000; Shibata *et al.*, 2005), and in other Pinaceae telomere sequences are confined to the chromosome ends (Lubaretz *et al.*, 1996). These cytogenetic studies reveal a more complex pattern of chromosome variation between both the genera and the species within genera, perhaps reflecting their ancient lineage.

The New Zealand species of *Lobelia* were formerly placed in the genus *Pratia* but have recently been reassigned to *Lobelia* (Knox *et al.*, 2008) and their new names will be used here. These species provide one of the best examples of a detailed cytogenetic analysis that involved extensive population sampling, meiotic analysis and *in situ* hybridisation coupled with the production of artificial hybrids and phylogenetic reconstruction (Murray *et al.*, 2004). The species studied by Murray *et al.* (2004) showed five different ploidy levels: *L. macrodon* had $2n = 2x = 14$, *L. perpusilla* had $2n = 6x = 42$, *L. arenaria* had $2n = 10x = 70$ and the most widespread species, *L. angulata*, was found to have two chromosome races, one with $2n = 10x = 70$ and the other with $2n = 20x = 140$. These two chromosome races had different and largely non-overlapping distributions and the phylogenetic analysis of both chloroplast and nuclear DNA sequences indicated clearly that the $20x$ plants are autoploids of the $10x$ ones. In the light of the discussion above on chromosome pairing in purported autoploids in New Zealand, the $20x$ plants were found to be exclusively bivalent forming, despite having a sufficiently high chiasma frequency to allow for multivalent formation. This is a further example of the problem of using chromosome pairing configurations to determine the type of polyploidy being studied.

In addition to these two chromosome races, plants initially collected as *L. angulata* were found to have $2n = 11x = 77$, $2n = 12x = 84$ and $2n = 13x = 91$. On closer examination these plants showed many intermediate characters between *L. angulata, L. perpusilla* and *L. macrodon* and it was proposed that they were interspecific hybrids. Artificial hybridisation experiments showed that fully viable hybrids could be made between all combinations of *L. perpusilla*, *L. arenaria* and the $10x$ and $20x$ races of *L. angulata*, indicating that there were no intrinsic barriers to hybridisation between these species. Confirmation of the hybrid origin of the $11x$ and $13x$ plants was obtained from the application of genomic *in situ* hybridisation (GISH) using whole genomic DNA from the putative parents as a probe on the chromosomes of the $11x$ and $13x$ plants. In both cases, as well as in artificial hybrids between *L. perpusilla* and $10x$ *L. angulata*, there was differential hybridisation to the total chromosome complement with the clear differentiation of genomes in the hybrids (Fig. 11.5f). When the number of chromosomes identified by the *in situ* hybridisation with the two probes was counted, the results suggested that unreduced (diploid) gametes must have been involved in the genesis of the hybrids having different chromosome numbers. The $2n = 11x = 77$ plants were suggested to have arisen following the fusion of a normal, reduced gamete from 10$x$ *L. angulata* (with 35 chromosomes) with an unreduced gamete from *L. perpusilla* (with 42 chromosomes). The mode of origin of the $2n = 13x = 91$ plants was less certain as they could have arisen from an unreduced gamete (with 70 chromosomes) from a 10$x$ or a reduced gamete (also with 70 chromosomes) from a 20$x$ *L. angulata* combined with a reduced gamete from *L. perpusilla* (with 21 chromosomes). With the availability of sequence information from the chloroplast genome, and the assumption that the chloroplasts were maternally inherited, it was shown that the 11$x$ hybrid had *L. perpusilla* as its maternal parent whereas the 13$x$ hybrid had *L. angulata* as its maternal parent, and it is probable that these hybrids can be produced in either direction as reciprocal artificial hybridisations were equally successful in producing progeny. The origin of the $2n = 12x = 84$ plants was not studied but it is possible that they represent hybrids between *L. angulata* and *L. macrodon*, the diploid species.

A second population-based study that involved two threatened and apparently distinct species of *Crassula*, *C. hunua* and *C. ruamahanga*, revealed an even more complex pattern of chromosome variation than seen in *Lobelia* in these two taxa and highlighted the importance of chromosome variation in the conservation of rare and endangered plants both in New Zealand and worldwide (de Lange *et al.*, 2008). Until the publication of that paper, the two species were accepted as being distinct entities though the differences between them were small (Allan, 1961; Webb *et al.*, 1988). However, in more recent years new collections were made that widened the range of locations where they were found and this led to a re-evaluation of their conservation status and also raised problems with the delimitation of the two species (de Lange, 2000; Sykes, 2005). With this background, more detailed studies were initiated that involved both morphological and chromosomal investigations as well as DNA sequencing for phylogenetic reconstructions. The morphological studies included 12 key characters that had traditionally been used to differentiate the species but analysis of this larger sample of plants showed that it was impossible to differentiate them one from another.

The chromosomal analysis of populations from throughout the range of both species showed extreme variation with 11 different chromosome numbers. Chromosome numbers ranged from $2n = 42$ to $2n = 100$ in plants from 16 populations and in many cases there did not appear to be any obvious numerical relationship between the different numbers. Some of the numbers, $2n = 42$, 70, 84, 91, 98, suggest a polyploid series based on $x = 7$ but others, such as $2n = 64$, 68, 78 and 94, suggest that aneuploidy is also involved. Population samples also showed variations in chromosome number within populations; for example, at the type locality for *C. hunua*, six different chromosome numbers were obtained. In most plants encompassing a wide range of chromosome numbers, meiosis appeared to be regular and bivalents were seen clearly aligned at the metaphase plate (Fig. 11.5h) but in others a mixture of bivalents and univalents was seen (Fig. 11.5g). Irregular meiotic pairing with bivalents and univalents was seen both in plants with expected euploid numbers (such as $2n = 70$) as well as in those that appeared to be aneuploids. No multivalents were observed in any meiotic cells despite the chromosomes having a sufficiently high chiasma frequency to permit multivalent formation. The pollen fertility of these *Crassula* plants was also very variable, ranging from 2% to 98% stainability with Cotton Blue, but this was not correlated with the regularity of meiotic metaphase pairing. A phylogenetic analysis using both nuclear (ITS) and chloroplast (*trnL* intron and *trnL-F* spacer) did not result in a clearly resolved phylogeny; many plants grouped together with the same sequences but in other cases plants collected as *C. hunua* or *C. ruamahanga* clustered into different clades that included other *Crassula* species such as *C. helmsii* and *C, sinclairii*. Overall, these results suggested that these plants are a species complex that has probably evolved through hybridisation and polyploidy, and the high levels of chromosomal and DNA sequence variation present a problem for their conservation.

In conclusion, the application of new methods and techniques has improved our knowledge and understanding of many evolutionary processes in the New Zealand flora though several important questions remain to be answered. Notable among them are the age of the key elements in the flora, the extent of hybridism and the role of hybridism in the origin of the many polyploid species that are so characteristic of the flora.

## Acknowledgements

We thank Kathy Ruggiero and Jeremy Rolfe for their help with the preparation of the figures and Joli Braggins and Jeremy Rolfe for the use of their images in Figures 11.1. and 11.2.

## References

Allan, H. H. (1924). On the hybridity of *Coprosma cunninghamii* Hook.f. *New Zeal. J. Sci. Technol.* 6: 310–318.

Allan, H. H. (1961). *Flora of New Zealand, Volume I.* Wellington, Government Printer.

APG III. (2009). An update of the Angiosperm Phylogeny Group classification for the orders and families of flowering plants: APG III. *Botanical Journal of the Linnean Society* 161: 105–121.

Bennett, M. D. (2004). Perspectives on polyploidy in plants – ancient and neo. *Biol. J. Linn. Soc.* 82: 411–423.

Brownsey, P. J. (2001). New Zealand's pteridophyte flora – plants of ancient lineage but recent arrival? *Brittonia* 53: 284–303.

Campbell, H. & Hutching, G. (2007). *In Search of Ancient New Zealand*. London, Penguin Books.

Cockayne, L. (1923). Hybridism in the New Zealand flora. *New Phytol.* 22: 105–127.

Cockayne, L. (1929). Hybridism in the forests of New Zealand. *Acta Forest. Fenn.* 34: 23.

Cockayne, L. & Allan, H. H. (1934). An annotated list of groups of wild hybrids in the New Zealand flora. *Ann. Bot.* 48: 1–55.

Connor, H. E. (1967). Interspecific hybrids in *Chionochloa* (Gramineae). *New Zeal. J. Bot.* 5: 3–16.

Connor, H. E. (1985). Biosystematics of higher plants in New Zealand 1965–1984. *New Zeal. J. Bot.* 23: 613–643.

Cross, E. W., Quinn, C. J. & Wagstaff, S. J. (2002). Molecular evidence for the polyphyly of *Olearia* (Astereae: Asteraceae). *Plant Syst. Evol.* 235: 99–120.

Davies, B. J., O'Brien, I. E. W. & Murray, B. G. (1997). Karyotypes, chromosome bands and genome size variation in New Zealand endemic gymnosperms. *Plant Syst. Evol.* 208: 169–185.

Dawson, M. I. & Beuzenberg, E. J. (2000). Contributions to a chromosome atlas of the New Zealand flora – 36. Miscellaneous families. *New Zeal. J. Bot.* 38: 1–23.

Dawson, M. I., Molloy, B. P. J. & Beuzenberg, E. J. (2007). Contributions to a chromosome atlas of the New Zealand flora – 39. Orchidaceae. *New Zeal. J. Bot.* 45: 611–684.

de Lange, P. J. (2000). A new northern limit for *Crassula ruamahanga*. *New Zeal. Bot. Soc. Newslett.* 60: 20–21.

de Lange, P. J. & Murray, B. G. (2008). *Ranunculus ranceorum*, a new name and rank for *Ranunculus recens* var. *lacustris* G. Simpson, an elusive, rarely seen buttercup of the Fiordland lakes, South Island, New Zealand. *New Zeal. J. Bot.* 46: 1–11.

de Lange, P. J., Datson, P. M., Murray, B. G. & Toelken, H. R. (2005). Hybridisim in the *Kunzea ericoides* complex (Myrtaceae): an analysis of artificial crosses. *Aust. Syst. Bot.* 18: 117–131.

de Lange, P. J., Sawyer, J. W. D. & Rolfe, J. R. (2006). *New Zealand Indigenous Vascular Plant Checklist*. Wellington, New Zealand Plant Conservation Network.

de Lange, P. J., Heenan, P. B., Keeling, D. J., Murray, B. G., Smissen, R. & Sykes, W. R. (2008). Biosystematics and conservation: a case study of two enigmatic and uncommon species of *Crassula* from New Zealand. *Ann. Bot.* 101: 881–899.

de Laubenfels, D. (1969). A revision of the Malesian and Pacific rainforest conifers 1. Podocarpaceae. *J. Arnold Arbor.* 50: 274–369.

Ehrendorfer, F. (1980). Polyploidy and distribution. In W. H. Lewis (ed.), *Polyploidy: Biological Relevance*. New York, Plenum.

Fleming, C. A. (1949). The geological history of New Zealand (with reference to the origin and history of the flora and fauna). *Tuatara* 2: 72–90.

Frankel, O. H. & Hair, J. B. (1937). Studies on the cytology, genetics, and taxonomy of New Zealand *Hebe* and *Veronica* (Part 1). *New Zeal. J. Sci. Technol.* 18: 669–687.

Gardner, R. C., de Lange, P. J., Keeling, J. K., Bowala, T., Brown, H. A. & Wright, S. D. (2004). A late Quaternary phylogeography for *Metrosideros* (Myrtaceae) in New Zealand inferred from chloroplast DNA haplotypes. *Biol. J. Linn. Soc.* 83: 399–412.

Gibbs, G. (2007). *Ghosts of Gondwana: The History of Life in New Zealand.* Nelson, New Zealand, Craig Potton Publishing.

Goldblatt, P. (1980). Polyploidy in angiosperms: monocotyledons. In W. H. Lewis (ed.), *Polyploidy: Biological Relevance.* New York, Plenum.

Guerra, M. (2008). Chromosome numbers in plant cytotaxonomy: concepts and implications. *Cytogenet. Genome Res.* 120: 339–350.

Hair, J. B. (1966). Biosystematics of the New Zealand Flora 1945–1964. *New Zeal. J. Bot.* 4: 559–595.

Hair, J. B. & Beuzenberg, E. J. (1958a). Contributions to a chromosome atlas of the New Zealand flora – 1. *New Zeal. J. Sci.* 1: 617–628.

Hair, J. B. & Beuzenberg, E. J. (1958b). Chromosomal evolution in the Podocarpaceae. *Nature* 181: 1584–1586.

Hauber, D. P. (1986). Autotetraploidy in *Haplopappus spinulosus* hybrids: evidence from natural and synthetic tetraploids. *Am. J. Bot.* 73: 1595–1606.

Hizume, M., Ishida, F. & Murata, M. (1992). Multiple locations of the rDNA genes in pines, *Pinus densiflora* and *P. thunbergii. Jpn. J. Genet.* 67: 389–396.

Jacobs, M. D., Gardner, R. C. & Murray, B. G. (2000). Cytological characterization of heterochromatin and rDNA in *Pinus radiata* and *P. taeda. Plant Syst. Evol.* 223: 71–79.

Knapp, M., Stöckler, K., Havell, D., Delsuc, F., Sebastiani, F. & Lockhart, P. J. (2005). Relaxed molecular clock provides evidence for long-distance dispersal of *Nothofagus* (Southern Beech). *Public Lib. Sci. Biol.* 3: 38–43.

Knapp, M., Mudaliar, R., Havell, D., Wagstaff, S. J. & Lockhart, P. J. (2007). The drowning of New Zealand and the problem of *Agathis. Syst. Biol.* 56: 862–870.

Knox, E. B., Heenan, P. B. & Muasya, A. M. & Murray, B. G. (2008). Phylogenetic position and relationships of *Lobelia glaberrima* (Lobeliaceae), a new alpine species from southern South Island (New Zealand). *New Zeal. J. Bot.* 47: 77–85.

Lange, P. J., Heenan, P. B., Keeling, D. J., Murray, B. G., Smissen, R., Sykes, W. R. (2008). Bio systematics and conservation: a case study with two enigmatic and uncommon species of *Crassula* from New Zealand. *Annals of Botany* 101: 881–899.

Lee, D. E., Bannister, J. M. & Lindqvist, J. K. (2007). Late Oligocene-Early Miocene leaf macrofossils confirm a long history of *Agathis* in New Zealand. *New Zeal. J. Bot.* 45: 565–578.

Lockhart, P. J., McLenachan, P. A., Havell, D., Glenny, D, Huson, D. & Jensen, U. (2001). Phylogeny, dispersal and radiation of New Zealand alpine buttercups: molecular evidence under split decomposition. *Ann. Mo. Bot. Gard.* 88: 458–477.

Lubaretz, O., Fuchs, J., Ahne, R. & Meister, A. (1996). Karyotyping of three Pinaceae species via fluorescent in situ hybridization and computer aided chromosome analysis. *Theor. Appl. Genet.* 92: 411–416.

McGlone, M. S., Duncan, R. P. & Heenan, P. B. (2001). Endemisim, species selection and the origin and distribution of the vascular flora of New Zealand. *J. Biogeogr.* 28: 199–216.

Molloy, B. P. J. (1995). *Manoao* (Podocarpaceae), a new monotypic conifer genus endemic to New Zealand. *New Zeal. J. Bot.* 33: 183–201.

Molloy, B. P. J. & Dawson, M. I. (1998). Speciation in *Thelymitra* (Orchidaceae) by natural hybridism and amphidiploidy. In R. Lynch (ed.), 'Ecosystems, entomology and plants: proceedings of a symposium held at Lincoln University to mark the retirement of Bryony Macmillan, John Dugdale, Peter Wardle and Brian Molloy', 1 September 1995. *R. Soc. New Zeal. Misc. Ser.* 48: 103–113.

Morgan-Richards, M., Smissen, R. D., Shepherd, L. D., Wallis, G. P., Hayward, J. J., Chi-hang, C. *et al.* (2009). A review of genetic analyses of hybridisation in New Zealand. *J. R. Soc. New Zeal.* 39: 15–34.

Murray, B. G., Sieber, V. K. & Jackson, R. C. (1984). Further evidence for the presence of meiotic pairing control genes in *Alopecurus* L. (Gramineae). *Genetica* 63: 13–19.

Murray, B. G., Braggins, J. E. & Newman, P. D. (1989). Intraspecific polyploidy in *Hebe diosmifolia* (Cunn.) Cockayne et Allan (Scrophulariaceae). *New Zeal. J. Bot.* 27: 587–589.

Murray, B. G., Friesen, N. & Heslop-Harrison, J. S. (2002). Molecular cytogenetic analysis of *Podocarpus* and comparison with other gymnosperm species. *Ann. Bot.* 89: 483–489.

Murray, B. G., Datson, P. M., Lai, E. L. Y., Sheath, K. M. & Cameron, E. K. (2004). Polyploidy, hybridization and evolution in *Pratia* (Campanulaceae). *New Zeal. J. Bot.* 42: 905–920.

Oliver, W. R. B. (1935). The genus *Coprosma. Bernice P. Bishop Mus. Bull.* 132: 1–207.

Otto, S. P. & Whitton, J. (2000). Polyploid incidence and evolution. *Annu. Rev. Genet.* 34: 401–437.

Page, C. N. (1990). *Podocarpaceae.* In K. Kubitzki (ed.), *The Families and Genera of Vascular Plants. 1. Pteridophytes and Gymnosperms*, pp. 332–346. Berlin, Springer.

Paun, O., Forest, F., Fay, M. F. & Chase, M. W. (2009). Hybrid speciation in angiosperms: parental divergence drives ploidy. *New Phytol.* 182: 507–518.

Pole, M. (1994). The New Zealand flora – entirely long-distance dispersal? *J. Biogeogr.* 21: 625–635.

Pole, M. (2008). The record of Araucariaceae macrofossils in New Zealand. *Alcheringa: An Australas. J. Palaeontol.* 32: 405–426.

Quinn, C. J. (1982). Taxonomy of *Dacrydium* Sol. ex Lamb. emend. de Laub. (Podocarpaceae). *Aust. J. Bot.* 30: 311–320.

Raven, P. H. & Raven, T. (1976). The genus *Epilobium* (Onagraceae) in Australasia: a systematic and evolutionary study. *New Zeal. DSIR Bull.* 216: 1–312.

Rendle, H. & Murray, B. G. (1989). Chromosome relationships and breeding barriers in New Zealand species of *Ranunculus. New Zeal. J. Bot.* 27: 437–448.

Riley, R. & Chapman, V. (1958). Genetic control of the cytologically diploid behaviour of hexaploid wheat. *Nature* 182: 713–715.

Schmidt, A., Doudrick, R. J., Heslop-Harrison, J. S. & Schmidt, T. (2000). The contribution of short repeats of low sequence complexity to large conifer genomes. *Theor. Appl. Genet.* 101: 7–14.

Shibata, F., Matsusaki, Y. & Hizume, M. (2005). AT-rich sequences containing *Arabidopsis*-type telomere sequences and their chromosomal distribution in *Pinus densiflora. Theor. Appl. Genet.* 110: 1253–1258.

Smissen, R. D. & Breitwieser, I. (2008). Species relationships and genetic variation in the New Zealand endemic *Leucogenes* (Asteraceae: Gnaphalieae). *New Zeal. J. Bot.* 46: 65–76.

Smissen, R. D., Breitwieser, I. & Ward, J. M. (2004). Phylogenetic implications of trans-specific chloroplast DNA sequence polymorphism in New Zealand Gnaphalieae (Asteraceae). *Plant Syst. Evol.* 239: 79–94.

Smith, A. C. (1996). *Flora Vitiensis Nova – A New Flora of Fiji (Spermatophytes Only).* Honolulu, National Tropical Botanical Garden, Heritage Graphics.

Stebbins, G. L. (1971). *Chromosomal Evolution in Higher Plants.* London, Arnold.

Stuessey, T. F. & Crawford, D. J. (1998). Chromosomal stasis during speciation in angiosperms of oceanic islands. In T. F. Stuessey & M. Ono (eds.), *Evolution and Speciation of Island Plants*, pp. 307–324. Cambridge, Cambridge University Press.

Sykes, W. R. (2005). Notes on *Euphorbia* and *Crassula* with a revised key to the latter wild in New Zealand. *New Zeal. Bot. Soc. Newslett.* 79: 8–16.

Trewick, S. A., Paterson, A. M. & Campbell, H. J. (2006). Guest Editorial: Hello New Zealand. *J. Biogeogr.* 34: 1–6.

Wagner, W. L., Herbst, D. R. & Sohmer, S. H. (1990). *Manual of the Flowering Plants of Hawaii, Volume I.* Honolulu, Bishop Museum Special Publication 83, University of Hawaii Press.

Wagstaff, S. J. & Garnock-Jones, P. J. (1998). Evolution and biogeography of the *Hebe* complex (Scrophulariaceae) inferred from ITS sequences. *New Zeal. J. Bot.* 36: 425–437.

Wagstaff, S. J., Bayly, M. J., Garnock-Jones, P. J. & Albach, D. G. (2002). Classification, origin, and diversification of the New Zealand *Hebes* (Scrophulariaceae). *Ann. Mo. Bot. Gard.* 89: 38–63.

Ward, J. M. (1997). *Raoulia* and its hybrids. In J. S. Sheppard (ed.), *Southern Alpines 1996. Conference Proceedings, New Zeal. Gard.* 1: 40–44.

Webb, C. J., Sykes, W. R. & Garnock-Jones, P. J. (1988). *Flora of New Zealand. Volume IV.* Christchurch, Botany Division, DSIR.

Wichman, S. R., Wright, S. D., Cameron, E. K. & Gardner, R. C. (2002). Elevated genetic heterogeneity and Pleistocene climatic instability: inferences from nrDNA in *Coprosma* (Rubiaceae). *J. Biogeogr.* 29: 943–954.

Winkworth, R. C., Robertson, A. W., Ehrendorfer, F. & Lockhart, P. J. (1999). The importance of dispersal and recent speciation in the flora of New Zealand. *J. Biogeogr.* 26: 1323–1325.

Winkworth, R. C., Wagstaff, S. J., Glenny, D. & Lockhart, P. J. (2002). Plant dispersal NWS from New Zealand. *Trends Ecol. Evol.* 17: 514–520.

# 12

# Jesters, red queens, boomerangs and surfers: a molecular outlook on the diversity of the Canarian endemic flora

JULI CAUJAPÉ-CASTELLS

The Canarian archipelago has been sampling biodiversity from other areas, possibly since shortly after the formation of its oldest islands in the Miocene (Fig. 12.1). The sources, and intensity, of such a prolonged sampling process have varied in different epochs, largely conditioning the origins, makeup and success of the colonising stocks (see Figs 12.1 and 12.2). Because of the bias inherent in any sampling, and the variations in extinction and colonisation rates through time and space, the current Canarian endemic flora is an imperfect representation of the historical floristic links between the archipelago and different geographic enclaves.

The shifting availability of sources of plant colonisers through time seems to have provided different 'windows of opportunity' (Carine, 2005) that, together with the vagaries of within-island ecological and geographic history, have shaped an extremely rich endemic flora, estimated to consist at present of about 607 species (Arnoldo Santos-Guerra, unpubl. data). Because the steep and inaccessible geography of the archipelago frequently hinders botanical exploration, this currently accepted figure is possibly an underestimate.

During the last *c.*25 years, molecular genetic investigations have mushroomed for the Canarian endemic flora. Up to April 2009, the phylogenetic information available consisted of *c.*133 papers encompassing 107 genera and *c.* 335 taxa, representing roughly 50% of the presently estimated Canarian endemic plant species. These investigations have used a total of *c.*30 different molecular markers (Fig. 12.3A), mostly sequences of different DNA regions, but also non-sequences (AFLPs, ISSRs and microsatellites, to mention only the three showing the highest utility in recent times). For the most part, they are biased towards groups that have radiated in the archipelago, such as *Echium* (Boraginaceae, Böhle *et al.*,

*The Biology of Island Floras*, eds. David Bramwell and Juli Caujapé-Castells. Published by Cambridge University Press. 

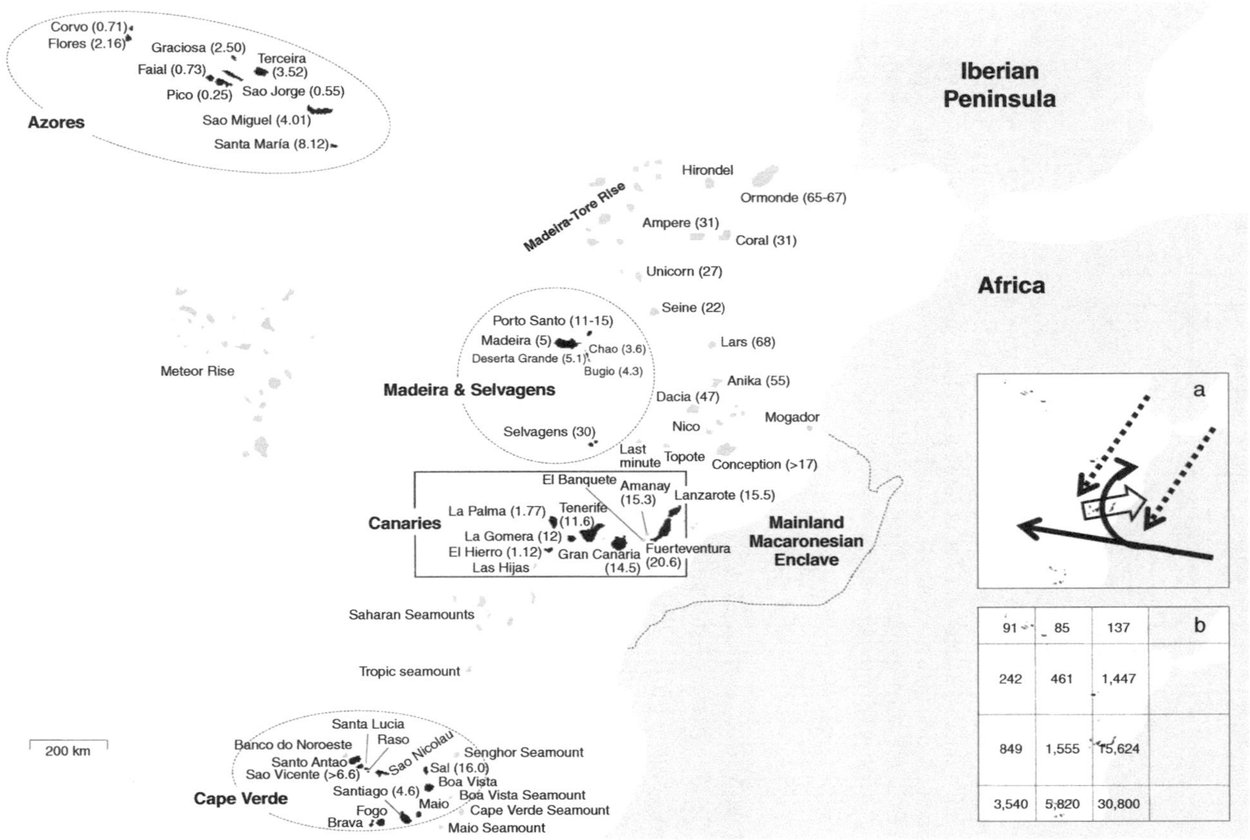


Figure 12.1 Physico-geographical setting of Macaronesia, including the mainland 'Macaronesian enclave' (Sunding, 1979). Numbers in parentheses indicate the maximum estimated ages (in Ma) of the emerged islands

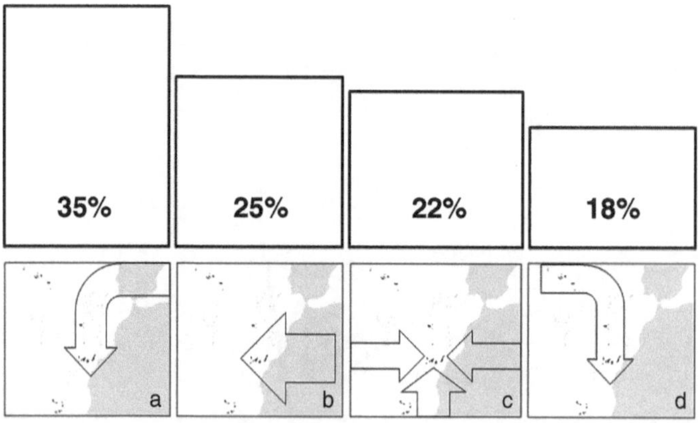


Figure 12.2 Bar plot of the most probable geographical origins for genera of the Canarian flora investigated using molecular markers, as estimated from the author's database (see text for comments). Only the probable origins of the Canarian endemic taxa are considered, according to the topology of the phylogenetic trees in the corresponding papers. (a) Mediterranean; (b) northwest African; (c) East African, South African or New World; (d) Macaronesia (including the Canaries).

1996; García-Maroto *et al.*, 2009), *Argyranthemum* (Asteraceae, Francisco-Ortega *et al.*, 1997), *Pericallis* (Asteraceae, Panero *et al.*, 1999), *Crambe* (Brassicaceae, Francisco-Ortega *et al.*, 2002), *Aeonium* (Crassulaceae, Mort *et al.*, 2002), *Convolvulus* (Convolvulaceae, Carine *et al.*, 2004), *Sideritis* (Lamiaceae, Barber *et al.*, 2007), or *Sonchus* (Asteraceae, Kim *et al.*, 2007), whereas non-radiating groups have been considered much more rarely, e.g. *Asplenium* (Aspleniaceae, van den Heede *et al.*, 2004), *Reseda* (Resedaceae, Martín-Bravo *et al.*, 2007), *Matthiola* (Brassicaceae, Jaén-Molina *et al.*, 2009), or *Festuca* and *Lolium* (Poaceae, Inda *et al.*, 2008).

Caption for Fig 12.1 (*cont.*)
(in black), and of presently submerged seamounts and rises (in grey). Data for the five Macaronesian archipelagos correspond to Carracedo *et al.* (2002), Holm *et al.* (2008), Borges and Hortal (2009), Geldmacher *et al.* (2001, 2005), and Schwartz *et al.* (2005). The approximate positions and maximum age estimates of the submarine land masses mapped were adapted from Rihm *et al.* (1998), Geldmacher *et al.* (2005), Patriat and Labails (2006) and Holm *et al.* (2008). Inset maps correspond to: (a) inferred wind influence and direction in Quaternary humid episodes (trade winds, dotted arrows) and dry episodes (Saharan winds and westerlies, solid and white arrows, respectively) in this geographical area (modified after Rognon & Coudé-Gaussen, 1996; Ortiz *et al.*, 2006); and (b) annual aerosol deposition rates over the North Atlantic, in $g/m^2 \times 10^3$ (after Goudie & Middleton, 2001).

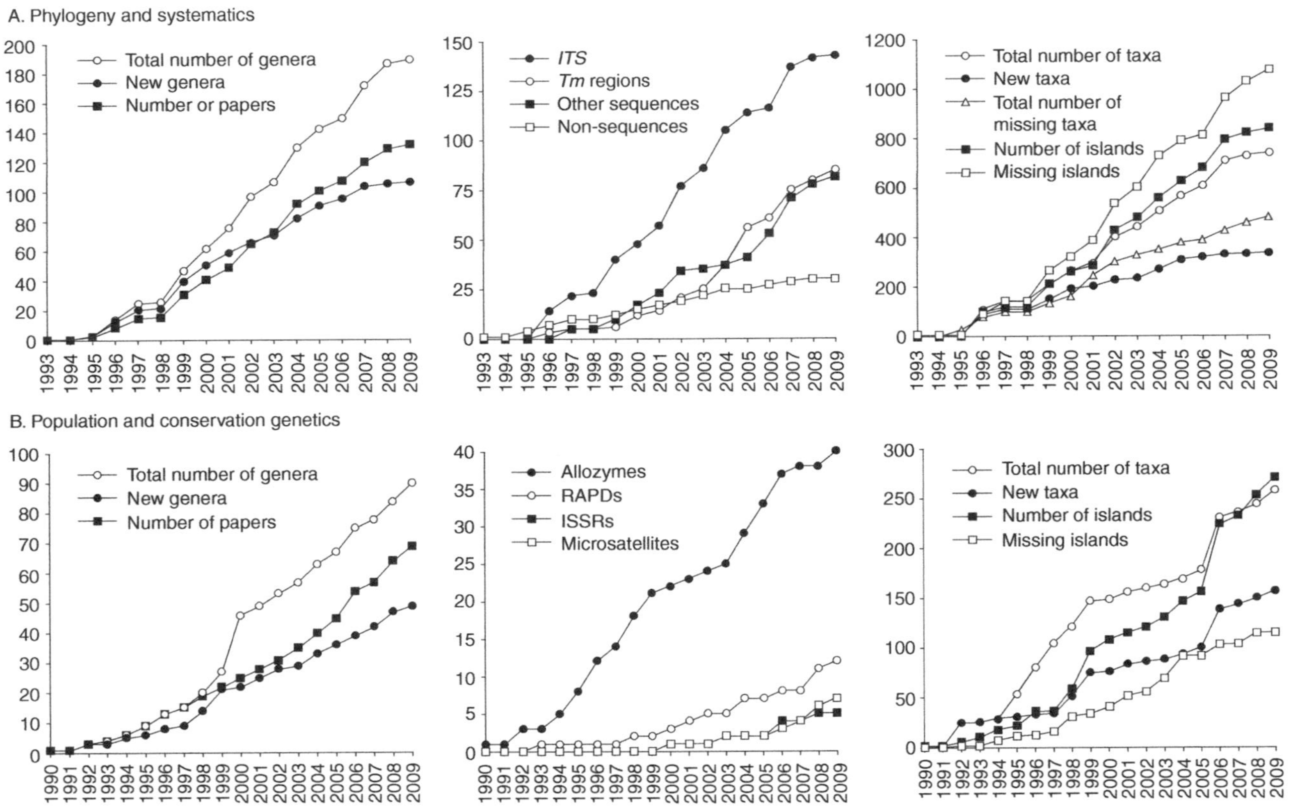


Figure 12.3 Cumulative plots of the evolution of phylogenetic (A) and population genetic (B) knowledge for the Canarian flora up to April 2009, as estimated from the author's database (see text for comments). In (A) 'other sequences' include *atpB-rbcL*, *ETS*, *FPS2*, *matK*, *ndhC-trnV*, *ndhF*, *ndhF-rpl32*, *nepGS*, *psbA-trnH*, *psbZ-trnfM*,

The currently available population genetic data for the Canarian flora is made up of *c*.67 papers, plus a much lower number of reports and results awaiting publication in scientific journals. These have considered 49 genera encompassing 162 species and infra-specific categories (*c*.25% of the flora at the species level) using seven different molecular markers (allozymes, RAPDs, ISSRs, microsatellites, AFLPs, PCR-RFLPs and endosperm proteins, in order of increasing prevalence; Fig. 12.3B). Because of the fragility inherent in insular species, many population genetic studies have focused on the conservation of extremely endangered elements and, less frequently, on phylogeographic aspects. Relevant examples include *Viola palmensis* (Violaceae, Batista & Sosa, 2002), *Parolinia* (Brassicaceae, Fernández-Palacios *et al.*, 2004), *Sonchus gandongeri* (Asteraceae, Kim *et al.*, 2005), several species of *Lotus* (Fabaceae, Oliva-Tejera *et al.*, 2005, 2006), *Solanum vespertilio* and *S. lidii* (Solanaceae, Prohens *et al.*, 2007), *Atractylis preauxiana* and *A. arbuscula* (Asteraceae, Caujapé-Castells *et al.*, 2008a, 2008b), *Gnaphalium teydeum* (Asteraceae, González-Pérez *et al.*, 2009), *Limonium dendroides* (Plumbaginaceae, Suárez-García *et al.*, 2009), some taxa of *Echium* (Boraginaceae, Mora-Vicente *et al.*, 2009), and *Olea guanchica* (García-Verdugo *et al.*, 2009). Population genetic studies of groups widespread in the archipelago are fewer (but see e.g. Francisco-Ortega *et al.*, 1993; Borgen 1996; Kim *et al.*, 1999; Korol *et al.*, 1999; Schiller *et al.*, 1999; González-Pérez *et al.*, 2004; Crawford *et al.*, 2006; Archibald *et al.*, 2006; Meimberg *et al.*, 2006).

Despite the heterogeneous limitations of these databases, the available molecular information has often provided important discrete, testable data to rebut or reinforce long-standing tenets on the origins of the Canarian endemic flora (e.g. Panero *et al.*, 1999; Andrus *et al.*, 2004; Trusty *et al.*, 2004; Vargas, 2007), and to address post-colonising evolution in some groups (e.g. Navascués *et al.*, 2006; Sánchez-Doreste *et al.*, 2006; Díaz-Pérez *et al.*, 2008).

Importantly, molecular markers have also triggered questions that could not be asked before their inception, and whose answer is still the subject of heated

Caption for Fig 12.3 (*cont.*)

*rbcL*, *RNApol2 intron 23*, *rpl20-rps18*, *rpoB-trnC*, *rps16*, *rps17*, *rpsL6* intron, *waxy*, and the 'more axillary' gene; 'non sequences' pools allozymes, chloroplastic and nuclear microsatellites, AFLPs, ISSRs, RFLPs and RAPDs. For clarity, only the four markers predominantly used in population genetic studies are plotted in (B). 'New taxa' plots the number of taxa for which new phylogenetic information is provided each year. Total number of missing taxa is the sum of unsampled taxa for the groups considered in the papers (estimated by comparison with Acebes-Ginovés *et al.*, 2004). 'Number of islands' is the sum of the sampled islands for all the taxa included in the papers. 'Number of missing islands' is the sum of unsampled islands for all the taxa included in the papers for the corresponding year (estimated by comparison with Acebes-Ginovés *et al.*, 2004).

discussion. One of these burning debates affects reconciling population genetic and phylogenetic evidence: while the levels of population genetic variation are relatively high for an insular enclave (Francisco-Ortega *et al.*, 2000; Pérez de Paz & Caujapé-Castells, 2010), the polymorphism of the sequence markers used so far for phylogenetic reconstruction at the species level is in general extremely limited (Crawford & Mort, 2004), and indicates that most groups examined in the Canarian flora are monophyletic (Silvertown *et al.*, 2005). Though remarkable exceptions exist to the predominant monophyly, the discussion hinges upon how a few colonisations might have led to the high levels of population genetic variation featured by the endemic flora. Consideration of this apparent paradox seems one pertinent approach to take in this chapter. First, we will review the existing molecular knowledge of the exceptionally rich Canarian flora in the light of our present understanding of the geographical-historical setting of the archipelago. Second, we will endeavour to guide the reader through the strengths and weaknesses of this database.

## Jesters and red queens

Tectonic and climatic events in the Atlantic basin have largely conditioned the dispersal of terrestrial plants from the mainland to the Canary Islands and among the different archipelagos of Macaronesia. Understanding the importance of these largely unpredictable abiotic factors in the colonisation of the Canaries can yield important insight into the evolutionary mechanisms that led to endemic diversity.

Different factors support the expectation that northwest Africa should be an important source of biodiversity for the Canaries. First, the closeness of the Canarian archipelago to Africa (barely 95 km at present, Fig. 12.1) underlies the remarkable botanical and ecological similarities between the two easternmost islands and the continental 'Macaronesian enclave' (Peltier, 1973; Sunding 1979). Second, the palaeo-wind regimes reconstructed for the area indicate that, during arid Quaternary episodes, the weakening of the monsoons produced a major influence of NEE-NNW Saharan winds in the Canaries (Fig. 12.1a, solid arrows), promoting propagule transport from northwest Africa (Ortiz *et al.*, 2006). And third, the three islands closest to the mainland are also the oldest in the archipelago, and probably have been hosting founders since their formation in the mid-Miocene; notably, data on Saharan dust storms over the North Atlantic (Fig. 12.1b) show that the easternmost Canary islands also display the largest annual aerosol deposition rates in Macaronesia (Goudie & Middleton, 2001).

However, molecular phylogenies converge with this expectation only to a certain extent, as they resolve a northwest African origin for the Canarian representatives of only *c.*25% of the genera surveyed (Fig. 12.2b). At odds with the current geographical setting of the region, a putative Mediterranean origin explains a majority of the plant colonisations examined with molecular data (roughly 35%, Fig. 12.2a) (see also Carine *et al.*, 2010). These two estimates have to be compared with caution, as they are surely distorted by extinction and/or defective sampling in some of the analysed cases, and by the fact that the present phylogenetic database does not include all the Canarian endemics. Nevertheless, they represent a substantial sample of the endemic flora, and they strongly suggest that factors other than proximity to the mainland may have prevailed in its origins.

Geologic and climatic data indicate that insular Macaronesia was substantially different in the Pleistocene and Holocene. All the submarine seamounts bridging the gap to Iberia already existed by the time of formation of the older islands (see the geological dating in Fig. 12.1, and references in the figure caption). Importantly, many of these also emerged during the last Quaternary glacial maximum *c.*20 000 Ma (Fernández-Palacios, 2010), and could have acted as stepping-stones, facilitating migration of biodiversity to the Canaries from the Mediterranean and Iberia (see Fig. 12.1), and vice versa (e.g. Carine *et al.*, 2004). A quantitative idea of the potential importance of these seamounts for dispersal is provided by the estimate of endemic plants per unit area in the Madeira-Selvagens archipelago, which far outnumbers that in the Canaries (0.17 vs 0.08, respectively; Caujapé-Castells *et al.*, 2010), despite its present isolation. According to Ortiz *et al.* (2006), the inferred palaeo-wind directions during Quaternary humid episodes may have favoured colonisation processes from the Mediterranean (Fig. 12.1a, dotted lines).

This high number of potential stepping stones could also have facilitated multiple geographic origins for congeneric taxa. However, except for *Asplenium* (van den Heede *et al.*, 2004), *Limonium* (Lledó *et al.*, 2005), *Reseda* (Martín-Bravo *et al.*, 2007) and *Rhamnus* (Bolmgren & Oxelman, 2004) and other cases mentioned in Sanmartín *et al.* (2008), most Canarian endemics included so far in phylogenies seem to have originated via a single introduction, and subsequently dispersed and diverged on the islands. The populations of *Monanthes muralis* from El Hierro and La Palma are indicated to be polyphyletic in Mes *et al.* (1997), but such a relationship could not be consistently resolved by the molecular data used (ITS); Canarian *Isoplexis* also appears as polyphyletic in Kelly and Culham (2008), but the markers used (*MAX4*-like genes) are considered highly questionable by these authors. Other cases of multiple introductions are seen when considering all of Macaronesia; *Festuca* (Díaz-Pérez *et al.*, 2008), *Matthiola* (Jaén-Molina *et al.*,

2009) and, possibly, *Tolpis* (Moore *et al.*, 2002) and *Marcetella* (Helfgott *et al.*, 2000) could be added to the 10 cases reported in Carine *et al.* (2004), though resolution is poor in the latter two.

Geographical proximity does not account for the origins of *c*.22% of Canarian endemics (Fig. 12.2c). In most of these cases, the present distribution of the phylogenetically closest extant congeners may be highly divergent from the ancestral area of origin; and discriminating between vicariance or long-range dispersal to explain the origins of these groups is complicated by the role of extinction in the mainland and the islands, especially during climatic changes (Andrus *et al.*, 2004). Consequently, both explanations have been favoured by different authors (Andrus *et al.*, 2004; Carine, 2005). The recent development of new methods of biogeographical analysis that allow the integration of chronological information (e.g. Sanmartín *et al.*, 2008) probably heralds an increase in the number of studies addressing the explanation of these and other intriguing patterns in the Canarian flora.

The contrast between the favourable climate of the Canarian archipelago and the harshness of the mainland during recurrent climatic changes in different epochs (see Rodríguez-Sánchez & Arroyo, 2008, for the Pleistocene) is one major factor conditioning the remarkable distinctiveness of the Canarian flora. For example, Quaternary glaciations played havoc with biodiversity on the mainland (Hewitt, 1996) and perhaps in some Macaronesian archipelagos (e.g. the Azores, Ávila *et al.*, 2008). However, fossil localities throughout North Africa (Jacobs & Herendeen, 2004) and in the islands (Anderson *et al.*, 2009) indicate that laurel forest elements, now extinct or rare on the mainland, could have thrived in Macaronesia in the late Miocene–early Pliocene. These records, and molecular phylogenetic evidence assessed in Vargas (2007), lend strong support to the contention (Bramwell, 1976, 1985) that some lineages that colonised the Canaries prior to climatic shifts (or during interglacials) could have survived there until the present. This emphasises the role of the archipelago as a biodiversity haven for a mainly tropical continental Tertiary flora.

Nevertheless, molecular phylogenies also suggest that Canarian endemics from at least 17 genera may have originated in Macaronesia (Fig. 12.2d, Table 12.1), thereby evidencing a considerable role for intra-archipelago diversification in shaping floristic singularity. Eight of these cases correspond to representatives of the 23 known endemic genera. The putative closest extant ancestors of other Canarian endemic genera included in molecular phylogenies are placed in the Mediterranean (*Vierea*, Andrus *et al.*, 2004; *Spartocitysus*, Cubas *et al.*, 2002; *Ixanthus*, Gentianaceae, Mansion & Struwe, 2004), East Africa (*Parolinia*, Brassicaceae, Warwick *et al.*, 2007), South Africa (*Kunkeliella*, Santalaceae, Der & Nickrent, 2008), northwest Africa (*Tinguarra*, Apiaceae, Downie *et al.*, 2000) or the New World (*Ceballosia*, Boraginaceae, Diane *et al.*, 2002). The origins of

Table 12.1 *Plant genera where the Canarian endemic species putatively originated in the Canaries or Macaronesia, according to molecular phylogenies. SPs: number of Canarian endemic species sampled. SPm: number of endemic species missing in the phylogeny. Is: sum of the number of islands of distribution sampled per each endemic taxon included in the corresponding phylogeny. Im: number of islands of distribution per each endemic taxon that are missing in the phylogeny. The number of missing species and islands was estimated by comparing the sampling in the references with distributional data in Acebes-Ginovés et al. (2004). Asterisks indicate Canarian endemic genera.*

| Genus | Family | SPs | SPm | Is | Im | Reference |
|---|---|---|---|---|---|---|
| *Aeonium* | Crassulaceae | 31 | 0 | 41 | 5 | Mort *et al.* (2002) |
| *Andryala* | Asteraceae | 1 | 3 | 1 | 6 | Fehrer *et al.* (2007) |
| *Babcockia** | Asteraceae | 1 | 0 | 1 | 0 | Lee *et al.* (2005) |
| *Bencomia* | Rosaceae | 4 | 0 | 4 | 4 | Helfgott *et al.* (2000) |
| *Chrysoprenanthes** | Crassulaceae | 1 | 0 | 1 | 0 | Kim *et al.* (2007) |
| *Dendriopoterium** | Rosaceae | 2 | 0 | 2 | 0 | Helfgott *et al.* (2000) |
| *Gonospermum** | Asteraceae | 4 | 0 | 6 | 2 | Francisco-Ortega *et al.* (2001) |
| *Greenovia** | Crassulaceae | 4 | 0 | 12 | 0 | Mort *et al.* (2002) |
| *Isoplexis* | Scrophulariaceae | 3 | 0 | 3 | 2 | Bräuchler *et al.* (2004) |
| *Lactucosonchus** | Asteraceae | 1 | 0 | 1 | 0 | Kim *et al.* (2007) |
| *Lugoa** | Asteraceae | 1 | 1 | 1 | 1 | Himmelreich *et al.* (2008) |
| *Schyzogyne* | Asteraceae | 1 | 0 | 1 | 0 | Liu *et al.* (2004) |
| *Sonchus* | Asteraceae | 16 | 3 | 22 | 4 | Lee *et al.* (2005) |
| *Sventenia** | Asteraceae | 1 | 0 | 1 | 0 | Lee *et al.* (2005) |
| *Taeckholmia* | Asteraceae | 3 | 3 | 3 | 8 | Lee *et al.* (2005) |
| *Tolpis* | Asteraceae | 10 | 4 | 11 | 10 | Moore *et al.* (2002) |

*Pleiomeris* (Myrsinaceae) are uncertain; though it is associated with the predominantly New World genus *Lysimachia* (Martins *et al.*, 2003), which also has narrow endemics in the Mediterranean region (Ibáñez *et al.*, 2002), this relationship is doubtful and could be an artifact associated with taxonomic sampling (Alfredo Reyes-Betancort, pers. comm.).

Divergence time estimates based on phylogenies indicate that plant colonisation of the Canaries may not have been continuous, but constrained to separate periods (from *c.*16 Ma to 2.5 Ma), coincident with radical climatic shifts that transformed the character of the vegetation of Africa and the Mediterranean (Kim *et al.*, 2008). Therefore, the Canarian flora must also contain a large proportion of neo-endemics and schizoendemics. As in the Hawaiian archipelago (Price & Clague, 2002), the fact that most phylogenetic studies that support recent origins for the Canarian endemics are biased towards large radiations reinforces, rather than weakens, the hypothesis of a non-relict condition for part of the flora.

The contingent processes entailing climatic and tectonic fluctuations that provided the windows of opportunity to colonise the Canaries adapt to the concept of a 'court jester' evolution, as they recall 'the capricious behavior of the licensed fool of Medieval times' (Benton, 2009). By contrast, post-colonisation evolution within the archipelago, where the dominant evolutionary processes include competition and adaptation to each island's ecological and geographic heterogeneity, can be viewed as belonging to an eminently Darwinian, 'red queen' evolutionary landscape (van Valen, 1973), that refers to the painstakingly dynamic struggle of organisms to maintain their fitness relative to other organisms evolving in their environment.

Both worlds have coexisted at different times throughout the history of the archipelago (either in the same or in different islands), both have played a significant role in the origins, composition and evolution of the flora, and both are still continually occurring. On the whole, 'red queen' evolution could only predominate once 'jester' processes had set the geographic and climatic stages for bringing biodiversity into the islands and for promoting microevolutionary change. However, after the islands' biotic colonisations, red queen processes were frequently overwhelmed by jester events such as giant landslides (e.g., Masson 1996; Urgeles *et al.*, 1997; Whelan & Kelletat, 2003) or tectonic collapses (e.g. Pérez-Torrado, 1997; Ancochea *et al.*, 2004) which drove many taxa to extinction, or fragmented their distributions. For instance, Marrero and Francisco-Ortega (2001) estimate that the volcanic events that collapsed a large part of Gran Canaria *c.*14 and 3.5 Ma (Pérez-Torrado, 1997) could bring about the extinction of *c.*80% and 50%, respectively, of this island's flora if they happened today. Tenerife has also been a very active volcanic arena since its formation, combining geological structures of considerable antiquity with other much more recent ones (Ancochea *et al.*, 2004). Undoubtedly, these tectonic contingencies have had profound implications for the biodiversity of the affected islands (e.g. De Nascimento *et al.*, 2009); although they aid in some studies by putting maximum ages on some populations, they also provoked extinctions that impinge on our present capability to assess post-colonisation microevolution in the archipelago.

Because cladistic biogeography considered dispersal a rare phenomenon (e.g. Humphries & Parenti, 1999), island floras were earlier assumed to have an improbable origin, and only from mainland areas. Certainly, many islands offer very small targets for organismic colonisation; nonetheless, the Canaries and Macaronesia have had a long history of opportunity to host the evolution of plant biota incoming from different vegetation types and from various sources, not necessarily close geographically, or mainland in nature. This evidence, the coexistence of relict species and recent ones, and the recurrent climatic variations between cool and warm conditions in the archipelago (Ortiz *et al.*, 2006), intimates that the region's flora should be envisioned as a heterogeneous mosaic of taxa whose evolutionary strategies may be sharply divergent (Dansereau, 1965; Barquín, 1984; Carine & Schaefer, 2010). Thus, the origins and post-colonising divergence of the Canarian flora circumscribe within an extremely dynamic scenario of recurrent island colonisation, dispersal, speciation and extinction, mainly developed in several discrete time periods after the formation of the first islands. Overall, the interpretation of molecular phylogenetic data in the face of the known tectonic-climatic changes compellingly suggests that part of the present endemic flora is recent, and that dispersal to the Canaries from the mainland or from other Macaronesian islands has not been rare at all.

## Boomerangs

Until recently, the colonisation of islands was also formally considered irreversible, so that the target archipelagos were largely equated to evolutionary cul-de-sacs, where only some of the 'ambushed' colonising lineages could thrive and diversify, while many others died off. Figure 16 in Sunding (1979) clearly illustrates this prevailing view of the Canarian flora.

As highlighted in Caujapé-Castells (2004) for the Canaries, and developed in a more general context by Bellemain and Ricklefs (2008), the irreversibility component that beset island biogeography theory was notably due both to biased interpretations of some of its tenets, and to the difficulty of detecting back-colonisations to the mainland. The term 'boomerang' was used by Caujapé-Castells (2004) as a general abstraction referring to the back-colonisation of biodiversity elements from an island to a mainland area, not necessarily near the source of the earlier island colonists.

Based on pollen analysis of deep-sea drillings off the NW coast of Morocco, Hooghiemstra *et al.* (1987) and Hooghiemstra (1989) had posited an increasing influence of NNE trade winds during the last glacial maximum in the Macaronesian area (see dotted arrows in Fig. 12.1a), and explicitly ruled out a west-to-east direction for palaeo-wind regimes. This interpretation represented a major hurdle arguing against the general occurrence of back-colonisations from the Canaries, unless the propagules were transported by birds, certainly not a predominant pattern for the present endemic flora (*c.*12% of the species; Manuel Nogales & Beatriz Rumeu, unpubl. data).

However, continental records analysed by Rognon and Coudé-Gaussen (1996) radically disagree with that earlier picture, and highlight a prevalence of the westerlies (Fig. 12.1a, solid arrow), involving wind-mediated propagule transport from the Canaries toward the nearby mainland, and eastwards inland. Albeit strikingly omitted in all phylogenetic literature on Canarian endemics to date, these continental analyses (but see also the winter coastal upwelling measurements in Machín *et al.*, 2006) seem to afford a better reconstruction of the real direction and strength of surface winds. Furthermore, they provide compelling evidence that the formerly construed dominance of NNE trade winds was an oversimplification of a complex palaeo-wind dynamics over northwest Africa during the Quaternary, and that the westerlies blew recurrently during a wide time window in this epoch, potentially fostering several waves of back-colonisation from the Canaries to (at least) the nearby continent. According to these data, trade winds could have been prevalent in the Canaries and Morocco much more recently than previously assumed (i.e. at the end of the deglaciation and during the Holocene).

One of the general tenets of island biology is the loss of dispersal power in successful colonists that have spent sufficient evolutionary time in the islands (Carlquist, 1966; MacArthur & Wilson, 1967). The basal position of numerous Canarian endemics in molecular phylogenies (see Vargas, 2007, for a recent review) indicates a considerable relict contingent within the present Canarian endemic flora. These older elements, however, coexist with other lineages that colonised the islands much more recently (e.g. Kim *et al.*, 2008). Admittedly, today, most of the older lineages probably have much less dispersal power than their ancestors. However, many of those lineages also colonised adjacent islands in the past; therefore, they should have been able to successfully back-colonise the continent, given the proper conditions. Analogously, many Canarian endemics that are recent in time must still retain the dispersal capabilities of their ancestors, so that most of them should be able to make the journey back. Two recent papers on archipelagos from the Pacific also suggest that, while loss of dispersability might be common in island endemics, in some cases they retain the capability for long-range dispersal (Harbaugh & Baldwin, 2007; Harbaugh *et al.*, 2009).

As pointed out by Bellemain and Ricklefs (2008), another element against the idea of back-colonisations was the generalisation that the ecological space of continental biota is more densely packed, and should be more difficult to invade by species from islands, adapted to lower-competition environments. In the Canaries, this is of doubtful relevance for two main reasons. First, extinction waves due to climatic changes helped to considerably reduce the biotic density of the nearby mainland during different epochs (especially the Quaternary). And second, the presently arid ecological conditions in the easternmost islands of Lanzarote and Fuerteventura could have been similar to those in mainland northwest Africa since the Pleistocene. Desert conditions seem to have been recurrent on the continent, but records at the Chott Rharsa Basin in southern Tunisia circumscribe the onset of the climatic shifts that gave rise to the Sahara at least 86 000 BP (Swezey, 1999; Schuster *et al.*, 2006). Furthermore, Fuerteventura and Lanzarote were much closer to Africa some 20 000 BP ago by virtue of the eustatic sea-level lowering associated with the glacial maximum (García-Talavera, 1997), a time window when the area occupied by a growing Sahara was probably already quite large, and when the westerlies are supposed to have been predominant (Rognon & Coudé-Gaussen 1996).

Overall, these multiple considerations support the view that 'biodiversity boomerangs' from the Canaries to the mainland may have been far more frequent than assumed at present, and were probably not as 'uncommon' or 'occasional' as suggested in some recent reviews covering the Canaries and other enclaves (Emerson, 2002; Heaney, 2007). Thus, mainland northwest Africa is not only one of the most important sources of the present flora (Fig. 12.2b), but it could be also a substantial historical sink for the archipelago's biodiversity.

Table 12.2 lists 30 genera where molecular phylogenies might substantiate a back-colonisation from the Canaries to the mainland; except for some cases, most of them should be recent in time. It is worth noting that only 12 of the listed cases (40%) entail a real boomerang (i.e. one that returns to its 'pitcher'), and all these represent floristic links between northwest Africa and the Canaries. In contrast to the diversity of potential source areas for insular colonisation (Fig. 12.2), the target of more than 68% of putative boomerangs (19 cases) is northwest Africa. Hence, while closeness to the mainland was not of outstanding importance for originating the current Canarian endemic flora (Fig. 12.2), it seems to have been determinant for back-colonisation success (or detection). Only *Convolvulus* (Carine *et al.*, 2004) shows a consistent pattern of back-colonisation to Iberia, and extinction or defective sampling might be influencing the examples in other cases. Increased botanical and molecular knowledge of the under-explored mainland area east of the Canaries (see Peltier, 1973; Sunding, 1979) could help discard or reinforce most of these examples, but probably it would also reveal others not listed.

Table 12.2 *Putative boomerangs suggested by molecular evidence for the Canarian flora (a boomerang is hypothesised when mainland taxa are in a derived or sister position relative to taxa distributed in the Canaries, even if it was not explicitly suggested in the corresponding reference). Origin: putative origin of the Canarian circumscription of the genus, according to the sampling and resolution of the corresponding phylogenetic tree(s). Target: nearest probable geographic target of the putative boomerang, as inferred from the sampling in the corresponding papers (in parentheses, mainland taxa that presumably originated after the boomerang). SPs: number of Canarian endemic species sampled. SPm: number of missing endemic species. Is: sum of islands of distribution sampled per each endemic taxon included in the corresponding phylogeny. Im: sum of the number of islands of distribution missing per each endemic taxon in the corresponding genus. Number of missing species and islands was estimated by comparison to distributional data in Acebes-Ginovés et al. (2004). Although mainland sampling and extinction have likely influenced the inferences in most cases, clade support was considered a priority to assess the likelihood of the boomerang. The dark shadings indicate the cases best supported by the tree topologies. Unshaded taxa are those for which topological resolution and/or sampling only provide flimsy evidence for a boomerang. Superscripts after the genus names indicate the major factors that might be undermining resolution power in each case.*

| Genus | Family | SPs | SPm | Is | Im | Origin | Target | Reference |
|---|---|---|---|---|---|---|---|---|
| *Aeonium* $^{E,S}$ | Crassulaceae | 31 | 0 | 41 | 5 | Macaronesia[1] | NE Africa (*A. leucoblepharum*) | Mort *et al.* 2002 |
| *Androcymbium* $^{P,S}$ | Colchicaceae | 2 | 0 | 5 | 0 | NW Africa | NW Africa (*A. wyssianum*) | Caujapé-Castells 2004 |
| *Andryala* $^{S}$ | Asteraceae | 1 | 3 | 1 | 6 | Madeira | Mediterranean (*A. ragusina, A. integrifolia*) | Fehrer *et al.* 2007 |
| *Arbutus* $^{E,S}$ | Arbutoideae | 1 | 0 | 1 | 4 | New World | Mediterranean (*A. andrachne, A. unedo*) | Hileman *et al.* 2001 |
| *Brachypodium* $^{S,P}$ | Poaceae | 1 | 0 | 1 | 2 | New World | Mediterranean (*B. retusum*) | Catalán & Olmstead 2000 |
| *Bupleurum* $^{P}$ | Apiacae | 1 | 0 | 1 | 4 | NW Africa | NW Africa (*B. canescens*) | Neves & Watson 2004 |
| *Ceballosia* $^{E,P}$ | Boraginaceae | 1 | 0 | 1 | 6 | New World | Old world (*Heliotropium*) | Diane *et al.* 2002 |

Table 12.2 (*cont.*)

| Genus | Family | SPs | SPm | Is | Im | Origin | Target | Reference |
|---|---|---|---|---|---|---|---|---|
| *Chamaecitysus* [P] | Fabaceae | 1 | 2 | 1 | 6 | Macaronesia[1] | Iberia and Mediterranean (several sps.) | Cubas *et al.* 2002 |
| *Convolvulus* [S] | Convolvulaceae | 8 | 2 | 9 | 20 | Mediterranean | Iberia (*C. fernandezii*) | Carine *et al.* 2004 |
| *Dactylis* [P] | Poaceae | 1 | 1 | 8 | 6 | NW Africa | NW Africa (*D. glomerata* var. *hispanica*) | Sahuquillo & Lumaret 1999 |
| *Euphorbia* [S,P] | Euphorbiaceae | 7 | 2 | 9 | 13 | Mediterranean | Iberia (*E. pedroi*) | Molero *et al.* 2002 |
| *Hedera* [S,P] | Araliaceae | 1 | 0 | 3 | 2 | Madeira[1] | NW Africa (*H. maroccana*) | Valcárcel *et al.* 2003 |
| *Helichryssum* | Asteraceae | 3 | 0 | 3 | 0 | Madeira[1] | Mediterranean (ancestor of several spp.) | Galbany-Casals *et al.* 2009 |
| *Ilex* [E] | Aquifoliaceae | 3 | 1 | 2 | 7 | Multiple | E Asia (*I. leucoclada, I. latifolia, I. rugosa*) | Manen *et al.* 2002 |
| *Kleinia* [S] | Asteraceae | 1 | 0 | 1 | 6 | E Asia | NW Africa (*K. anteuphorbium*) | Pelser *et al.* 2007 |
| *Limonium* [S,P] | Plumbaginaceae | 6 | 12 | 7 | 17 | Mediterranean | NW Africa (*L. pectinatum, L. jovi-barba*) | Lledó *et al.* 2005 |
| *Lotus* [S] | Fabaceae | 11 | 6 | 12 | 15 | NW Africa | NW Africa (*L. assakensis*) | Allan *et al.* 2004 |
| *Matthiola* [S] | Brassicaceae | 3 | 0 | 3 | 1 | NW Africa | NW Africa (*M. bolleana*) | Jaén-Molina *et al.* 2009 |
| *Olea* [P] | Oleaceae | 1 | 0 | 6 | 1 | NW Africa | NW Africa (*O. maroccana*) | Besnard *et al.* 2009 |
| *Navaea* | Malvaceae | 1 | 0 | 1 | 0 | S Africa | Eurasia (*Lavatera olbia*) | Fuertes-Aguilar *et al.* 2002 |
| *Phoenix* [S,P] | Arecaceae | 1 | 0 | 1 | 6 | NW Africa | NW Africa (*P. canariensis?*) | Asmussen *et al.* 2006 |
| *Plantago* [S,P] | Asteraceae | 2 | 2 | 2 | 11 | NW Africa | NW Africa (*P. mauritanica*) | Ronsted *et al.* 2002 |

| | | | | | | | | |
|---|---|---|---|---|---|---|---|---|
| *Pulicaria* [E,P] | Asteraceae | 1 | 1 | 1 | 2 | NW Africa | NW Africa (*P. odora*) and Mediterranean (*P. arabica*) | Andrus *et al.* 2004 |
| *Reichardia* [P] | Asteraceae | 1 | 2 | 1 | 9 | Mediterranean | NW Africa (*R. crystallina*) | Kim *et al.* 2007 |
| *Reseda* [P] | Crassulaceae | 3 | 0 | 7 | 7 | NW Africa | NW Africa (*R. lutea* ssp. *neglecta*) | Martín-Bravo *et al.* 2007 |
| *Sideroxylon* [E] | Sapotaceae | 1 | 0 | 1 | 5 | E Asia | Cape Verde (*S. marginatum*), South Africa (*S. inerme*) | Smedmark *et al.* 2006 |
| *Spartocytisus* [E,P] | Fabaceae | 1 | 1 | 2 | 4 | Mediterranean | NW Africa (*Cytisus valdesii*) | Cubas *et al.* 2002 |
| *Teline* [S] | Fabaceae | 1 | 1 | 2 | 4 | NW Africa | NW Africa (*T. monspessulana*) | Percy & Cronk 2002 |
| *Tinguarra* [E] | Apiaceae | 1 | 0 | 1 | 3 | NW Africa | NW Africa (*T. sicula*) | Downie *et al.* 2000 |
| *Tolpis* [S,P] | Asteraceae | 1 | 13 | 1 | 20 | Azores[1] | NW Africa (*T. virgata*) | Park *et al.* 2001 |
| *Tolpis* [S] | Asteraceae | 10 | 4 | 11 | 10 | Madeira[1] | NW Africa (*T. virgata*) | Moore *et al.* 2002 |

[E] Possible major influence of extinction.

[P] Phylogenetic tree poorly resolved (i.e. low support for the derived position of the mainland taxa, or polytomic topology combining mainland and island populations).

[S] Sampling deficiencies, especially in mainland areas near the Canaries (i.e. the 'Macaronesian enclave' and its surroundings).

[1] Canarian taxa with putative origin in Macaronesia have been included in this list on the grounds on their ultimate origin via a continental ancestor.

## Surfers

Sanmartín *et al.* (2008) developed a sophisticated Bayesian, phylogeny-based approach to island biogeography under a homogeneous, time-reversible Markov process. This approach ignores temporal changes within the islands, and assumes both constancy in the relationships among islands through time, and that islands are older than the organisms inhabiting them. In this conceptual framework, these authors estimate the islands' carrying capacity (i.e. the proportion of the total island lineages expected on a given island at equilibrium conditions) and apply several stochastic dispersal models to 13 molecular phylogenetic studies of Canarian organisms (six plant lineages and seven animal lineages) to infer the most probable inter-island dispersal routes in the Canaries. Based on their quantitative results, they propose a general biogeographical hypothesis whereby (1) inter-island dispersal is the major driver of diversification within lineages, and (2) the three 'central' islands (Gran Canaria, Tenerife and La Gomera) are the predominant diversification and dispersal centres in the archipelago and show the highest rate of exchange with the mainland (even though this is rare in comparison with inter-island dispersal).

The availability of different population genetic databases for Canarian plant endemics offers the possibility of assessing the degree of overlap between these biogeographic predictions and genetic diversity patterns. Published information on allozyme variation for the Canarian endemic flora encompasses useful genetic diversity data for 320 populations (out of a total of 540 populations in the database; Pérez de Paz & Caujapé-Castells, 2010) that belong to 115 endemic taxa (species, subspecies and varieties), and is today the largest population genetic database for the archipelago (see Fig. 12.3B). The overall levels of allozyme variation detected in the Canarian flora are higher than those estimated for other oceanic archipelagos, or for endemics in general (Francisco-Ortega *et al.*, 2000; Pérez de Paz & Caujapé-Castells, 2010).

For the purpose of comparison, it is convenient to view each island as an 'allele sink', whose genetic variation can be estimated by the average population values of four basic indicators: the number of alleles per locus ($A$), proportion of polymorphic loci ($P$), observed heterozygosity ($H_o$) and expected heterozygosity ($H_e$). The philosophy of this approach is analogous to Sanmartín *et al.*'s (2008) in that islands are simply considered as a background against which genetic diversity patterns are recorded; yet, this simplicity can be used as an asset to interpret test results incorporating phylogenetic and historical considerations.

Population expansions and colonisations have been modelled mathematically as advancing genetic wave fronts where alleles can surf and eventually spread out (Edmonds *et al.*, 2007; Hallatschek *et al.*, 2007; Excoffier & Ray, 2008). As predicted by population genetics theory, any geographic expansion or colonisation would lead to a reduction of average genetic diversity in the populations at the wave front relative to those at the wave origin, because the gene pool for the new habitats is provided by only a small number of individuals (Hallatschek & Nelson, 2008). The pattern of decreasing south–north gradient in genetic diversity in the Northern Hemisphere (Hewitt, 1996) is one example consistent with this prediction.

If, according to Sanmartín *et al.* (2008), the three central islands (Gran Canaria, Tenerife and La Gomera) are the diversification and dispersal centre of the Canarian archipelago, we would expect them to exhibit a higher genetic diversity than the older, eastern islands (Fuerteventura and Lanzarote) or the much younger westernmost ones (La Palma and El Hierro). In these latter two groups, random sweeps of alleles in founders from the central islands would have promoted further reductions of genetic diversity, or fixation, due to drift and eventual selection. The expected genetic diversity distribution in the archipelago under this scenario is represented in Fig. 12.4a. Likewise, if the eastern islands were the centre of diversification and dispersal, then we should expect an east–west decreasing genetic diversity gradient (Fig. 12.4b). In some cases, these patterns could be altered by drift or selection (which would be more likely as more time passed since the time of dispersal), and also by human disturbance, ecological diversity enabling diversification (or not), and so on.

The detected patterns of genetic diversity (Fig. 12.4, Table 12.3) do not fit the biogeographic predictions in Sanmartín *et al.* (2008). First, the three central islands defined by these authors are not those with highest genetic diversity in the archipelago. And second, there are radical differences in this group; most notably, Gran Canaria exhibits significantly higher population genetic variation than Tenerife in all indicators examined (Table 12.3). The results obtained agree much better with Fig. 12.4b, though a consistent genetic diversity wave is only detectable for two groups of islands: an eastern one (the wave origin), with uniformly high values of all indicators (F, L, C); and a western one (at the wave front), with significantly lower values (T, G, P, H). In the western group, La Gomera contains much lower genetic diversity than the other islands, despite being much older than El Hierro or La Palma (Table 12.3). Incidentally, this island is situated in the area of the Atlantic where the aerosol deposition rates decrease abruptly (Fig. 12.1b). It might thus be possible that the emergence of Tenerife (the next youngest island) between La Gomera and the much older eastern group was necessary for successful plant colonisation. The elicited east–west genetic diversity gradient is consistent with both the geological history of the archipelago (Fig. 12.4, Carracedo, 1999) and with the pattern of shared endemism between adjacent islands (Whittaker *et al.*, 2008).

A criticism that can be made of this test is that considering the islands as genetic diversity sinks neglects the diverging phylogenetic histories of the non-congeneric populations pooled. However, the overall patterns detected in the congeners

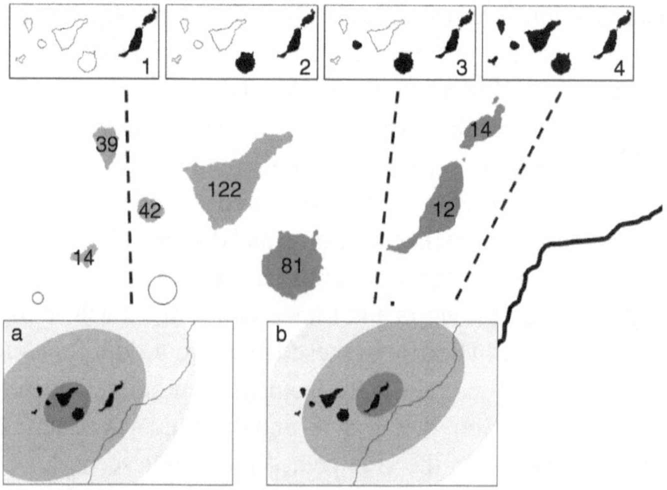


Figure 12.4 The Canaries as genetic variation sinks for the terrestrial flora. Top: sequential surfacing of the Canary Islands, simplified from Carracedo (1999) (1: 20–15 Ma, 2: 15–12 Ma, 3: 12–7.5 Ma, 3: 7.5–0.0 Ma; emerged islands at each stage are shown in black). Middle: qualitative representation of the genetic diversity gradient among islands according to the Kruskal-Wallis tests in Table 12.1 (a more intense grey colour corresponds to significantly higher population genetic variation contents in the corresponding island). Numbers within the island area are exclusive island endemics according to Acebes-Ginovés *et al.* (2004). The three island groups according to the carrying capacities estimated by Sanmartín *et al.* (2008) are indicated by dotted lines (the size of the circles at the bottom left of each subdivision roughly represents the relative carrying capacity for each island group, according to these researchers); note the sharply divergent genetic diversity contents in the three central islands. Bottom: conceptual representation of the expected shapes of the genetic diversity gradients for a hypothetical situation of the centres of dispersal and diversification in the central islands (a), or the eastern islands (b). More intense grey areas correspond to expected higher levels of population genetic diversity.

for which allozyme data in the two significant island groups are available (see Table 12.4) are also consistent with an east–west genetic diversity gradient. The single discrepancy is *Lobularia canariensis* (Brassicaceae), where the genetic diversity centre seems to be in the western islands, and for which a molecular phylogeny is not available. Other population genetic data for groups distributed in most islands are also coincident with an east–west genetic diversity gradient (e.g. AFLPs in *Olea guanchica*; García-Verdugo *et al.*, 2009).

Considering the current historical-geographical setting of the archipelago, a feasible explanation of how the eastern islands attained and maintained their much higher levels of variation is that they could be acting as genetic variation melting pots, where multiple colonisations from the mainland gave rise to generalised

Table 12.3 *Sample sizes, group averages and significances for the allozyme genetic diversity indicators subjected to Kruskal-Wallis tests. Island codes correspond to the first letter in the islands' name, except for Gran Canaria (C).*

| Subdivisions tested | $A$ | $P$ | $H_o$ | $H_e$ |
|---|---|---|---|---|
| [WEST], [CENTRAL], [EAST] | | | | |
| [P+H], [G+T+C], [F+L] | | | | |
| sample sizes | (51, 213, 47) | (49, 213, 51) | (40, 196, 32) | (51, 218, 49) |
| averages | (1.67, 1.84, 1.81) | (0.43, 0.47, 0.51) | (0.11, 0.14, 0.17)** | (1.81, 1.84, 1.68) |
| [WEST], [EAST] | | | | |
| [P+H+G+T], [C+F+L] | | | | |
| sample sizes | (130, 181) | (129, 181) | (110, 158) | (134, 179) |
| averages | (1.65, 1.9,)*** | (0.41, 0.51)*** | (0.12, 0.15)* | (0.15, 0.20)*** |
| OTHER SUBDIVISIONS | | | | |
| [C], [T] | | | | |
| sample sizes | (130, 51) | (130, 52) | (126, 43) | (128, 53) |
| averages | (1.98, 1.77)* | (0.52, 0.44)* | (0.15, 0.14) | (0.20, 0.17)* |
| [G], [P+H] | | | | |
| sample sizes | (33, 46) | (29, 48) | (28, 39) | (33, 48) |
| averages | (1.42, 1.68)* | (0.30, 0.43) | (0.12, 0.11) | (0.12, 0.15) |
| [P+H], [F+L] | | | | |
| sample sizes | (51, 47) | (49, 51) | (40, 32) | (51, 49) |
| averages | (1.67, 1.81) | (0.43, 0.51) | (0.11, 0.17)** | (1.81, 1.68) |
| [F], [L], [C] | | | | |
| sample sizes | (25, 26, 130) | (25, 26, 130) | (16, 16, 126) | (25, 26, 128) |
| averages | (1.94, 1.68, 1.98) | (0.55, 0.47, 0.52) | (0.19, 0.15, 0.15) | (0.23, 0.17, 0.20) |
| [T], [P], [G], [H] | | | | |
| sample sizes | (51, 43, 33, 3) | (52, 44, 29, 4) | (43, 37, 28, 2) | (53, 44, 33, 4) |
| averages | (1.76, 1.68, 1.43, 1.77)* | (0.44, 0.43, 0.30, 0.45) | (0.14, 0.11, 0.12, 0.10) | (0.17, 0.15, 0.12, 0.13) |

*$P < 0.05$; **$P < 0.01$; ***$P < 0.001$

Table 12.4 *Genetic variation indicators in the four lineages in the Canarian flora where allozyme data from exhaustive population sampling are available for the groups of islands considered in the genetic diversity tests (between brackets, numbers of populations considered per group). Except for Gran Canaria (C), island abbreviations correspond to the first letter of the island name. Numbers in parentheses after the lineages' names are the number of populations included in each island group. Shadings indicate groups where intra-population sampling was severely limited respect to estimated population sizes.*

| | Eastern [F+L+C] | | | | Western [T+G+P+H] | | | |
|---|---|---|---|---|---|---|---|---|
| Island groups | $A$ | $P$ | $H_o$ | $H_e$ | $A$ | $P$ | $H_o$ | $H_e$ |
| The woody *Sonchus* alliance[1] (5, 11) | 1.448 | 0.420 | 0.097 | 0.161 | 1.240 | 0.214 | 0.066 | 0.073 |
| *Androcymbium*[2] (2, 3) | 1.240 | 0.235 | 0.206 | 0.108 | 1.120 | 0.133 | 0.107 | 0.058 |
| *Lobularia canariensis*[3] (6, 12) | 2.150 | 0.683 | 0.184 | 0.242 | 2.485 | 0.762 | 0.227 | 0.296 |
| *Lotus*[4] (23, 16) | 2.209 | 0.604 | 0.122 | 0.223 | 1.545 | 0.375 | 0.060 | 0.135 |
| *Pinus canariensis*[5] (4, 17) | 1.820 | 0.500 | 0.130 | 0.127 | 1.744 | 0.511 | 0.114 | 0.126 |
| *Phoenix canariensis*[6] (2, 7) | 1.750 | 0.583 | 0.178 | 0.192 | 1.547 | 0.371 | 0.179 | 0.148 |
| Averages | 1.770 | 0.504 | 0.153 | 0.176 | 1.381 | 0.394 | 0.125 | 0.139 |

[1] Kim *et al.* (1999)
[2] Pedrola-Monfort & Caujapé-Castells (1994)
[3] Borgen (1996)
[4] Oliva-Tejera *et al.* (2005, 2006 and unpubl. data)
[5] Schiller *et al.* (1999)
[6] González-Pérez *et al.* (2004)

genetic admixture processes. Sanmartín *et al.* (2008) estimate low colonisation rates from the mainland to the eastern islands, but they also admit a substantial influence of extinction warping their results. Although founder events are also likely to impact other islands (especially Gran Canaria, given its relative closeness to the mainland), the possibility of successful colonisation is higher in Fuerteventura and Lanzarote, due to their proximity and ecological similarity to the mainland, and their antiquity. Genetic admixture in these islands is bound to have been additionally fostered by their relatively shallow relief and ecological uniformity, two traits that are expected to increase gene flow and erode any sharp genetic differences among founder populations (Barton & Bengtsson, 1986).

Given enough phylogenetic closeness between the colonising genomes, these processes likely generated syngameons (*sensu* Grant, 1957) derived from gene flow among genotypes previously isolated in the mainland, and consisting of morpho-species, or of incipient species that in some cases still retain crossing compatibility. When these multiple colonisations involved self-incompatible taxa, it is feasible that highly variable colonisers could have been able to establish sexually reproducing populations, because propagules from different dispersal events were cross-compatible (Daniel J. Crawford, pers. comm.). This predominance of gene flow in Fuerteventura and Lanzarote could be an additional factor to consider (together with island age and extinction) in explaining the low number of endemics in these islands (Fig. 12.4); as expected, highest endemic species richness and range-size rarities in these two islands concentrate in their only remarkable geographical features, where isolation can effectively reduce gene flow: the massifs of Jandía and Famara, respectively (Reyes-Betancort *et al.*, 2008). The back-colonisation dynamics discussed earlier entails gene flow from the eastern islands to the mainland; therefore, we might expect to detect syngameons also in the flat extensions of northwest Africa facing the Canaries.

Much higher ecological heterogeneity and isolation promoted by the more rugged geography of the western islands would have induced substantial population genetic diversity loss following colonisation from the eastern islands, while increasing the degree of interpopulation differentiation. Although there are other geographic regions from where propagules can colonise the western islands, sufficient genetic variation furnished by founder stocks from syngameons in the eastern islands (or the mainland) may have been critically important in overcoming selective pressures and the deleterious consequences of further genetic drift, as hypothesised in other enclaves (Fuertes-Aguilar *et al.*, 1999; Rieseberg *et al.*, 2003); subsequently, the populations could develop adaptations to a particular environment, and colonise similar habitats within the islands, or in other islands. Since the predominant evolutionary processes within the western islands entail speciation and cessation of gene flow, we would expect a much higher number of endemics in these islands than in Fuerteventura and Lanzarote, as is the case (Fig. 12.4).

This 'surfing syngameon' hypothesis does not conflict with eventual competition preventing multiple colonisations in most radiating monophyletic lineages (i.e. the 'niche preemption' hypothesis; Silvertown *et al.*, 2005), because most colonisers would be close genetically, and admixture processes would have happened earlier than radiations. More precisely, it predicts that moderate to high levels of gene flow within the eastern islands, and between them and the mainland, should stall and blur evolutionary change in these regions. By contrast, the predominance of drift and selection in the more rugged western islands would tend to exacerbate the genetic differences with respect to their easternmost congeners. This hypothesis also introduces an additional new component to interpret the lower number of endemic species in the eastern Canaries in terms of the predominance of gene flow and admixture in these islands, which is compatible with the prevalent pattern of monophyly in the Canarian flora. Indeed, recent reviews of introgression (Currat *et al.*, 2008; Pétit & Excoffier, 2009) find a significant negative correlation between intra- and interspecific hybridisation. This result emphasises a major role for enhanced gene flow in the maintenance of species cohesion and in promoting rapid monophyly, as opposed to previous interpretations that equated hybridisation with fuzziness in species boundaries (Rieseberg *et al.*, 2006).

The general possibility of multiple colonisations and hybridisation before radiations took place was first suggested by Francisco-Ortega *et al.* (2000) to explain the contrast between the high population genetic diversity in Canarian plant endemics and their predominantly monophyletic origin. Subsequently, other theories (Herbén *et al.*, 2005; Saunders & Gibson, 2005) have favoured hybridisation over interspecific competition (Silvertown, 2004) to account for the monophyly landslide. However, none of these authors considered the relevance of the eastern islands in hosting admixture processes, or their possible role as shuttles to colonisation success in the western islands. Although there are still few empirical studies to substantiate this hypothesis, artificial crossings between the Canarian endemic *Androcymbium* and their closest mainland ancestor according to the molecular phylogenies (*A. wyssianum*, Del Hoyo and Pedrola-Monfort, 2008) agree with this hypothesis in showing a greater seed germination percentage from crossings between *A. wyssianum* and the eastern islands endemic *A. psammophyllum* (32.5%) than between *A. wyssianum* and the western islands endemic, *A. hierrense* (15.5%) (Núria Membrives, pers. comm.; Ardanuy, 1997).

Two predictions of this model cannot be tested due to various limitations of the present population genetic database for the Canarian flora. First, we would expect the average inter-population genetic differentiation (as measured, e.g., by Nei's (1973) $G_{ST}$ values) to be greater within each western island, where higher ecological heterogeneity and isolation have likely promoted a much larger population differentiation than in the eastern islands. In close connection, the total genetic diversity *per genus* in the western islands (particularly the number of alleles, *A*) might equal or even exceed that in the eastern islands, because ecological speciation and isolation may have induced sharp differences in the qualitative allele composition of congeneric populations. The latter expectation is however tentative, as the markers used to assess genetic variation are neutral, and therefore do not have to reflect adaptation.

This proposed interpretation of the east–west genetic diversity gradient is not incompatible with Sanmartín *et al.*'s (2008) contention that the three 'central' Canary Islands are the centre of diversification in the archipelago. Rather, it reinforces their role as active speciation grounds with a causal explanation; namely, that the success of diversification events in these ecologically and geographically more complex islands is bound to be higher when the founder stocks originate from syngameons in the eastern islands or the mainland. It also converges with Sanmartín *et al.*'s (2008) expectation that dispersal to the much younger El Hierro and La Palma predominantly originates in Tenerife and La Gomera.

The general geographic patterns elicited for population genetic diversity are, however, difficult to reconcile with the prediction of highest dispersal from central islands to eastern islands and to the mainland. It can be argued that genetic admixture in the eastern islands following multiple introductions from the central islands could also have given rise to the detected genetic diversity gradient. However, if inter-island colonisation of similar ecological zones is indeed the primary mode of species diversification in Canarian plants (Francisco-Ortega *et al.*, 2002; Allan *et al.*, 2004; Silvertown *et al.*, 2005), the success of these colonisation routes would be compromised by the radical differences between the putative source and target islands, and by the predictably low genetic variation of the founder stocks. Furthermore, the long-term population viability of most well-known interspecific hybrids between Canarian plant endemics (for a comprehensive list see Francisco-Ortega & Santos-Guerra, 2001) is at least questionable.

There are several possible reasons for the observed mismatch between the biogeographic and the population genetic predictions. First, some of the biogeographic models used, or some general assumptions of the Bayesian approach (e.g. ignoring island changes through time) might just not be of general application in plants. Second, given the low resolution of some plant phylogenies considered (e.g. *Tolpis*, Moore *et al.*, 2002 or *Lotus*, Allan *et al.*, 2004), and the glaring incongruence between nuclear and plastid phylogenies in some cases (e.g. in *Sideritis*, Barber *et al.*, 2007), the accuracy of the biogeographical reconstructions for the different plant groups used could be questionable. And third, the sampling of genetic diversity in the Canarian flora, while extensive, has not been done in the same rigorous manner at the population level for all groups (e.g. Caujapé-Castells, 2010, and some cases in Table 12.4).

The complex historical-geographic setting of the Canarian archipelago makes it simplistic to assume that there can be a single general explanatory model for the post-colonisation sequence of endemic plants. For instance, Fuerteventura could have reached a height of more than 3000 m during the shield stage (Stillman, 1999; Acosta *et al.*, 2003) and likely hosted many ecological zones that, at present, are only found in some of the westernmost islands (Fernández-Palacios, pers. comm.). Because these radical island changes in the past are bound to have influenced the current distribution of many contemporary endemic lineages, they cannot be ignored in any general biogeographic or microevolutionary hypothesis.

Furthermore, in many cases, the detection of an eventual genetic diversity gradient is bound to be blurred by the impact of the recent geological history of the islands on different biological groups, by particular adaptations and reproductive syndromes that could increase or diminish their colonising capabilities, by extinction/fragmentation followed by recolonisation, or by poor resolution of the molecular markers used. The surfing syngameon could be used as a null hypothesis for a large part of the current Canarian endemic flora, i.e. the groups that colonised the Canaries from northwest Africa when the easternmost islands were already ecologically uniform and practically flat, as at present. However, a fair test can be made only when there are well-resolved molecular phylogenies, dynamic biogeographical island models that allow the inclusion of more variables (as envisioned by Sanmartín *et al.*, 2008) and adequate genetic diversity data for the same lineages.

## A whole forest behind the trees

The outstanding environmental, historical, geographical, taxonomic and genetic diversities of the Canarian flora conceal a vast heterogeneity of evolutionary processes that serve as a constant reminder of why oceanic archipelagos continue to fascinate evolutionary biologists. The excellent methodological and scientific achievements in the molecular realm in only two decades support the expectation of continuing progress in the coming years. Nevertheless, care should be exerted in mitigating some basic knowledge gaps that emerge from the present database.

Undoubtedly, molecular phylogenies have represented a fruitful approach to understanding speciation in the Canarian flora. However, incomplete phylogenetic trees have been used, with the risk of error in comparisons of subclades, calculations of evolutionary rates, or monophyly inferences. Tables 12.1 and 12.2 illustrate this point by highlighting a considerable number of unsampled species and islands of occurrence in the corresponding phylogenies. For the whole phylogenetic database of the Canarian flora, a similar pattern emerges (Fig. 12.3A): each year, the number of missing taxa per genus included in phylogenies is higher than the number of new taxa sampled, that almost plateaus from 2005 on. Admittedly, these statistics include even very general papers whose focus is not the Canarian circumscription of the groups, and there are significant exceptions in speciose genera where taxonomic and geographic sampling have been exhaustive, e.g. *Sideritis* (Barber *et al.*, 2007), *Descurainia* (Goodson *et al.*, 2006 new), *Aeonium* (Mort *et al.*, 2002), *Sonchus* (Lee *et al.*, 2005; Kim *et al.*, 2007), *Crambe* (Francisco-Ortega *et al.*, 2002), or *Lotus* (Allan *et al.*, 2004). However, by and large, sampling gaps exist which impede a thorough understanding of the origins and colonisation sequences of the Canarian endemic flora. Paradoxically, the increasing accuracy and economy achieved by DNA sequencing since the turn of the twenty-first century have not been paralleled by a thorough sampling in most phylogenies of Canarian groups. It must be noted, however, that in many of these cases taxon sampling has been much more strongly influenced by difficulties in locating the populations, rather than by not having the funds or dedication to collect and analyse the samples. Although these floristic difficulties do hinder resolution in many cases, it is more meaningful to use a partial phylogeny than nothing at all. This is particularly true considering the likelihood that all phylogenies are incomplete to some extent, due to extinction and cryptic diversity.

The ITS region currently forms the largest phylogenetic database (Fig. 12.3A), but its general lack of polymorphism (e.g. Böhle *et al.*, 1996; Allan *et al.*, 2004 ; Francisco-Ortega *et al.*, 2002) has motivated the quest for other sequences (see e.g. Crawford & Mort, 2004). Poor polymorphism is probably related to lineage recentness, and has triggered the application of hypervariable non-sequence markers (e.g. ISSRs, or AFLPs, Fig. 12.3A) on the grounds that they are appropriate below the level where ITS or similar DNA regions are useful for resolving relationships (Bussell *et al.*, 2005; Archibald *et al.*, 2006). Although the use of these markers for phylogenetic inference is still a minority, it has experienced an upsurge since 2005 (Fig. 12.3A).

Resolution problems notwithstanding, a thorough test of the monophyly of an insular group can only be achieved by sampling all closely related species from all neighbouring continental areas. As, for example, Emerson (2002) points out, inclusion or exclusion of different combinations of mainland congeners in a phylogeny of insular taxa can lead to radically different topologies, some of them entailing paraphyly or monophyly of the same island group. In the Canarian phylogenetic database, the sampling of mainland congeners from northwest Africa is limited, even for taxa that are well sampled in the islands. Notably, previously assumed monophyly in the polyploid Hawaiian silversword alliance was revealed to be spurious just by enhancing the genetic sampling (Barrier *et al.*, 1999). However, Silvertown *et al.* (2005) rightly point out that there is no single example where monophyly of a Macaronesian group has been refuted by additional character sampling. It seems thus of utmost importance to extend mainland sampling to assess the consistency of monophyly in the Canarian flora.

Critically, most published phylogenies of Canarian endemics represent cases of conspicuous morphological radiations in Macaronesia where either (1) only a few (if any) distantly related extant taxa are known on the near mainland or in more remote areas, or (2) a poor sampling of extant congeners in the continental enclave has been carried out, for example, *Echium* (Boraginaceae, García-Maroto *et al.*, 2009), *Lotus* (Fabaceae, Allan *et al.*, 2004), *Cheirolophus* (Asteraceae, Susanna *et al.*, 1999) and *Euphorbia* (Euphorbiaceae, Molero *et al.*, 2002). Furthermore, only some taxa for which phylogenies are available have an exhaustive population genetic counterpart. Even in these cases (see e.g. Pedrola-Monfort & Caujapé-Castells, 1994, 1996, Caujapé-Castells *et al.*, 1999 and Del Hoyo & Pedrola-Monfort, 2008 with *Androcymbium*, Moore *et al.*, 2002, Crawford *et al.*, 2006 and Archibald *et al.*, 2006 with *Tolpis*, or Bräuchler *et al.*, 2005 and Meimberg *et al.*, 2006 with *Micromeria*), problems related to a limited intra-population sampling might apply. Clearly, there is a lack of complementarity between population genetic and phylogenetic knowledge in the Canarian flora that impinges on our understanding of the floristic link between this archipelago and the mainland. This problem is not unique to the Canaries, as many studies across the world do not have the advantage of detailed phylogenetic and population genetic studies on the same taxa.

Insight on the colonisation sequence of the Canaries, the patterns and levels of gene flow within the archipelago and with the mainland, and the role played by the Canaries in the floristic makeup of the African continent, is best afforded by combining population genetic and phylogenetic studies based on exhaustive geographic sampling of recent groups (Jaén-Molina *et al.*, 2009); these have not yet radiated, and are less prone to the warping influences of 'jester' historical processes than older groups.

Because, in general, population genetic investigations have a much narrower taxonomic focus than phylogenetic studies, geographic sampling per island of distribution has been less of an issue (Fig. 12.3B). Many of the problems in this field arise from (1) defective intra-population sampling, (2) scarcity of comparisons with mainland congeners, and (3) the lack of a publicly accessible database.

A biased intra-population sampling in many papers is prone to have inflated the estimates of inter-population genetic differentiation, so that the previously estimated low levels of gene flow in the Canarian flora (Francisco-Ortega *et al.*, 2000) might not reflect the underlying biological reality; as discussed in Caujapé-Castells (2010) and substantiated by this contribution, the emerging picture for many endemics is one of frequent inter-island colonisation and considerable levels of local gene flow.

Comparisons of genetic variation between Canarian endemics and mainland relatives are extremely rare except for some cases, e.g. in the genera *Androcymbium* (Colchicaceae, Pedrola-Monfort & Caujapé-Castells, 1994, 1996), *Argyranthemum*

(Asteraceae, Francisco-Ortega *et al.*, 1995), *Dactylis* (Poaceae, Sahuquillo & Lumaret, 1995), *Laurus* (Lauraceae, Arroyo-García *et al.*, 2001) and *Olea* (Oleaceae, Lumaret *et al.*, 2004; García-Verdugo *et al.*, 2009). Therefore, we lack a proper gauge for the high levels of genetic variation detected in the Canarian flora (i.e. are the insular populations *really* less diverse than their mainland congeners?), which also prevents testing of some general hypotheses on the levels and directions of gene flow between the archipelago and northwest Africa.

Another important weakness of the present population genetic database is the lack of a matrix repository containing the raw data (i.e. genotypes) of the investigated Canarian endemics. Excellent population genetic data swiftly become obsolete because they cannot be used in a versatile way to test novel hypotheses, to properly standardise parameters for meta-analyses, or to establish meaningful comparisons with more recent data (Caujapé-Castells, 2007). These and other inconveniencies prevent the consideration of many alternative analytical approaches that do not exist today, but which will surely appear in the future. Some activities in a recently funded project (Demiurgo, MAC/1/C020) aim at mitigating this weakness through collecting, preserving and disseminating genotype matrices in the format of the computer program Transformer-3 (Caujapé-Castells & Baccarani-Rosas, 2005) and its future upgrade, T-4. Hopefully, the projected public matrix repository DEMIURGE (www.demiurge-project.org/) will soon begin to fulfil its main mission: to collect and preserve molecular population genetic genotype matrices, not to freeze them in a given conceptual mould but to allow their use in the progressive construction of a better understanding of biodiversity through unrestricted access to all the dimensions of knowledge that they contain.

DNA banking is of urgent general importance for all these evolutionary issues and for conservation research problems, partly motivated by the predicted effects of climate change on biodiversity (Savolainen & Reeves, 2004; Gran Canaria Group, 2006). The DNA bank of the Canarian flora was started in 2005 at the Jardín Botánico Canario 'Viera y Clavijo' (Caujapé-Castells *et al.*, 2006), to complement the research on the origins, microevolution, conservation and taxonomic identification developed at its different research departments. A vast majority of the bank's collections are funded by institutional projects at the 'Viera y Clavijo', but it also hosts samples sent by other collaborating botanists and institutions; overall, these samples now represent *c.*50% of the Canarian endemic flora at the species level, and a similar proportion of the Cape Verdean endemic flora (whose samples are kept as a 'trust fund'). A great deal of effort is still needed to consolidate these collections but, in parallel with other lines of action, they will, it is hoped, allow us to estimate the phylogenetic diversity (PD) of these floras (see Faith *et al.*, 2004; Forest *et al.*, 2007) with a view to broadening conservation guidelines based on taxonomic diversities per unit area (Reyes-Betancort *et al.*, 2008). Molecular population and phylogenetic studies based on exhaustive sampling in Macaronesia *sensu lato* are also under way with the genera *Androcymbium*, *Asparagus*, *Bupleurum*, *Crambe*, *Dracaena*, *Heberdenia*, *Juniperus*, *Lotus*, *Matthiola*, *Minuartia*, *Neochamaelea*, *Parolinia*, *Pleiomeris* and *Ruta* among others; some of these entail collaboration with other research institutions.

Thanks to the insightful outlooks of several research groups and individuals, new hypotheses are emerging that challenge the accepted mainstream knowledge on the current Canarian endemic flora, and which need proper multidisciplinary testing in a geographical context encompassing not only the Canaries, but also the other Macaronesian archipelagos, mainland northwest Africa and the Mediterranean basin. Any further approach to enhance our scientific understanding of the evolution, diversity and conservation of this region's terrestrial floras will only deliver if based on the synergies among (at least) taxonomical, reproductive and molecular investigations, and on the progressive implementation of more modern techniques (e.g. FISH, Rosato *et al.*, 2008).

Quite clearly, Macaronesia is a burgeoning discussion arena encompassing heterogeneous facts and theories on the evolution, diversity and conservation of oceanic island floras. If we want this scientific knowledge to gel into a consistent whole, we need to foster (and enhance) the existing collaboration among the five botanic gardens in this multinational oceanic region, and other research institutions or universities with full programmes on the archipelagos' terrestrial plants. The natural and political-legal threats faced by the closely related Macaronesian endemic floras can only be contained through a coordinated scientific effort of the same magnitude.

## Acknowledgements

I thank David Bramwell for his continued support, and for offering me this challenge; all the biologists at the Jardín Botánico Canario 'Viera y Clavijo' – Unidad Asociada CSIC for their many impressive multidisciplinary contributions to the knowledge of the Canarian flora, their stimulus and trust, especially Julia Pérez de Paz, Águedo Marrero Rodríguez, Bernardo Navarro, Ruth Jaén-Molina, Alicia Roca, Rosa Febles, José Naranjo, Felicia Oliva-Tejera, Olga Fernández-Palacios, Magui Olangua and Moisés Soto; and the following persons for additional meaningful insight on this manuscript: Gonzalo Nieto-Feliner and Javier Fuertes-Aguilar (at the Real Jardín Botánico de Madrid), Dan Crawford and Jenny Archibald (at the University of Kansas), Alfredo Reyes-Betancort (at the Jardín de Aclimatación de La Orotava in Tenerife), Mark Carine (at the Natural History Museum, London), Núria Membrives and Pep Toni Rosselló (at the Jardí Botànic Marimurtra-Fundació Karl Faust, Blanes, Girona), José María

Fernández-Palacios (at the Universidad de la Laguna in Tenerife) and Juan Carlos Moreno Saiz (Universidad Autónoma de Madrid). This chapter would not have been possible without the contributions of the researchers who have shaped the present molecular knowledge of the Canarian endemic flora; however, the scientific perspective and insight in the pioneering investigations of Javier Francisco-Ortega, Arnoldo Santos-Guerra, Bob Jansen, Dan Crawford and their various collaborators have been always a source of inspiration. I am especially indebted to the Cabildo de Gran Canaria, the CSIC, the Instituto Tecnológico de Canarias, the Consortium for the Barcode of Life, the Programa de Cooperación Transnacional MAC 2007–2013, the Spanish Ministry of Science, the Canarian Government and the Fundación Amurga-Maspalomas for providing financial support to many ideas that lie at the core of our scientific missions at the Jardín Botánico Canario 'Viera y Clavijo' – Unidad Asociada CSIC of the Cabildo de Gran Canaria. I also acknowledge the support of the UNESCO-Unitwin chair on 'Biodiversity conservation and climate change in Macaronesia and the West of Africa', based at the Jardín Botánico Canario 'Viera y Clavijo'.

## References

Acebes-Ginovés, J. R., del Arco-Aguilar, M., García-Gallo, A., León-Arencibia, M. C., Pérez de Paz, P. L. *et al.* (2004). *Pteridophyta, Spermatophyta.* In I. Izquierdo, J. L. Martín, N. Zurita & M. Arechavaleta (eds.), *Lista de especies silvestres de Canarias (hongos, plantas y animales terrestres)*, pp. 96–143. Santa Cruz de Tenerife, Consejería de Política Territorial y Medio Ambiente de Canarias.

Acosta, J., Uchupi, E., Muñoz, A., Herranz, P., Palomo, C., Ballesteros, M. & ZEE Working Group (2003). Geologic evolution of the Canarian Islands of Lanzarote, Fuerteventura, Gran Canaria and La Gomera and comparison of landslides at these islands with those at Tenerife, La Palma and El Hierro. *Mar. Geophys. Res.* 24: 1–40.

Allan, G., Francisco-Ortega, J., Santos-Guerra, A., Boerner, E. & Zimmer, E. A. (2004). Molecular phylogenetic evidence for the geographic origin and classification of Canary Island *Lotus* (Fabaceae: Loteae). *Mol. Phylogenet. Evol.* 32: 123–138.

Ancochea, E., Barrera, J. L., Bellido, F. *et al.* (2004). Canarias y el vulcanismo neógeno peninsular. In E. A. Vera (ed.), *Geología de España*, pp. 635–682. Madrid, SGE-IGME.

Anderson, C. L., Channing, A. & Zamuner, A. B. (2009). Life, death and fossilization on Gran Canaria: implications for Macaronesian biogeography and molecular dating. *J. Biogeogr.* 36: 2189–2201.

Andrus, N., Trusty, J., Santos-Guerra, A., Jansen, R. K. & Francisco-Ortega, J. (2004). Using molecular phylogenies to test phytogeographical links between E/S Africa, southern Arabia and the Macaronesian islands: a review, and the case of *Vieraea* and *Pulicaria* sect. *Vieraeopsis. Taxon* 53: 333–346.

Archibald, J. K., Crawford, D. J., Santos-Guerra, A. & Mort, M. E. (2006). The utility of automated analysis of inter-simple sequence repeat (ISSR) loci for resolving relationships in the Canary Island species of *Tolpis* (Asteraceae). *Am. J. Bot.* 93: 1154–1162.

Ardanuy, A. (1977). Estudis bàsics per la gestió agronòmica i la conservació de sis espècies endèmiques del gènere *Androcymbium* Willd. Treball fi de carrera. Enginyeria Tècnica Agrícola, especialitat Explotacions Agropecuàries. Escola Politécnica Superior, Universitat de Girona.

Arroyo-García, R., Martínez-Zapater, J. M., Fernández Prieto, J. A. & Álvarez-Arbesu, R. (2001). AFLP evaluation of genetic similarity among laurel populations (*Laurus L.*). *Euphytica* 122: 155–164.

Asmussen, C. B., Dransfield, J., Deickmann, V., Barfod, A. S., Pintaud, J. C. & Baker, W. J. (2006). A new subfamily classification of the palm family (Arecaceae): evidence from plastid DNA phylogeny. *Bot. J. Linn. Soc.* 151: 15–38.

Ávila, S. P., Madeira, P., Marques da Silva, C., Cachaõ, M., Landau, B., Quartau, R. et al. (2008). Local disappearance of bivalves in the Azores during the last glaciations. *J. Quat. Sci.* 23: 777–785.

Barber, J. C., Finch, C. C., Francisco-Ortega, J., Santos-Guerra, A. & Jansen, R. K. (2007). Hybridization in Macaronesian *Sideritis* (Lamiaceae): evidence from incongruence of multiple independent nuclear and chloroplast sequence datasets. *Taxon* 56: 74–88.

Barquín, E. (1984). Matorrales de la transición entre el piso basal y el montano de la isla de Tenerife, Canarias. PhD dissertation: Universidad de la Laguna, Tenerife, Spain.

Barrier, M., Baldwin, B. G., Robichaux, R. H. & Purugganan, M. D. (1999). Interspecific hybrid ancestry of a plant adaptive radiation: allopolyploidy of the Hawaiian silversword alliance (Asteraceae) inferred from floral homeotic gene duplications. *Mol. Biol. Evol.* 16: 1105–1113.

Barton, N. & Bengtsson, B. O. (1986). The barrier to genetic exchange between hybridising populations. *Heredity* 57: 357–376.

Batista, F. & Sosa, P. (2002). Allozyme diversity in natural populations of *Viola palmensis* Webb & Berth. (Violaceae) from La Palma (Canary Islands): implications for conservation genetics. *Ann. Bot.* 90: 725–733.

Bellemain, E. & Ricklefs, R. E. (2008). Are islands the end of the colonization road? *Trends Ecol. Evol.* 23: 461–468.

Benton, M. J. (2009). The Red Queen and the Court Jester: species diversity and the role of biotic and abiotic factors through time. *Science* 323: 728–732.

Besnard, G., Rubio de Casas, R., Cristin, P. A. & Vargas, P. (2009). Phylogenetics of *Olea* (Oleaceae) based on plastid and nuclear ribosomal DNA sequences: Tertiary climatic shifts and lineage differentiation times. *Ann. Bot.* 104: 143–160.

Böhle, U. R., Hilger, H. H. & Martin, W. F. (1996). Island colonization and evolution of the insular woody habit in *Echium* L. (Boraginaceae). *Proc. Natl. Acad. Sci. USA* 93: 11740–11745.

Bolmgren, K. & Oxelman, B. (2004). Generic limits in *Rhamnus* L. s.l. (Rhamnaceae) inferred from nuclear and chloroplast DNA sequence phylogenies. *Taxon* 53: 383–390.

Borgen, L. (1996). Genetic differentiation in endemic *Lobularia* (Brassicaceae) in the Canary Islands. *Nord. J. Bot.* 16: 487–503.

Borges, P. A. V. & Hortal, J. (2009). Time, area and isolation: factors driving the diversification of Azorean arthropods. *J. Biogeogr.* 36: 178–191.

Bramwell, D. (1976). The endemic flora of the Canary Islands. In G. Kunkel (ed.), *Biogeography and Ecology in the Canary Islands*, pp. 207–240. The Hague, Junk.

Bramwell, D. (1985). Contribución a la biogeografía de las Islas Canarias. *Bot. Mac.* 14: 3–34.

Bräuchler, C., Meimberg, H. & Heubl, G. (2004). Molecular phylogeny of the genera *Digitalis* L. and *Isoplexis* (Lindley) Loudon (Veronicaceae) based on ITS- and trnL-F sequences. *Plant Syst. Evol.* 248: 111–128.

Bräuchler, C., Meimberg, H., Abele, T. & Heubl, G. (2005). Polyphyly of the genus *Micromeria* (Lamiaceae): evidence from cpDNA sequence data. *Taxon* 54: 639–650.

Bussell, J. D., Waycott, M. & Chappill, J. A. (2005). Arbitrarily amplified DNA markers as characters for phylogenetic inference. *Perspect. Plant Ecol. Evol. Syst.* 7: 3–26.

Carine, M. A. (2005). Spatio-temporal relationships of the Macaronesian endemic flora: a relictual series or window of opportunity? *Taxon* 54: 895–903.

Carine, M. A. & Schaefer, H. (2010). The Azores diversity enigma: why are there so few Azorean endemic flowering plants and why are they so widespread? *J. Biogeogr.* 37: 77–89.

Carine, M. A., Russell, S. J., Santos-Guerra, A. & Francisco-Ortega, J. (2004). Relationships of the Macaronesian and Mediterranean floras: molecular evidence for multiple colonizations into Macaronesia and back-colonization of the continent in *Convolvulus* (Convolvulaceae). *Am. J. Bot.* 91: 1070–1085.

Carine, M. A., Santos-Guerra, A., Guma, I. R. & Reyes-Betancort, J. A. (2010). Endemism and evolution of the Macaronesian Flora. In: *Beyond Cladistics*, pp. 101–124. Berkeley, CA, University of California Press.

Carlquist, S. (1966). The biota of long-distance dispersal. III. Loss of dispersability in the Hawaiian flora. *Brittonia* 18: 310–335.

Carracedo, J. C. (1999). Growth, structure, instability and collapse of Canarian volcanoes and comparisons with Hawaiian volcanoes. *J. Volcanol. Geoth. Res.* 94: 1–19.

Carracedo, J. C., Pérez, F. J., Ancochea, E., Meco, J., Hernán, F., Cubas, C. R. *et al.* (2002). Cenozoic volcanism II: The Canary Islands. In W. Gibbons & T. Moreno (eds.), *The Geology of Spain*, pp. 439–472. London, Geological Society of London.

Catalán, P. & Olmstead, R. G. (2000). Phylogenetic reconstruction of the genus *Brachypodium* P. Beauv. (Poaceae) from combined sequences of chloroplast *ndhF* gene and nuclear ITS. *Plant Syst. Evol.* 220: 1–19.

Caujapé-Castells, J. (2004). Boomerangs of biodiversity? The interchange of biodiversity between mainland north Africa and the Canary Islands as inferred from cpDNA RFLPs in genus *Androcymbium*. *Bot. Mac.* 25: 53–69.

Caujapé-Castells, J. (2007). SAGE: una base de datos moleculares poblacionales con capacidades analíticas inherentes. *Conserv. Veget.* 11: 40–41.

Caujapé-Castells, J. (2010). General *GST* and $\theta$ inflation due to biased intra-population sampling, and its consequences for the conservation of the Canarian flora. *Conserv. Genet.* 11: 709–720.

Caujapé-Castells, J. & Baccarani-Rosas, M. (2005). Transformer-3: a program for the analysis of molecular population genetic data. EXEGEN software & Jardín Botánico Canario 'Viera y Clavijo' (distributed by the first author).

Caujapé-Castells, J., Jansen, R. K., Pedrola-Monfort, J. & Membrives, N. (1999). Chloroplast DNA restriction site phylogeny of the genus *Androcymbium* (Colchicaceae). *Syst. Bot.* 24: 581–597.

Caujapé-Castells, J., Jaén-Molina, J. & Cabrera-García, N. (2006). El banco de ADN de la flora canaria: creación, progresos y líneas futuras de desarrollo. *Bot. Mac.* 26: 3–16.

Caujapé-Castells, J., Naranjo-Suárez, J., Santana, I., Baccarani-Rosas, M., Cabrera-García, N., Marrero, M. et al. (2008a). Population genetic suggestions to offset the extinction ratchet in the endangered Canarian endemic *Atractylis preauxiana* (Asteraceae). *Plant Syst. Evol.* 273: 191–199.

Caujapé-Castells, J., Marrero-Rodríguez, A., Cabrera-García, N., Baccarani-Rosas, M. & Vilches, B. (2008b). Population genetics of the endangered Canarian endemic *Atractylis arbuscula* (Asteraceae): implications for taxonomy and conservation. *Plant Syst. Evol.* 274: 99–109.

Caujapé-Castells, J., Tye, A., Crawford, D., Santos-Guerra, A., Sakai, A., Beaver, K. et al. (2010). Conservation of island floras: present assets and future global challenges. *Perspect. Plant Evol. Ecol. Syst.* 12: 107–129.

Crawford, D. J. & Mort, M. E. (2004). Single-locus molecular markers for inferring relationships at lower taxonomic levels: observations and comments. *Taxon* 53: 631–635.

Crawford, D. J., Archibald, J. K., Santos-Guerra, A. & Mort, M. E. (2006). Allozyme diversity within and divergence among species of *Tolpis* (Asteraceae-Lactuceae) in the Canary islands: systematic, evolutionary, and biogeographical implications. *Am. J. Bot.* 93: 656–664.

Cubas, P., Pardo, C. & Tahiri, H. (2002). Molecular approach to the phylogeny and systematics of *Cytisus* (Leguminosae) and related genera based on nucleotide sequences of nrDNA (ITS region) and cpDNA (trnL-trnF intergenic spacer). *Plant Syst. Evol.* 233: 223–242.

Currat, M., Ruedi, M., Pétit, R. & Excoffier, L. (2008). The hidden side of invasions: massive introgression by local genes. *Evolution* 62: 1908–1920.

Dansereau, P. (1965). Le contrôle de la végétation dans les îles oceaniques. Simposio: Aportación de las investigaciones ecológicas y agrícolas en la lucha del mundo contra el hambre. CSIC, Patronato 'Alonso de Herrera', Madrid, Spain.

de Nascimento, L., Willis, K. J., Fernádez-Palacios, J. M., Criado, C. & Whittaker, R. J. (2009). The long-term ecology of the lost forests of La Laguna, Tenerife (Canary Islands). *J. Biogeogr.* 36: 499–514.

del Hoyo, A. & Pedrola-Monfort, J. (2008). Phylogeny of *Androcymbium* (Colchicaceae) based on morphology and DNA sequences. *Plant Syst. Evol.* 273: 151–167.

Der, J. P. & Nickrent, D. L. (2008). A molecular phylogeny of Santalaceae (Santalales). *Syst. Bot.* 33: 107–116.

Diane, N., Förther, H. & Hilger, H. H. (2002). A systematic analysis of *Heliotropium*, *Tournefortia*, and allied taxa of the Heliotropiaceae (Boraginales) based on ITS1 sequences and morphological data. *Am. J. Bot.* 89: 287–295.

Díaz-Pérez, A., Sequeira, M., Santos-Guerra, A. & Catalán, P. (2008). Multiple colonizations, in situ speciation, and volcanism-associated stepping-stone dispersals shaped the phylogeography of the Macaronesian red fescues (*Festuca* L., Gramineae). *Syst. Biol.* 57: 732–749.

Downie, S. R., Katz-Downie, D. S. & Spalik, K. (2000). A phylogeny of Apiaceae tribe Scandiceae: evidence from nuclear ribosomal DNA internal transcribed spacer sequences. *Am. J. Bot.* 87: 76–95.

Edmonds, C. A., Lillie, A. S. & Cavalli-Sforza, L. L. (2007). Mutations arising in the wave front of an expanding population. *Proc. Natl. Acad. Sci. USA* 101: 975–979.

Emerson, B. C. (2002). Evolution on oceanic islands: molecular phylogenetic approaches to understanding pattern and process. *Mol. Ecol.* 11: 951–966.

Excoffier, L. & Ray, N. (2008). Surfing during population expansions promotes genetic revolutions and structuration. *Trends Ecol. Evol.* 23: 347–351.

Faith, D. P., Reid, C. A. M. & Hunter, J. (2004). Integrating phylogenetic diversity, complementarity, and endemism. *Conserv. Biol.* 18: 255–261.

Fehrer, J., Gemeinholzer, B., Chrtek, J. & Braütigam, S. (2007). Incongruent plastid and nuclear DNA phylogenies reveal ancient intergeneric hybridization in *Pilosella* hawkweeds (*Hieracium*, Cichorieae, Asteraceae). *Mol. Phylogenet. Evol.* 42: 347–361.

Fernández-Palacios, J. M. (2010). The islands of Macaronesia. In A. Serrano, P. Borges, M. Boieiro & P. Oromí (eds.), *Terrestrial Arthropods of Macaronesia: Evolution, Ecology and Conservation*. doi: 10.1111/j.1365-2699.2010.02427.

Fernández-Palacios, O., Pérez de Paz, J., Febles, R. & Caujapé-Castells, J. (2004). Diversidad genética en *Parolinia*: *P. glabriuscula* y *P. platypetala* (Brassicaceae: Matthioleae). *Bot. Mac.* 25: 143–160.

Forest, F., Grenyer, R., Rouget, M., Davies, J., Cowling, R. M. *et al.* (2007). Preserving the evolutionary potential of floras in biodiversity hotspots. *Nature* 445: 757–760.

Francisco-Ortega, J. & Santos-Guerra, A. (2001). Genes y conservación de plantas vasculares. In J. M. Fernández-Palacios & J. L. Martín-Esquivel (eds.), *Naturaleza de las Islas Canarias. Ecología y Conservación*, pp. 357–365. Santa Cruz de Tenerife, Publicaciones Turquesa.

Francisco-Ortega, J., Newbury, H. J. & Ford-Lloyd, B. V. (1993). Numerical analyses of RAPD data highlight the origin of cultivated tagasaste (*Chamaecytisus proliferus* ssp. *palmensis*) in the Canary Islands. *Theoret. Appl. Genet.* 87: 264–270.

Francisco-Ortega, J., Crawford, D. J., Santos-Guerra, A. & Safontinha, S. (1995). Genetic diversity among Mediterranean and Macaronesian genera of the subtribe Chrysantheminae (Asteraceae). *Am. J. Bot.* 82: 1321–1328.

Francisco-Ortega, J., Crawford, D. J., Santos-Guerra, A. & Jansen, R. K. (1997). Origin and evolution of *Argyranthemum* (Asteraceae: Anthemideae) in Macaronesia. In T. J. Givnish & K. J. Sytsma (eds.), *Molecular Evolution and Adaptive Radiation*, pp. 407–431. Cambridge, Cambridge University Press.

Francisco-Ortega, J., Santos-Guerra, A., Kim, S. C. & Crawford, D. J. (2000). Plant genetic diversity in the Canary Islands: a conservation perspective. *Am. J. Bot.* 87: 909–919.

Francisco-Ortega, J., Barber, J. C., Santos-Guerra, A., Febles-Hernández, R. & Jansen, R. K. (2001). Origin and evolution of the endemic genera of Gonosperminae (Asteraceae: Anthemideae) from the Canary Islands: evidence from nucleotide sequences of the internal transcribed spacers of the nuclear ribosomal DNA. *Am. J. Bot.* 88: 161–169.

Francisco-Ortega, J., Fuertes-Aguilar, J., Kim, S. C., Santos-Guerra, A., Crawford, D. J. & Jansen, R. K. (2002). Phylogeny of the Macaronesian endemic *Crambe* section *Dendrocrambe* (Brassicaceae) based on internal transcribed spacer sequences of nuclear ribosomal DNA. *Am. J. Bot.* 89: 1984–1990.

Fuertes-Aguilar, J., Rosselló-Picornell, J. A. & Nieto-Feliner, G. (1999). Molecular evidence for the compilospecies model of reticulate evolution in *Armeria* (Plumbaginaceae). *Syst. Biol.* 48: 735–754.

Fuertes-Aguilar, J., Ray, M. F., Francisco-Ortega, J., Santos-Guerra, A. & Jansen, R. K. (2002). Evidence from chloroplast and nuclear markers for multiple colonizations of *Lavatera* (Malvaceae) in the Canary Islands. *Syst. Bot.* 27: 74–83.

Galbany-Casals, M., García-Jacas, N., Sáez, L. L., Benedí, C. & Susanna, A. (2009). Phylogeny, biogeography, and character evolution in Mediterranean, Asiatic, and Macaronesian *Helichrysum* (Asteraceae, Gnaphalieae) inferred from nuclear phylogenetic analyses. *Int. J. Plant Sci.* 170: 365–380.

García-Maroto, F., Mañas-Fernández, A., Garrido-Cárdenas, J. A., López Alonso, D., Guil-Guerrero, J. L., Guzmán, B. et al. (2009). $\Delta^6$-Desaturase sequence evidence for explosive Pliocene radiations within the adaptive radiation of Macaronesian *Echium* (Boraginaceae). *Mol. Phylogenet. Evol.* 52: 563–574.

García-Talavera, F. (1997). Las Canarias orientales y vecina costa africana en el Holoceno. *Eres* 7: 55–63.

García-Verdugo, C., Fay, M. F., Granado-Yela, C., Rubio de Casas, R., Balaguer, L., Besnard, G. & Vargas, P. (2009). Genetic diversity and differentiation processes in the ploidy series of *Olea europaea* L.: a multiscale approach from subspecies to insular populations. *Mol. Ecol.* 18: 454–467.

Geldmacher, J., Hoernle, K., van den Bogaard, P., Zankl, G. & Garbe-Schönberg, D. (2001). Earlier history of the ≥70-Ma-old Canarian hotspot based on the temporal and geochemical evolution of the Selvagen Archipelago and neighboring seamounts in the Eastern North Atlantic. *J. Volcanol. Geoth. Res.* 111: 55–87.

Geldmacher, J., Hoernle, K., van den Bogaard, P., Duggen, S., Werner, R. *et al.* (2005). New $^{40}$Ar/ $^{39}$Ar age and geochemical data from seamounts in the Canary and Madeira volcanic provinces: Support for the mantle plume hypothesis. *Earth Planet. Sci. Lett.* 237: 85–101.

González-Pérez, M. A., Caujapé-Castells, J. & Sosa, P. A. (2004). Allozyme variation and structure of the Canarian endemic palm tree *Phoenix canariensis* (Arecaceae): implications for conservation. *Heredity* 93: 307–315.

González-Pérez, M. A., Sosa, P. A., González-González, E. A., Bañares, A., Marrero, M., Carqué, E. *et al.* (2009). *Gnaphalium teydeum* and *Gnaphalium luteo-album*: two taxa of the Canary Islands with different genetic histories. *Plant Syst. Evol.* 276: 39–49.

Goodson, B. E., Santos-Guerra, A. & Jansen, R. K. (2006). Molecular systematics of *Descurainia* (Brassicaceae) in the Canary Islands: biogeographic and taxonomic implications. *Taxon* 55: 671–682.

Goudie, A. S. & Middleton, N. J. (2001). Saharan dust storms: nature and consequences. *Earth-Sci. Rev.* 56: 179–204.

Gran Canaria Group (2006). The Gran Canaria Declaration II on climate change and plant conservation (www.bgci.org/ourwork/gcdccpc/).

Grant, V. (1957). The plant species in theory and practice. In E. Mayr (ed.), *The Species Problem*. Washington, DC, American Association for the Advancement of Science.

Hallatschek, O., Hersen, P., Ramanathan, S. & Nelson, D. R. (2007). Genetic drift at expanding frontiers promotes gene segregation. *Proc. Natl. Acad. Sci. USA* 104: 19926–19930.

Hallatschek, O. & Nelson, D. R. (2008). Gene surfing in expanding populations. *Theoret. Pop. Biol.* 73: 158–170.

Harbaugh, D. T. & Baldwin, B. G. (2007). Phylogeny and biogeography of the sandalwoods (*Santalum*, Santalaceae): repeated dispersals throughout the Pacific. *Am. J. Bot.* 94: 1028–1040.

Harbaugh, D. T., Wagner, W. L., Allen, G. J. & Zimmer, E. A. (2009). The Hawaiian Archipelago is a stepping-stone for dispersal in the Pacific: an example from the plant genus *Melicope* (Rutaceae). *J. Biogeogr.* 36: 230–241.

Heaney, L. R. (2007). Is a new paradigm emerging for oceanic island biogeography? *J. Biogeogr.* 34: 753–757.

Helfgott, D. M., Francisco-Ortega, J., Santos-Guerra, A., Jansen, R. K. & Simpson, B. B. (2000). The origin of the *Bencomia* alliance (Rosaceae) in Macaronesia. *Syst. Bot.* 25: 82–97.

Herben, T., Suda, J. & Munclinger, P. (2005). The ghost of hybridization past: niche preemption is not the only explanation of apparent monophyly in island endemics. *J. Ecol.* 93: 572–575.

Hewitt, G. M. (1996). Some genetic consequences of ice ages, and their role in divergence and speciation. *Biol. J. Linn. Soc.* 58: 247–276.

Hileman, L. C., Vasey, M. C. & Parker, V. T. (2001). Phylogeny and biogeography of the Arbutoideae (Ericaceae): implications for the Madrean-Tethyan hypothesis. *Syst. Bot.* 26: 131–143.

Himmelreich, S., Källersjö, M., Eldenäs, P. & Oberprieler, C. (2008). Phylogeny of southern hemisphere Compositae-Anthemideae based on nrDNA ITS and cpDNA ndhF sequence information. *Plant Syst. Evol.* 272: 131–153.

Holm, P. M., Grandvuinet, T., Friis, J., Wilson, J. R., Barker, A. K. & Plesner, S. (2008). An $^{40}Ar$-$^{39}Ar$ study of the Cape Verde hot spot: temporal evolution in a semistationary plate environment. *J. Geophys. Res.* 113: B08201.

Hooghiemstra, H. (1989). Variations of the NW African trade wind regime during the last 140,000 y: changes in pollen flux evidenced by marine sediment records. In M. Leinen & M. Sarnthein (eds.), *Paleoclimatology and Paleometeorology*, pp. 733–770. Washington, DC, NATO-ASI series 282.

Hooghiemstra, H., Bechler, A. & Beug, H. J. (1987). Isopollen maps for 18,000 y BP of the Atlantic offshore of NW Africa: evidence for paleowind circulation. *Paleoceanography* 2: 561–582.

Humphries, C. J. & Parenti, L. (1999). *Cladistic Biogeography: Interpreting Patterns of Plant and Animal Distribution.* New York, Oxford University Press.

Ibáñez, O., Calero, C., Mayol, M. & Rosselló, J. A. (2002). Isozyme uniformity in a wild extinct insular plant, *Lysimachia minoricensis* J. J. Rodr. (Primulaceae). *Mol. Ecol.* 8: 813–817.

Inda, L. A., Segarra-Moragues, J. G., Müller, J., Peterson, P. M. & Catalán, P. (2008). Dated historical biogeography of the temperate Loliinae (Poaceae, Pooideae) grasses in the northern and southern hemispheres. *Mol. Phylogenet. Evol.* 46: 932–957.

Jacobs, B. F. & Herendeen, P. S. (2004). Eocene dry climate and woodland vegetation in tropical Africa reconstructed from fossil leaves from northern Tanzania. *Palaeogeogr. Palaeoclim. Palaeoecol.* 213: 115–123.

Jaén-Molina, R., Caujapé-Castells, J., Reyes-Betancort, J. A., Akhani, H., Fernández-Palacios, O., Pérez de Paz, J. *et al.* (2009). Molecular phylogeny of *Matthiola* R. Br. (Brassicaceae) inferred from ITS sequences, with special emphasis on the Macaronesian endemics. *Mol. Phylogenet. Evol.* 53: 97–981.

Kelly, L. J. & Culham, A. (2008). Phylogenetic utility of MORE AXILLARY GROWTH4 (MAX4)-like genes: a case study in *Digitalis/Isoplexis* (Plantaginaceae). *Plant Syst. Evol.* 273: 133–149.

Kim, S. C., Crawford, D. J., Francisco-Ortega, J. & Santos-Guerra, A. (1999). Adaptive radiation and genetic differentiation in the woody *Sonchus* alliance (Asteraceae: Sonchinae) in the Canary Islands. *Plant Syst. Evol.* 215: 101–118.

Kim, S. C., Lee, C. & Santos-Guerra, A. (2005). Genetic analysis and conservation of the endangered Canary Island woody sow-thistle, *Sonchus gandogeri* (Asteraceae). *J. Plant Res.* 118: 147–153.

Kim, S. C., Lee, C. & Mejías, J. A. (2007). Phylogenetic analysis of chloroplast DNA matK gene and ITS of nrDNA sequences reveals polyphyly of the genus *Sonchus* and new relationships among the subtribe Sonchinae (Asteraceae: Cichorieae). *Mol. Phylogenet. Evol.* 44: 578–597.

Kim, S. C., McGowen, M. R., Lubinsky, P., Barber, J. C., Mort, M. E. & Santos-Guerra, A. (2008). Timing and tempo of early and successive adaptive radiations in Macaronesia. *PLoS ONE* 3: e2131.

Korol, L., Gil, S. L., Climent, M. J., Zehavi, A. & Schiller, G. (1999). Canary Islands pine (*Pinus canariensis* Chr. Sm. ex DC.) 2. Gene flow among native populations. *Forest Genet.* 6: 277–282.

Lee, C., Kim, S. C., Lundy, K. & Santos-Guerra, A. (2005). Chloroplast DNA phylogeny of the woody *Sonchus* alliance (Asteraceae: Sonchinae) in the Macaronesian islands. *Am. J. Bot.* 92: 2072–2085.

Liu, H., Trusty, J., Oviedo, R., Anderberg, A. & Francisco-Ortega, J. (2004). Molecular phylogenetics of the Caribbean genera *Rhodogeron* and *Sachsia* (Asteraceae). *Int. J. Plant Sci.* 165: 209–217.

Lledó, M. D., Crespo, M. B., Fay, M. F. & Chase, M. W. (2005). Molecular phylogenetics of *Limonium* and related genera (Plumbaginaceae): biogeographical and systematic implications. *Am. J. Bot.* 92: 1189–1198.

Lumaret, R., Ouazzani, N., Michaud, H., Vivier, G., Deguilloux, M. F. & di Giusto, F. (2004). Allozyme variation of oleaster populations (wild olive tree) (*Olea europaea* L.) in the Mediterranean Basin. *Heredity* 92: 343–351.

MacArthur, R. H. & Wilson, E. O. (1967). *The Theory of Island Biogeography.* Princeton, NJ, Princeton University Pressn.

Machín, F., Hernández-Guerra, A. & Pelegrí, J. L. (2006). Mass fluxes in the Canary Basin. *Progr. Oceanogr.* 70: 416–447.

Manen, J. F., Boulter, M. C. & Naciri-Graven, Y. (2002). The complex history of the genus Ilex L. (Aquifoliaceae): evidence from the comparison of plastid and nuclear DNA sequences and from fossil data. *Plant Syst. Evol.* 235: 79–98.

Mansion, G. & Struwe, L. (2004). Generic delimitation and phylogenetic relationships within the subtribe Chironiinae (Chironieae: Gentianaceae), with special reference to *Centaurium*: evidence from nrDNA and cpDNA sequences. *Mol. Phylogenet. Evol.* 32: 951–977.

Marrero, A. & Francisco-Ortega, J. (2001). Evolución en islas: la metáfora espacio-tiempo-forma. In J. M. Fernández-Palacios & J. L. Martín-Esquivel (eds.), *Naturaleza de las Islas Canarias. Ecología y Conservación*, pp. 133–140. Santa Cruz de Tenerife, Publicaciones Turquesa.

Martín-Bravo, S., Meimberg, H., Luceño, M., Märkl, W., Valcárcel, V., Bräuchler, C. *et al.* (2007). Molecular systematics and biogeography of Resedaceae based on ITS and *trnL-F* sequences. *Mol. Phylogenet. Evol.* 44: 1105–1120.

Martins, L., Oberprieler, C. & Hellwig, F. H. (2003). A phylogenetic analysis of Primulaceae s.l. based on internal transcribed spacer (ITS) DNA sequence data. *Plant Syst. Evol.* 237: 75–85.

Masson, D. G. (1996). Catastrophic collapse of the volcanic island of Hierro 15 Ka ago and the history of landslides in the Canary Islands. *Geology* 24: 231–234.

Meimberg, H., Abele, T., Bräuchler, C., McKay, J. K., Pérez de Paz, P. L. & Heubl, G. (2006). Molecular evidence for adaptive radiation of *Micromeria* Benth. (Lamiaceae) on the Canary Islands as inferred from chloroplast and nuclear DNA sequences and ISSR fingerprint data. *Mol. Phylogenet. Evol.* 41: 566–578.

Mes, T. H. M., Wijers, G. J. & 't Hart, H. (1997). Phylogenetic relationships in *Monanthes* (Crassulaceae) based on morphological, chloroplast and nuclear DNA variation. *J. Evol. Biol.* 10: 193–216.

Molero, J., Garnatje, T., Rovira, A., Garcia-Jacas, N. & Susanna, A. (2002). Karyological evolution and molecular phylogeny in Macaronesian dendroid spurges (*Euphorbia* subsect. *Pachycladae*). *Plant Syst. Evol.* 231: 109–132.

Moore, M. J., Francisco-Ortega, J., Santos-Guerra, A. & Jansen, R. K. (2002). Chloroplast DNA evidence for the roles of island colonization and extinction in *Tolpis* (Asteraceae: Lactuceae). *Am. J. Bot.* 89: 518–526.

Mora-Vicente, S., Caujapé-Castells, J., Pérez de Paz, J., Febles, R. & Malo, J. (2009). Isozyme diversity in some Canarian woody endemics of the genus *Echium* L. (Boraginaceae). *Plant Syst. Evol.* 279: 139–149.

Mort, M. E., Soltis, D. E., Soltis, P. S., Francisco-Ortega, J. & Santos-Guerra, A. (2002). Phylogenetics and evolution of the Macaronesian clade of Crassulaceae inferred from nuclear and chloroplast sequence data. *Syst. Bot.* 27: 271–288.

Navascues, M., Vaxevanidou, Z., González-Martínez, S. C., Climent, J., Gil, L. & Emerson, B. C. (2006). Chloroplast microsatellites reveal colonization and metapopulation dynamics in the Canary Island pine. *Mol. Ecol.* 15: 2691–2698.

Nei, M. (1973). Analysis of gene diversity in subdivided populations. *Proc. Natl. Acad. Sci. USA* 70: 3321–3323.

Neves, S. S. & Watson, M. F. (2004). Phylogenetic relationships in *Bupleurum* (Apiaceae) based on nuclear ribosomal DNA ITS sequence data. *Ann. Bot.* 93: 379–398.

Oliva-Tejera, F., Caujapé-Castells, J., Naranjo-Suárez, J., Navarro-Déniz, J., Acebes-Ginovés, J. R. & Bramwell, D. (2005). Population genetic differentiation in taxa of *Lotus* (Fabaceae: Loteae) endemic to the Gran Canarian pine forest. *Heredity* 94: 199–206.

Oliva-Tejera, F., Caujapé-Castells, J., Navarro-Déniz, J., Reyes-Betancort, A., Baccarani-Rosas, M. & Cabrera-García, N. (2006). Population genetics of three Canarian endemic *Lotus*: levels and structuring of genetic variation and implications for the taxonomy of the endangered *L. kunkelii*. *Am. J. Bot.* 93: 1116–1124.

Ortiz, J. E., Torres, T., Yanes, Y., Castillo, C., de la Nuez, J., Ibáñez, M. *et al.* (2006). Climatic cycles inferred from the aminostratigraphy and aminochronology of Quaternary dunes and palaeosols from the eastern islands of the Canary Archipelago. *J. Quat. Sci.* 21: 287–306.

Panero, J., Francisco-Ortega, J., Jansen, R. K. & Santos-Guerra, A. (1999). Molecular evidence for multiple origins of woodiness and a New World biogeographic connection of the Macaronesian Island endemic *Pericallis* (Asteraceae: Senecioneae). *Proc. Natl. Acad. Sci. USA* 96: 13886–13891.

Park, S. J., Korompai, E. J., Francisco-Ortega, J., Santos-Guerra, A. & Jansen, R. K. (2001). Phylogenetic relationships of *Tolpis* (Asteraceae: Lactuceae) based on *ndhF* sequence data. *Plant Syst. Evol.* 226: 23–33.

Patriat, M. & Labails, C. (2006). Linking the Canary and Cape-Verde Hot-Spots, Northwest Africa. *Mar. Geophys. Res.* 27: 201–215.

Pedrola-Monfort, J. & Caujapé-Castells, J. (1994). Allozymic and morphological relationships among *Androcymbium gramineum*, *A. europaeum*, and *A. psammophilum* (Colchicaceae). *Plant Syst. Evol.* 191: 111–126.

Pedrola-Monfort, J. & Caujapé-Castells, J. (1996). Genetic and morphological divergence in the *Androcymbium gramineum* complex (Colchicaceae). *Plant. Syst. Evol.* 201: 149–162.

Peltier, J. P. (1973). Endémiques Macaronésiennes au Maroc. *Monogr. Biol. Canarienses* 4: 134–142.

Pelser, P. B., Nordenstam, B., Kadereit, J. W. & Watson, L. E. (2007). An ITS phylogeny of tribe Senecioneae (Asteraceae) and a new delimitation of *Senecio* L. *Taxon* 56: 1077–1104.

Percy, D. M. & Cronk, Q. C. B. (2002). Different fates of island brooms: contrasting evolution in *Adenocarpus*, *Genista*, and *Teline* (Genisteae, Fabaceae) in the Canary Islands and Madeira. *Am. J. Bot.* 89: 854–864.

Pérez de Paz, J. & Caujapé-Castells, J. (2010). The allozyme genetic diversity of the Canarian flora: relationships with reproductive, biotic and abiotic factors. In *Proceedings of the International Symposium FloraMac2010*, pp. 47. Ponta Delgada, Azores, Portugal.

Pérez-Torrado, F. J. (1997). Análisis de la evolución geológica y la formación de los paisajes característicos de la isla de Gran Canaria. In P. L. Pérez de Paz (ed.), *Máster en gestión ambiental*, pp. 85–90. Santa Cruz de Tenerife, Gobierno Canario, Univ. de La Laguna y de Las Palmas de Gran Canaria.

Pétit, R. & Excoffier, L. (2009). Gene flow and species delimitation. *Trends Ecol. Evol.* 24: 386–393.

Price, J. P. & Clague, D. A. (2002). How old is the Hawaiian biota? Geology and phylogeny suggest recent divergence. *Proc. R. Soc. Lond. B* 269: 2429–2435.

Prohens, J., Anderson, G., Herraiz, F., Bernardello, G., Santos-Guerra, A., Crawford, D. J. *et al.* (2007). Genetic diversity and conservation of the endangered Canary Island endemics: *Solanum vespertilio* and *S. lidii* (Solanaceae). *Genet. Resour. Crop. Evol.* 54: 451–464.

Reyes-Betancort, J. A., Santos-Guerra, A., Guma, I. R., Humphries, C. J. & Carine, M. A. (2008). Diversity, rarity and the evolution and conservation of the Canary Islands endemic flora. *Anales del Jardín Botánico de Madrid* 65: 25–45.

Rieseberg, L. H., Raymond, O., Rosenthal, D. M., Lai, Z., Livingstone, K. *et al.* (2003). Major ecological transitions in wild sunflowers facilitated by hybridization. *Science* 301: 1211–1216.

Rieseberg, L. H., Wood, T. E. & Baak, E. J. (2006). The nature of plant species. *Nature* 440: 524–527.

Rihm, R., Jacobs, C. L., Krastel, S., Schmincke, H. U. & Alibes, B. (1998). Las Hijas seamounts-the next Canary Island? *Terra Nova* 10: 121–125.

Rodríguez-Sánchez, F. & Arroyo, J. (2008). Reconstructing the demise of Tethyan plants: climate-driven range dynamics of *Laurus* since the Pliocene. *Global Ecol. Biogeogr.* 17: 685–695.

Rognon, P. & Coudé-Gaussen, G. (1996). Paleoclimates off northwest Africa (28º-35º N) about 18,000 yr B. P. based on continental eolian deposits. *Quat. Res.* 46: 118–126.

Ronsted, N., Chase, M. W., Albach, D. C. & Bello, M. A. (2002). Phylogenetic relationships within *Plantago* (Plantaginaceae): evidence from nuclear ribosomal ITS and plastid *trnL-F* sequence data. *Bot. J. Linn. Soc.* 139: 323–338.

Rosato, M., Castro, M. & Rosselló, J. A. (2008). Relationships of the woody *Medicago* species (Section *Dendrotelis*) assessed by molecular cytogenetic analyses. *Ann. Bot.* 112: 15–22.

Sahuquillo, E. & Lumaret, R. (1995). Variation in the subtropical group *Dactylis glomerata* L. 1. Evidence from allozyme polymorphism. *Bioch. Syst. Ecol.* 23: 407–418.

Sahuquillo, E. & Lumaret, R. (1999). Chloroplast DNA variation in *Dactylis glomerata* L. taxa endemic to the Macaronesian islands. *Mol. Ecol.* 8: 1797–1803.

Sánchez-Doreste, J. L., Caujapé-Castells, J., Reyes Betancort, A. & Scholz, S. (2006). Population genetic variation of *Matthiola bolleana* (Brassicaceae), an endemic to the eastern Canary Islands of Fuerteventura and Lanzarote. *Plant Syst. Evol.* 262: 139–151.

Sanmartín, I., van der Mark, P. & Ronquist, F. (2008). Inferring dispersal: a Bayesian approach to phylogeny-based island biogeography, with special reference to the Canary Islands. *J. Biogeogr.* 35: 428–449.

Saunders, N. E. & Gibson, D. J. (2005). Breeding system, branching processes, hybrid swarm theory, and the humped-back diversity relationship as additional explanations for apparent monophyly in the Macaronesian island flora. *J. Ecol.* 93: 649–652.

Savolainen, V. & Reeves, G. (2004). A plea for DNA banking. *Science* 304: 1445.

Schiller, G., Korol, L., Ungar, E. D., Zehavi, A., Gil, S. L. & Climent, M. J. (1999). Canary Islands pine (*Pinus canariensis* Chr. Sm. ex DC.) 1. Differentiation among native populations in their isoenzymes. *Forest Genet.* 6: 257–276.

Schuster, M., Duringer, P., Ghienne, J. F., Vignaud, P., Taisso-MacKaye, H., Likius, A. *et al.* (2006). The Age of the Sahara Desert. *Science* 311: 281.

Schwarz, S., Klügel, A., van den Bogaard, P. & Geldmacher, J. (2005). Internal structure and evolution of a volcanic rift system in the eastern North Atlantic: the Desertas rift zone, Madeira archipelago. *J. Volcanol. Geoth. Res.* 141: 123–155.

Silvertown, J. (2004). The ghost of competition past in the phylogeny of island endemic plants. *J. Ecol.* 92: 168–173.

Silvertown, J., Carine, M. A. & Francisco-Ortega, J. (2005). The monophyly of island radiations: an evaluation of niche pre-emption and some alternative explanations. *J. Ecol.* 93: 653–657.

Smedmark, J. E. E., Swenson, U. & Anderberg, A. A. (2006). Accounting for variation of substitution rates through time in Bayesian phylogeny reconstruction of Sapotoideae (Sapotaceae). *Mol. Phylogenet. Evol.* 39: 706–721.

Stillman, C. J. (1999). Giant Miocene landslides and evolution of Fuerteventura, Canary Islands. *J. Volcanol. Geotherm. Res.* 94: 89–104.

Suárez-García, C., Pérez de Paz, J., Febles, R. & Caujapé-Castells, J. (2009). Genetic diversity and floral dimorphism in *Limonium dendroides* (Plumbaginaceae), a woody Canarian species on the way of extinction. *Plant Syst. Evol.* 280: 105–117.

Sunding, P. (1979). Origins of the Macaronesian flora. In D. Bramwell (ed.), *Plants and Islands*, pp. 13–40. London, Academic Press.

Susanna, A., Garnatje, T. & Garcia-Jacas, N. (1999). Molecular phylogeny of *Cheirolophus* (Asteraceae: Cardueae-Centaureinae) based on ITS sequences of nuclear ribosomal DNA. *Plant Syst. Evol.* 214: 147–160.

Swezey, C. (1999). A very old Sahara. *Science* 286: 239.

Trusty, J., Olmstead, R. G., Bogler, D. J., Santos-Guerra, A. & Francisco-Ortega, J. (2004). Using molecular data to test a biogeographic connection of the Macaronesian genus *Bystropogon* (Lamiaceae) to the New World: a case of conflicting phylogenies. *Syst. Bot.* 29: 702–715.

Urgeles, R., Canals, M., Baraza, J., Alonso, B. & Masson, D. (1997). The most recent mega-landslides of the Canary Islands: El Golfo debris avalanche and Canary debris flow, west El Hierro Island. *J. Geophys. Res.* 102: 20305–20323.

Valcárcel, V., Fiz, O. & Vargas, P. (2003). Chloroplast and nuclear evidence for multiple origins of polyploids and diploids of *Hedera* (Araliaceae) in the Mediterranean basin. *Mol. Phylogenet. Evol.* 27: 1–20.

van den Heede, J., Pajarón, S., Pangua, E. & Viane, R. L. L. (2004). *Asplenium ceterach* and *A. octoploideum* on the Canary Islands (Aspleniaceae, Pteridophyta). *Am. Fern J.* 94: 81–111.

van Valen, L. (1973). A new evolutionary law. *Evol. Theor.* 1: 1–30.

Vargas, P. (2007). Are Macaronesian islands refugia of relict plant lineages?: a molecular survey. In S. J. Weiss & N. Ferrand (eds.), *Phylogeography in Southern European Refugia: Evolutionary Perspectives on the Origins and Conservation of European biodiversity*, pp. 297–314. Dordrecht, Netherlands, Kluwer Academic Publishers.

Warwick, S. I., Sauder, C. A., al-Shehbaz, I. A. & Jacquemoud, F. (2007). Phylogenetic relationships in the tribes Anchonieae, Chorisporeae, Euclidieae, and Hesperideae (Brassicaceae) based on nuclear ribosomal ITS DNA sequences. *Ann. Mo. Bot. Gard.* 94: 56–78.

Whelan, F. & Kelletat, D. (2003). Submarine slides on volcanic islands – a source for mega-tsunamis in the Quaternary. *Progr. Phys. Geogr.* 27: 198–216.

Whittaker, R. J., Triantis, K. A. & Ladle, R. J. (2008). A general dynamic theory of oceanic island biogeography. *J. Biogeogr.* 35: 977–994.

# 13

# Endemism and evolution in Macaronesian and Mediterranean *Limonium* taxa

MARÍA DOLORES LLEDÓ, PER O. KARIS, MANUEL B. CRESPO, MICHAEL F. FAY AND MARK W. CHASE

The study of plant speciation on oceanic islands has improved enormously with the help of molecular systematics. Studies have targeted groups present on both the mainland and islands with the aim of understanding plant migration and evolution in isolation. In addition, relatively young volcanic islands give the opportunity to place the evolutionary process in a time frame, by dating molecular trees according to the age of the islands or by relying on the fossil record. Molecular phylogenetics can also be valuable in helping to reconstruct character evolution and understand the syndrome of characters diagnosing oceanic species.

Finding the perfect group to study is always a challenge. Islands too close to the continent may not be isolated and migration and gene flow may be too common to consider them independent units. Thus, the plant group needs to be chosen carefully. Island clades with distinctive traits (e.g. woodiness, different karyology) are especially interesting. Needless to say, the perfect scenario can be difficult to find, and we are left with systems less than perfect but not necessarily less interesting.

Taxa of the Canary Islands have been the focus of many evolutionary studies. Their geology and volcanic evolution are well documented (Juan *et al.*, 2000), and the islands form part of a biogeographic unit, Macaronesia, which also includes the Azores, Madeira, the Salvage or Selvagem Islands and the Cape Verde Islands (Sunding, 1979). In geographic terms, the Canary Islands are too close to the continent to be considered a truly oceanic archipelago. The closest island to western Africa, Fuerteventura, is only 110 km away (Juan *et al.*, 2000). Nevertheless, from a biological point of view, the archipelago shows high levels of endemism (40% according to Santos-Guerra, 1999), with a high percentage of woody taxa (70% of endemics; Aldridge, 1979) compared with the closest relatives from the continent. It has been argued that many Canarian endemics can be considered

*The Biology of Island Floras*, eds. David Bramwell and Juli Caujapé-Castells. Published by Cambridge University Press. 

palaeo-endemics that migrated in the Tertiary and Cretaceous (Sunding, 1979) and retain ancestral characters. Another line of evidence suggests that the retention of plesiomorphic features cannot be considered as proof of ancestry, and speciation within the islands is more recent (Humphries, 1979). Focussing on the flora rather than the fauna, the Canary Islands have affinities mainly with the Mediterranean region, although elements from more remote regions can also be found (Sunding, 1979).

*Limonium* Mill., the generic focus of this study, although mainly Mediterranean, is present in Macaronesia with representatives from three clades that can be considered products of at least three migrations. Molecular phylogenetic studies and DNA dating have been carried out for this group (Lledó *et al.*, 2005); their morphology is also well documented (Karis, 2004).

## Material and methods

The phylogenetic trees based on molecular characters and molecular clock are taken from Lledó *et al.* (2005). Information of plant material, sequencing and analysis can also be found there. The tree based on morphological characters is taken from Karis (2004).

## *Placing* **Limonium** *in context: family Plumbaginaceae*

Since publication of the first broadly sampled phylogenetic tree of flowering plants based on molecular data (Chase *et al.*, 1993), the family Plumbaginaceae has been undoubtedly placed within the broadly defined order Caryophyllales (APG II, 2003). Although mainly present in coastal or arid and saline zones of the Northern Hemisphere, members of Plumbaginaceae are native to all continents, excluding Antarctica (Kubitzki, 1993).

Phylogenetic studies of Plumbaginaceae based on plastid DNA sequences and morphology have shown that they are classified in two subfamilies, Plumbaginoideae and Staticoideae, well differentiated by morphological, chemical and molecular characters (Lledó *et al.*, 1998, 2001, 2005). Staticoideae are the largest and most diverse. More than 85% of the species belong in three genera, *Limonium*, *Armeria* and *Acantholimon*, and the remaining species belong to monotypic or small genera split from these (e.g. *Myriolimon*, Lledó *et al.*, 2003, 2005). Although mainly found in the Old World, especially around the Mediterranean basin, members of Plumbaginaceae are cosmopolitan. Migrations into the Southern Hemisphere have occurred several times, and the existence of genera with peculiar morphological features is well documented in Australia and South America. The genus *Aegialitis*, with two closely related species native of Southeast Asia and Australia, can be considered as the oldest migration into the Southern Hemisphere. It has reproductive features similar to subfamily Plumbaginoideae but chemical characters closer to Staticoideae, and it exhibits a large number of autapomorphies, probably developed as adaptations to life in mangrove swamps (Lledó *et al.*, 2001).

According to Lledó *et al.* (2005), this migration occurred about 16 mya. The migration of another Australian genus, *Muellerolimon*, is much recent (around 7 mya according to Lledó *et al.*, 2005). The colonisation of the Americas also occurred at different times. The Argentinean genus *Bakerolimon* is an older migration (around 8 mya), whereas species of *Limonium* related to *L. vulgare* present in both continents probably migrated during the late Miocene. Unexpectedly, the former genus *Afrolimon* from the Cape region of South Africa, although having many autapomorphies, is now considered a member of *Limonium* related to *L. vulgare* from Europe (Lledó *et al.*, 2005).

*Limonium* is the largest genus of Plumbaginaceae, although the number of species is difficult to establish. Their unusual reproductive strategies (apomixis, hybridisation) linked to the fragmented habitats where they grow (coastal cliffs and salt marshes) make this genus taxonomically complex (Erben, 1979; Pignatti, 1971, 1972: 302; Cowan *et al.*, 1998). Even establishing an infrageneric arrangement has proved to be difficult.

## A bit of history

Much of the taxonomic work done on *Limonium* has been restricted to sections or specific geographical areas. Since Boissier published his accounts for the family (1848, 1859), subsequent authors have followed his lead with small variations (Hooker, 1876; Pax, 1889). The former *Statice* (*Statice*, *nom. rej.* vs *Armeria*; Greuter *et al.*, 2000: 268) was divided into 13 sections belonging to two main groups: *'corolla polypetala'* with eight sections, and *'corolla gamopetala'* with four sections (for a comprehensive study of the *'corolla gamopetala'* group, see Lledó *et al.*, 2003, 2005). Only a handful of these sections are currently recognised as taxonomically stable units. Some of them, mainly from the *'corolla gamopetala'* group, are now accepted as separate genera (e.g. *Bakerolimon, Muellerolimon, Myriolimon*), whereas other sections have been reincorporated into *Limonium* (e.g. the former *Afrolimon*).

At least for European species, most authors follow the infrageneric arrangement published by Pignatti (1972) in *Flora Europaea*, in which *Limonium* was divided into three subgenera. *Limonium* subgen. *Myriolepis pro parte* is now accepted as a distict genus *Myriolimon* (Lledó *et al.*, 2003, 2005). *Limonium* subgen. *Pteroclados* (equivalent to the section with the same name) has about 20 taxa, mostly endemic to the Canary Islands. The largest subgenus, *L.* subgen.

*Limonium*, includes many agamospermous species. Relationships to other species and sections of *Limonium* have not been clearly established yet. Whether to name *Pteroclados* and *Limonium* as subgenera or sections is a matter of taste. Lledó *et al.* (2005) opted to accept them as subgenera, leaving taxa under this rank unclassified. From their point of view, both groups show enough morphological, chemical and molecular divergence to be accepted as subgenera. Nevertheless, as Karis (2004) pointed out, sections and series have not been formally combined within these two subgenera. For example, Boissier divided *L.* section *Pteroclados* in two series, *Odontolepidae* and *Nobiles*, that were never been formally recombined as sections under the subgenus. As this group is the focus of the present study, we have opted to informally call these two groups *'Odontolepidae'* and *'Nobiles'* without going any deeper into their taxonomic status. For *L.* subgenus *Limonium*, the taxonomic situation is even more complicated, and the proliferation of agamospermous species from the Mediterranean only confuses the matter more. It is not our aim to clarify this complex group here. When discussing relationships within subgenus *Limonium*, the sectional names correspond to Boissier's sections, unless otherwise stated.

Phylogenetic studies based on plastid sequences (*rbcL* and *trnL* intron and *trnL-F* intergene spacer – hereafter *trnL-F*) have clarified several questions, but left many unanswered (Lledó *et al.*, 2005). The genus *Limonium*, excluding former *L.* sect. *Myriolepis* – now recognised as genus *Myriolimon* – is monophyletic (Fig. 13.1). Three groups – *L.* subgen. *Pteroclados*, sections *Limoniodendron*, *Jovibarba* and *Ctenostachys* – form a grade relative to the rest of the genus (core *Limonium*). This grade is primarily found in northern and western Africa and Macaronesia. Sampling within core-*Limonium* is more sparse, and relationships within the large groups are more difficult to interpret.

## *Limonium* subgenus *Pteroclados*

This subgenus is sister to the rest of *Limonium*, and it is easily recognised by a syndrome of characters that includes the presence of winged stems and/or spikelets, pyxidiate fruit, a coloured extended calyx limb and pinnate leaf venation (Karis, 2004). Its monophyly has never been questioned, despite the fact that it lacks unique synapomorphies. All morphological characters are present in other genera or sections of *Limonium*. For example, the winged stems are also present in other Staticoideae (*Goniolimon, Psylliostachys* and *L.* sect. *Ctenostachys*). However the presence of these homoplasies in not closely related terminals can be safely considered as independent gains, and therefore they can be interpreted as local synapomorphies for each group (Karis, 2004). Chemically, *L.* subgen. *Pteroclados* shows a combination of characters also present in subfamily Plumbaginoideae and

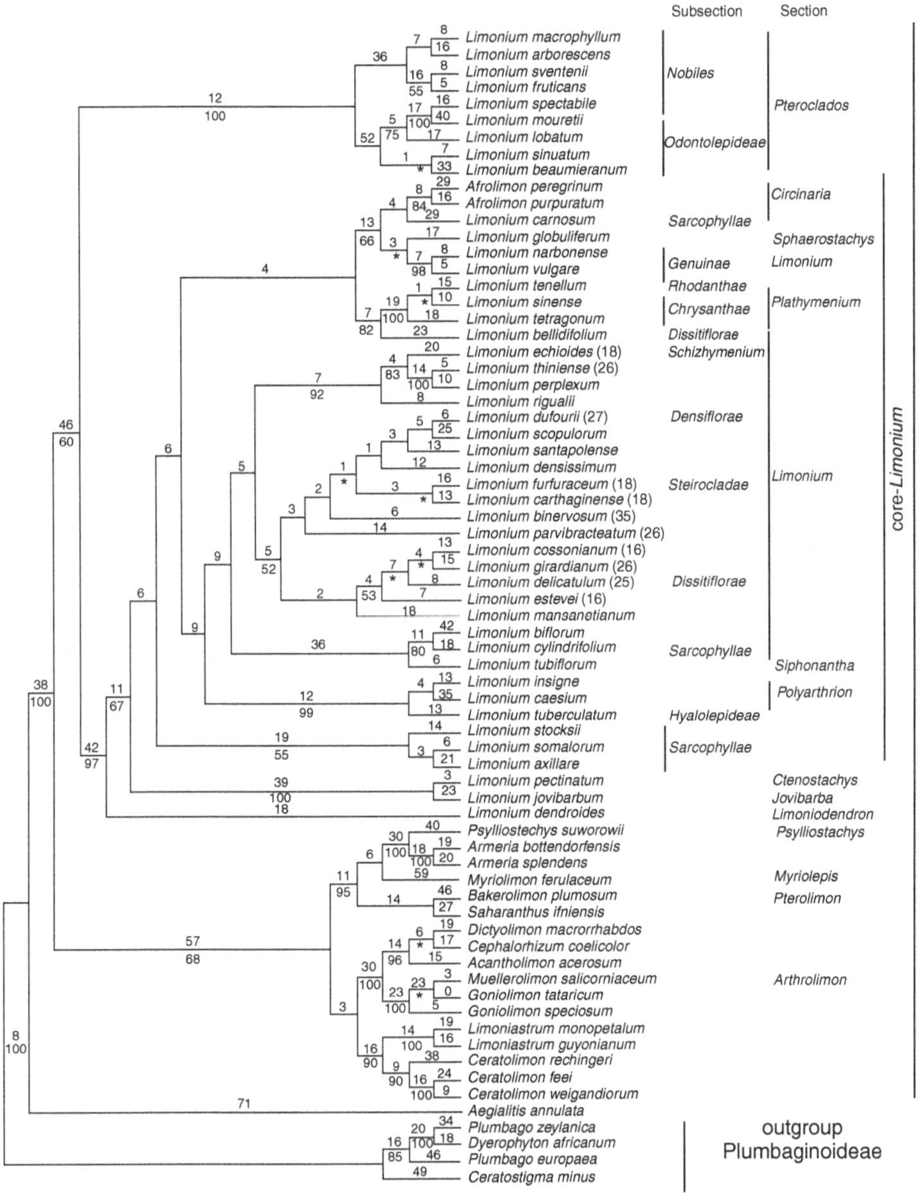


Figure 13.1 Phylogeny of Plumbaginaceae based on *trnL-F* and *rbcL* sequences. Taken form Lledó *et al.* (2005). Sectional names correspond mostly to Boissier's classification.

*Aegialitis*: it produces glycine betaine, but lacks β-alanine betaine, a chemical present in core-*Limonium* (Hanson *et al.*, 1994).

The two series of *L.* subgen. *Pteroclados* are well differentiated by morphological characters (Karis, 2004), i.e. conspicuous wing extensions at the stem nodes, an often two-pointed bract to the first spicule, β-prophylls with lateral appendages and a three-cleft apex (*'Odontolepidae'*), and diaspores with one to four spicules, apically entire α1 and β prophylls, and a white corolla (*'Nobiles'*). *'Odontolepidae'* contains two species, *L. sinuatum* (L.) Mill. and *L. lobatum* (L.f.) Kuntze, both widespread in the Mediterranean region. *Limonium Bonduelei* (T. Lestib.) Kuntze and *L. beaumierianum* Maire, both North African taxa, are sometimes accepted as subspecies of *L. sinuatum. Limonium mouretii* (Pit.) Maire is endemic to the Atlas mountains, Morocco, of North Africa and even if accepted as a member of *'Odontolepidae'*, it shares synapomorphies with both series. In Karis's morphological analysis (2004), this taxon was sister to the *'Nobiles'* clade, although this relationship was not strongly supported, and in Lledó *et al.* (2005), was included in *'Odontolepidae'* with strong support (Fig. 13.1). Series *'Nobiles'* contains about 18 taxa, all endemic to the Canary Islands. Most of them have narrow distributions, and in most cases they are restricted to a single island, except for *L. imbricatum* (Webb ex Girard) C. E. Hubb. which is present on both Tenerife and La Palma, and *L. puberulum* (Webb) Kuntze and *L. bourgaei* (Webb ex Boiss.) Kuntze, both native to Fuerteventura and Lanzarote (Febles & Pérez-Rodríguez, 2004). Ecologically they have broad habitat preferences, including coastal deserts to laurel and pine forests and lowland scrubs. The allopatric and fragmented distribution does not allow the production of natural hybrids. Hybrids, however, can be produced under controlled conditions.

Lledó *et al.* (2005) included nine species of *Pteroclados* in their analysis. Except for the position of *L. spectabile* (Svent.) G. Kunkel & Sunding, sister to *L. mouretii* in the *'Odontolepidae'*, both groups were monophyletic. In Karis's (2004) morphological analysis of the group, *L. spectabile* was placed in the *Nobiles* clade. In order to confirm this conflicting result from the molecular analyses, new DNA from *L. spectabile* plants growing at the Royal Botanic Gardens, Kew, was extracted, and new *rbcL* and *trnL-F* sequences were obtained. The methods for DNA extraction and sequencing are explained in Lledó *et al.* (2005). An analysis was performed including all *Pteroclados* species but reducing the terminals in other groups (e.g. including only one species of the other genera and reducing the number of terminals in core *Limonium*). The methodology is the same as used in Lledó *et al.* (2005). The new sequences confirm the position of *L. spectabile* within the *'Nobiles'* clade (Fig. 13.2). Its misplaced position as a member of the *'Odontolepidae'* clade can probably be regarded as a contamination or sample handling error in the previous analyses.

## The Atlantic grade

The next nodes in the cladogram (Fig. 13.1) are supported by distinctive morphological characters that separate them from core *Limonium*. In addition, *L.* sections *Limoniodendron, Jovibarbum* and *Ctenostachys* have their own autapomorphies and peculiarities, and all of them are endemic to Macaronesia or western Africa.

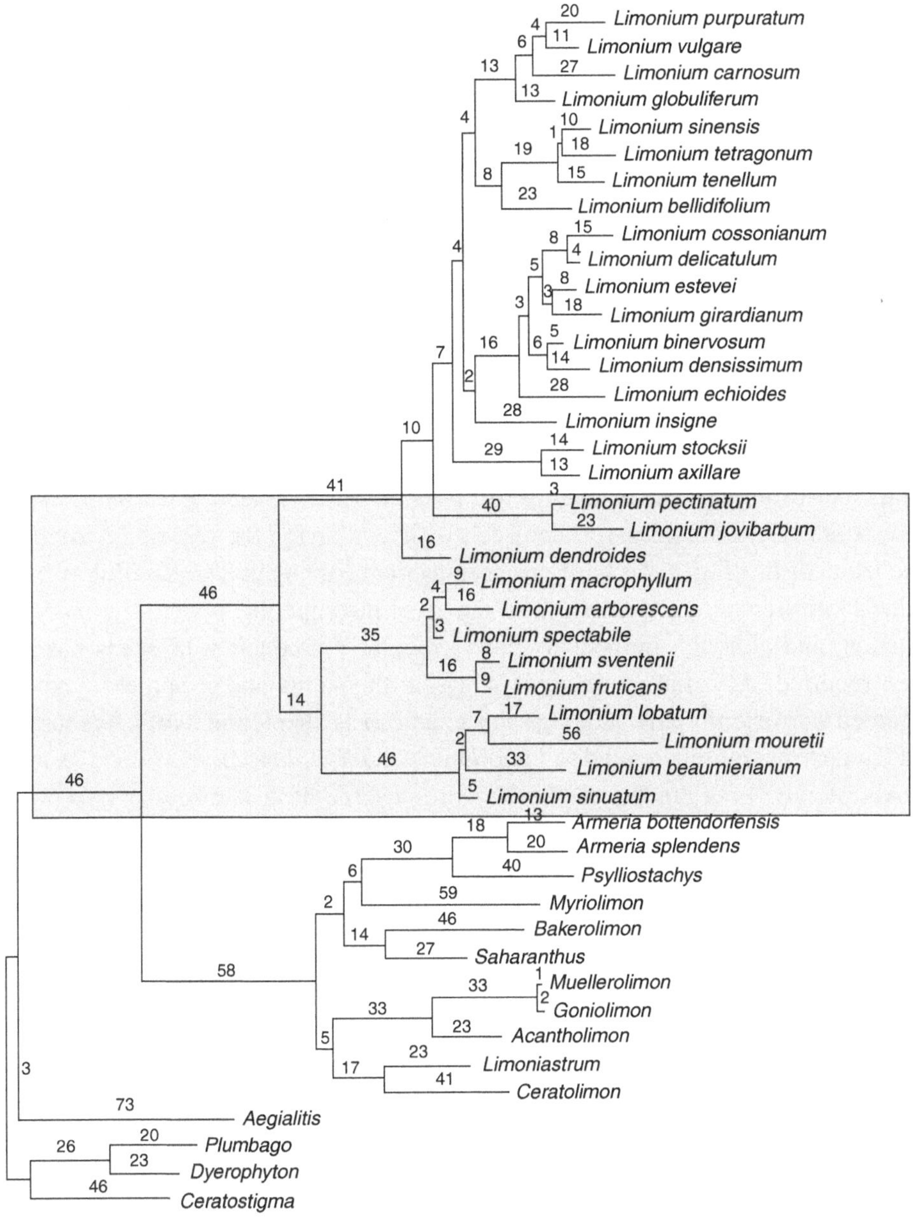


Figure 13.2 One of the most parsimonius trees found with the new *trnL-F* and *rbcL* sequences for *L. spectabile*. The Macaronesian clades are highlighted in the tinted box.

The first clade successively sister to the rest is composed of a single species *L. dendroides* Svent., known only from some inaccessible cliffs on La Gomera (Figs 13.1 and 13.2). It was described in 1960 by Sventenius, and similarly to *L.* subgen. *Pteroclados* it has an arborescent habit – reaching 3 m in height. It shows the same chemical patterns as the rest of the genus (Hanson *et al.*, 1994), but autapomorphies such as presence of salt-secreting glands in the spikelet instead of on the leaves and its phylogenetic isolation support its placement in a monotypic section *Limoniodendron*, as originally described by Sventenius (1960). The last clade of the Atlantic grade consists of *L.* sections *Jovibarba* (*L. jovibarbum* (Webb ex Boiss.) Kuntze) and *Ctenostachys* (Figs 13.1 and 13.2). *Limonium jovibarbum*, the only representative of *L.* sect. *Jovibarba*, is endemic to Cape Verde. Species of *L.* sect. *Ctenostachys* are native to Morocco, the Canary Islands, the Salvage or Selvagem Islands and Cape Verde. This clade is then sister to core *Limonium*, which includes the rest of the sections recognised by Bossier, although most of them have not been demonstrated to be monophyletic, with the exception of those currently recognised as independent genera.

Some characters of the Atlantic grade can be regarded as plesiomorphic for the genus. For example, taxa belonging to *Pteroclados*, *Jovibarba* and *Ctenostachys* have fruits with circumscissile dehiscence (irregular for the rest of *Limonium*). The fruits of *L. dendroides* have not been available thus far. The handful of individuals known from La Gomera show the same incompatible pollen–stigma combination, and although individuals flower regularly production of seeds has not been reported. Also the presence of wings in the stems and/or spikelets can be regarded as plesiomorphic, although this character is also found in other genera of Staticoideae not closely related to *Limonium* (e.g. *Goniolimon*).

All Atlantic taxa are diploids, with chromosome base numbers of $x = 7$ for '*Nobiles*', $x = 6$ or 8 for '*Odontolepidae*', $x = 9$ for *L. dendroides* and $x = 6$ for *L.* sect. *Ctenostachys* (Febles & Pérez-Rodríguez, 2004). For the rest of the genus, polyploidy is common, and base numbers $x = 6$, 8 and 9 can be found in multiple combinations to give a complete range of even and odd chromosome numbers. Hybrid populations that otherwise would be lost due to reproductive incompatibility are fixed due to apomixis and agamospermy. Karyotypes of the Atlantic and Mediterranean groups are different, even if their base number is the same. For example, in *L.* subgen. *Limonium*, the base number, $x = 8$, has a large metacentric chromosome absent in species with different base numbers. Erben (1979) suggested that this large chromosome was due to the fusion of two smaller chromosomes. However, this large metacentric chromosome is absent in *Pteroclados* taxa with $x = 8$, so it appears likely that $x = 8$ has been derived in more than one manner.

## Atlantic *versus* Mediterranean: the nature of the Macaronesian *Limonium* taxa

The composition of the Macaronesian flora has been extensively studied (Sunding, 1979). The level of endemism is high, especially in the Canary Islands, and the closest relatives of Macaronesian plant groups are diverse. The major component is Mediterranean (Sunding, 1979; Carine *et al.*, 2004). Examples of plant groups with Mediterranean affinities are abundant (e.g. *Argyranthemun*, Francisco-Ortega *et al.*, 1999; *Sideritis*, Barber *et al.*, 2002; *Androcymbium*, Caujapé-Castells *et al.*, 1999). A northern Atlantic element can be found in many pteridophytes. In addition, examples of relationships to plant groups from East Africa, the Sahara, South Africa and even East Asia have also been proposed (Andrus *et al.*, 2004). In this scenario, it is difficult to explain the diversity of the Macaronesian flora purely by long-distance dispersal and relatively recent migrations. The molecular data obtained for several plant groups, including *Limonium*, suggest a scenario in which Macaronesia has acted in many cases as refuges for a tropical flora that was present in Africa and Europe in the late Tertiary, although the data obtained for many groups suggest that many models of migration/colonisation and radiation in Macaronesia have occurred. Macaronesian species display a wide range of phytogeographic events, such as single colonisations (or *in situ* radiation; Vargas *et al.*, 1999; Francisco-Ortega *et al.*, 1996), multiple colonisations (Bohs & Olmstead, 2001; Carine *et al.*, 2004), dispersals among islands (Mort *et al.*, 2002), or even as sources for continental recolonisation (Goertzen *et al.*, 2002).

Lledó *et al.* (2005) estimated the ages of lineages within Plumbaginaceae using *L. dendroides* as a calibration point and produced a chronogram. This species is unique in many ways. Its phylogenetic isolation and presence on an island that has not undergone much volcanic activity since it was formed about 10 million years ago (mya; Juan *et al.*, 2000) made it the best choice for calibrating the phylogenetic tree, using the age of the island as a maximum age. According to this chronogram (Fig. 13.3, Table 13.1), speciation of *L.* subgen. *Pteroclados* took place around 7.5 mya. *Limonium jovibarbum* and *L. pectinatum* (Aiton) Kuntze were not included in this analysis as some of the data were missing, which could produce discrepancies in the analysis. It is more difficult to discuss the radiation within this group as only species from two islands were included. The chronogram was mainly produced to explain the main events that Plumbaginaceae have undergone, and it was not intended to examine in detail the Atlantic grade. When discussing ages of lineages in *Limonium* many things need to be taken into consideration. For example, if *L. dendroides* was older than La Gomera (e.g. the current distribution of *L. dendroides* could be a relict from a much wider area where now it is extinct), the dates given in this chronogram would be younger than in reality.

In the case of *Limonium*, speciation in Atlantic and Mediterranean groups has been driven by different factors. Diploid, woody, Atlantic taxa could be regarded as ancient groups, probably reminiscent of a more abundant group widespread when climatic conditions in Africa were more benign. Nevertheless, the presence of plesiomorphic characters does not prove an ancestral condition (Humphries, 1979).

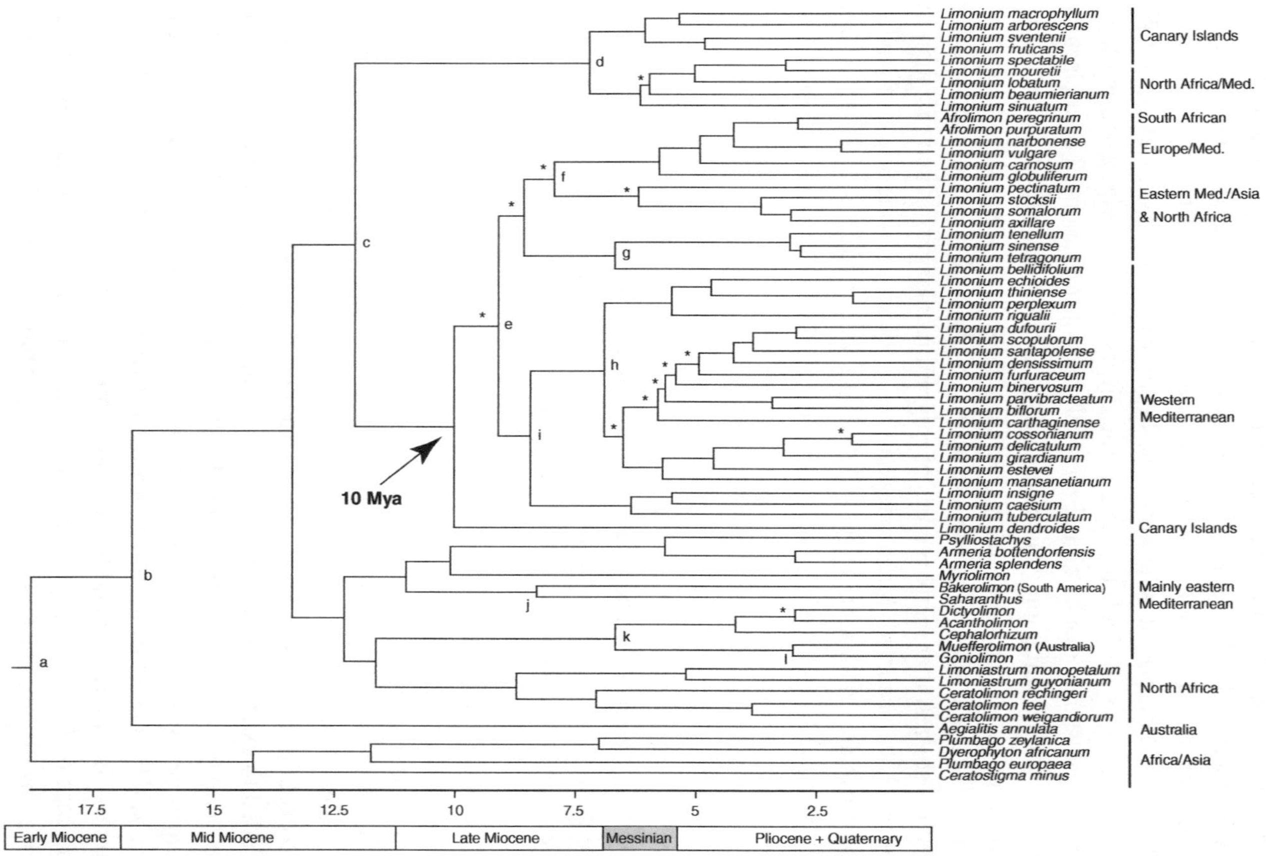


Figure 13.3 Chronogram of family Plumbaginaceae taken from Lledó *et al.* (2005). The ages of the marked nodes are shown in Table 13.1. Asterisks correspond to branches that collapse in the strict consensus tree.

Table 13.1 *Estimated ages of the marked nodes in Fig. 13.3. Source: (Lledó et al., 2005).*

| | Parsimony | Maximum likelihood |
|---|---|---|
| Node a | 18.84 | 16.94 |
| Node b | 16.68 | 15.99 |
| Node c | 12.09 | 11.95 |
| Node d | 9.10 | 9.06 |
| Node e | 7.16 | 6.74 |
| Node f | 7.89 | 8.50 |
| Node g | 6.65 | 7.46 |
| Node h | 8.42 | 8.88 |
| Node i | 6.68 | 7.18 |
| Node j | 11.02 | 11.60 |
| Node k | 8.31 | 9.55 |
| Node l | 6.59 | 7.54 |
| Node m | 2.94 | 3.56 |
| Node n | 8.71 | 9.47 |

In order to confirm the ages for the Macaronesian lineages, more data are needed, and, more importantly, the chronogram should be calibrated using a calibration point outside the clades that are the focus of this study.

Speciation in '*Nobiles*' is related to geographical isolation on islands from the Canarian Archipelago. In contrast, diversification in the Mediterranean groups of *L.* subgen. *Limonium* is more recent, according to the chronogram (Fig. 13.3) in the Messinian period about 5.5–4.5 mya (Hsü *et al.*, 1973, 1977; Bocquet *et al.*, 1978), when conditions in the Mediterranean basin were undergoing constant change due to fluctuations in sea level. Often polyploid populations of hybrid origin became isolated in newly emerged zones and were able to thrive as a result of agamospermy.

## References

Aldridge, A. E. (1979). Evolution within a single genus: *Sonchus* in Macaronesia. In D. Bramwell (ed.), *Plants and Islands*, pp. 279–291. London & New York, Academic Press.

Andrus, N., Trusty, J., Santos-Guerra, A., Jansen, R. K. & Francisco-Ortega, J. (2004). Using molecular phylogenies to test phytogeographical links between East/South Africa-southern Arabia and the Macaronesian islands: a review and the case of *Vierea* and *Pulicaria* section *Vieraeopsis* (Asteraceae). *Taxon* 53: 333–346.

APG II (2003). An update of the Angiosperm Phylogeny Group classification for the orders and families of flowering plants: APG II. *Bot. J. Linn. Soc.* 141: 399–436.

Barber, J., Francisco-Ortega, J., Santos-Guerra, A., Turner, K. G. & Jansen, R. J. (2002). Origin of Macaronesian *Sideritis* (Lamioideae: Lamiaceae) inferrend from nuclear and chloroplast sequence datasets. *Mol. Phylogenet. Evol.* 23: 293–306.

Bocquet, G., Wilder, B. & Kiefer, H. (1978). The Messinian Model – a new outlook for the floristics and systematics of the Mediterranean area. *Candollea* 33: 269–287.

Bohs, L. & Olmstead, R. G. (2001). A reassessment of *Normania* and *Triquera* (Solanaceae). *Plant Sys. Evol.* 228: 33–48.

Boissier, E. (1848). Plumbaginales. In: A.P. de Candolle (ed.), *Prodromus systematis naturalis regni vegetabilis* 12: 617–696. Paris, Treuttel et Wurz.

Boissier, E. (1859). *Diagnoses plantarum orientalium novarum.* Ser. 2, 4: 61–71. Genèva, Leipzig, Paris.

Carine, M. A., Russell, S. J., Santos-Guerra, A. & Francisco-Ortega, J. (2004). Relationships of the Macaronesian and Mediterranean floras: molecular evidence for multiple colonizations into Macaronesia and back-colonization of the continent in *Convolvulus* (Convolvulaceae). *Amer. J. Bot.* 91: 1070–1085.

Caujapé-Castells, J., Jansen, R. K., Pedrola-Monfort, J. & Membrives, M. (1999). Chloroplast restriction site phylogeny of the genus *Androcymbium* (Colchicaceae). *Syst. Bot.* 24: 581–597.

Chase, M. W., Soltis, D. E., Olmstead, R. G., Morgan, D., Les, D. H., Mishler, B. D. *et al.* (1993). DNA sequence phylogenetics of seed plants: an analysis of the plastid gene *rbcL*. *Ann. Missouri Bot. Gard.* 80: 528–580.

Cowan, R., Ingrouille, M. J. & Lledó, M. D. (1998). The taxonomic treatment of agamosperms in the genus *Limonium* Mill. (Plumbaginaceae). *Folia Geobotanica* 33: 353–366.

Erben, M. (1979). Karyotype differentiation and its consequences in Mediterranean *Limonium*. *Webbia* 34: 409–417.

Febles, R & Pérez-Rodríguez, E. (2004). Análisis cariotípico de *Limonium* Mill. sección *Pteroclados* Boiss. subsección *Nobiles* Boiss. (Plumbaginaceae). *Bot. Mac.* 25: 79–94.

Francisco-Ortega, J., Jansen, R. K. & Santos-Guerra, A. (1996). Chloroplast DNA evidence of colonisation, adaptative radiation, and hybridization in the evolution of the Macaronesian flora. *Proc. Natl. Acad. Sci. USA* 93: 4085–4090.

Francisco-Ortega, J., Santos-Guerra, A., Hines, A. & Jansen, R. J. (1999). Molecular evidence for a Mediterranean origin of the Macaronesian endemic genus *Argyranthemun (Asteraceae)*. *Amer. J. Bot.* 84: 1595–1613.

Goertzen, L. R., Francisco-Ortega, J., Santos-Guerra, A., Mower, J. P., Linder, C. R. & Jansen, R. K. (2002). Molecular systematics of the *Asteriscus* alliance (Asteraceae: Inuleae) II: combined nuclear and chloroplast data. *Syst. Bot.* 27: 815–823.

Greuter, W., McNeill, J., Barrie, F. R., Burdet, H. M., Demoulin, V., Filgueiras, T. S. *et al.* (eds.) (2000). *International Code of Botanical Nomenclature (Saint Louis Code).* Koeltz Scientific Books, Königstein, Germany [*Regnum Vegetabile* 138].

Hanson, A. D., Rathinassabapathi, B., Rivoal, J., Burnet, M., Dillon, M. O. & Gage, D. A. (1994). Osmoprotective compounds in the Plumbaginaceae: a natural experiment in metabolic engineering of stress tolerance. *Proc. Natl. Acad. Sci. USA* 91: 306–310.

Hooker, J. D. (1876). Plumbaginaceae. In G. Bentham & J. D. Hooker (eds.), *Genera Plantarum ad Exemplaria Imprimis in Herbariis Kewensibus Servata*, vol. 2, 623–628. London, Reeve & Co.

Hsü, J. K., Ryan, W. B. F. & Cita, M. B. (1973). Late Miocene desiccation of the Mediterranean. *Nature* 242: 240–244.

Hsü, J. K., Montadert, L., Bernoulli, D., Cita, M. B., Erickson, A., Garrison, R. E. *et al.* (1977). History of the Mediterranean salinity crisis. *Nature* 267: 399–403.

Humphries, C. J. (1979). Endemism and evolution in Macaronesia. In D. Bramwell (ed.), *Plants and Islands*, pp. 171–199. London, Academic Press.

Juan, C., Emerson, C. E., Oromí, P. & Hewitt, G. M. (2000). Colonization and diversification: towards a phylogeographic synthesis for the Canary Islands. *Trends Ecol. Evol.* 15: 104–109.

Karis, P. O. (2004). Taxonomy, phylogeny and biogeography of *Limonium* sect. *Pteroclados* (Plumbaginaceae), based on morphological data. *Bot. J. Linn. Soc.* 144: 461–482.

Kubitzki, K. (1993). Plumbaginaceae. In K. Kubitzki, J. G. Rohwer & V. Bittrich (eds.), *The Families and Genera of Vascular Plants*. Vol. 2, 523–530. Berlin, Springer.

Lledó, M. D., Crespo, M. B., Cameron, K. M., Fay, M. F. & Chase, M. W. (1998). Systematics of Plumbaginaceae based upon cladistic analysis of *rbcL* sequence data. *Syst. Bot.* 23: 21–29.

Lledó, M. D., Karis, P. O., Crespo, M. B., Fay, M. F. & Chase, M. W. (2001). Phylogenetic position and taxonomic status of the genus *Aegialitis* and subfamilies Staticoideae and Plumbaginoideae (Plumbaginaceae): evidence from plastid DNA sequences and morphology. *P. Syst. Evol.* 229: 107–124.

Lledó, M. D., Erben, M. & Crespo, M. B. (2003). *Myriolepis*, a new genus segregated from *Limonium* (Plumbaginaceae). *Taxon* 52: 67–73.

Lledó, M. D., Erben, M. & Crespo, M. B. (2005). *Myriolimon*, a new name for the recently described *Myriolepis*. *Taxon* 54: 811–812.

Mort, M. E., Soltis, D. E., Soltis, P. S., Francisco-Ortega, J. & Santos-Guerra, A. (2002). Phylogenetics and evolution of the Macaronesian clade of Crassulaceae inferred from nuclear and chloroplast sequence data. *Syst. Bot.* 27: 271–288.

Pax, F. (1889). Plumbaginaceae. In A. Engler & K. Prantl (eds.), *Die Natürlichen Pflanzenfamilien* Vol. 4, 116–125. Leipzig, W. Englemann.

Pignatti, S. (1971). Studi sui *Limonium*, VIII. In V. H. Heywood (ed.), *Florae Europaea. Notulae Systematicae ad Flora Europaeam Spectantes. Bot. J. Linn. Soc.* 64: 353–381.

Pignatti, S. (1972). *Limonium*. In T. G. Tutin, V. H. Heywood, N. A. Burges, *et al.* (eds.), *Flora Europaea* Vol. 3, 38–50. Cambridge, Cambridge University Press.

Santos-Guerra, A. (1999). Origen y evolución de la flora de Canarias. In J. M. Fernández-Palacios, J. J. Bacallado & J. A. Belmonte (eds.), *Ecología y Cultura en Canarias*, pp. 107–129. Santa Cruz de Tenerife, Organismo Autónomo de Museos y Centros. Cabildo de Tenerife.

Sunding, P. (1979). Origins of the Macaronesian flora. In D. Bramwell (ed.), *Plants and Islands*, pp. 13–40. London, Academic Press.

Sventenius, E. R. S. (1960). *Additamentum ad Floram Canariensem*. Madrid, Matritri Hispanicum Institutum.

Vargas, P., Morton, C. M. & Jury, S. L. (1999). Biogeographic patterns in Mediterranean and Macaronesian species of *Saxifraga* (Saxifragaceae) inferred from phylogenetic analysis of ITS sequences. *Amer. J. Bot.* 86: 724–734.

# 14

# Dispersal, diversity and evolution of the Macaronesian cryptogamic floras

ALAIN VANDERPOORTEN, BENJAMIN LAENEN, ROSALINA GABRIEL, JUANA M. GONZÁLEZ-MANCEBO, FRED J. RUMSEY AND MARK A. CARINE

The Macaronesian region comprises the volcanic, oceanic archipelagos of the Cape Verdes, the Canaries, the Savage or Selvagem Islands, Madeira and the Azores, located in the Atlantic Ocean between 15 and 30°N (Fig. 14.1). In keeping with other volcanic oceanic archipelagos, Macaronesia is characterised by high levels of endemicity and recent years have seen the rapid development of novel hypotheses to explain aspects of the evolution of diversity in the region. The availability of checklists for the biota for each of the Macaronesian archipelagos (Izquierdo *et al.*, 2004; Arechavaleta *et al.*, 2005; Borges *et al.*, 2005a, 2008) has led to the development of new hypotheses to explain diversity patterns in the flora in relation to abiotic factors (Emerson & Kolm, 2005; Whittaker *et al.*, 2007, 2008; Borges & Hortal, 2009). Silvertown (2004; see also Silvertown *et al.*, 2005; and for other views, Herben *et al.*, 2005; Saunders & Gibson, 2005) reviewed the results of phylogenetic analyses and suggested that 'niche pre-emption' may explain the radiation of some groups but not others. Carine *et al.* (2010) also used phylogenetic data to demonstrate that angiosperm groups which showed island woodiness (i.e. the evolution of woodiness linked to the colonisation of islands) were typically more species-rich than groups with the same habit on both continent and islands and hypothesised that the evolution of woodiness promoted diversification. Data from checklists together with phylogenetic information led Carine and Schaefer (2010) to suggest that differences in the climatic histories of the Canaries and Azores may explain the marked differences observed in patterns of endemicity in the angiosperm floras of the two archipelagos.

The angiosperm flora of Macaronesia has not only served as an important focus for the recent development of hypotheses to explain the origins of island diversity but has been an important focus for biodiversity research since the Enlightenment

*The Biology of Island Floras*, eds. David Bramwell and Juli Caujapé-Castells. Published by Cambridge University Press. 

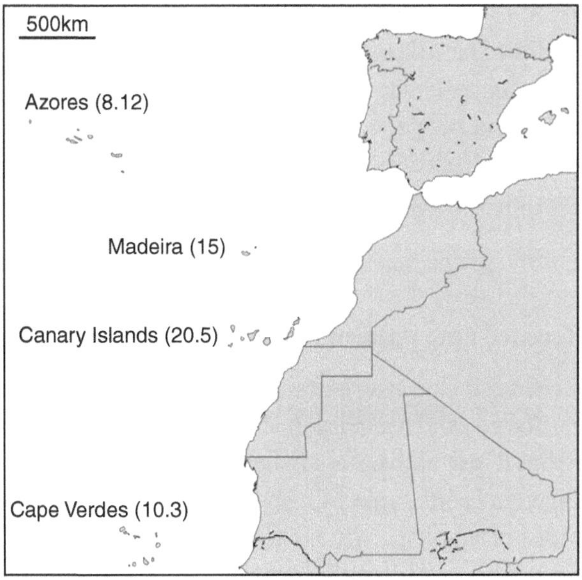


Figure 14.1 The Macaronesian region. Maximum ages for each archipelago (in mya) are given in parentheses.

period (Francisco-Ortega *et al.*, 2010). In contrast, the cryptogamic flora has received much less attention. In part, this is the result of the 'island laboratory' paradigm that has tended to focus the attention, at least of island evolutionary biologists, on those groups that have undergone evolutionary radiations on islands (e.g. arthropods, terrestrial molluscs and vascular plants), a phenomenon largely absent from the cryptogamic flora. Nevertheless, cryptogamic plants constitute a significant component of Macaronesian ecosystems, both in terms of numbers and biomass. In the Azorean hyper-humid native forests, for example, bryophyte diversity can reach more than 25 species within 30 × 30 cm plots, which is comparable to diversity patterns traditionally found in tropical rainforests (Gabriel & Bates, 2005). The region itself is recognised as a globally important centre for both pteridophyte and bryophyte diversity (Sérgio *et al.*, 1995; Hallingbäck & Hodgetts, 2000). Furthermore, given the differences in the biology of cryptogams and angiosperms, cryptogamic plants may offer novel insights into the processes underlying the evolution of insular plant diversity.

In this paper, we focus specifically on the archipelagos of the Canary Islands, Madeira and Azores, and highlight three areas in which the patterns of cryptogamic plant diversity differ markedly from angiosperm diversity patterns: (1) regional levels of diversity and the extent of evolutionary radiations; (2) the floristic relationships of the archipelago floras; and (3) the distribution of sister taxa of Macaronesian endemic lineages. We consider these differences in the context of hypotheses developed to explain the evolution of the Macaronesian biota. The patterns observed highlight the role of both extinction and dispersal in shaping extant distributions and we suggest that intense dispersal in cryptogamic groups, both between archipelagos and between continents and the archipelagos, may be a key factor in explaining the differences observed between angiosperms and cryptogams. Finally, we review molecular studies of cryptogams to support this hypothesis.

## Difference one: patterns of diversity and endemism

Table 14.1 summarises the characteristics of the cryptogamic and angiosperm floras of the Canaries, Azores and Madeira. It shows that for angiosperms and mosses, the Canaries have the richest native flora, whereas for pteridophytes and liverworts, Madeira is most diverse. From Table 14.1 it is also apparent that, for each archipelago, the total number of angiosperms exceeds the number of species within each of the cryptogamic groups considered except in the case of the Azores, where the moss flora is considerably more diverse than the angiosperm flora.

Levels of endemism in the native angiosperm floras range from 22.5% for Madeira to 45.5% for the Canaries (Table 14.1). Among cryptogams, endemism is much lower, ranging from 0.7% in the case of Canary Island liverworts to 9.7% in the case of Madeiran pteridophytes. The bryophyte floras exhibit very low levels of endemism in comparison with both pteridophytes and angiosperms, with Azorean liverworts (1.8%) showing the highest level of endemism among the bryophyte groups. Madeira harbours the highest levels of endemic pteridophyte diversity (9.7%) and the Canaries exhibits the lowest (5.5%), a pattern that is in marked contrast to that observed in angiosperms.

Table 14.2 summarises levels of regional endemism in the floras of the archipelagos, i.e. for each archipelago, the proportion of taxa that occur on at least two of the three archipelagos considered here (Canaries, Madeira and Azores) but do not occur outside of those archipelagos. Such taxa constitute a small proportion of the angiosperm floras (between 4.6–8.4%, Table 14.2) and one that is much smaller than the proportion of archipelago endemics in the flora (22.5–45.5%, Table 14.1). In contrast, regional endemics (2.4–12.9%; Table 14.2) constitute a larger proportion of the native cryptogamic floras than do archipelago endemics (0.7–9.7%; Table 14.1).

The intra-archipelago distribution of endemic diversity also differs between angiosperms and cryptogams. Recent analyses of patterns of endemicity in the Canary Islands angiosperm flora highlighted the highly localised concentration of endemic angiosperms (Reyes-Betancort *et al.*, 2008), facilitating the recognition of intra-island areas of endemism (Carine *et al.*, 2009). The restricted distributions of Canary Island

Table 14.1 *Rates of strict endemism (i.e. species that are strict endemics for single archipelagos) for the Canary Islands, the Azores and Madeira: comparative data for mosses, liverworts (including hornworts), pteridophytes and angiosperms, expressed as a percentage of the native flora.*

| | Canary Islands | Azores | Madeira |
|---|---|---|---|
| Mosses, endemics/total/% endemism | 6/353/1.7 | 5/292/1.7 | 6/346/1.7 |
| Liverworts, endemics/total/% endemism | 1/146/0.7 | 3/163/1.8 | 3/180/1.6 |
| Bryophytes, endemics/total/% endemism | 7/499/1.4 | 8/455/1.7 | 9/526/1.7 |
| Pteridophytes, endemics/total/% endemism | 3/57/5.3 | 5/53/9.4 | 6/62/9.7 |
| Angiosperms, endemics/total/% endemism | 567/1246/45.5 | 70/197/35.5 | 145/645/22.5 |

Data from González-Mancebo *et al.*, 2008, 2009; Sérgio & González-Mancebo, 2009, for the Canary Islands, Sérgio *et al.*, 2008, Sim-Sim *et al.*, 2008 for Madeira and Gabriel *et al.*, 2005 and Gabriel (unpubl. data) for the Azores. The endemics *Cyrto-hypnum montei, Fissidens luisieri, Thamnobryum rudolphianum, T. fernandesii, Tylimanthus azoricus* and *T. madeirensis*, were excluded for reasons detailed in the text. The non-endemics *Jungermannia caespiticia, Lejeunea mandonii, Plagiochila denticulata, Porella pinnata, Petalophyllum ralfsii, Preissia quadrata, Bryoxiphium norvegicum, Bryum kunzei, Campylopus subulatus, Drapanocladus aduncus, Entosthodon fascicularis, Eurhynchiastrum pulchellum, Fissidens monguillonii, Haplocladium virginianum, Pleuridium subulatum, Pohlia elongata* and *Zygodon rupestris* were included in the Azores checklist based upon Gabriel (unpubl. data). *Aloina ambigua* (=*A. aloides*), *Bryum muehlenbeckii* (=*B. subapiculatum*), *Chiloscyphus pallescens* (=*C. polyanthos*), *Dichodontium flavescens* (=*D. pellucidum*), *Dicranum canariense* (=*D. scottianum*), *Dialytrichia fragilifolia* (=*D. mucronata*), *Didymodon insulanus* (=*D. vinealis*), *H. stellata* (=*H. ciliata*), *Heterocladium flaccidum* and H. *wulfsbergii* (=*H. heteropterum*), *Hygroamblystegium* spp. (=*H. varium*), *Hypnum andoi* (=*H. cupressiforme*), *Plagiothecium succulentum* (=*P. nemorale*), *Rhynchostegiella litorea* (=*R. tenella*), *Sphagnum inundatum* (=*S. auriculatum*) and *Thamnobryum maderense* were excluded from the checklists because the circumscription and/or acceptation of those species has been an area of controversy or because those species were recently described or re-erected at the species level. As a result, their occurrence and distribution in the different archipelagos remain difficult to ascertain and we decided to conservatively exclude them from the lists. *Radula complanata*, included with doubts in the Madeiran checklist (Sérgio *et al.*, 2008), was excluded based on most recent phylogenetic evidence (Laenen *et al.*, 2011).

endemic angiosperms is further illustrated by the PKI (per-K-island) curve for the endemic flora which is strongly right skewed because most endemics are restricted to a single island (Fig. 14.2e). Such a pattern is common across the endemic angiosperm floras of oceanic islands more generally (Carine & Schaefer, 2009) and indeed the only documented exception to date is the Azores where the flora shows a left-skewed

Table 14.2 *Rates of Macaronesian endemism (i.e. species that are restricted to Macaronesia and occur in two or three archipelagos, excluding single-archipelago endemics and Cape Verde) within the Canary Islands, the Azores, Madeira and Macaronesia: comparative data for mosses, liverworts (including hornworts), pteridophytes and angiosperms.*

| | Canary Islands | Azores | Madeira | Macaronesia s.s. |
|---|---|---|---|---|
| Mosses, endemics/total/% endemism | 11/353/3.1 | 7/292/2.4 | 13/346/3.6 | 13/484/2.7 |
| Liverworts, endemics/total/% endemism | 8/146/5.5 | 5/163/3.1 | 9/180/5 | 10/227/4.4 |
| Bryophytes, endemics/total/% endemism | 19/499/3.8 | 12/455/2.6 | 22/526/4.2 | 23/711/3.2 |
| Pteridophytes, endemics/total/% endemism | 5/57/8.7 | 6/53/11.3 | 8/62/12.9 | 9/88/10.2 |
| Angiosperms, endemics/total/% endemism | 57/1246/4.6 | 8/197/4.1 | 54/645/8.4 | |

distribution, indicating that most endemics are widespread across the archipelago (Fig. 14.2e; see also Borges *et al.*, 2005b).

PKI curves for archipelago endemic bryophytes (Fig. 14.2a–c) show that, in contrast to most archipelago endemic angiosperm floras, endemic species either tend to be restricted to single islands or are widespread across islands within an archipelago. PKI curves for regional endemic cryptogams are left skewed (Fig. 14.3), indicating that such taxa have wide distribution ranges in terms of number of islands occupied, a pattern comparable with that observed in Azorean endemic angiosperms (Fig. 14.2e).

A further difference between diversity among Macaronesian cryptogams and angiosperms concerns the extent of intra-regional evolutionary radiations. In angiosperms, Carine *et al.* (2010) noted that 75% of endemic angiosperm lineages studied using molecular data contained two or more species and there are numerous examples of endemic groups that have undergone extensive intra-regional radiation following colonisation of the region (e.g. *Argyranthemum*, the *Aeonium* alliance and Macaronesian *Echium*).

In bryophytes, Vanderpoorten and Long (2006) reported that 77% of Macaronesian endemic bryophytes are single-genus endemics, i.e. are the sole representative of the

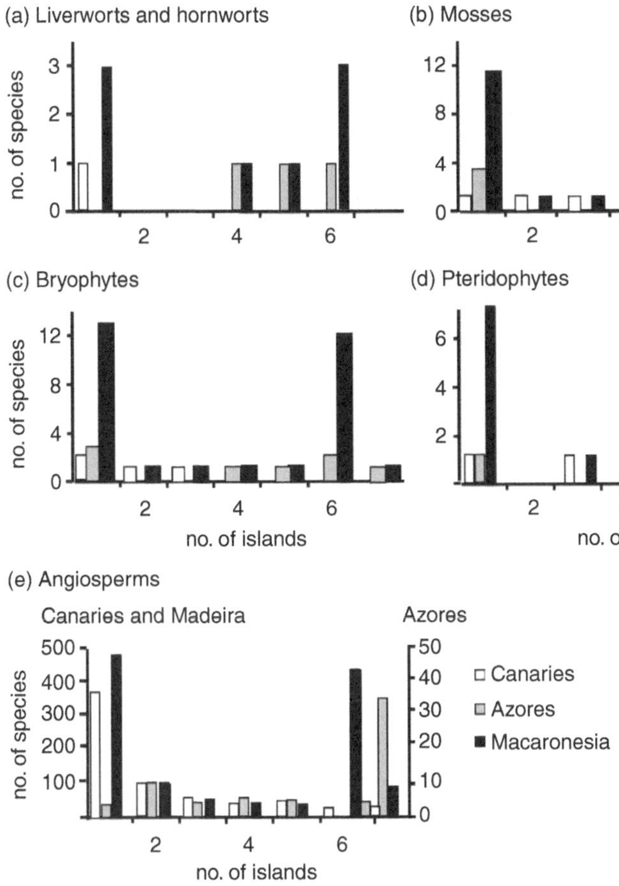


Figure 14.2 Number of islands occupied by strict archipelago endemic species (i.e. endemic species to a single archipelago): comparison between (a) liverworts and hornworts; (b) mosses; (c) all bryophytes; (d) pteridophytes; and (e) angiosperms.

genus in the region. Furthermore, among the few bryophyte genera thought to include more than one endemic, recent taxonomic and phylogenetic evidence suggests that, in some cases, Macaronesian endemics are conspecific with widespread species (*Tylimanthus*; Burghardt & Gradstein, 2008) or that the taxa have a polyphyletic origin in the region, with the extent of radiation in the region therefore more limited than previously thought (*Echinodium*, wherein only three of four Macaronesian endemics are a monophyletic group, Stech *et al.*, 2008; *Rhynchostegiella*, wherein only two of three endemics are sister taxa, Aigoin *et al.*, 2009a).

In pteridophytes, at the regional scale, the only lineages containing more than one endemic, or near endemic species (i.e. with one or few occurrences in continental enclaves) show very limited radiation with taxa partitioned between

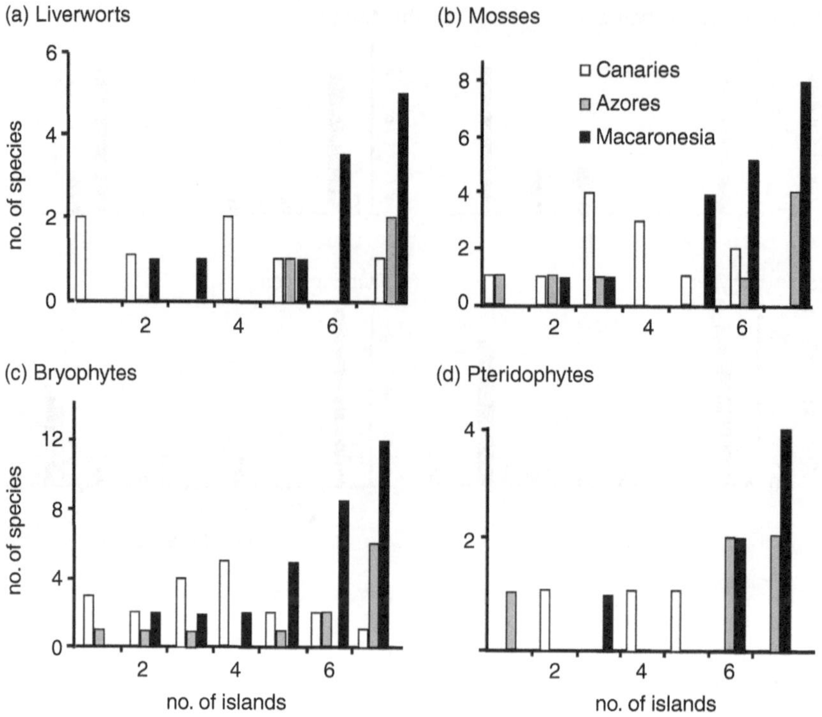


Figure 14.3 Number of islands occupied by regional Macaronesian endemic species (i.e. Macaronesian species that occur on two or three archipelagos): comparison between (a) liverworts; (b) mosses; (c) all bryophytes; and (d) pteridophytes.

archipelagos (e.g. *Polypodium macaronesicum* (Canaries, Madeira)/*P. azoricum* (= *P. macaronesicum* subsp. *azoricum*) (Azores)) or speciation through hybridisation between taxa which have colonised the region independently (e.g. *Hymenophyllum maderense*, Gibby & Lovis, 1989; *Asplenium* subgenus *Ceterach*, van den Heede *et al.*, 2004), or a combination of both (e.g. *Dryopteris*, Gibby *et al.*, 1977; Gibby, 1983, 1985). Allopolyploidy has played an important role in pteridophyte speciation in the region. It is notable that, while allopolyploidy is prevalent among many angiosperm groups globally, it is less significant among Macaronesian angiosperms as evidenced by the limited variation in ploidy level observed among congeneric endemic taxa (e.g. Baltisberger & Widmer, 2006).

In comparison with angiosperms, the taxonomy of the cryptogamic groups may be considered less well worked. This suggests that the species-level diversity has not yet been fully described and catalogued and that the knowledge of species' global, regional and even local distributions is still insufficient. Thus, one possible explanation for the extremely low rates of endemism and evolutionary radiation in the bryophyte and, to a lesser extent, pteridophyte floras of the Macaronesian archipelagos is that additional Macaronesian endemic taxa remain to be described or re-assessed.

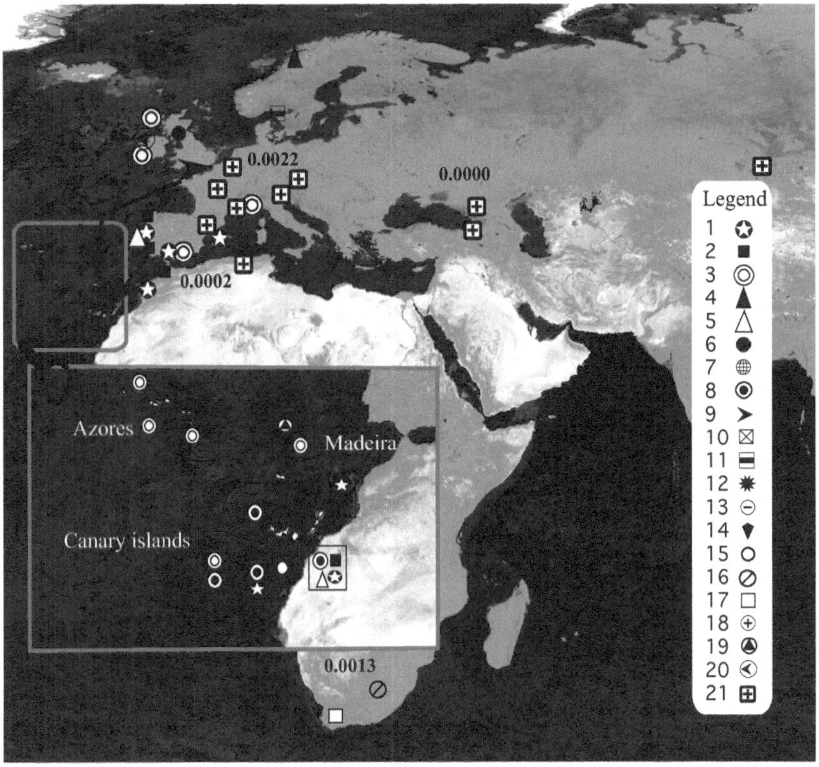


Figure 14.4 Nucleotidic diversity (numbers) and dispersal events (plain arrows = colonisation of Macaronesia from a European ancestor; dashed arrows = back-colonisation of the European continent from the Macaronesian refugium) in the liverwort *Radula lindenbergiana* in the Azores, Madeira, the Canaries, Iberian Peninsula and mid-western Europe, inferred from the analysis of the variation of a representative sampling of specimens genotyped at four cpDNA regions. Circles correspond to different haplotypes. Instances of colonisation of Macaronesia and back-colonisation to the continents were inferred from a reconstruction of ancestral distribution areas on the phylogeography (from Laenen *et al.*, unpubl.).

Within bryophytes, the recent transfer of the Madeira endemic *Brachythecium percurrens* to the monotypic endemic genus *Hedenasiastrum* (Aigoin *et al.*, 2009b) suggests that the highly divergent nature of some lineages remains to be discovered while several endemic species have been described during the past 20 years: e.g. *Aloina humilis* (Gallego *et al.*, 1998) and *Orthotrichum handiense* (Lara *et al.*, 1999) from the Canary Islands and *Frullania sergiae* (Sim-Sim *et al.*, 2000) and *Riccia atlantica* (Sérgio & Perold, 1992) from Madeira. Recent phylogenetic and chemical work also 'resurrected' the Madeiran endemic liverwort *Plagiochila madeirensis* (Rycroft *et al.*, 2004), which had been reduced to synonymy on morphological grounds. *Neckera intermedia* was identified as a Macaronesian endemic since continental specimens attributed to that species have been accommodated within recently described *N. baethica* (Draper *et al.*, 2011). Furthermore, other

Macaronesian endemic taxa have been resolved as monophyletic in molecular phylogenetic analyses (*Homalothecium mandonii*, Huttunen *et al*., 2008; *Leptoscyphus azoricus*, Vanderpoorten & Long, 2006) but, owing to the lack of any morphological synapomorphies, they have either not been recognised as distinct pending further taxonomic investigations (Huttunen *et al*., 2008) or they have been treated as subspecies (Vanderpoorten *et al*., 2010). In the liverwort *Radula lindenbergiana*, molecular investigations recently revealed the existence of multiple endemic haplotypes in Macaronesia despite the lack of any apparent morphological differentiation (Fig. 14.4). Bryophytes exhibit reduced morphologies in comparison with angiosperms and their diversification is not necessarily paralleled by morphological differentiation. This phenomenon, known as cryptic speciation, has increasingly been reported in bryophytes (see Vanderpoorten *et al*., 2010 for review) and might offer, to a certain extent, an explanation for the apparent low rates of radiation among Macaronesian mosses.

The increase in the diversity of the Macaronesian endemic bryophyte flora resulting from the discovery of new taxa (including cryptic taxa) is counterbalanced both by the discovery that species previously regarded as endemics also occur outside the region, specifically along the Atlantic fringe of the European and African continents (*Tetrastichium fontanum*, Rumsey & Vogel, 1999; *T. virens*, Gallego *et al*., 2004; *Thamnobryum maderense*, Jimenez *et al*., 2000), and by the reduction to synonymy of many endemics in light of recent phylogenetic evidence and taxonomic revisions (e.g. *Platyhypnidium torrenticola*, Werner *et al*., 2007; *Entosthodon krausei*, Bruges & Sérgio, 2010; *Tylimanthus azoricus* and *T. madeirensis*, Burghardt & Gradstein, 2008; *Fissidens luisieri*, Werner *et al*., 2009). The Madeiran endemic *Cyrto-hypnum montei* is suspected to be conspecific with *Thuidiopsis sparsa* and probably of introduced origin with ornamental plants (Touw, 2001). The inclusion of the endemic mosses *Thamnobryum fernandesii* and *T. rudolphianum* within a clade of the widespread *T. alopecurum* (Olsson *et al*., 2009) further suggests that these two species should be considered as aquatic forms of the latter. Three other endemics, namely *Sphagnum nitidulum, Bryoxyphium maderense* and *Trematodon perssoniorum*, are of doubtful taxonomic status and are likely to be conspecific with the widespread *S. rubellum* and *B. norvegicum* (B. Goffinet, pers. comm.) and *T. longicollis* (Rumsey, unpubl. data), respectively.

Critical study of several pteridophyte genera, initially from a cytological perspective by Manton and co-workers, revealed several intriguing examples of intra-regional variation in cytotype distribution. *Adiantum reniforme*, for example, was found to be tetraploid in the Canaries and decaploid in Madeira. *Asplenium aethiopicum sensu lato* is hexaploid in the Canaries and dodecaploid in Madeira (Manton *et al*., 1986). Other taxa exhibit different cytotypes in Macaronesia and on the continent (e.g. *Arachniodes webbianum*, Gibby *et al*., 1992). Molecular phylogenetic studies

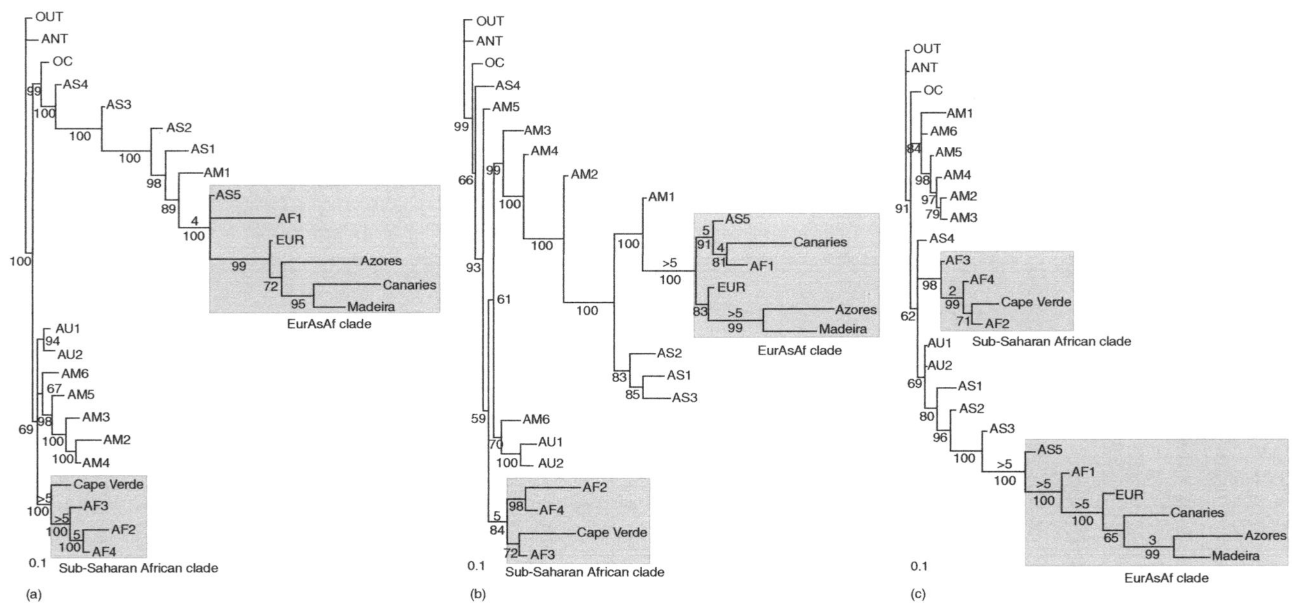


Figure 14.5 Floristic affinities of the Macaronesian liverwort (a), moss (b) and pteridophyte (c) floras inferred from the analysis of a data matrix documenting the presence of each species present on each archipelago in Macaronesia together with its occurrence within each of the twenty world biogeographic regions defined by van der Wijk *et al*. (1959): Europe (EUR), northern Africa (AF1), continental sub-Saharan Africa (AF2), Mascarene Islands (AF3), southern Africa (AF4), northern Asia (AS1), eastern Asia (AS2), southern Asia (AS3), southwestern Asia (AS4), western Asia (AS5), North America (AM1), Central America (AM2), Caribbean islands (AM3), northern South America (AM4), Brazil (AM5), southern South America (AM6), Australia (AU1), New Zealand (AU2), Antarctica (ANT), and Oceania (OC). The trees are the 50% majority-rule consensus sampled from the posterior probability distribution from a Bayesian analysis implementing a model of species 'gains' and 'losses' among geographic areas. Numbers below the branches correspond to their posterior probabilities (adapted from Vanderpoorten *et al*., 2007).

are now necessary to elucidate the evolutionary history of these taxa within the region, answering the critical question: are these distinct lineages the product of radiation within the region, or the result of independent colonisations?

Despite some taxonomic uncertainty, it is clear that differences in diversity patterns between angiosperms and cryptogams are not simply attributable to taxonomic artifact alone. The evolution of endemic lineages and the intra-regional radiation of those lineages have been much less common in cryptogams than in angiosperms and, in the case of pteridophytes, has involved different evolutionary processes (allopolyploidy). The distribution of diversity across the region differs from group to group. It is notable that while the Canarian flora is the most diverse and endemic rich for angiosperms, this is not the case for all groups, with the Madeiran flora, for example, the most diverse for pteridophytes and liverworts and the most endemic-rich archipelago for liverworts. Furthermore, while angiosperms dominate the floras of the Canaries and Madeira in terms of species numbers, this is not the case in the Azores where mosses constitute the most diverse group (Borges *et al.*, 2005b).

## Difference two: floristic relationships

Vanderpoorten *et al.* (2007) recently examined the relationships of the Macaronesian archipelago cryptogamic floras in a global context. The results of their analyses (Fig. 14.5) demonstrated that for all cryptogamic groups examined (mosses, liverworts and pteridophytes), the Cape Verdes flora is more closely related to the flora of tropical Africa than to floras of the other Macaronesian archipelagos. This is perhaps not surprising, given that the Cape Verdes are remote from the rest of Macaronesia and much closer to tropical Africa than to any of the other archipelagos in the region, and a similar relationship was proposed for angiosperms by Lobin (1982).

A strong affinity between Europe and the Macaronesian islands was also evident for all three cryptogamic groups analysed by Vanderpoorten *et al.* (2007). For both pteridophytes and liverworts, the three northern archipelagos were resolved as a group (Macaronesia s.s.) that was placed as sister to Europe; in the case of the mosses, Madeira and the Azores were resolved as a group sister to Europe with the Canary Islands resolved as sister to North Africa. The relationship between the archipelagos and Europe is supported in all cases by an Atlantic fringe element, i.e. a suite of taxa that occur only on the archipelagos and the western seaboard of Europe. A number of liverwort species exhibit such a distribution (see Vanderpoorten *et al.*, 2007 for details) as do the pteridophytes, *Ophioglossum azoricum* and *Hymenophyllum wilsonii*. Several further pteridophytes have distributions heavily centred on Macaronesia but with disjunct populations on the continent, many co-occurring with other largely Macaronesian cryptogams in small climatically defined areas which have been considered to be Macaronesian enclaves (Rumsey & Vogel, 1999). The most abundant and widespread Macaronesian pteridophyte species show varying degrees of continental disjunction: *Diplazium caudatum* only occurs in the sierras just north of Algeciras in southernmost Spain, where it is associated with several other Macaronesian species, notably the mosses *Tetrastichium fontanum*, (Rumsey & Vogel, 1999) and *T. virens* (Gallego *et al.*, 2004) and the pteridophytes *Culcita macrocarpa*, *Christella dentata*, *Davallia canariensis*, *Dryopteris guanchica* (possibly now locally extinct; Fraser-Jenkins, 1982), *Polypodium macaronesicum* (Neuroth, 1996), *Trichomanes speciosum* and *Pteris incompleta* (Rumsey & Vogel, 1998). The latter has also been recorded in the Sintra area of Portugal along with *Asplenium hemionitis*, *Dryopteris guanchica* and *Trichomanes speciosum* and near Cap Spartel in northernmost Morocco, along with *Asplenium hemionitis* and *Davallia canariensis*. *Culcita macrocarpa* is absent from North Africa but does occur in the Iberian enclaves, further extending to the northern coastal region where it grows with *Hymenophyllum tunbrigense*, *Trichomanes speciosum*, *Cystopteris diaphana*, *Davallia canariensis* and *Stegnogramma pozoi* – the last curiously absent from southern Iberia and North Africa but frequent in southern Africa. *Asplenium hemionitis*, which occurs on all Macaronesian archipelagos, has a single European locality, near Sintra in Portugal. However, it does occur in coastal enclave areas in North Africa.

A particularly notable relationship documented by Vanderpoorten *et al.* (2007) is the Azores-Madeira grouping resolved in the moss and pteridophyte floras. This relationship reflects a substantial New World element in the native pteridophyte and moss floras of the archipelagos. In the pteridophytes, it is supported by taxa such as *Ceradenia jungermannioides* and *Grammitis marginella* that are disjunct in the Azores and Madeira but have their main distribution area on the American continent. Another example, again a cloud-forest epiphyte, is *Elaphoglossum semicylindricum*, treated by some as a Macaronesian endemic but perhaps conspecific with material from the Antilles and Central America (as *E. paleaceum*; see Proctor, 1985). Among mosses, five of the nine species of *Campylopus* present in the Azores have a distribution range that spans Macaronesia and the Caribbean islands (Frahm, 1999) while in the liverwort genus, *Plagiochila*, more than 50% of the species found on Madeira also occur in the neotropics (Sim-Sim *et al.*, 2005). Such a New World element is almost entirely absent from the angiosperm floras of the archipelagos that rather show a close affinity with the European and Mediterranean floras (e.g. Schaefer, 2003 ; Carine *et al.*, 2004, 2010).

## Difference three: spatio-temporal relationships of the endemic floras

Molecular phylogenetic data suggest that the closest relatives of most Macaronesian endemic angiosperm groups are distributed in the near-continent (Europe and the

Mediterranean; e.g. Carine *et al.*, 2004, 2010), although notable exceptions exist (e.g. *Kunkeliella* with a South African sister group; Der & Nickrent, 2008). Among bryophytes, phylogenetic evidence points to various biogeographic origins for Macaronesian endemics, including Europe (*Homalothecium mandonii*, Huttunen *et al.*, 2008; *Rhynchostegiella* spp., Aigoin *et al.*, 2009a; *Isothecium prolixum*, Stech *et al.*, 2008), Southeast Asia (*Porella inaequalis*, Hentschel *et al.*, 2007), tropical America and Africa (*Echinodium* spp., Stech *et al.*, 2008), North America (*Radula jonesii*, Devos *et al.*, unpublished results) and South America (*Plagiochila madeirensis*, Rycroft *et al.*, 2004; *Leptoscyphus azoricus*, Vanderpoorten & Long, 2006). Strong neotropical affinities are also evident in the endemic pteridophyte flora. *Huperzia dentata* is the only Old World representative of the otherwise New World section *Reflexa*. Cytological data has shown that is a hexaploid and is therefore likely to be a derived representative of this distinct lineage that evolved following colonisation of Macaronesia from the New World (Vanderpoorten *et al.*, 2007). As already noted, *Elaphoglossum semicylindricum* is closely related to if not conspecific with neotropical taxa (Proctor, 1985) and a further New World link is provided by the Azorean endemic aquatic *Isoetes azorica*, which has been shown to be most closely related to the North American *I. tuckermannii* (Britton & Brunton, 1996).

Few of the strictly endemic pteridophytes might be considered palaeoendemic or relictual; indeed, many are allopolyploids of presumed recent local origin, their parents known, local and sympatric. In bryophytes, molecular phlyogenetic data indicate that the origin of endemics includes a mixture of palaeo- and neo-endemism. An analysis of divergence times for the Madeiran endemic moss bryophyte *Hedenasiastrum percurrens* suggested that this monotypic genus diverged from its closest extant relative *c.*40 Ma (Aigoin *et al.*, 2009a). *Hedenasiastrum* is characterised by the retention of plesiomorphic traits, making it one of the rare examples of 'living fossils' (Aigoin *et al.*, 2009a). The basal phylogenetic position of *Alophosia azorica* within the Polytrichaceae (Hyvönen *et al.*, 2004) also suggests that this taxon may be a palaeo-endemic. In other instances, Macaronesian endemics appear to have originated recently. For instance, the Azorean endemic liverwort *Leptoscyphus azoricus* shares alleles with some neotropical *L. porphyrius* that have not yet been eliminated by drift, rendering the latter paraphyletic (Vanderpoorten & Long, 2006) and suggesting a recent origin for the former. Other endemics, such as *Homalothecium mandonii* (Huttunen *et al.*, 2008), are morphologically similar to their close relatives on the continent, suggesting limited time since divergence.

While the temporal relationships hypothesised for Macaronesian endemic bryophytes are comparable with those observed in angiosperms wherein the colonisation of the region has taken place over a long period (albeit with colonisations possibly confined to discrete windows of opportunity; Carine, 2005; Kim *et al.*, 2008), the sister group distributions observed among Macaronesian endemic cryptogams are much more diverse than the mostly Mediterranean and European affinities of angiosperm endemics (*c*.90%; see Carine *et al.*, 2004, 2010).

## Explanations

Engler (1879) proposed that the distinctive endemic element of the Macaronesian flora was, for the most part, a relict of a formerly widespread subtropical flora that covered southern Europe and North Africa during the Tertiary (hereafter the Engler refugium model). This hypothesis was later supported by Bramwell (1972, 1985) and Sunding (1979), among others. Dispersalist island biogeographic models, emphasising the interchange of species between continental and island areas, were first developed for the Macaronesian archipelagos by Wallace (1881) and suggested the more recent and frequent interchange of diversity between islands and continent. Such ideas were later developed for island floras more generally by authors including MacArthur and Wilson (1967) and Carlquist (1974), with the latter emphasising the importance of both dispersal and establishment in the successful colonisation of islands.

In the context of the temporal evolution of the Macaronesian flora, recent molecular phylogenetic work on angiosperms suggests that the patterns of relationship and evolution of the Macaronesian endemic angiosperm flora are much more complex than Engler's 'refugium model' would predict (see Emerson, 2002; Carine *et al.*, 2004; and Carine, 2005 for a review). As we have already noted, for cryptogams there also appear to be both palaeo-endemics (*Hedenasiastrum*, Aigoin *et al.*, 2009a; *Alophosia*, Hyvönen *et al.*, 2004) consistent with Engler's view on the origin of the endemic flora together with taxa that are neo-endemics and of recent origin (e.g. *Hymenophyllum maderense* and *Asplenium trichomanes* subsp. *maderense*, Gibby & Lovis, 1989; *Dryopteris crispifolia*, Gibby *et al.*, 1977; *Homalothecium mandonii*, Huttunen *et al.*, 2008; *Leptoscyphus azoricus*, Vanderpoorten & Long, 2006).

Recently, hypotheses have been developed to explain diversity patterns in the Macaronesia in relation to abiotic factors (Emerson & Kolm, 2005; Whittaker *et al.*, 2007, 2008; Borges & Hortal, 2009), climatic history (Carine & Schaefer, 2009), 'niche pre-emption' (Silvertown, 2004; Silvertown *et al.*, 2005; but see Herben *et al.*, 2005; Saunders & Gibson, 2005) and key innovations, notably the evolution of woodiness (Carine *et al.*, 2010). Each of these may help to explain the patterns of cryptogamic diversity observed in Macaronesia. However, we would suggest that the ideas developed by Carlquist (1974) that emphasise dispersal and establishment barriers, together with the historically dynamic nature of distribution ranges, are key to understanding the differences in diversity patterns between angiosperms and cryptogams documented in this chapter.

The relationships of several endemic pteridophytes demonstrate that extinction and dispersal have played a key role in shaping extant distribution ranges of taxa.

For example, both the polyploid endemics *Asplenium aureum* from the Canaries (van den Heede *et al.*, 2003) and *A. azoricum* from the Azores (Rumsey *et al.*, 2004) have one diploid parent not extant in the region and indeed not known from elsewhere and thus presumably extinct.

*Cheilanthes pulchella*, considered a Canarian endemic – although an ancient Galician specimen is recorded by Prelli (2004) – has given rise to *C. guanchica* through hybridisation with the diploid *C. maderensis* and subsequent polyploidy. *Cheilanthes guanchica* is known from all the Macaronesian archipelagos and the European and African continents. The differing distributions of parent and hybrid indicate either range contraction in the case of *C. pulchella* and/or range expansion in the case of *C. guanchica. Cheilanthes pulchella* is also parent to *C. tinaei*, derived from hybridisation with *C. hispanica*, a diploid not currently known from Macaronesia. A few records for this tetraploid exist for the Canaries and Madeira but it is more abundant and widespread in Europe and North Africa. The latter would certainly argue for an extra-Macaronesian origin and hence a once wider distribution for *C. pulchella* with subsequent range contraction. In *Dryopteris*, the endemic diploid species *D. maderensis* (syn. *D. intermedia* subsp. *maderensis*) from Madeira, and the Azorean *D. azorica*, are both genomically indistinguishable from one another and from *D. intermedia*, which is widely distributed in North America (Gibby *et al.*, 1977). Both Macaronesian taxa have given rise to allopolyploid derivatives: *D. maderensis*, through hybridisation with *D. aemula*, to *D. guanchica*, which does not occur where the parental taxa are sympatric on Madeira but is found on the Canaries and in Portugal, northwestern Spain and was once collected in the Algeciras area of Southern Spain in the nineteenth century (Fraser-Jenkins, 1982). The origin(s) of this plant and the past distribution of its Macaronesian diploid progenitor are therefore open to speculation. *Asplenium lolegnamense* was previously thought to be a Madeiran endemic hexaploid derived from two Canarian endemic taxa, the tetraploid *A. aureum* and the octoploid *A. octoploideum* (to which *aureum* is part parental). Hexaploid plants have now been found where the parents are sympatric (van den Heede *et al.*, 2004). However, the mutually exclusive distributions of parental taxa and hybrid is intriguing. Why only the hexaploid has dispersed and not the parental taxa is unclear.

In bryophytes the significance of dispersal in explaining current distribution patterns is highlighted by: (1) the left-skewed frequency distributions of regional endemics (Fig. 14.3), which suggests that species tend to be widespread within the region; (2) the highly homoplastic distribution patterns of bryophyte species; and (3) the biogeographic relationships of endemics. Vanderpoorten and Long (2006) hypothesised that most Macaronesian endemic cryptogams are of non-European/North African origin because of extensive gene flow between Macaronesian and populations on the near-continent. Dispersal between the Macaronesian archipelagos and between the archipelagos and the Atlantic and Mediterranean seaboards of Europe and North Africa would promote gene flow and inhibit genetic isolation and speciation, resulting in the low levels of endemism observed. In such a scenario, and in contrast to angiosperms, endemics are most likely to evolve following rare and stochastic long-distance dispersal events or extinction in continental areas. This is consistent with the observation that most sister groups of endemic cryptogamic taxa are not distributed on the near-continent and most endemic bryophytes are laurel forest species, a habitat type largely absent from the near-continent.

Range disjunctions between Macaronesia and the neotropics in the cryptogamic floras are also consistent with the the greater dispersal ability of cryptogams. In addition to the examples already referred to (*Campylopus*, Frahm, 1999; Stech *et al.*, 2007; *Plagiochila*, Groth *et al.*, 2003) it is notable that, at the molecular level, populations of the moss *Scleopodium touretii* from Tenerife share a mutation that is otherwise limited to New World plants, suggesting the existence of transatlantic dispersal events (Shaw *et al.*, 2003). In the context of Southern Hemisphere bryophyte distributions, Muñoz *et al.* (2004) suggested that wind connectivity rather than geographic proximity is the main force driving current bryophyte distributions. Frequent depressions moving rapidly eastwards at relatively low altitude (3000 m) from the American coasts and the existence of tropical cyclones of west Caribbean origin that can carry even large propagules across the Atlantic (Schaefer, 2003) would certainly facilitate dispersal of cryptogams from the New World to the Macaronesian archipelagos, which may also have served as stepping stones for the eastward spread (and occasional isolation in the case of endemics) of American taxa (Sim-Sim *et al.*, 2005).

Cladistic analyses of Canary Island endemic angiosperm groups suggest that allopatric speciation through the isolation of lineages on different islands (or isolation in different areas within the same island) has been an important process in generating endemic diversity (Baldwin *et al.*, 1999; see also Carine & Schaefer, 2009). This conclusion is supported by analyses of endemic angiosperm distribution patterns (Reyes-Betancort *et al.*, 2008) that demonstrated a close correspondence between range size rarity and phylogenetic diversity in the endemic angiosperm flora, suggesting that the centres of endemism observed are not the result of the localised radiation of lineages but, rather, the evolution of endemics from different lineages through geographical isolation. If cryptogams do indeed exhibit higher dispersal ability, then radiation through geographical isolation is unlikely, particularly at the intra-regional scale, and this is again consistent with the very low levels of endemism and the widespread distribution of endemics observed in the cryptogamic floras.

Molecular analyses of non-endemic elements of the Macaronesian cryptogamic floras provide strong evidence for the lack of geographical isolation and strong connectivity both among archipelagos and between archipelagos and continents. Phylogeographic structure in the moss *Grimmia montana* (Vanderpoorten *et al.*,

2008) suggests that gene flow is restricted at the intercontinental scale but repeated sister group relationships observed among accessions from different regions strongly suggest multiple instances of transatlantic dispersal. Furthermore, the low value of $F_{ST}$ (a measure of the variance of allele frequences among populations) between Europe and North America, albeit significantly different from 0, suggests the existence of at least some transatlantic gene flow. This observation is consistent with recent and mounting evidence from correlative analyses between species distributions and air currents (Muñoz *et al.*, 2004), interpretations of species distributions in a phylogenetic context (e.g. Shaw *et al.*, 2003; Heinrichs *et al.*, 2005, 2006; Vanderpoorten & Long, 2006) and genetic inferences of dispersal (McDaniel & Shaw, 2005), which indicate that the intercontinental disjunct distributions exhibited by many bryophyte species result from long-distance dispersal rather than continental drift. The importance of dispersal to the phylogeography of *G. montana* is particularly obvious in Macaronesia. None of the Macaronesian island haplotype groups is monophyletic, which strongly suggest multiple colonisation events. $F_{ST}$ values between Madeira, western Mediterranean islands, North America and Europe are not significantly different from 0 and suggest that Madeira and the southwestern Mediterranean are subject to substantial transatlantic gene flow. In contrast, haplotypes found in the Canary Islands are otherwise distributed in southwestern Europe or southern Africa. Such a pattern could either result from relictualism or from more recent dispersal. Although these competing hypotheses cannot be distinguished in the absence of an absolute time scale, previous analyses in the liverwort *Porella canariensis* suggested substantial gene flow between the Canary Islands and southwestern Europe (Freitas & Brehm, 2001).

Durães (2005) showed that, for the fern *Asplenium hemionitis*, private cpDNA haplotype and allozyme variation existed in all of the Macaronesian archipelagos and in the Sintra region of Portugal, suggesting a degree of isolation between regions and a long presence in the latter. However, the data also suggested that low levels of inter-regional gene flow must exist, maintaining species integrity but not sufficient for homogenisation.

In the fern *Trichomanes speciosum*, cpDNA haplotype data (Rumsey *et al.*, 1996) provided evidence for dispersal from the respective Atlantic archipelagos to continental Europe but no evidence of dispersal between the Azores and Madeira/the more southerly archipelagos, or vice versa. A relationship between the Canaries and the southern Iberian enclaves was indicated. Such a pattern may indicate that the Macaronesian archipelagos served as refugia for the post-glacial recolonisation of the Atlantic fringes of Europe.

This interpretation is further supported by the most recent phylogeography of the liverwort *Radula lindenbergiana* (Laenen, *et al.*, 2011). In fact, despite substantial dispersal among archipelagos and continents resulting in the homogenisation of haplotype frequencies among regions, the species underwent at least two cryptic radiations in Macaronesia, making the region the richest hotspot of genetic variation for the species across its entire distribution range. None of the endemic haplotypes is entirely restricted to a single island, again emphasising that isolation by distance is not the driver of diversification in cryptogams. Rather, the observed endemic diversification might result from sympatric, adaptive speciation along the very wide range of habitats where the species thrives in Macaronesia and North Africa.

The patterns observed in both *T. speciosum* and *R. lindenbergiana* are consistent with Engler's refugium hypothesis in so far as they suggest that the islands have served as refugia. However, in contrast to Engler's views, the history of these taxa appears to be highly dynamic. The Macaronesian islands have not been the 'end of the colonisation road' (Bellemain & Ricklefs, 2008) for these species. In fact, in the case of *R. lindenbergiana*, reconstruction of ancestral distributions areas strongly suggest that Macaronesia was colonised at least twice from European ancestors, after which Macaronesia served as a reservoir of genetic diversity from which Europe was back-colonised after the glaciations. Conceivably, the islands may have served not only as sinks but also as source areas for the re-population of the near-continent and particularly the Atlantic fringe in many cryptogamic groups. Indeed, the Atlantic fringe element in the cryptogamic flora of the archipelagos offers a rich suite of examples to further investigate the nature of the relationships between continental and island populations and to evaluate the contribution of dispersal and extinction to the generation of extant patterns of distributions.

The greater dispersal ability of cryptogams may explain why back-colonisation, so rarely documented robustly in angiosperms (e.g. see Carine *et al.*, 2004; see also Chapter 12, this volume), may be much more common in cryptogams. Indeed, dispersal ability may be the key to understanding the markedly different patterns of diversity observed between angiosperms and cryptogams more generally.

## Acknowledgements

AV acknowledges financial support from the Belgian Funds for Scientific Research (FNRS) (grants 2.4557.11 and 1.5036.11), the University of Liège (grant C-11/32) and the Léopold III Fund.

## Appendix 14.1

List and distribution of strict archipelago endemics and Macaronesian endemics (i.e. species that are restricted to Macaronesia and occur on at least two archipelagos) in the moss and liverwort floras. Island abbreviations: C: Corvo; L: Flores; F: Faial; G: Graciosa; P: Pico; S: Santa Maria; J: Sao Jorge; M: Sao Miguel; T: Terceira; D: Madeira; H: El Hierro; P: La Palma; G: La Gomera; T: Tenerife; C: Gran Canaria; F: Fuerteventura; L: Lanzarote.

| | C | L | F | G | P | S | J | M | T | D | H | P | G | T | C | F | L |
|---|---|---|---|---|---|---|---|---|---|---|---|---|---|---|---|---|---|
| **Liverworts** | | | | | | | | | | | | | | | | | |
| *Aphanolejeunea madeirensis* | | 1 | 1 | | 1 | | 1 | | 1 | 1 | | | | | | | |
| *Bazzania azorica* | | 1 | 1 | | 1 | | 1 | 1 | 1 | | | | | | | | |
| *Calypogeia azorica* | | 1 | 1 | | 1 | | 1 | 1 | 1 | 1 | | | | | | | |
| *Cheilolejeunea cedercreutzii* | 1 | 1 | | 1 | | | 1 | | | | | | | | | | |
| *Cololejeunea schaeferi* | | | | | | | | | | 1 | 1 | 1 | 1 | 1 | | 1 | |
| *Frullania polysticta* | | | | | | | | | | 1 | 1 | 1 | 1 | 1 | 1 | 1 | 1 |
| *Frullania sergiae* | | | | | | | | | | 1 | | | | | | | |
| *Heteroscyphus denticulatus* | 1 | 1 | 1 | 1 | 1 | | 1 | 1 | 1 | 1 | | 1 | 1 | 1 | 1 | | |
| *Lejeunea canariensis* | | | | | | | | | | 1 | 1 | 1 | | 1 | 1 | | |
| *Leptoscyphus azoricus* | | 1 | | | 1 | | 1 | 1 | 1 | | | | | | | | |
| *Plagiochila madeirensis* | | | | | | | | | | 1 | | | 1 | 1 | | | |
| *Porella inaequalis* | | | | | | | | | | 1 | | | | | | | |
| *Radula jonesii* | | | | | | | | | | 1 | | | | 1 | | | |
| *Radula wichurae* | 1 | 1 | 1 | 1 | 1 | 1 | 1 | 1 | 1 | 1 | | | | 1 | | | |
| *Riccia atlantica* | | | | | | | | | | 1 | | | | | | | |
| *Riccia teneriffae* | | | | | | | | | | | | | | 1 | | | |
| *Telaranea azorica* | 1 | 1 | | | 1 | | 1 | 1 | 1 | | | | | 1 | | | |
| **Mosses** | | | | | | | | | | | | | | | | | |
| *Aloina humilis* | | | | | | | | | | | 1 | | | 1 | | 1 | |
| *Alophosia azorica* | 1 | 1 | 1 | | 1 | | 1 | 1 | 1 | 1 | | | | | | | |
| *Andoa berthelotiana* | 1 | 1 | 1 | 1 | 1 | 1 | 1 | 1 | 1 | 1 | | 1 | | 1 | | 1 | |
| *Breutelia azorica* | | 1 | 1 | | 1 | | 1 | 1 | 1 | | | | | | | | |

| | | | | | | | | | | | | | | | | | |
|---|---|---|---|---|---|---|---|---|---|---|---|---|---|---|---|---|---|
| *Bryoxyphium maderense* | | | | | | | | | | 1 | | | | | | | |
| *Cryptoleptodon longisetus* | | | | | | | | | | 1 | 1 | 1 | 1 | 1 | 1 | 1 | |
| *Echinodium prolixum* | 1 | 1 | 1 | 1 | 1 | 1 | 1 | 1 | 1 | 1 | | | | | | | |
| *Echinodium renauldii* | 1 | 1 | 1 | | 1 | | 1 | 1 | 1 | | | | | | | | |
| *Echinodium setigerum* | | | | | | | | | | 1 | | | | | | | |
| *Echinodium spinosum* | | | | | | | | | | 1 | | 1 | | | | | |
| *Fissidens azoricus* | | | 1 | | | | | | | | | | | | | | |
| *Fissidens coacervatus* | | | | | | 1 | | | 1 | 1 | | | 1 | 1 | 1 | | |
| *Fissidens microstictus* | | | | | | | | | | 1 | | | | | | | |
| *Fissidens nobreganus* | | | | | | | | | | 1 | | | | | | | |
| *Fissidens sublinaefolius* | | 1 | 1 | | 1 | | 1 | 1 | 1 | 1 | | | | | | | |
| *Grimmia curviseta* | | | | | | | | | | | | 1 | | 1 | | | |
| *Hedenasiastrum percurrens* | | | | | | | | | | 1 | | | | | | | |
| *Leucodon canariensis* | | 1 | | 1 | | | | 1 | | 1 | 1 | 1 | | 1 | 1 | 1 | 1 |
| *Leucodon treleasei* | 1 | 1 | 1 | 1 | 1 | 1 | 1 | 1 | 1 | 1 | | 1 | | 1 | 1 | | |
| *Neckera intermedia* | 1 | 1 | 1 | 1 | 1 | 1 | 1 | 1 | 1 | | 1 | 1 | | 1 | 1 | 1 | |
| *Nobregaea latinervis* | | | | | | | | | | 1 | | | | | | | |

| | C | L | F | G | P | S | J | M | T | D | H | P | G | T | C | F | L |
|---|---|---|---|---|---|---|---|---|---|---|---|---|---|---|---|---|---|
| *Orthotrichum handiense* | | | | | | | | | | | | | | | | 1 | |
| *Pelekium atlanticum* | | | | | | | | | | 1 | 1 | 1 | 1 | 1 | | | |
| *Rhynchostegiella bourgeana* | | | | | | | | 1 | | | 1 | | | 1 | 1 | | |
| *Rhynchostegiella macilenta* | | | | | | | | | | 1 | | 1 | 1 | 1 | 1 | | |
| *Rhynchostegiella trichophylla* | | | | | | | | | | | 1 | 1 | 1 | 1 | 1 | | 1 |
| *Sphagnum nitidulum* | | | | | | | | | 1 | | | | | | | | |
| *Tortella limbata* | | | | | | | | | | 1 | | | 1 | | 1 | | |
| *Tortula ampliretis* | | | | | | | | | | | 1 | 1 | 1 | 1 | 1 | | 1 |
| *Trematodon perssonorum* | | | | | | | | 1 | | | | | | | | | |

## References

Aigoin, D. A., Devos, N., Huttunen, S., Ignatov, M. S., González-Mancebo, J. M. & Vanderpoorten, A. (2009a). And if Engler was not completely wrong? Evidence for multiple evolutionary origins in the moss flora of Macaronesia. *Evolution* 63: 3248–3257.

Aigoin, D. A., Huttunen, S., Ignatov, M. S., Dirkse, G. M. & Vanderpoorten, A. (2009b). *Rhynchostegiella* (Brachytheciaceae): molecular re-circumscription of a convenient taxonomic repository. *J. Bryol.* 31: 213–221.

Arechavaleta, M., Zurita, N., Marrero, M. C. & Martín, J. L. (eds.) (2005). *Lista preliminar de especies silvestres de Cabo Verde (hongos, plantas y animales terrestres).* La Laguna, Tenerife, Consejería de Medio Ambiente y Ordenación Territorial, Gobierno de Canarias.

Baldwin, B. G., Crawford, D. J., Francisco-Ortega, J., Kim, S-C., Sang, T. & Stuessy, T. F. (1999). Molecular phylogenetics and island plants. In D. E. Soltis, P. S. Soltis & J. J. Doyle (eds.), *Molecular Systematics of Plants II: DNA Sequencing*, pp. 410–441. Boston, MA, Kluwer Academic Press.

Baltisberger, M. & Widmer, A. (2006). Chromosome numbers of plant species from the Canary Islands. *Bot. Helvet.* 116: 9–30.

Bellemain, E. & Ricklefs, R. E. (2008). Are islands the end of the colonisation road? *Trends Ecol. Evol.* 23: 461–468.

Borges, P. A. V. & Hortal, J. (2009). Time, area, and isolation: factors driving the diversification of Azorean arthropods. *J. Biogeogr.* 36: 178–191.

Borges, P. A. V., Cunha, R., Gabriel, R., Martins, A. M. F., Silva, L. & Vieira, V. (2005a). *A list of the terrestrial fauna (Mollusca and Arthropoda) and flora (Bryophyta, Pteridophyta and Spermatophyta) from the Azores.* Horta, Angra do Heroísmo and Ponta Delgada: Direcção Regional do Ambiente and Universidade dos Açores.

Borges, P. A. V., Cunha, R., Gabriel, R., Martins, A. F., Silva, L., Vieira, V. *et al.* (2005b). Description of the terrestrial Azorean biodiversity. In P. A. V. Borges, R. Cunha, R. Gabriel *et al.* (eds.), *A list of the terrestrial fauna (Mollusca and Arthropoda) and flora (Bryophyta, Pteridophyta and Spermatophyta) from the Azores*, pp. 21–68. Horta, Angra do Heroísmo and Ponta Delgada: Direcção Regional de Ambiente and Universidade dos Açores.

Borges, P. A. V., Abreu, C., Aguiar, A. M. F., Carvalho, P., Jardim, R., Melo, I. *et al.* (2008). *A list of the terrestrial fungi, flora and fauna of Madeira and Selvagens archipelagos.* Funchal and Angra do Heroísmo: Direcção Regional do Ambiente da Madeira and Universidade dos Açores.

Bramwell, D. (1972). Endemism in the flora of the Canary Islands. In D. H. Valentine (ed.), *Taxonomy, Phytogeography and Evolution*, pp. 141–159. London, Academic Press.

Bramwell, D. (1985). Contribución a la biogeografía de las Islas Canarias. *Bot. Mac.* 14: 3–34.

Britton, D. M. & Brunton, D. F. (1996). Spore morphology and cytology of *Isoetes azorica* (Pteridophyta, Isoetaceae) and its affinity with North America. *Fern Gazette* 15: 113–118.

Bruges, M. & Sérgio, C. (2010). On the identity of some *Entosthodon* species endemic to Macaronesia or the North of Africa. *Cryptog., Briol.* 31: 277–280.

Burghardt, M. & Gradstein, S. R. (2008). A Revision of *Tylimanthus* (Acrobolbaceae, Marchantiophyta) in Tropical America, Africa, and Macaronesia. *Fieldiana Botany* 47: 199–210.

Carine, M. A. (2005). Spatio-temporal relationships of the Macaronesian endemic flora: a relictual series or window of opportunity? *Taxon* 54: 895–903.

Carine, M. A. & Schaefer, H. (2009). The Azores diversity enigma: why are there so few Azorean endemic flowering plants and why are they so widespread? *J. Biogeogr.* 37: 77–89.

Carine, M. A., Francisco-Ortega, J., Santos-Guerra, A. & Russell, S. J. (2004). Relationships of island and continental floras: molecular evidence for multiple colonisations into Macaronesia and subsequent back-colonisation of the continent in *Convolvulus* L. *Am. J. Bot.* 91: 1070–1085.

Carine, M. A., Humphries, C. J., Guma, I. R., Reyes-Betancort, J. A. & Santos Guerra, A. (2009). Areas and algorithms: numerical approaches to the delimitation of areas of endemism in the Canary Islands archipelago. *J. Biogeogr.* 36: 593–611.

Carine, M. A., Santos Guerra, A., Guma, I. R. & Reyes-Betancort, J. A. (2010). Endemism and evolution of the Macaronesian Flora. In D. M. Williams & S. Knapp. *Beyond Cladistics: The Branching of a Paradigm* pp. 101–124. Berkeley, CA, University of California Press.

Carlquist, S. J. (1974). *Island Biology*. New York, Columbia University Press.

Der, J. P. & Nickrent, D. L. (2008). A molecular phylogeny of Santalaceae (Santalales). *Syst. Bot.* 33: 107–116.

Draper, I., González-Mancebo, J. M., Werner, O., Patio, J., Ros, R. M. (2011). Phylogeographic relationships between the mosses *Exsertotheca intermedia* from Macaronesian islands and *Neckera baetica* from southern glacial refugia of the Iberian Peninsula. *Acta Bot. Fenn.* 48, in press.

Durães, C. J. R. (2005). Genetic variation and evolutionary history of the Macaronesian fern *Asplenium hemionitis* L. Unpublished PhD thesis: University of Edinburgh.

Emerson, B. C. (2002). Evolution on oceanic islands: molecular phylogenetic approaches to understanding pattern and process. *Mol. Ecol.* 11: 951–966.

Emerson, B. C. & Kolm, N. (2005). Species diversity can drive speciation. *Nature* 434: 1015–1017.

Engler, A. (1879). *Versuch einer Eintwicklungsgeschichte, insbesondere der Florengebiete seit der Tertiärperiode. I. Die extra-tropischen Gebiete der nördlischen Hemisphäre.* Leipzig, Engelmann.

Frahm, J. P. (1999). A survey of the *Campylopus* species from the Azores. *Cryptogam. Bryol.* 20: 145–152.

Francisco-Ortega, J., Santos Guerra, A., Jarvis, C. E., Carine, M. A., Menezes de Sequeira, M. & Maunder, M. (2010). Early British collectors and observers of the Macaronesian flora: from Sloane to Darwin. In D. M. Williams & S. Knapp. *Beyond Cladistics: The Branching of a Paradigm*. Berkeley, CA, University of California Press.

Fraser-Jenkins, C. R. (1982). *Dryopteris* in Spain, Portugal and Macaronesia. *Boletim da Sociedade Broteriana, Sér. 2* 55: 175–336.

Freitas, H. & Brehm, A. (2001). Genetic diversity of the Macaronesian leafy liverwort *Porella canariensis* inferred from RAPD markers. *J. Hered.* 92: 339–345.

Gabriel, R. & Bates, J. W. (2005). Bryophyte community composition and habitat specificity in the natural forests of Terceira, Azores. *Plant Ecol.* 177: 125–144.

Gabriel, R., Sjögren, E., Schumacker, R., Sérgio, C., Frahm, J. P. & Sousa, E. (2005). List of bryophytes. In P. A. V. Borges, R. Cunha, R. Gabriel, *et al.* (eds.), *A list of terrestrial fauna (Mollusca and Arthropoda) from the Azores*, pp. 117–129. Horta, Angra do Heroísmo and Ponta Delgada: Direcçao Regional Do Ambiente, Governo Regional dos Açores.

Gallego, M. T., Cano, M. J., Ros, R. M., Guerra, J. & Dirkse, G. M. (1998). *Aloina humilis* sp. nov. (Bryopsida, Pottiaceae) from the Canary Islands. *Nova Hedwigia* 67: 9–124.

Gallego, M. T., Jimenez, J. A. & Guerra, J. (2004). *Tetrastichium virens* with sporophytes in mainland Europe: an SEM study. *J. Bryol.* 26: 224–226.

Gibby, M. (1983). The *Dryopteris dilatata* complex in Macaronesia and the Iberian peninsula. *Acta Bot. Malac.* 8: 59–72.

Gibby, M. (1985). Hybridization and speciation in the genus *Dryopteris* (Pteridophyta: Dryopteridaceae) on Pico Island in the Azores. *Plant Syst. Evol.* 149: 241–252.

Gibby, M. & Lovis, J. D. (1989). New ferns of Madeira. *Fern Gazette* 13: 285–290.

Gibby, M., Jermy, A. C., Rasbach, H., Rasbach, K., Reichstein, T. & Vida, G. (1977). The genus *Dryopteris* in the Canary Islands and Azores and the description of two new tetraploid species. *Bot. J. Linn. Soc.* 74: 251–277.

Gibby, M., Rasbach, T., Reichstein, T., Widén, C.-J. & Viane, R. L. L. (1992). Micromorphology, chromosome numbers and phloroglucinols of *Arachniodes foliosa* and *A. webbiana* (Dryopteridaceae, Pteridophyta). *Bot. Helvet.* 102: 229–245.

González-Mancebo, J. M., Romaguera, F., Ros, R. M., Patino, J. & Werner, O. (2007). Bryophyte flora of the Canary Islands: an updated compilation of the species list with an analysis of distribution patterns in the context of the Macaronesian Region. *Cryptogam. Bryol.* 29: 315–357.

González-Mancebo, J. M., Draper, I., Lara, F., Marraero, J. D., Muñoz, J., Patino, J. *et al.* (2009). Amendments to the bryophyte flora of the Cape Verde and Canary Islands. *Cryptogam. Bryol.* 30(4): 433–441.

Groth, H., Lindner, M., Wilson, R., Hartmann, F. A., Schmull, M., Gradstein, S. R. et al. (2003). Biogeography of *Plagiochila* (Hepaticae): natural species groups span several floristic kingdoms. *J. Biogeogr.* 30: 965–978.

Hallingbäck, T. & Hodgetts, N. (2000). *Mosses, Liverworts and Hornworts. Status Survey and Conservation Action Plan for Bryophytes.* Gland, Switzerland, IUCN/SSC Bryophyte Specialist Group.

Heinrichs, J., Lindner, M., Gradstein, S. R., Groth, H., Buchbender, V., Solga, A. *et al.* (2005). Origin and subdivision of *Plagiochila* (Jungermanniidae: Plagiochilaceae) in tropical Africa based on evidence from nuclear and chloroplast DNA sequences and morphology. *Taxon* 54, 317–333.

Heinrichs, J., Lindner, M., Groth, H., Hentschel, J., Feldberg, K., Renker, C. et al. (2006). Goodbye or welcome Gondwana? – Insights into the phylogenetic biogeography of the leafy liverwort *Plagiochila* with a description of *Proskauera*, gen. nov. (Plagiochilaceae, Jungermanniales). *Plant Syst. Evol.* 258: 227–250.

Hentschel, J., Zhu, R. L., Long, D. G., Davison, P. G., Schneider, H., Gradstein, S. R. *et al.* (2007). A phylogeny of *Porella* (Porellaceae, Jungermanniopsida) based on nuclear and chloroplast DNA sequences. *Mol. Phylogenet. Evol.* 45: 693–705.

Herben, T., Suda, J. & Munclinger, P. (2005). The ghost of hybridization past: niche preemption is not the only explanation of apparent monophyly in island endemics. *J. Ecol.* 93: 572–575.

Huttunen, S., Hedenäs, L., Ignatov, M. S., Devos, N. & Vanderpoorten, A. (2008). Origin and evolution of the northern hemisphere disjunction in the moss genus *Homalothecium* (Brachytheciaceae). *Am. J. Bot.* 95: 220–250.

Hyvönen, J., Koskinen, S., Merrill, G. L. S., Hedderson, T. A. & Stenroos, S. (2004). Phylogeny of the Polytrichales (Bryophyta) based on simultaneous analysis of molecular and morphological data. *Mol. Phylogenet. Evol.* 31: 915–928.

Izquierdo, I., Martín, J. L., Zurita, N. & Arechavaleta, M. (eds.) (2004). *Lista de especies silvestres de Canarias (hongos, plantas, y animales terrestres).* La Laguna, Tenerife: Consejería de Medio Ambiente y Ordenación Territorial, Gobierno de Canarias.

Jimenez, J. A., Ros, R. M., Cano, M. J. & Gallego, M. T. (2000). *Thamnobryum maderense* (Kindb.) Hedenäs in Spain and northern Africa. *Nova Hedwigia* 71: 439–445.

Kim, S-C., McGowen, M. R., Lubinsky, P., Barber, J. C. & Mort, M. E. (2008). Timing and tempo of early and successive adaptive radiations in Macaronesia. *PLoS ONE* 3: e2139. doi:10.1371/journal.pone.0002139

Laenen, B., Désamoré, A., Devos, N., Shaw, A. J., Carine, M. A., González-Mancebo, J. M., Vanderpoorten, A. (2011). Macaronesia: a source of hidden genetic diversity for post-glacial recolonization of western Europe in the leafy liverwort *Radula lindenbergiana. J. Biogeogr.* 38: 631–639.

Lara, F., Mazimpaka, V., Garilleti, R. & Brugues, M. (1999). *Orthotrichum handiense*, sp. nov. from Fuerteventura, Canary Islands. *Bryologist* 102: 619–622.

Lobin, W. (1982). Untersuchung über Flora, Vegetation und biogeographische Beziehungen der Kapverdischen Inseln. *Courier Forschungsinstitut Senckenberg* 53: 1–112.

MacArthur, R. H. & Wilson, E. O. (1967). *The Theory of Island Biogeography.* Princeton, NJ, Princeton University Press.

Manton, I., Lovis, J. D., Vida, G. & Gibby, M. (1986). Cytology of the fern flora of Madeira. *Bull. Br. Mus. Nat. Hist. (Bot.)* 15: 123–161.

McDaniel, S. F. & Shaw, A. J. (2005). Selective sweeps and intercontinental migration in the cosmopolitan moss *Ceratodon purpureus* (Hedw.) Brid. *Mol. Ecol.* 14: 1121–1132.

Muñoz, J., Felicisimo, A. M., Cabezas, F., Burgaz, A. R. & Martinez, I. (2004). Wind as a long-distance dispersal vehicle in the southern hemisphere. *Science* 304: 1144–1147.

Neuroth, R. (1996). Biosystematik und Evolution des *Polypodium vulgare*-Komplexes (Polypodiaceae; Pteridophyta). *Dissertationes. Botanicae* 256: 1–209.

Olsson, S., Rumsey, F., Grundmann, M., Russell, S., Enroth, J. & Quandt, D. (2009). Origin of British and Macaronesian endemic *Thamnobryum* species (Neckeraceae). *J. Bryol.* 31: 1–10.

Prelli, R. (2004). *Les Fougères et plantes alliées de France et d'Europe occidentale.* Paris: Belin.

Proctor, G. R. (1985). *Ferns of Jamaica.* London, British Museum (Natural History).

Reyes-Betancort, J. A., Santos Guerra, A., Guma, I. R., Humphries, C. J. & Carine, M. A. (2008). Diversity, rarity and the evolution and conservation of the Canary Islands endemic flora. *Anales Jard. Bot. Madrid* 65: 25–45.

Rumsey, F. J. & Vogel, J. C. (1998). *Trichomanes speciosum* Willd. (Hymenophyllaceae) in southern Spain. *Fern Gaz.* 15: 197–203.

Rumsey, F. J. & Vogel, J. C. (1999). *Tetrastichium fontanum* (Mitt.) Card. new to mainland Europe. *J. Bryol.* 21: 72–73.

Rumsey, F. J., Russell, S. J., Ji, J., Barrett, J. A. & Gibby, M. (1996). Genetic variation in the endangered filmy fern *Trichomanes speciosum* Willd. In J. M. Camus, M. Gibby

& R. J. Johns (eds.). *Pteridology in Perspective*, pp. 161–165. London, Royal Botanic Gardens Kew.

Rumsey, F. J., Russell, S. J., Schäfer, H. & Rasbach, H. (2004). Distribution, ecology and cytology of *Asplenium azoricum* Lovis, Rasbach & Reichstein (Aspleniaceae, Pteridophyta) and its hybrids. *Am. Fern J.* 94: 113–125.

Rycroft, D. S., Groth, H. & Heinrichs, J. (2004). Reinstatement of *Plagiochila maderensis* (Jungermanniopsida: Plagiochilaceae) based on chemical evidence and nrDNA ITS sequences. *J. Bryol.* 26, 37–45.

Saunders, N. E. & Gibson, D. J. (2005). Breeding system, branching processes, hybrid swarm theory, and the hump-back diversity relationship as additional explanations for apparent monophyly in the Macaronesian island flora. *J. Ecol.* 93: 649–652.

Schaefer, H. (2003). Chorology and diversity of the Azorean flora. *Dissertationes Botanicae* 374: 1–130.

Sérgio, C. & González-Mancebo, J. M. (2009). 3. Alguns briófitos novos ou pouco conhecidos para a Ilha de Gomera e Lanzarote (Ilhas Canárias). *Portugaliae Acta Biol.* 23: 280–282.

Sérgio, C. & Perold, S. (1992). A new species of *Riccia* L., from the island of Madeira, *Riccia atlantica*, sp. nov. *J. Bryol.* 17: 127–132.

Sérgio, C., Schumacker, R. & Martiny, P. (1995). Threatened bryophytes in Europe including Macaronesia. In *Red Data Book of European Bryophytes*. Trondheim, Norway, E.C.C.B.

Sérgio, C., Sim-Sim, M., Fontinha, S. & Figueira, R. (2008). The bryophytes (Bryophyta) of the Madeira and Selvagems archipelagos. In P. A. V. Borges, C. Abreu, A. M. Franquinho Aguiar *et al.* (eds.), *A list of terrestrial fungi, flora and fauna of Madeira and Selvagems archipelagos*, pp. 143–156. Funchal and Angra do Heroísmo: Direcção Regional do Ambiente da Madeira and Universidade dos Açores.

Shaw, A. J., Werner, O. & Ros, R. M. (2003). Intercontinental Mediterranean disjunct mosses: morphological and molecular patterns. *Am. J. Bot.* 90: 540–550.

Silvertown, J. (2004). The ghost of competition past in the phylogeny of island endemic plants. *J. Ecol.* 92: 168–173.

Silvertown, J., Francisco-Ortega, J. & Carine, M. (2005). The monophyly of island radiations: an evaluation of niche pre-emption and some alternative explanations. *J. Ecol.* 93: 653–657.

Sim-Sim, M., Fontinha, S., Mues, R. Lion, U. (2000). A new *Frullania* species (subg. *Frullania*) from Deserta Grande, Madeira archipelago, *Frullania sergiae* sp. nov. *Nova Hedwigia* 71: 185–193.

Sim-Sim, M., Esquivel, M. D., Fontinha, S. & Stech. M. (2005). The genus *Plagiochila* (Dumort.) Dumort. (Plagiochilaceae, Hepaticophytina) in Madeira Archipelago: molecular relationships, ecology, and biogeographic affinities. *Nova Hedwigia* 81: 449–461.

Sim-Sim, M., Luís, L., Garcia, C., Fontinha, S., Lobo, C. Martins, S. & Stech, M. (2008). New data on the status of threatened bryophytes of Madeira Island. *J. Bryol.* 28: 63–70.

Stech, M., Sim-Sim, M. &. Frahm, J. P. (2007). *Campylopus* (Leucobryaceae, Bryopsida) on Madeira island: molecular relationships and biogeographic affinities. *Nova Hedwigia* 131 (Suppl.): 91–100.

Stech, M., Sim-Sim, M., Equivel, M. G., Fontinha, S., Tangney, R., Lobo, C. *et al.* (2008). Explaining the 'anomalous' distribution of *Echinodium* (Bryopsida:

Echinodiaceae): independent evolution in Macaronesia and Australasia. *Organisms Divers. Evol.* 8: 282–292.

Sunding, P. (1979). Origins of the Macaronesian flora. In D. Bramwell (ed.), *Plants and Islands*, pp. 13–40. London, Academic Press.

Touw, A. (2001). A review of the Thuidiaceae (Musci) and a realignment of taxa traditionally accommodated in *Thuidium* sensu amplo (*Thuidium* Schimp., *Thuidiopsis* (Broth.) M. Fleisch., and *Pelekium* Mitt.), including *Aequatoriella* gen. nov. and *Indothuidium* gen. nov. *J. Hatt. Bot. Lab.* 90: 1–136.

van den Heede, C. J., Viane, R. L. L. & Chase, M. W. (2003). Phylogenetic analysis of *Asplenium* subgenus *Ceterach* (Pteridophyta: Aspleniaceae) based on plastid and nuclear ribosomal ITS DNA sequences. *Am. J. Bot.* 90: 481–495.

van den Heede, C. J., Pajarón, S., Pangua, E. & Viane, R. L. L. (2004). *Asplenium ceterach* and *A. octoploideum* on the Canary Islands (Aspleniaceae, Pteridophyta) *Am. Fern J.* 94: 81–111.

Vanderpoorten, A. & Long, D. G. (2006). Budding speciation and Neotropical origin of the Azorean endemic liverwort, *Leptoscyphus azoricus*. *Mol. Phylogenet. Evol.* 40: 73–83.

Vanderpoorten, A., Rumsey, F. J. & Carine, M. A. (2007). Does Macaronesia exist? Conflicting signal in the bryophyte and pteridophyte floras. *Am. J. Bot.* 94: 625–639.

Vanderpoorten, A., Devos, N., Goffinet, B., Hardy, O. & Shaw, A. J. (2008). The barriers to oceanic island radiation in bryophytes: insights from the phylogeography of the moss *Grimmia montana*. *J. Biogeogr.* 35: 654–663.

Vanderpoorten, A., Schäfer-Verwimp, A., Heinrichs, J., Devos, N. & Long, D. G. (2010). The taxonomy of the liverwort genus *Leptoscyphus* revisited. *Taxon* 59(1): 176–186.

van der Wijk, R., Margadant, W. D. & Florschütz, P. A. (1959). *Index Muscorum*. Utrecht: International Association of Plant Taxonomists.

Wallace, A. R. (1881). *Island Life*. New York: Harper & Bros.

Werner, O., Patino, J., González-Mancebo, J. M. & Ros, R. M. (2007). The taxonomic status of *Platyhypnidium torrenticola* based on ITS sequence data. *Cryptogam. Bryol.* 28: 187–195.

Werner, O., Patino, J., González-Mancebo, J. M., Gabriel, R. M. & Ros, R. M. (2009). The taxonomic status and the geographical relationships of the Macaronesian endemic moss *Fissidens luisieri* (Fissidentaceae) based on DNA sequence data. *Bryologist* 112: 315–324.

Whittaker, R. J., Ladle, R. J., Araújo, M. B., Fernández-Palacios, J. M., Domingo Delgado, J. & Arévalo, J. R. (2007). The island immaturity-speciation pulse model of island evolution: an alternative to the 'diversity begets diversity' model. *Ecography* 30: 321–327.

Whittaker, R. J., Triantis, K. A. & Ladle, R. J. (2008). A general dynamic theory of oceanic island biogeography. *J. Biogeogr.* 35: 977–994.

# 15

# Invasive alien species and islands

MICHAEL KIEHN

Invasive alien species are considered to be a major contributor to the high extinction rates of species native to islands compared with other habitats (Sherley *et al.*, 2000). In this respect, the effect of invasive alien species is second only to the destruction or modification of natural environments caused by humans. Island ecosystems, especially on small or remote islands, seem to be especially vulnerable to alien plant species invasions with negative impacts on their native floras. Numerous reasons for this phenomenon have been put forward in the literature. They include assessments of 'unbalanced island floras' or different aspects of lower competitiveness of 'island-born' species in comparison to the non-native 'newcomers' (Denslow, 2003) as well as correlations with abiotic factors such as island size, number of native species, rate of disturbance and distance to continental areas (Barney & Whitlow, 2008).

Documented negative impacts of invasive alien plant species relate to whole-island ecosystems and plant community levels. For example, invasive alien plant species can transform formerly treeless areas into woodland. Such a situation is found on the Galapagos Islands, where the quinine tree, *Cinchona pubescens* (Rubiaceae), has invaded higher-elevation areas on Santa Cruz Island (Figs 15.1, 15.2) (Soria *et al.*, 2002). In the fruiting season, each tree of this species produces thousands of wind-dispersed seeds which have high germination rates (Kiehn, pers. obs.). By 2002, *c.*11 000 hectares of original grass- and shrubland on Santa Cruz Island were covered by quinine trees (Soria *et al.*, 2002). Not only is the native vegetation displaced, but ground-nesting birds lose their nesting sites (Schofield, 1989; Baskin, 2002).

Equally, native woodlands on islands can be negatively affected by invasive alien plant species which bring about changes in hydrology, nutrient cycles, light or

*The Biology of Island Floras*, eds. David Bramwell and Juli Caujapé-Castells. Published by Cambridge University Press. 

Figure 15.1 Galapagos Islands, Santa Cruz: mountain top region of Cerro Puntudo. Situation in 1972: treeless vegetation. Photo: Ole Hamann, Denmark.

Figure 15.2 Same area as shown in Fig. 15.1, in 2007, covered with *Cinchona pubescens* trees. Photo: Ole Hamann, Denmark.

fire regimes and soil erosion, causing large-scale disturbance and destruction. Even smaller-scale alterations, such as changes in the microclimate or the soil chemistry, can be triggered by new species in a habitat, especially when they form local monotypic stands. Beyond that, invasive aliens often have direct negative impacts on the population level of native species when outcompeting elements of indigenous island floras (Meyer & Florence, 1996; Sherley, 2000). As a result of climate change effects, the critical situation for many native island species can be expected to become even worse in the future (Low, 2008).

What characterises an 'invasive alien species'? It is a non-native or non-indigenous species; introduced intentionally or accidentally, usually by humans, in

Table 15.1 *Ratio of native plant species (endemic and indigenous species) to introduced naturalised species for some tropical oceanic islands.*[1]

| Island or island group | Native species | Introduced species | Introduced/ native |
|---|---|---|---|
| French Polynesia (Pacific Ocean) | 690 | 520 | 0.75 |
| Hawaiian Islands (Pacific Ocean) | 956 | 869 | 0.91 |
| La Réunion (Indian Ocean) | 675 | 628 | 0.93 |
| Cook Islands (Pacific Ocean) | 278 | 286 | 0.97 |
| Seychelles (Indian Ocean) | 233 | 247 | 1.06 |
| Mauritius (Indian Ocean) | 685 | 731 | 1.07 |
| Juan Fernández archipelago (Pacific Ocean) | 209 | 232 | 1.11 |
| Nauru (Pacific Ocean) | 45 | 81 | 1.8 |
| St Helena (Atlantic Ocean) | 50 | 270 | 5.4 |
| Ascension (Atlantic Ocean) | 20 | 200 | 10.0 |

[1] Data from Cronk, 2000; Greimler *et al.*, 2002; Merlin & Juvik, 1992; Meyer, 2004; Meyer & Lavergne, 2004.

an area where it has never occurred or evolved before; naturalised or established in that area, i.e. being able to reproduce sexually or vegetatively without human assistance; expanding its distribution and/or increasing its abundance; occurring in natural and/or semi-natural habitats; and showing significant ecological impacts (Meyer, 2000: 86) or causing economic damage besides creating environmental harm for ecosystems, habitats or species (McNeely *et al.*, 2001).

As result of alien species invasions, many island floras today are dominated by naturalised non-native species (Sax *et al.*, 2002). The situation of the floristic composition for some tropical oceanic islands and island groups is shown in Table 15.1.

The aims of the present chapter are to exemplify mechanisms of arrival and establishment of invasive alien plant species on islands, to illustrate their effects on island biota, to give examples of management strategies to mitigate their influence and to propose actions related to invasions still to come.

**There are more to come...**

The translocation of plant species from one geographic area to another by humans, either intentionally (as a result of human needs or desires for plants) or unintentionally (as an unnoticed 'travelling companion' or as a 'blind passenger' along with an intentionally transferred plant) is as old as human history. 'Amelioration' of islands harbouring comparatively few native plant species suited for human use was a common concept of sea-travelling people wishing to 'improve' the utility of the island for human uses. This resulted in a number of intentional plant introductions to islands (Mack, 2001). An early example of such island-related transfer activities of non-native plants is the 'canoe-plant' – plants the Polynesians carried with them on their voyages in search of new land to colonise. Out of *c*.60 species they travelled with in southern Polynesia, 27 were introduced and became established on the Hawaiian Islands (Whistler, 2009). Other well-documented numbers for introduced species relate to the ornamental plant trade. More than 250 non-native plant species were introduced to Tahiti between 1921 and 1944 as ornamentals for a private 'botanical garden' (Meyer, 2003) – this is considered to be a high number for that time. In recent years, with increasing human mobility, the number of species brought to islands has grown exponentially. A quite recent exponential increase in the number of introduced plant species is documented for the Galapagos Islands: 77 such species were known in 1971 and 471 in 1999. In 2001, more than 600 non-native species were reported (Tye *et al.*, 2002; Baskin, 2002) and a recent survey (Guézou *et al.*, 2010) lists 754 alien taxa for the archipelago. A more continuous increase of non-native plant species is documented for the Juan Fernández archipelago (Greimler *et al.*, 2002).

A major source of potentially invasive plant species on islands are introduced 'useful' plants including ornamentals. Taking the absolute numbers of such introductions into account, this is no surprise. Serious estimates indicate, for example, that more than 1700 non-native plant species were intentionally brought to French Polynesia (Florence, 2003), over 2000 to La Réunion (Lavergne *et al.*, 1999) and over 8000 to the Hawaiian Islands (Staples *et al.*, 2000), and a single large nursery in Hawaii today might hold a stock of some hundreds of non-native species (Kiehn, pers. obs.). Even if only a small percentage of these introductions does (or has the potential to) become established as invasive in its new environment (as discussed by Williamson & Fitter, 1996), the resulting numbers are quite high.

### ... and more to stay ...

A trend in horticulture contributing to the spread of alien species on islands was analysed by Meyer and Lavergne (2004). They studied the potential invasiveness of non-native members of the Acanthaceae in the Indo-Pacific. This plant family contains a considerable number of showy species suited as ornamentals in tropical and subtropical regions, and this is the reason for a high demand for Acanthaceae in horticulture. The authors documented 52 species of the family as being naturalised in the Indo-Pacific or in other tropical regions; 16 of those can be rated as, at least regionally, invasive plant species in the Indo-Pacific region. Eight of the listed species, with growth forms from creeping herbs to woody vines, have meanwhile been

Figure 15.3 *Miconia calvescens* infestation at the Vaihiria valley region on Tahiti, French Polynesia. Photo: Jean-Yves Meyer, Délégation à la Recherche, Tahiti.

proven to be problematic invasive species on Indo-Pacific islands: *Justicia carnea, Odontonema strictum, Phlogacanthus turgidus, Sanchezia speciosa, Strobilanthes hamiltonianus, Thunbergia grandiflora, Hemigraphis alternate* and *Ruellia brevifolia.* Most of them impact forest understoreys on the islands they have reached by forming dense monospecific groundcovers or thickets, but they can even negatively affect forest canopies by overgrowing native trees (as is the case with *Thunbergia grandiflora;* Fig. 15.3b).

Looking at the situation in the Pacific region, Meyer (2000, 2004) identified several other taxa which are already invasive on some island groups, and are present (but not invasive yet) or likely to reach additional islands in the near future. This holds true for the octopus or umbrella tree, *Schefflera actinophylla* (Araliaceae), a popular ornamental tree in the tropics. It is well known as an invasive plant in the Hawaiian Islands (Staples *et al.*, 2000). However, the species is commonly cultivated on the Society Islands and on the Cook Islands, and is reported now as being naturalised in the understorey of mesic forests of Tahiti at between 300 and 400 m elevation (Meyer, 2004).

Another example of a potential invasion ahead, mentioned by Meyer (2004), is the quinine tree, *Cinchona pubescens* (Rubiaceae). This species, with proven high invasive capacity on the Galapagos Islands (see above), was introduced to Tahiti in the 1940s for commercial purposes. Meyer (2004) reported the presence of isolated fruiting trees with seedling carpets in the cloud forest above 950 m elevation on Tahiti – the 'perfect' starting point for a successful colonisation of the island. As *Cinchona pubescens* has been cultivated on numerous other tropical islands since the 1940s (e.g. on Hawaii), similar problems as have occurred on Galapagos are to be expected in other Pacific regions as well.

## ... and some might even start after a delay

Problems associated with non-native plant species have been increasing due to their constantly growing numbers and the many disturbances on islands caused by human activities (especially related to land use) which favour the establishment and spread of non-native plant species. However, not all species immediately become invasive in a new environment. Some 'newcomers' seem to start their area expansion or increase in numbers only after a time lag. In the literature, such delayed colonisation behaviour is, at least in part, interpreted as being a result of evolutionary processes (Henderson *et al.*, 2006; Prentis *et al.*, 2008).

Henderson *et al.* (2006) report an example for the Galapagos Islands: *Rubus niveus* (hill raspberry, Rosaceae) was introduced to San Cristóbal Island in 1983 as a fruit plant. The species initially needed careful weeding, fertilisation and watering for survival. Only a short time after the first ripe fruits were produced, however, young plants appeared quite far away from the parent plants. In the following years the species spread constantly and now covers several hundred hectares.

Another example of the time lag between naturalisation and becoming invasive is *Strobilanthes hamiltonianus* (Acanthaceae), a species native to India and Sri Lanka. Introduced to the Mascarenes as an ornamental, it was first collected as a naturalised plant on La Réunion and Mauritius in 1956 but, for a long time, was not considered to be invasive. In 2004, however, a survey showed that the species had already covered *c.*1000 hectares of wet mountain areas on La Réunion at between 900 m and 1500 m elevation. Initially, *Strobilanthes hamiltonianus* on La Réunion and Mauritius spread only by vegetative propagation. But in 2002 and 2003, the first plants with fruits were found, indicating a potential for sexual reproduction and for dispersal through seeds (Meyer & Lavergne, 2004).

## Successful establishment on tropical islands

Hundreds of alien plant species have become established on tropical islands. Surveys of such species are available, e.g. for the South Pacific (Meyer, 2000), for Eastern Polynesia (Meyer, 2004), for the Austral-Pacific region (Shine *et al.*, 2003), for the Juan Fernández archipelago (Greimler *et al.*, 2002) and for Hawaii (Staples *et al.*, 2000). Recently, Kueffer *et al.* (2010) provided a global comparison of plant invasions in oceanic islands. A few examples shall illustrate the history of some of these plant invaders.

The cinnamon tree, *Cinnamomum verum* (Lauraceae), is today the most widespread tree in the Seychelles, found from sea level to the highest elevations.

Figure 15.4 *Morella faya* dominated site at the Volcanoes National Park, Big Island, Hawaii. Photo: Michael Kiehn.

Introduced in 1772, it spread so rapidly that by the end of the nineteenth century it formed monotypic forests (Vielle, 2003). For French Polynesia, Meyer (2003, 2004) reported quite typical examples of alien invasive species, their means of arrival, their patterns of establishment and their negative effects on island floras: *Psidium guajava* (common guava, Myrtaceae) was intentionally introduced to French Polynesia in 1815. In November 1835, during his visit to Tahiti on the *Beagle*, Charles Darwin observed that 'the species is so abundant that it is quite a weed'. Today, the common guava is invasive on many Pacific islands, mostly in forest ecosystems (Meyer, 2000). Several of the *c.*250 non-native ornamental plant species introduced by a private botanical garden on Tahiti between 1921 and 1944 are now invasive in French Polynesia, e.g. *Cecropia peltata* (trumpet tree, Cecropiaceae), *Schinus terebinthifolius* (Brazilian pepper, Anacardiaceae), *Spathodea campanulata* (African tulip tree, Bignoniaceae; Fig. 15.3a), *Cestrum nocturnum* (night-blooming jasmine, Solanaceae), *Ardisia elliptica* (shoebutton ardisia, Myrsinaceae; Fig. 15.3c) and *Miconia calvescens* (miconia, Melastomataceae; Fig. 15.4) (Meyer, 2004).

The latter species, a small tree with showy flowers and foliage, originally native to Central and tropical America, turned out to be especially detrimental for the ecosystems it invaded: introduced to French Polynesia as an ornamental in 1937, the first populations of the species on natural sites on Tahiti were only documented in the early 1970s. Subsequently however, the species quickly expanded its distribution area on Tahiti (Meyer & Florence, 1996). Only three decades later, in 2003, more than 80 000 hectares on Tahiti had been invaded by *Miconia calvescens*, and the species covered about 3500 hectares, *c.*35% of the island surface, on Moorea (Meyer, 2003).

One would have expected that *Miconia calvescens*, due to its known invasiveness, would be the subject of strict controls to avoid its further spread to other islands in the region. But since the mid-1990s, the species has been found on the Marquesas Islands (Nuku Hiva, Fatu Hiva) and the Austral Islands (Rurutu and Rapa Iti), most probably brought there with contaminated soil (Meyer, 2003). This underlines the general observation made by Mauremootoo and Boudjelas (2003), that multi-island states, because of the frequent inter-island travel together with a lack of inter-island quarantine, are extremely vulnerable to invasive alien species transfers from island to island.

*Miconia calvescens* shows most characteristics typical of successful plant colonisers on islands: competitiveness, adaptability to different habitats, capability of surviving and establishment in different habitats, tolerance of variable environmental conditions (light, temperature, moisture), tolerance of disturbance, easily dispersible to new localities (Staples *et al.*, 2000; Meyer, 2003). Framework pattern which make a coloniser successful are currently the subject of intensive ongoing research (see, e.g., Barney & Whitlow, 2008; Henderson *et al.*, 2006; Prentis *et al.*, 2008).

## Effects on unique island ecosystems

Of special concern are the negative impacts of alien plants related to so-called 'endemic' species, elements of the native floras having evolved uniquely on one single island or archipelago and quite often exhibiting a very restricted area of distribution. Each case of a considerable reduction in numbers of individuals of such an endemic species or, in the worst case, an extinction is an irreversible loss. Competition with invasive alien species is documented to have caused severe impacts on some species endemic to islands, as the following examples show. On Tahiti, 40–50 native plant species are directly affected by the invasion of the Melastomataceae *Miconia calvescens* (Meyer & Florence, 1996). Several endemic species have become seriously endangered, including, among others, *Psychotria speciosa* (Rubiaceae) and *Lepinia taitensis* (Apocynaceae) (Meyer, 2004). The strawberry guava, *Psidium cattleianum* (Myrtaceae), has invaded forests on most islands in the Austral-Pacific region (Shine *et al.*, 2003) and on Hawaii (Wagner *et al.*, 1999). On Tubuai (Austral Islands), stands of *P. cattleianum* threaten the rare native species *Charpenteria australis* (Amaranthaceae) and *Meryta brachypoda* (Araliaceae) (Meyer, 2004). On Hawaii, *Psidium cattleianum* has resulted in a drastic decline in numbers of species and individuals of rare endemic understorey plants (e.g. of the genus *Cyrtandra*, Gesneriaceae), probably by reducing the success rate of seedling establishment due to altered light and soil conditions (Kiehn, unpubl.). As result of competition with a number of invasive alien species including *Ardisia elliptica* (Myrsinaceae), *Cestrum nocturnum* (Solanaceae), *Mikania micrantha* (Asteraceae) and *Passiflora rubra* (Passifloraceae), the Euphorbiaceae

Figure 15.5 Examples of ornamental plants with proven invasive capacity on (sub)tropical islands: (a) *Spathodea campanulata*; (b) *Thunbergia grandiflora*; (c) *Ardisia elliptica* – all photos from natural sites on the Hawaiian Islands. Photos: Michael Kiehn.

*Acalypha wilderi*, endemic to Rarotonga (Cook Islands), meanwhile might have gone extinct (Meyer, 2004). The firetree *Morella or Myrica faya* (Myricaceae; Fig. 15.5), originally native to the Azores, Madeira and the Canary Islands, and introduced to Hawaii in the 1890s, was first observed at Hawaii Volcanoes National Park in 1961. Due to its seeds being dispersed by birds and its ability to fix nitrogen from the air with the help of bacteria living in its roots, it has spread considerably, in spite of eradication efforts and some dieback (Adler *et al.*, 1998). At Hawaii Volcanoes National Park, *Morella faya* now covers large areas and is seriously threatening native species only occurring there (Baskin, 2002). In addition, the species is also found on other Hawaiian Islands (Whiteaker & Gardner, 1992).

## Strategies and actions

In recent years, there has been an increasing awareness of the negative impacts of alien plant invaders on island ecosystems, and of the need for actions to mitigate these effects. Activities to reduce the number of potentially invasive plant species transferred to new geographical areas have been started. For example, in Article 8(h) of the *Convention on Biological Diversity* (CBD), Parties have committed themselves to 'prevent the introduction of, control or eradicate those alien species which threaten ecosystems, habitats and species'. Measures include strategies and legislation related to alien invasive species at national and/or regional levels, and reports of best practice examples (www.cbd.int/invasive/about.shtml; McNeely *et al.*, 2001). Coordination bodies for such activities include the IUCN Invasive Species Specialist Group (ISSG) and the Global Invasive Species Programme (GISP). The latter was established in 1996 as a partnership between the Scientific Committee on Problems of the Environment (SCOPE), the World Conservation Union (IUCN) and the Centre for Agricultural and Bioscience International (CABI) with the mission to conserve biodiversity and sustain human livelihoods by minimising the spread and impact of invasive alien species. The GISP aims at providing support to the implementation of Article 8(h) of the CBD, and has three overarching objectives: (1) to facilitate the development, adoption and implementation of science-based policies and systems of governance; (2) to raise awareness of threats posed by, and solutions for, addressing the issue; (3) to promote the exchange of information for broader uptake by decision makers (GISP, 2008).

### *Collaborations and networks*

Shortcomings and obstacles hampering effective mitigation of negative impacts of invasive aliens to island ecosystems include the lack (or inaccessibility) of scientific information needed as a basis for risk-assessment and management of invasive species, or insufficient networking mechanisms and collaboration beyond island and/or national borders for the dissemination of information. Also the existing legislation and its enforcement are sometimes not suited to fully address problems created by alien invasive plants. Trained personnel to monitor sites and to detect newly arrived potentially invasive species are lacking, as well as sufficient funding for such training, the establishment of the necessary research and information infrastructure, the development of risk assessment procedures, and the management of and research on invasive species (Sherley, 2000).

To take a global approach and to combine efforts against the negative impacts of alien invaders in island ecosystems, a Cooperative Initiative on Invasive Alien Species on Islands (CII) was launched in April 2002, at the sixth Conference of the Parties of the CBD, under the umbrella of GISP. Initiated by the group of small island developing states and developed as a joint initiative involving the New Zealand Government and the Invasive Species Specialist Group (ISSG) of the IUCN, the CII aims to facilitate cooperation and build capacity to manage invasive alien species on islands (Clout, 2004).

There is general agreement that mitigation of the effects of invasive species on biodiversity is best coordinated regionally (Sherley, 2000). Therefore, networks at local and regional level to improve the exchange of scientific information, to foster scientific training and to coordinate activities related to invasive alien species on islands are being set up. In the Pacific region the CII created the Pacific Invasives Initiative (PII). The GISP started a network for the Austral-Pacific region (Shine *et al.*, 2003) and another network was set up in the context of the South Pacific Regional Environment Programme (SPREP), with the goal to 'promote the efforts of Pacific island countries and territories in protecting and maintaining the rich and fragile natural heritage of the Pacific islands from the impacts of invasive species through cooperative efforts' (Sherley *et al.*, 2000). This network organised a number of conferences, resulting in a technical review of invasive species and a draft regional strategy for the Pacific island region. The aim of this strategy is: 'To promote the efforts of Pacific island countries and territories in protecting and maintaining the rich and fragile natural heritage of the Pacific islands from the impacts of invasive species through operative efforts to: (1) develop and maintain an effective, coordinated network of information and technical expertise; (2) prevent the introduction of new invasive species; (3) reduce the impact of existing invasive species; (4) raise awareness; and (5) build the capacity required to manage the threats posed by invasive species' (Sherley, 2000). Steps proposed to reduce or avoid negative effects of invasive alien plants on islands thus include: raising of awareness, creation of information and infrastructure, exchange of knowledge, precautionary measures to prevent new invasions (including early warning systems and legislative regulations) and mitigation of effects of taxa already present (including attempts to control or to eradicate such taxa). In 2010, a Pacific Island Specialist Group (PIPSG) was established within the IUCN Species Survival Commission, with one focus on dealing with invasive alien plant species.

In the Caribbean region CABI, with a number of national, regional and international partners, has been since July 2006 involved in a pilot project aiming at 'mitigating the threats of invasive alien species in the insular Caribbean'. This project intends to (1) develop national strategies; (2) establish a Caribbean-wide cooperation and strategy; (3) improve information management; (4) prevent introductions; and (5) facilitate early detection and management. Project activities were initially focused on six pilot countries: the Bahamas, Cuba, the Dominican Republic, Jamaica, St Lucia and Trinidad & Tobago. In-country and international workshops were organised and gap analyses were carried out as the basis for a full-size project, which is expected to last from 2008 to 2012 (GISP, 2007).

### *Capacity building and training*

An important capacity-building process is training in identification for early detection of known potential invasive plant species in areas not yet affected by them. Such initiatives have already begun in various regions. The Pacific Invasive Learning Network (PILN), a body established in 2006 with the mission 'to empower effective invasive species management through a participant-driven network that meets priority needs, rapidly shares skills and resources, provides links to technical expertise, increases information exchange, and accelerates on-the-ground action', provides this type of expertise (GISP, 2006). A training workshop for the design, implementation, monitoring and evaluation of priority invasive plant management projects in Micronesia was held in Pohnpei (Federal States of Micronesia) in February-March 2009, in the context of the PII (Pacific Invasives Initiative, 2009).

### *Prevention and monitoring*

Preventing the arrival and establishment of new alien invasive plant species should be given highest priority (McNeely *et al.*, 2001; Wittenberg & Cock, 2001). Only early detection allows for early control and improves the chances of successful eradication. Detection, at least, will leave options for measures to reduce the negative impacts on biodiversity. Control of invasive plant species will become more difficult and expensive if founder populations cannot be efficiently detected and destroyed early.

An increasing number of publications deal with preventive measures, early detection, risk assessments and control methods for invasive alien plant species on islands, or present optimised approaches for monitoring of areas and/or species of potential concern (e.g. Cronk & Fuller, 1995; Daehler & Carino, 2000; Daehler *et al.*, 2004; Meyer, 2000, 2004; Sherley & Lowe, 2000; Timmins & Braithwaite, 2002). For example, proposals for precautionary measures are provided in relation to anticipated invasions of taxa already known for their invasive potential. Success of an invasive species depends not only upon its biological characteristics but also on the traits of the habitat invaded. Thus, looking at the 'behaviour' of a potential alien invasive species in other ecosystems helps to gain a more realistic risk analysis, not only providing insights on potential invasiveness, but also on habitat and species characteristics susceptible for that potential invasion. This additional information allows optimising of management strategies to reduce the invader's effect (Daehler *et al.*, 2004; Traveset *et al.*, 2008). As regards taxa of supposingly high impact, based on observations in already invaded regions, Sherley and Lowe (2000) describe miconia (*Miconia* calvescens, Melastomataceae) and the 'mile-a-minute' vine (*Mikania micrantha*, Asteraceae) as 'devastating groups of organisms for which there should be a special watch'. Meyer (2000, 2004) lists species which are already found as invasives on a few islands in Polynesia, and proposes careful monitoring for these species on other Polynesian islands to avoid their introduction or, at least, their naturalisation. Taxa explicitly mentioned by Meyer (2004) include the African rubber tree, *Funtumia elastica* (Apocynaceae; actually only reported as invasive on the island of 'Upolu, Samoa), *Mikania micrantha* (Asteraceae; invasive in Rarotonga), and Koster's curse, *Clidemia hirta* (Melastomataceae; highly invasive on Samoa, Fiji, Wallis and Futuna, and on the Hawaiian Islands, but still absent from eastern Polynesia).

The example of *Clidemia hirta* in Hawaii illustrates how a delay in taking appropriate measures leads to a situation where a species gets beyond control (Mack & Lonsdale, 2002; Smith, 1992): *C. hirta* was first collected in the wild on Oahu in 1949. Immediate action should have been taken at that time as it was already known as an invasive weed from Fiji. In 1952, *C. hirta* only covered *c.*100 hectares on Oahu and eradication, most probably, would still have been possible. But the action taken against its spreading in 1954 was not an eradication plan, but an attempt to control it through the release of a thrip. This did not prevent *C. hirta*, whose seeds are dispersed by birds, from spreading into native forests on the island. By the 1970s, the species covered more than 30 000 hectares on Oahu; further activities to manage the species were unsuccessful, and by 1992, *C. hirta* was reported to occupy *c.*100 000 hectares on Oahu and had spread to five other main islands in the Hawaiian Islands, with the tendency to continue its area expansion.

Meyer and Lavergne (2004), in their paper on invasive and potentially invasive alien species of the family Acanthaceae, critically remark that 'a large number of websites dedicated to ornamental garden plants and gardening books still recommend these invasive acanths as ornamental plants in the tropics'. As a precautionary measure, they recommend banning or discouraging the introduction, cultivation and transportation of at least the 16 Acanthaceae species identified as currently or/and potentially invasive within the Pacific region.

## *Eradication*

If prevention has not been successful, eradication of an invasive species with proven negative impacts should be given high priority. This is also expressed in the CBD, Article 8(h), which encourages the Parties to the convention to 'prevent the introduction of, control or eradicate those alien species which threaten ecosystems, habitats or species'. With good planning it is, in many cases, feasible to eradicate invasive species, especially in the early stages of an invasion, or where a population is confined to an island or small habitat. However, eradication requires well-planned strategies (e.g. to tackle the whole population of the species to be removed), adequate techniques (to remove individuals of the targeted species faster than they reproduce, and to prevent re-invasion) and sustained effort to complete the eradication (including monitoring afterwards). Thus eradication measures are likely to be quite labour and cost intensive (Clout & Veitch, 2002). To implement eradication programmes successfully it is important to collect as much biological information as possible about the species concerned (including, e.g., data about reproduction cycles, means of dispersal and habitat preferences), to exchange experiences about details of earlier eradication or management measures, to demonstrate the benefits of the eradication programme (for the habitats and the species concerned, as well as for the livelihood of the local people, if applicable) and to ensure the involvement and support of local people.

In recent years, a numbers of scientific programmes, conferences and papers have dealt with the eradication of invasive alien species on islands and have provided baseline information, best practice examples and benchmarks (e.g. Stone *et al.*, 1992; Sherley *et al.*, 2000; Veitch & Clout, 2002; Mauremootoo, 2003; Shine *et al.*, 2003; Clout, 2004). General lessons learned from eradication projects have been summarised by Mack and Lonsdale (2002):

1 The ideal eradication campaign would see all individuals of a potentially-invasive species destroyed immediately upon their arrival. Few immigrants meet this fate; more usually there is a failure to act until damage has been inflicted by the invader.
2 Failing the destruction of all immigrants upon their entry, maximum effort should be lodged against the immigrants' small, isolated foci ...
3 Eradication or even effective control of invasive species requires repeatedly surveying the same area for surviving plants. Virtual eradication of *Schinus terebinthifolius* and other invasive species on tiny islands in Bermuda has clearly succeeded through such diligence.
4 Control or even eradication of a single invasive species may ultimately produce little benefit, if its demise only sparks the rise of another non-indigenous species.

Eradication campaigns for invasive plant species typically need to last for 10 years or more. If the goal in eradication is environmental conservation, attention should focus on creating a zone sanitaire, rather than on undertaking single species removals, and on establishing the desired native vegetation as rapidly as possible.

*(Mack & Lonsdale, 2002)*

Rejmánek and Pitcairn (2002) provide data about areas of infestation likely to by successfully eradicated. They conclude that eradication of alien plant invasions on areas smaller than one hectare are usually possible and that measures on larger areas are less successful, while the costs rise drastically. They state that 'with a realistic amount of resources, it is very unlikely that infestations larger than 1000 ha can be eradicated', but emphasise that even for areas where eradication of a plant invader will not be possible it might be worthwhile and beneficial for the native species to carry out an ongoing programme to control the invasive alien.

Such baseline data and experiences are used for feasibility studies (for example, by the CII to plan the eradication of invasive weeds on Cocos Island, Costa Rica; Clout, 2004), or form the basis for concrete actions, e.g. by the Pacific Invasive Learning Network (PILN) to eradicate strawberry guava (*Psidium cattleyanum*,

Myrtaceae) on American Samoa or the mile-a-minute vine (*Mikania micrantha*, Asteraceae) on Pohnpei (GISP, 2006).

## *Controls*

Different options for control measures have been utilised against alien invasive plant species. They range from regular mechanical removal to the use of herbicides or the release of animals (mostly specialised insects) that damage the invasive plant species. Numerous case studies have been published, and research and documentation of such approaches is ongoing (see, e.g., www.invasive.org/). A well-documented example of the successful control of a single invasive plant species in a limited area is provided by Flint and Rehkemper (2002). They describe the measures taken against the sandbur, *Cenchrus echinatus* (Poaceae), on Laysan Island, a Hawaiian Islands National Wildlife Refuge. This annual grass species native to Central America was first documented on Laysan in 1961. By 1991 the species was dominant on 30% of the vegetated area of the island, with negative impacts on the native bunchgrass, *Eragrostis variabilis*, and also on the breeding habitats for indigenous and endemic animals including the endangered Laysan finch (*Telespiza cantans*) and the Laysan duck (*Anas laysanensis*). An eradication programme against the species started in 1991, using a herbicide (glyphosate) and mechanical control (hand pulling). Ongoing monitoring indicated that, as a result of these efforts, the number of *Cenchrus echinatus* plants on Laysan has continuously declined, and in 1999 the species was so rare that it was no longer considered to have a negative effect on the ecosystems of the island. However, the authors stated the need for ongoing monitoring to avoid the re-establishment of the species.

## Perspectives

In addition to the serious threats to island native plant species posed by alien invaders, there are a number of other interesting facts to be mentioned. For example, the arrival of new species on islands has increased the number of plant species present on individual islands up to 10-fold (Table 15.1); even on a large island like New Zealand the number of plant species has doubled from 2000 to 4000 (Sax *et al.*, 2007). This provides opportunities for new research related to plant evolution and speciation which can contribute to better conservation methodologies (Sax *et al.*, 2007). Such studies indicate, for example, that competition from alien invaders can cause extinction of native species at a local scale, but is unlikely to have caused a complete extinction of a species up till now. Other factors are claimed to be the driving forces for such extinction events (Sax *et al.*, 2007).

The effects of climate change are likely to create new challenges for the future. This already can be seen on the southernmost outpost of Australia, Heard Island. This partially glaciated island was free of non-native plant invaders until the 1980s, when winter grass (*Poa annua*, Poaceae) was first reported. This species, originating from Europe, is advancing on sites where glaciers are withdrawing, at more than 100 metres a year, with unknown consequences for the island ecosystems (Low, 2008).

More and more tools are available to help decision makers and scientists to optimise their conservation efforts on islands. Studies of plant invasions on islands remain a necessary and rewarding challenge. They are helpful to evaluate aspects of global change. Their results play an important role in the development of sound conservation, management and policy strategies. By contributing to our understanding of mechanisms of species extinctions, species evolution and diversification, and ecosystem functioning, such studies help to safeguard the survival of threatened species and habitats.

## Acknowledgements

The author thanks Ole Hamann (Copenhagen, Denmark) and Jean-Yves Meyer (Délégation à la Recherche, Tahiti, French Polynesia) for providing photos and David Lorence (National Tropical Botanical Garden, Kauai'i, Hawaii, USA), Jean-Yves Meyer (Délégation à la Recherche, Tahiti, French Polynesia) and Josef Greimler (Department of Systematic and Evolutionary Botany, University of Vienna, Austria) for stimulating discussion and sharing information on the subject.

## References

Adler, P. B., d'Antonio, C. M. & Tunison, J. T. (1998). Understory succession following a dieback of *Myrica faya in* Hawaii Volcanoes National Park. *Pac. Sci.* 52: 69–78.

Barney, J. N. & Whitlow, T. H. (2008). A unifying framework for biological invasions: the state factor model. *Biol. Invas.* 10: 259–272.

Baskin, Y. (2002). *A Plague of Rats and Rubbervines: The Growing Threat of Species Invasions.* Washington, DC, Island Press.

Clout, M. N. (2004). The GISP partnership network in action: a focus on the Invasive Species Specialist Group (ISSG). *GISPNews* 3: 6–7.

Clout, M. N. & Veitch, C. R. (2002). Turning the tide of biological invasion: the potential for eradicating invasive species. In C. R. Veitch & M. N. Clout (eds.), *Turning the Tide: The Eradication of Invasive Species. Proceedings of the International Conference on Eradication of Island Invasives.* Occasional Papers of the IUCN Species Survival Commission 27. pp. 1–3. Gland, Switzerland, IUCN.

Cronk, Q. C. B. (2000). *The Endemic Flora of St Helena.* Oswestry, UK, Anthony Nelson.

Cronk, Q. C. B. & Fuller, J. L. (1995). *Plant Invaders: The Threat to Natural Ecosystems.* London, Chapman & Hall.

Daehler, C. C. & Carino, D. A. (2000). Predicting invasive plants: prospects for a general screening system based on current regional models. *Biol. Invas.* 2: 92–103.

Daehler, C. C., Denslow, J. S., Ansari, S. & Kuo, H-C. (2004). A risk-assessment system for screening out invasive pest plants from Hawaii and other Pacific Islands. *Conserv. Biol.* 18: 360–368.

Denslow, J. S. (2003). Weeds in paradise: thoughts on the invasibility of tropical islands. *Ann. Mo. Bot. Gard.* 90: 119–127.

Flint, E. & Rehkemper, C. (2002). Control and eradication of the introduced grass, *Cenchrus echinatus*, at Laysan Island, Central Pacific Ocean. In C. R. Veitch & M. N. Clout (eds.), *Turning the Tide: The Eradication of Invasive Species. Proceedings of the International Conference on Eradication of Island Invasives, Occasional Papers of the IUCN Species Survival Commission* 27: 110–115. Gland, Switzerland, IUCN.

Florence, J. (2003). La Polynésie française: environnement phytogéographique, l'homme et ses plantes. Archéologie en Oceanie Insulaire. In C. Orliac (ed.), *Peuplement, Société et Paysages*, pp. 130–149. Paris, Editions Artcom.

GISP (Global Invasive Species Programme) (2006). Palau. Pacific Invasive Learning Network meets. *GISPNews* 6: 17.

GISP (Global Invasive Species Programme) (2007). CABI spearheads Caribbean project. *GISPNews* 9: 8.

GISP (Global Invasive Species Programme) (2008). Global Strategy 2008–2010. Nairobi, Kenya, GISP.

Greimler, J., Stuessy, T. F., Swenson, U., Baeza, C. M. & Matthei, O. (2002). Plant invasions on an oceanic archipelago. *Biol. Invas.* 4: 72–85.

Guézou, A., Trueman, M., Buddenhagen, C. E., Chamorro, S., Guerrero, A. M., Poza, P. & Atkinson, R. (2010). An extensive alien plant inventory from the inhabited areas of Galapagos. *PLos ONE 5.doi:10.1371/journal.pone.0010276 (30.3.2011).*

Henderson, S., Dawson, T. P. & Whittaker, R. J. (2006). Progress in invasive plants research. *Prog. Phys. Geog.* 30: 25–46.

Kueffer, C., Daehler, C. C., Torres-Santana, C. W., Lavergne, C., Meyer, J.-Y., Otto, R. & Silva, L. (2010). A global comparison of plant invasions on oceanic islands. *Perspectives in Plant Ecology, Evolution and Systematics* 12: 145–161.

Lavergne, C., Rameau, J.-C. & Figier, J. (1999). The invasive woody weed *Ligustrum robustum* subsp. *walkeri* threatens native forests on La Réunion. *Biol. Invas.* 1: 377–392.

Low, T. (2008). *Climate Change & Invasive Species: A Review of Interactions. November 2006 Workshop Report.* Canberra, Department of the Environment, Water, Heritage and the Arts.

Mack, R. N. (2001). Motivations and consequences of the human dispersal of plants. In J. A. McNeely (ed.), *The Great Reshuffling: Human Dimensions in Invasive Alien Species*, pp. 23–34. Gland, Switzerland, IUCN.

Mack, R. N. & Lonsdale, W. M. (2002). Eradicating invasive plants: hard-won lessons for islands. In C. R. Veitch & M. N. Clout (eds.), *Turning the Tide: The Eradication of Invasive Species. Proceedings of the International Conference on Eradication of Island Invasives, Occasional Papers of the IUCN Species Survival Commission* 27: 164–172. Gland, Switzerland, IUCN.

Mauremootoo, J. (ed.) (2003). *Proceedings of the Regional Workshop on Invasive Alien Species and Terrestrial Ecosystem Rehabilitation in Western Indian Ocean Island States – Sharing Experience, Identifying Priorities and Defining Joint*

*Action. Seychelles 13–17 October 2003.* Quatre Bornes, Mauritius, Indian Ocean Commission.

Mauremootoo, J. & Boudjelas, S. (2003). The work of the IUCN in invasive species management. In J. Mauremootoo (ed.), *Proceedings of the Regional Workshop on Invasive Alien Species and Terrestrial Ecosystem Rehabilitation in Western Indian Ocean Island States – Sharing Experience, Identifying Priorities and Defining Joint Action. Seychelles 13–17 October 2003*, pp. 11–21. Quatre Bornes, Mauritius, Indian Ocean Commission.

McNeely, J. A., Mooney, H. A., Neville, L. E., Schei, P. & Waage, J. K. (eds.) (2001). *Global Strategy on Invasive Alien Species.* Gland, Switzerland, IUCN.

Merlin, M. & Juvik, J. (1992). Relationships between native and alien plants on Pacific islands with and without significant human disturbance and feral ungulates. In C. P. Stone, C. W. Smith & J. T. Tunison (eds.), *Alien Plant Invasions in Native Ecosystems of Hawaii: Management and Research*, pp. 597–625. Honolulu, University of Hawaii Press.

Meyer, J.-Y. (2000). Preliminary review of the invasive plants in the Pacific islands (SPREP Member Countries). In G. Sherley (ed.), *Invasive Species in the Pacific: A Technical Review and Draft Regional Strategy*, pp. 85–114. Apia, Samoa, South Pacific Regional Environment Programme.

Meyer, J.-Y. (2003). French Polynesia. In C. Shine, J. K. Reaser & A. T. Gutierrez (eds.), *Invasive Alien Species in the Austral-Pacific Region. National Reports & Directory of Resources*, pp. 22–34. Cape Town, South Africa, Global Invasive Species Programme.

Meyer, J.-Y. (2004). Threat of invasive alien plants to native flora and forest vegetation of eastern Polynesia. *Pac. Sci.* 58: 357–375.

Meyer, J.-Y. & Florence, J. (1996). Tahiti's native flora endangered by the invasion of *Miconia calvescens* DC. (Melastomataceae). *J. Biogeogr.* 23: 775–783.

Meyer, J.-Y. & Lavergne, C. (2004). Beautés fatales: Acanthaceae species as invasive alien plants on tropical Indo-Pacific Islands. *Diversity Distrib.* 10: 333–347.

Pacific Invasives Initiative (2009). Another successful training workshop for weed management project design and implementation. *Pacific Invasives Initiative: The News* May 2009.

Prentis, P. J., Wilson, J. R. U., Dormontt, E. E., David, M., Richardson, D. M. & Lowe, A. J. (2008). Adaptive evolution in invasive species. *Trends Plant Sci.* 13: 288–294.

Rejmánek, M. & Pitcairn, M. J. (2002). When is eradication of exotic pests a realistic goal? In C. R. Veitch & M. N. Clout (eds.), *Turning the Tide: The Eradication of Invasive Species. Proceedings of the International Conference on Eradication of Island Invasives. Occasional Papers of the IUCN Species Survival Commission* 27: 249–253. Gland, Switzerland, IUCN.

Sax, D. F., Gaines, S. D. & Brown, J. H. (2002). Species invasions exceed extinctions on islands worldwide: a comparative study of plants and birds. *Am. Nat.* 160: 766–783.

Sax, D. F., Stachowicz, J. J., Brown, J. H., Bruno, J. F., Dawson, M. N., Gaines, S. D. *et al.* (2007). Ecological and evolutionary insights from species invasions. *Trends Ecol. Evol.* 22: 465–471.

Schofield, E. K. (1989). Effects of introduced plants and animals on island vegetation: Examples from the Galapagos Archipelago. *Conserv. Biol.* 3: 227–238.

Sherley, G. (ed.) (2000). *Invasive Species in the Pacific: A Technical Review and Draft Regional Strategy*. Apia, Samoa, South Pacific Regional Environment Programme.

Sherley, G. & Lowe, S. (2000). Towards a regional invasive species strategy for the South Pacific: issues and options. In G. Sherley (ed.) *Invasive Species in the Pacific: A Technical Review and Draft Regional Strategy*, pp. 7–18. Apia, Samoa, South Pacific Regional Environment Programme.

Sherley, G., Timmins, S. & Lowe, S. (2000). Draft Invasive Species Strategy for the Pacific Islands Region. In G. Sherley (ed.) *Invasive Species in the Pacific: A Technical Review and Draft Regional Strategy*, pp. 1–5. Apia, Samoa, South Pacific Regional Environment Programme.

Shine, C., Reaser, J. K. & Gutierrez, A. T. (eds.) (2003). *Invasive Alien Species in the Austral-Pacific Region. National Reports & Directory of Resources*. Cape Town, South Africa, Global Invasive Species Programme.

Soria, M. C., Gardener, M. R. & Tye, A. (2002). Eradication of potentially invasive plants with limited distribution in the Galapagos Islands. In C. R. Veitch & M. N. Clout (eds.), *Turning the Tide: The Eradication of Invasive Species. Proceedings of the International Conference on Eradication of Island Invasives,. Occasional Papers of the IUCN Species Survival Commission* 27: 287–292. Gland, Switzerland, IUCN.

Smith, C. W. (1992). Distribution, status, phenology, rate of spread, and management of *Clidemia* in Hawaii. In C. P. Stone, C. W. Smith & J. T. Tunison (eds.), *Alien Plant Invasions in Native Ecosystems of Hawaii: Management and Research*, pp. 241–253. Honolulu, University of Hawaii Press.

Staples, G. W., Herbst, D. & Imada, C. T. (2000). Survey of invasive or potentially invasive cultivated plants in Hawaii. *Bishop Museum Occasional Papers* 65.

Stone, C. P., Smith, C. W. & Tunison, J. T. (eds.) (1992). *Alien Plant Invasions in Native Ecosystems of Hawaii: Management and Research*. Honolulu, University of Hawaii Press.

Timmins, S. M. & Braithwaite, H. (2002). Early detection of weeds on islands. In C. R. Veitch & M. N. Clout (eds.), *Turning the Tide: The Eradication of Invasive Species. Proceedings of the International Conference on Eradication of Island Invasives. Occasional Papers of the IUCN Species Survival Commission* 27: 311–318. Gland, Switzerland, IUCN.

Traveset, A., Brundu, G., Carta, L., Mprezetou, I., Lambdon, P., Manca, M. *et al.* (2008). Consistent performance of invasive plant species within and among islands of the Mediterranean basin. *Biol. Invas.* 10: 847–858.

Tye, A., Soria, M. C. & Gardener, M. R. (2002). A strategy for Galapagos weeds. In C. R. Veitch & M. N. Clout (eds.), *Turning the Tide: The Eradication of Invasive Species. Proceedings of the International Conference on Eradication of Island Invasives. Occasional Papers of the IUCN Species Survival Commission* 27: 336–341. Gland, Switzerland, IUCN.

Veitch, C. R. & Clout, M. N. (eds.) (2002). *Turning the Tide: The Eradication of Invasive Species. Proceedings of the International Conference on Eradication of Island Invasives. Occasional Papers of the IUCN Species Survival Commission* 27. Gland, Switzerland, IUCN.

Vielle, M. (2003). Management priorities for Seychelles – a basis for discussion. In J. Mauremootoo (ed.), *Proceedings of the Regional Workshop on Invasive Alien*

*Species and Terrestrial Ecosystem Rehabilitation in Western Indian Ocean Island States – Sharing Experience, Identifying Priorities and Defining Joint Action. Seychelles 13–17 October 2003*, pp. 99–104. Quatre Bornes, Mauritius, Indian Ocean Commission.

Wagner, W. L., Herbst, D. R. & Sohmer, S. H. (1999). *Manual of the Flowering Plants of Hawaii*, revised edition. Honolulu, University of Hawaii Press and Bishop Museum Press.

Whistler, W. A. (2009). *Plants of the Canoe People*. Lawai, Kauai, National Tropical Botanical Garden.

Whiteaker, L. D. & Gardner D. E. (1992). Firetree (*Myrica faya*) distribution in Hawaii. In C. P. Stone, C. W. Smith & J. T. Tunison (eds.), *Alien Plant Invasions in Native Ecosystems of Hawaii: Management and Research*, pp. 225–240. Honolulu, University of Hawaii Press.

Williamson, M. & Fitter, A. (1996). The varying success of invaders. *Ecology* 77: 1661–1666.

Wittenberg, R. & Cock, M. J. W. (2001). *Invasive Alien Species: A Toolkit of Best Prevention and Management Practices*. Wallingford, UK, CAB International.

# 16

# Ecology, demography and conservation in the Galapagos Islands flora

OLE HAMANN

## The islands

The Galapagos are true oceanic islands, which except for the two northernmost, Darwin and Wolf, rise from a relatively shallow submarine platform. The oldest islands are composed of uplifted lava flows and are about 4.5 million years old (Ma): they are found in the eastern part of the archipelago (Española, Santa Fe, Plazas, Baltra, Seymour and a narrow strip along the northeast coast of Santa Cruz). All the other islands are formed by great volcanoes and are much younger; no rocks older than 0.7 Ma have been found on Fernandina, Isabela, Santiago, Pinta, Genovesa, Marchena and Darwin, and Fernandina could even be as young as 60 000 BP, according to Geist (1996). The islands consist typically of one or several shield volcanoes culminating in craters or large calderas. Eruptions are frequent, particularly in the westernmost islands, Isabela and Fernandina, and the landscape displays a variety of craters, huge, barren fields of lava, ash or scoria, and vegetated areas.

The lower, gently sloping parts of the islands are predominantly composed of lava, and soils are poorly developed or lacking, while the higher, steeply sloped parts are predominantly composed of pyroclastic deposits. On the young islands, Fernandina and Isabela, large areas at high altitudes are covered with recent lava flows or pyroclastic material, which are not at all or only sparsely weathered, and soil development in such areas is mostly poor. On older islands, such as Santa Cruz and San Cristóbal, both lava and pyroclastic material have been exposed to weathering and erosion for a much longer time, and deeper soils have developed.

*The Biology of Island Floras*, eds. David Bramwell and Juli Caujapé-Castells. Published by Cambridge University Press. 

## Climate

The climate of the Galapagos Islands differs from most other equatorial islands, being situated in the Pacific dry zone where the principal ocean currents of the Eastern Pacific converge. From about May/June to November/December a cool season is determined by the southern equatorial current systems with cool waters and steady southeast trade winds. In this season the air temperatures are low, and the trade winds bring moisture-laden air to the islands. On the windward sides of the higher islands the air is forced uphill, resulting in overcast skies, fog, drizzle and rain, called *garúa*, whereas the leeward slopes receive much less moisture because of the rain-shadow effect. The *garúa* is light at low elevations, but becomes more intense with increasing elevation and steepness on southeastern exposed slopes, which thereby receive much moisture during this season; thus, in non-El Niño years (see below) the cool season at higher elevations is the wetter season, while it is the dry season at lower elevations. At about 1000–1300 m elevation an inversion layer is usually formed, which prevents the moist air from rising further. Consequently, both the lowlands and the high altitude areas above the inversion layer on Fernandina and Isabela receive very little moisture through *garúa* during the cool season.

From about January to May/June the archipelago experiences a hot, wet season when the trade winds weaken and warmer waters from the north prevail in the archipelago. The temperatures are tropical and the skies are usually clear. Large cumulus clouds form over the islands, resulting in heavy rain showers. Although the rain may fall all over the archipelago, the interior of the larger islands usually receives most.

Annual rainfall is extremely variable; in some years the rains in the hot season fail, and years of severe drought are common (Hamann, 1979a). The variation in annual precipitation in the cool season is smaller, because the moisture then is provided by *garúa*.

Every 2–4 years the Galapagos experiences a phenomenon known as the El Niño Southern Oscillation (ENSO) event. (The term ‘El Niño’ was originally used for the annual weakening of the southeast trade winds and arrival of warm, wet weather, which generally takes places around Christmas, hence El Niño in reference to the Christ child). However, when El Niño is coupled with a reversal of the atmospheric circulation (the Southern Oscillation) the result is an ENSO event.

The occurrence of the ENSO phenomenon dramatically alters the general pattern of seasonality. During an ENSO event, the hot, wet season becomes more intense and longer-lasting, because the usual circulation patterns of winds and oceanic currents are reversed. ENSO events often comprise an El Niño period followed by a period of drought, called La Niña, of 1–2 years’ duration or sometimes more

(Snell & Rea, 1999). A 'minor' ENSO event occurs every 2–4 years, while 'major' events are less frequent, in the past century around every 15–20 years (Snell & Rea, 1999).

During the El Niño period of an ENSO event, warmer waters remain around Galapagos for long periods of time, bringing high temperatures and very intense rainfalls during several months to two years. The amount of rain can increase a hundredfold as a result of a major El Niño (van der Werff, 1990). In consequence, the seasonal and annual variations in temperatures and rainfall in the Galapagos become extreme (Hamann, 1979a; Mueller-Dombois & Fosberg 1998; Snell & Rea, 1999; Tye *et al.*, 2002).

## Ecology and distribution of vegetation (spatial patterns)

The two main factors determining the spatial occurrence and distribution of unique and characteristic ecological conditions, habitats and vegetation formations in the Galapagos are the arid, volcanic nature of the islands and the local climatic conditions, especially rainfall.

The size of the area of an island suitable for plant growth, excluding fresh lava fields, gives the strongest correlation of plant species diversity and area. Other factors of importance for the distribution of plants and vegetation are the height and age of the island. In Galapagos, the lowlands are arid while the highlands are comparatively moist. The high islands have a wider spectrum in climatic conditions, and the older the island the more time there has been for soil formation and erosion. This leads to a rather clear distinction between arid and humid areas with respect to occurrence of plant species and vegetation. At the coast, the long-term mean annual precipitation ranges from about 0 to 300 mm, while the higher parts on the south-exposed slopes of the larger islands receive an average of 2000 mm per year or more. Consequently, small, low islands are predominantly arid, while larger, high islands receive much more rain, particularly on the southern slopes exposed to the prevailing winds, and they display vegetation types ranging from dry to humid.

Many earlier researchers of the Galapagos noted the obvious altitudinal differences in vegetation found on the higher islands. For example, four altitudinal regions of vegetation were recognised by Stewart (1911, 1915), i.e. the dry, the transition, the moist and the grassy region, in effect a zonation of vegetation caused by rainfall increasing from low to high altitudes. However, Stewart did note that not all four regions were present or prominent on several large islands. Furthermore, it was not possible for Stewart to spend much time exploring the high volcanoes on Isabela and Fernandina, which meant that the vegetation at high altitudes on these volcanoes was not included in his treatment of vegetation regions (Stewart, 1911, 1915).

Bowman (1961) recognised the following altitudinal vegetation zones on Santa Cruz Island, based on structure and floristics: Littoral, Arid coastal, Transition, *Scalesia* forest, Brown, *Miconia* belt and Upland zone. This was further developed by Wiggins and Porter (1971), based on their explorations on Floreana, San Cristóbal, Santa Cruz and Volcan Sierra Negra on Isabela. Wiggins and Porter recognised seven altitudinal vegetation zones: Littoral, Arid, Transition, *Scalesia*, Brown, *Miconia* and Fern-Sedge Zone. This pattern of altitudinal zonation of vegetation has been regarded as a standard feature in the Galapagos, and it is widely used as a frame of reference in the Galapagos literature, both in research papers and in more general works such as travel guides. However, Wiggins and Porter (1971) presented the scheme with a caveat that has largely been overlooked: they noted that there were hundreds of square miles on Isabela that (at that time) had not been visited by biologists, and that consequently little was known about the similarities or differences in the ecological communities of similar habitats on different islands, or even on different parts of the same island.

Since 1971 numerous studies of Galapagos vegetation and plant communities have been made, including from areas that previously were not well known (van der Werff, 1978, 1980, 1990; Hamann, 1981; Adsersen, 1989; Atkinson, 1990; Itow, 1990, 1992, 1995, 2003; van der Werff & Adsersen, 1993), and a broader picture of the Galapagos vegetation has emerged. In 2002, Tye *et al.* presented a more comprehensive classification of the Galapagos vegetation, which is applicable to all islands in the archipelago. The Tye *et al.* (2002) classification is here adopted as an outline for a redefined, general classification of Galapagos vegetation; however, in the present chapter the classification is modified with a focus on vegetation structure, i.e. on the most common vegetation formations and subformations (Table 16.1). Therefore, the presented classification:

- focuses on vegetation structure;
- is applicable to all the islands in Galapagos;
- formally recognises a High Altitude Dry Zone, occurring on the highest volcanoes of the western islands, above the cool-season inversion layer and the level at which the hot-season clouds commonly form, and compounded by rain-shadow effects;
- treats pioneer vegetation and freshwater aquatic vegetation as distinctive entities.

As with previous vegetation classifications, this one reflects that rainfall is the main determining factor for the spatial distribution of distinct vegetation formations and subformations in the Galapagos. The focus on structural formations and subformations provides an overview of the vegetation in the archipelago and, at the same time, provides the vernacular name used for the most common vegetation units,

Table 16.1 *Galapagos vegetation. The vegetation zones display a range of characteristic formations, from forest to formations without woody plants. The zonation is largely determined by rainfall and substrate. The azonal formations are largely determined by habitat type. In most cases, the type of formation in combination with the name of the common characteristic species will give the vernacular name and classification of the vegetation, for example* Bursera *forest,* Opuntia *woodland,* Scalesia *forest and* Miconia *scrub.*

| Climatic conditions | Zones | Habitat types | Common vegetation formations and subformations (structural classes) | Common characteristic species (see text for discussion on species) |
|---|---|---|---|---|
| Dry | Littoral | Rockland & cliffs<br>Sandy beaches<br>Dunes<br>Salt marshes<br>Salt flats<br>Brackish lagoons | Forest<br>Scrub<br>Scrubland<br>Herbfield<br>Grassland<br>Sedgefield | *Rhizophora mangle*<br>*Cryptocarpus pyriformis*<br>*Nolana galapagensis*<br>*Conocarpus erecta*<br>*Scalesia helleri*<br>*Salicornia fruticosa*<br>*Tiquilia* spp.<br>*Ruppia maritima*<br>*Sporobolus virginicus*<br>*Cyperus* spp.<br>*Eleocharis* spp. |
| Dry | Dry Lowland Zone | | Forest<br>Woodland<br>Scrubland<br>Scrub steppe savanna<br>Desert scrubland | *Bursera graveolens*<br>*Opuntia* spp.<br>*Cordia lutea*<br>*Lippia rosmarinifolia*<br>*Macraea laricifolia* |
| Dry to Humid | Transition Zone | | Forest<br>Woodland<br>Low forest and scrub | *Pisonia floribunda*<br>*Psidium galapageium*<br>*Zanthoxylum fagara* |

Table 16.1 (*cont.*)

| Climatic conditions | Zones | Habitat types | Common vegetation formations and subformations (structural classes) | Common characteristic species (see text for discussion on species) |
|---|---|---|---|---|
| Humid | *Scalesia* Zone | | Forest<br>Woodland | *Scalesia pedunculata*<br>*Psidium galapageium* |
| Humid | *Zanthoxylum* Zone (including Brown Zone) | | Low forest and woodland<br>Scrub | *Zanthoxylum fagara*<br>*Croton scouleri* |
| Humid | *Miconia* Zone | | Scrub<br>Tree-fern grove | *Miconia robinsoniana*<br>*Cyathea weatherbyana* |
| Humid | Fern-Sedge Zone | | Tree-fern grove<br>Heathland<br>Herbfield<br>Grassland<br>Sedgefield<br>Fernland<br>Mossfield | *Cyathea weatherbyana*<br>*Pernettya howellii*<br>*Jaegeria crassa*<br>*Paspalum conjugatum*<br>*Rhyncospora* spp.<br>*Pteridium arachnoideum*<br>*Sphagnum* spp. |
| Dry | High Altitude Dry Zone | | Low forest<br>Woodland<br>Scrub<br>Scrubland<br>Scrub steppe savanna<br>Grassland<br>Fernland | *Scalesia cordata*<br>*Scalesia microcephala*<br>*Macraea laricifolia*<br>*Lippia rosmarinifolia*<br>*Zanthoxylum fagara*<br>*Pennisetum pauperum*<br>*Polypodium tridens* |

| | | | | |
|---|---|---|---|---|
| Mainly humid | Aquatic | Freshwater springs | Herbfield | *Potamogeton pectinatus* |
| | | Freshwater temporary pools | Floating fernfield | *Azolla microphylla* |
| Mainly dry | Pioneer | Lava & rockland | Woodland | *Jasminocereus thouarsii* |
| | | Tufa & scoria | Scrub | *Macraea laricifolia* |
| | | Ash & gravel plains | Scrubland | *Lecocarpus* spp. |
| | | Fumaroles | Cactusfield | *Brachycereus nesioticus* |
| | | | Herbfield | *Tiquilia* spp. |
| | | | Sedgefield | *Fimbristylis dichotoma* |
| | | | Fernland | *Polypodium tridens* |

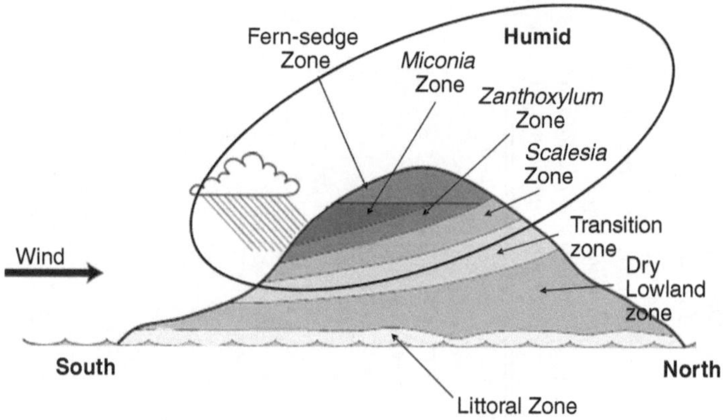


Figure 16.1 Vegetation Zones, Santa Cruz. The vegetation zones are mainly formed by the local climatic conditions and follow the gradient in increasing rainfall from low to high elevations. During most of the year the prevailing winds come from the southeast, so that south-exposed slopes receive more precipitation than north-exposed slopes. Consequently, the vegetation zones are shifted upwards on northern slopes. In addition, moisture is captured through condensation, particularly in high areas during the cool season, and large cumulus clouds form and produce frequent rain showers inland during the hot season. Thus, the high areas receive most precipitation.

such as *Bursera* forest, *Opuntia* woodland, *Scalesia* forest and *Miconia* scrub, by combining the type of formation or subformation with the name of common characteristic species.

## The Galapagos vegetation

The weathering of the volcanic substrate and the development and occurrence of soils are closely related to the altitudinal distribution of rainfall, the spatial and altitudinal distribution of substrate, and the age of the islands and the substrate. These factors determine the occurrence of defined vegetation zones (see Fig. 16.1), especially on the larger islands in the archipelago (van der Werff, 1979; Itow, 1992, 2003; Hamann, 1995; Mueller-Dombois & Fosberg, 1998; Tye *et al.*, 2002).

The constituent species often occur in a range of closely related vegetation formations (plant names are in accordance with Jørgensen and Léon-Yamez, 1999). For example: *Cryptocarpus pyriformis* can be dominant in scrub and is also a characteristic member in scrubland in the Littoral Zone as well as in various formations in the Dry Lowland Zone; *Opuntia echios* occurs in the Littoral Zone and in the whole range of vegetation formations in the Dry Lowland Zone, where it can be dominant in both forest and woodland, as well as being a characteristic member in scrubland, scrub steppe savanna and desert scrubland; *Pisonia floribunda* can be dominant in forest in the Transition Zone as well as being a member of forest and woodland in the *Scalesia* Zone; and vice versa, *Scalesia pedunculata* can be a member of the forest in the Transition Zone as well as being dominant in forest in the *Scalesia* Zone; *Zanthoxylum fagara* occurs in almost all vegetation zones, from the Dry Lowland to the High Altitude Dry Zone, but dominates in the *Zanthoxylum* Zone, and so forth. This means that the presentation below and in Table 16.1 is restricted to examples of recognisable and characteristic vegetation zones, to well-developed formations and subformations, and to their common characteristic species. There is no attempt to present a fully comprehensive description and table of all the various compositions of the Galapagos vegetation formations nor a treatment of the numerous (over 100 described) plant communities comprising the vegetation of the formations and subformations.

A brief characteristic of the vegetation zones and their vegetation formations is given below.

## Littoral Zone

The vegetation of the Littoral Zone is composed of structurally and floristically different salt-tolerant vegetation formations. The Littoral Zone forms a narrow fringe along the coast, and is not determined by rainfall, but by substrate, saltwater infiltration into groundwater, and salt spray. The larger part of the Galapagos coastline is characterised by barren cliffs and rocks, so plant life at the coast occurs in rather limited areas and in a limited number of different habitats. The littoral vegetation in the Galapagos is poorly developed compared to other tropical coasts (van der Werff & Adsersen, 1993).

The vegetation formations in the Littoral Zone range from forest, scrub and scrubland, to herbfield, grassfield and sedgefield (Table 16.1).

### *Forest, scrub and scrubland*

The littoral forest in the Galapagos is an evergreen mangrove forest, which is most fully developed in the tidal zone in sheltered bays and around lagoons on sandy clay, where sedimentation of fine material is possible (Wium-Andersen & Hamann, 1986; van der Werff & Adsersen, 1993). In Galapagos, the mangrove forest is composed of four species only, namely *Rhizophora mangle*, *Avicennia germinans*, *Laguncularia racemosa* and *Conocarpus erecta*, which occur in various combinations in different places. *Rhizophora mangle* grows to about 15 m or more in height and forms a closed forest. Usually *Rhizophora mangle* grows in a dense belt along the high tide mark, while *Avicennia germinans* and *Laguncularia racemosa* dominate farther inland. The mangroves can also establish themselves on lava rock, in fissures with very little sediment; in such places they form a low scrubland, where *Conocarpus erecta* is often the main constituent.

By far the most common and widespread littoral vegetation is a more or less evergreen scrub with *Cryptocarpus pyriformis* as the main species (saltbush scrub). The scrub occurs on almost all coastal substrates, from lava rock to sandy beaches. Besides *Cryptocarpus*, the main constituents are *Lycium minimum, Scutia spicata, Grabowskia boerhaaviaefolia, Maytenus octogona, Vallesia glabra, Alternanthera filifolia* and *Opuntia echios.* When best developed, the saltbush scrub is about 1.5–2 m high and forms a very dense belt 40–50 m. Some of the species in the scrub also occur, or even have their main occurrence, further inland in other vegetation zones, such as *Opuntia echios* and *Maytenus octogona.*

Other types of littoral scrub and scrubland are confined to specific substrates, for example, sand beaches and dunes. The main constituents in these habitats are *Nolana galapagensis* and *Scaevola plumieri*, which form a low, evergreen scrub or scrubland. Around shallow, hypersaline lagoons, *Batis maritima* often forms a low scrub, accompanied by *Salicornia fruticosa.*

Littoral scrubland is widespread on rock pavements and on cliffs, especially in wind-exposed areas. Succulent species such as *Sesuvium edmonstonei* and *Portulaca howellii* are more or less confined to rockland, while other common species such as *Opuntia echios* and *Jasminocereus thouarsii* also occur on cliffs as well as farther inland, with a main distribution in the Dry Lowland Zone. The scrubland on coastal cliffs in the Galapagos often has an unusual species composition: for example, *Scalesia helleri* ssp. *helleri* is confined to cliffs on Santa Fe Island, and *Scalesia crockeri* is only found on the cliffs on Baltra and northeastern Santa Cruz.

### *Herbfield, sedgefield and grassland*

Littoral herbfield and grassland occur on sandy beaches and dunes, as well as on salt flats and around lagoons. Among the common species on sandy substrates are *Tiquilia* spp., *Heliotropium curassavicum, Ipomoea pes-caprae* and *Sporobolus virginicus.* These species may occur in dense formations with no woody species present, but they can also form a ground layer in the coastal scrubland.

Salt flats and lagoons are widespread, mainly on the larger islands. In the lagoons, *Ruppia maritima* often form a dense, submerged herbfield, while a number of herbs, grasses and sedges constitute the fringes of the lagoons. Various species of *Cyperus* and *Eleocharis* are often dominant in sedgefields around the lagoons, accompanied by a few salt-tolerant herb and grass species, such as *Atriplex peruviana* and *Sporobolus virginicus.*

Figure 16.2 *Opuntia echios* var. *barringtonensis*, Santa Fe. *Opuntia echios* var. *barringtonensis* is endemic to Santa Fe Island. It is the most impressive of the Galapagos *Opuntia* taxa, growing to a height of 10–12 m and developing a trunk up to 1.25 m in diameter. It is a conspicuous and often dominant member of the prevailing Dry Lowland Zone on the island.

## Dry Lowland Zone

The vegetation in the Dry Lowland Zone is determined by the low annual precipitation, by the lava substrates and by poor soil development. The magnificent *Opuntia* (Fig. 16.2) and *Jasminocereus* cacti and the *Bursera graveolens* trees of this zone are the hallmark of the Galapagos landscape. Small and low islands have only the Littoral and Dry Lowland Zones of vegetation. On larger islands, the Dry Lowland Zone stretches inland from the coast, often many kilometres. The leeward slopes receive much less rain than the windward slopes because of the rain-shadow effect; consequently, the vegetation zones are shifted up on leeward slopes, so that extensive areas on the leeward sides of large islands are covered with the xerophytic vegetation of the Dry Lowland Zone (Fig. 16.1). Of all the vegetation zones, the Dry Lowland Zone covers by far the largest area in the islands.

The formations and subformations in the Dry Lowland Zone range from forest, woodland (steppe forest) and scrubland to desert steppe savanna and desert scrubland (Table 16.1).

### *Forest and woodland*

When best developed, the characteristic vegetation in the Dry Lowland Zone is dry-season deciduous forest. The forest occurs on all large and several medium-sized islands, where rainfall is adequate. The vegetation is often dense, with a low tree canopy, a dense shrub layer and a seasonal ground layer of herbs and vines. The canopy is dominated by *Bursera graveolens* and/or *Opuntia* spp. and tree species such as *Zanthoxylum fagara*, *Piscidia carthagenensis* and *Erythrina velutina*. Large *Jasminocereus thouarsii* are often conspicuous, emerging above the tree and shrub layer and reaching several metres in height. The shrub layer in the forest is formed by *Croton scouleri*, low-grown *Maytenus octogona*, *Acacia* spp. and *Prosopis juliflora*, plus numerous shrubby species such as *Scutia spicata*, *Cordia* spp., *Tournefortia* spp., *Lantana peduncularis*, *Castela galapageia*, *Alternanthera* spp., *Clerodendrum molle* and *Waltheria ovata*.

In areas with less rainfall, the vegetation takes the form of woodland, which in most places is formed by the same species as in the forest. Typically, *Bursera graveolens* and/or *Opuntia* spp. occur as the only tall species, while a variety of shrubby species form an open shrub layer. For example, vast areas on the leeward slopes of the larger islands are covered with *Bursera* woodland, in which *Opuntia* spp. is present, and where species such as *Waltheria ovata*, *Alternanthera* spp., *Cordia* spp., *Gossypium* spp. and *Tournefortia psilostachya* are common. On the lower slopes of the western islands, *Bursera* often dominates the woodland, whereas *Opuntia* spp. plays a minor role or may even be absent; the *Bursera* woodland in these areas may have *Scalesia affinis* or other shrub species such as *Waltheria ovata* as main constituents in an open shrub layer.

### *Scrubland, scrub steppe savanna and desert scrubland*

On the lowest slopes on the northern sides of the large islands, and on many small islands, where the rainfall is very low, the Dry Lowland Zone vegetation is sparse, taking the form of scrubland, scrub steppe savanna and desert scrubland. Tye *et al.* (2002) suggested that this part of the Dry Lowland Zone should be recognised as a distinct zone, because it perhaps is the most varied in terms of local community dominant species. However, the vegetation formations and subformations are rather consistent, whatever the species composition may be, and so the focus on structural features of the vegetation is maintained here. Dry Lowland Zone scrub is in some places dominated by species such as *Cordia lutea*, *Croton scouleri*, *Gossypium darwinii*, *Waltheria ovata* or *Scalesia* spp., particularly on flat lava substrates on older islands, while more irregular lava substrates on younger islands can carry scrubland or scrub steppe savanna dominated by, for example, *Lippia rosmarinifolia*. Desert scrubland is widespread in the archipelago on various substrates: on older islands *Castela galapageia* and *Opuntia* spp. are often the most conspicuous species, while *Macraea laricifolia* is the most common in such vegetation in many areas on younger islands.

## Transition Zone

The Transition Zone is typically found on larger islands and, as the name indicates, it occurs mostly at intermediate elevations as a transition between the Dry Lowland Zone and the humid zones above. However, Transition Zone vegetation is also found in places where there is no Dry Lowland Zone present, and even in places where neither Dry Lowland Zone nor humid vegetation types are present nearby. Despite the name, the Transition Zone is without doubt a distinct vegetation zone, determined by the level of rainfall and the substrate. The formations and subformations in the Transition Zone range from forest and woodland (steppe forest) to scrub (Table 16.1).

### *Forest, low forest, woodland and scrub*

The most widespread Transition Zone vegetation is dry-season deciduous to semi-evergreen forest with a closed canopy, and discernible shrub and ground layers. The forest is characterised by a mixture of species from the Dry Lowland Zone below and the *Scalesia* Zone above. The dominant species are *Pisonia floribunda*, *Psidium galapageium, Zanthoxylum fagara* and, in some places, *Hippomane mancinella*. In addition, such species as *Piscidia carthagenensis*, *Bursera graveolens* and *Opuntia* are common in this vegetation at lower altitudes, while *Scalesia* may occur at higher elevations. The shrub layer is fairly rich in species, with *Clerodendrum molle*, *Tournefortia* spp., *Cordia* spp., *Croton scouleri*, *Chiococca alba* and *Psychotria rufipes* the most common. *Ipomoea* spp. vines are quite common, and the seasonal ground layer is rather well developed.

On the older, central and eastern islands, Transition Zone woodland occurs with *Psidium galapageium* and *Zanthoxylum fagara* as dominant species, while the younger, western and northern islands often display Transition Zone low forest, woodland and scrub dominated by *Zanthoxylum fagara*.

The distinctness of the Transition Zone vegetation is illustrated by the following two examples.

At Caleta Iguana, Isabela, the slope of Volcan Cerro Azul rises steeply from the sea. There is no gentle coastal slope stretching inland, as in most other places in the Galapagos, and in this place there is no Dry Lowland Zone. The vegetation just inward from the coastline is dry-season deciduous to semi-evergreen forest and woodland, with *Pisonia floribunda* and *Hippomane mancinella* as dominant species. Among other characteristic woody species are *Psidium galapageium, Cordia leucophlyctis* and *Zanthoxylum fagara*. The occurrence of Transition Zone vegetation right at the coast is determined by the exceptionally high rainfall and humidity caused by the exposure and steepness of the volcano in this locality.

At about 500 m elevation on the southwestern slopes of Volcan Darwin, Isabela, dense Transition Zone forest is found on gently sloping areas with good soil development, which clearly has a major role in determining the vegetation here. The dominant woody species are *Pisonia floribunda, Bursera graveolens* and *Zanthoxylum fagara.* A rich shrub layer is formed by many species, such as *Croton scouleri, Tournefortia* spp., *Chiococca alba, Waltheria ovata, Gossypium darwinii* and *Solanum erianthum.* The slopes below and above these small forested areas have neither Dry Lowland Zone nor humid vegetation types, respectively, but are characterised by vast stretches of lava and pyroclastic deposits carrying pioneer vegetation only.

## Humid vegetation types

Only seven islands are high enough to carry humid vegetation types, viz. Fernandina, Floreana, Isabela, Pinta, Santa Cruz, San Cristóbal and Santiago. The humid regions are characterised by a relatively high annual precipitation and intense *garúa* in the cool season, and soils are generally well developed in the lower parts (*Scalesia* and *Zanthoxylum* Zones), whereas the areas at higher elevations have less well developed or poorly developed soils. An important part of water reaching the ground comes when *garúa* fog and drizzle hit the vegetation and condense, and the numerous epiphytic plants, in particular pteridophytes and bryophytes, accumulate much water through condensation. The fog-drip condensation plays an important role in maintaining the humidity and the plant diversity of the zone; this was demonstrated on Santiago Island and Volcan Alcedo, Isabela Island, where feral goats, pigs and donkeys destroyed the vegetation (until these animals were finally exterminated in 2006). Beneath surviving trees in the humid vegetation on Santiago and Alcedo, a condensation field, in the shape of a projection of the crown of the tree, displayed a ground cover vegetation with about 10 times as many species (mostly perennials) as outside the condensation field (mostly annuals) where no fog-drip condensation took place because of the lack of tree and shrub cover.

Large parts of the humid regions on the four large inhabited islands have lost their natural vegetation since human colonisation began in the nineteenth and twentieth centuries. It was these areas that proved to be suitable for agriculture, and most of them on Floreana, southern Isabela, Santa Cruz and San Cristóbal have been transformed for agricultural purposes, and are not included in the National Park.

The formations and subformations range from forest, woodland (steppe forest), scrub and tree-fern grove to heathland, herb- and sedgefield, grass- and fernland and mossfield (including vertical bogs) (Table 16.1).

Figure 16.3 In the *Scalesia* Zone on Santa Cruz Island. In the foreground is *Scalesia pedunculata*, the trunks of which are covered by epiphytic mosses and hepatics; in the middle, fernland vegetation is dominated by *Pteridium arachnoideum*; in the background is *Scalesia* forest, with *Scalesia pedunculata* as the mono-dominant tree.

## Scalesia Zone

### *Forest and woodland*

The main vegetation formation of the *Scalesia* Zone is a moist evergreen forest with a closed canopy, well-developed shrub and ground layers, and many vines and epiphytes. *Scalesia pedunculata* (Fig. 16.3) dominates (or dominated in former times) this forest on the central and eastern islands (Floreana, Santa Cruz, San Cristóbal and Santiago), while *S. microcephala* dominates on Fernandina and northern Isabela, and *S. cordata* on southern Isabela.

When best developed, *Scalesia* forms pure stands of even-aged trees, almost like a monoculture. In some areas, other woody species are prominent members of the *Scalesia* forest, such as *Psidium galapageium*, *Zanthoxylum fagara* and *Pisonia floribunda*. There is a rich shrub layer composed of many species, the most common of which are *Psychotria* spp., *Tournefortia rufo-sericea*, *Cordia* spp., *Clerodendrum molle*, *Chiococca alba*, *Darwiniothamnus* spp. and *Duranta repens*. The composition of the *Scalesia* forest differs between the central/eastern islands and the western islands, not only with respect to species of *Scalesia* in the tree layer, but also with respect to species dominance in the shrub layer. For example, *Psychotria rufipes* is generally the dominant shrub in *Scalesia pedunculata* forest, while *Darwiniothamnus tenuifolius* can be dominant in *Scalesia microcephala* forest and *D. lancifolius* in *S. cordata* forest. Vines such as *Cissampelos pareira*, *Cissus sicyoides*, *Ipomoea* spp. and *Passiflora* spp. are in places quite common, and generally there is an abundance of epiphytes.

Among the common epiphytic plants are *Racinaea insularis*, *Epidendrum spicatum*, *Ponthieva maculata*, *Peperomia* spp., numerous pteridophytes such as species of *Polypodium*, *Asplenium* and *Lycopodium*, and a wealth of mosses and liverworts. The ground layer is often dominated by a large number of ferns (*Thelypteris*, *Ctenitis*, *Polypodium*, *Adiantum* spp., *Asplenium* spp., etc.) and *Justicia galapagana*, *Alternanthera halimifoli* and *Spermacoce remota*. (According to A. Tye and C. Taylor (pers. comm.), *Spermacoce remota* (Rubiaceae) is the correct name for the species that has long been known in the Galapagos as *Borreria laevis*, and including *Diodia radula*; see e.g. Wiggins & Porter, 1971.)

On leeward slopes in the lower parts of the zone the vegetation is woodland with *Psidium galapageium* as the dominant species and *Scalesia* spp., *Zanthoxylum fagara* and *Pisonia floribunda* as other main constituents. As in the closed forest, different shrub species tend to dominate in different regions in the archipelago.

## *Zanthoxylum* Zone

### *Low forest, woodland and scrub*

The *Zanthoxylum* Zone occurs on at least five of the high islands (Isabela, Pinta, Santa Cruz, San Cristóbal and Santiago). This zone includes the previously described Brown Zone, which has practically disappeared from Santa Cruz and San Cristóbal because of human colonisation and agriculture, and only small, scattered remnants exist today. The Brown Zone was characterised (Bowman, 1961) by the many brown and dark epiphytic bryophytes that covered the shrubs and low-grown trees (mainly *Zanthoxylum fagara* and *Psidium galapageium*). However, in addition to the remnants found on Santa Cruz and San Cristóbal, a *Zanthoxylum* Zone, with the characteristic structure and appearance, occurs at relatively high elevations on the islands of Pinta, Santiago and Isabela. The vegetation in such places is an evergreen forest, dominated by *Zanthoxylum fagara*, which grows to a height of only 3–4 m. A few other woody species reach the same height, for example, *Solanum erianthum*. A dense shrub layer is present, usually formed by *Psychotria rufipes*, *Tournefortia rufo-sericea*, *Chiococca alba* and occasionally *Cordia scouleri*, *Acnistus ellipticus* and *Darwiniothamnus tenuifolius*. In general, an abundance of epiphytic bryophytes, in particular brown and dark liverworts, cover the trunks, branches and even leaves of the woody plants in the *Zanthoxylum* Zone, even more than in the *Scalesia* Zone.

In some places in the lower parts of the *Zanthoxylum* Zone the vegetation is woodland, where *Zanthoxylum fagara* dominates, but other tree species such as *Psidium galapageium* and *Pisonia floribunda* can occur, as well as a rather dense and tall shrub layer formed by the same species as in the closed forest.

On some high islands, scrub formations formed by *Zanthoxylum fagara* grow only to the same height as the commonly present shrub species *Cordia* spp., *Tournefortia* spp., *Chiococca alba* and *Psychotria rufipes*. Tree-fern groves of *Cyathea weatherbyana* are often found in association with such scrub. Scrub formations with other species compositions occur on leeward slopes, for example, scrub dominated by *Croton scouleri* (broad-leaved form) or *Acnistus ellipticus*.

## *Miconia* Zone

### *Scrub and tree-fern grove*

The *Miconia* scrub is found only at higher elevations on Santa Cruz and San Cristóbal, where it occurs in a broad belt on the windward slopes above the low *Zanthoxylum* Zone and below the Fern-Sedge Zone (Fig. 16.1). The main formation is evergreen scrub, dominated entirely by *Miconia robinsoniana*, which grows to a height of *c*.3 m. Only a few other woody species occur together with *Miconia*, such as *Acnistus ellipticus*, *Darwiniothamnus tenuifolius* and *Psychotria rufipes*. The tree-fern *Cyathea weatherbyana*, which also occurs in the Fern-Sedge Zone, tends to form conspicuous groves in sheltered places, ravines, small craters, etc. The *Miconia* scrub is rich in pteridophytes, both in the ground layer and as epiphytes (*Pteridium arachnoideum*, *Nephrolepis pectinata*, *Adiantum* spp., *Asplenium* spp., *Thelypteris* spp., *Blechnum* spp. and *Lycopodium* spp.); in addition, there is a rich epiphytic flora of mosses and hepatics (such as *Frullania* spp.).

## Fern-Sedge Zone

The Fern-Sedge Zone is found in areas on high islands above the elevation where woody vegetation types are prevalent. Such areas are wind-exposed and receive less rain that lower down, but they experience intense *garúa* in the cool season. On some islands, the Fern-Sedge Zone covers extensive, continuous areas above the *Zanthoxylum* and *Miconia* zones (e.g. on Santa Cruz and San Cristóbal). On other islands such as Fernandina, the Fern-Sedge Zone occurs in patches, in accordance with the patchy occurrence of suitable substrates, while the zone on Pinta Island only covers a very small but distinct area at the top of the island above the *Zanthoxylum* Zone. The soils are generally acidic and are well developed in some areas, but less so and shallow in other places.

The Fern-Sedge Zone is characterised by the almost total absence of indigenous trees, apart from tree-ferns (*Cyathea weatherbyana*), which tend to occur in groves in sheltered places, where low-grown *Zanthoxylum* trees and shrubby species such as *Acnistus ellipticus*, *Psychotria rufipes* and *Tournefortia rufo-sericea* also can be present.

The origin of this treeless vegetation has been the subject of much discussion. Several factors have been suggested to be of importance in preventing indigenous trees from growing in the zone: wind and its drying effect (Stewart, 1911; Bowman, 1961; Wiggins & Porter, 1971), or reduced evapotranspiration and reduced uptake of nutrients due to a saturated atmosphere (van der Werff, 1978), which could also explain the low growth of woody plants in the *Zanthoxylum* Zone (a stunting effect) (Hamann, 1979a). However, Itow (1990, 2003) demonstrated that the most likely reason is the effect of alterations between extremely dry periods caused by strong winds and extremely wet periods caused by heavy *garúa*, and that indigenous trees adapted at the same time to very dry and very wet conditions did not exist in the Galapagos. This observation is supported by the fact that large areas of both the Fern-Sedge Zone and the *Zanthoxylum* Zone on Santa Cruz today are massively invaded by the non-indigenous tree *Cinchona pubescens*. This species was introduced to the islands in 1946, but has turned out to be very invasive and well adapted to the Galapagos conditions (Hamann, 1974; Itow, 2003; Jäger *et al.*, 2007). Thus, the ecological conditions in the two zones do not, as such, prevent trees from growing there.

The plants of the Fern-Sedge Zone occur in a number of different formation types, and a considerable number of plant communities have been described from this zone. They are mostly characterised by dominant species, which represent different life-forms. Thus, the vegetation formations are to a large degree equivalent to the described communities: tree-fern groves are dominated by *Cyathea weatherbyana*, heathland by *Pernettya howellii*, and herbfield by species such as *Jaegeria crassa*, *Lobelia xalapensis*, *Pilea baurii*, *Hypericum uliginosum*, *Centella asiatica* and *Polygonum opelousanum*; grassland is mostly dominated by *Paspalum conjugatum* or *Trisetum howellii* and sedgefield by species of *Eleocharis*, *Cyperus*, *Scleria* and *Rhyncospora*. The most widespread vegetation formation in the subzone is fernland dominated by *Pteridium arachnoideum*, but with many other pteridophytes as constituents or even as dominants in some places, such as *Polypodium tridens*, *Blechnum polypodioides*, *B. occidentale*, *Nephrolepis pectinata*, *Pityrogramma calomelanos*, *Lycopodium clavatum*, *Lycopodiella cernua*, *Ctenitis* spp., *Thelypteris* spp., etc. According to Itow (1990, 2003), variations in soils, in aspect and in exposure determine the distribution of the communities, and hence the formations, too.

Mossfield vegetation, dominated by *Sphagnum*, occurs in the Fern-Sedge Zone on the large, high islands of Isabela, San Cristóbal, Santa Cruz and Santiago. Itow and Weber (1974) termed the *Sphagnum*-dominated vegetation 'fens and bogs'. Fens occur in shallow, wet or water-holding depressions with *Sphagnum cuspidatum*.

Numerous sedges and herbs are usually found in and around such fens. 'Raised bogs' are the distinct formations of *Sphagnum* vegetation in small crater bottoms. The *Sphagnum* layer can be quite thick, and is formed by either *S. cuspidatum* or *S. erythrocalyx*, the only *Sphagnum* species present in the islands. In such bogs numerous herbs, sedges and ferns (e.g. *Elaphoglossum* spp.) are often found.

'Vertical bogs' are found on exposed and wet, vertical sides of small cinder cones and craters. These vertical bogs are formed by thick layers of *Sphagnum erythrocalyx*, and a number of pteridophytes are usually growing there, too, for example the tree-fern *Cyathea weatherbyana* and species of *Elaphoglossum*. The common presence of filmy ferns (*Hymenophyllum* spp. and *Trichomanes reptans*) on the trunks of *Cyathea* in such areas indicates the extreme humidity of this habitat.

## High Altitude Dry Zone

Only the western islands, Fernandina and Isabela, have a High Altitude Dry Zone of vegetation. The zone is found above the levels at which the cool-season inversion layer and the hot-season rain-clouds commonly form. However, the pattern of vegetation at these high elevations is very complex because of volcanic activity, which creates a mosaic of lava flows, lava boulders, cinders and ash- and pumice-fields. Eruptions are frequent, and the vegetation in small or large areas can be totally destroyed by eruptions. Consequently, patches of young substrates with no vegetation at all intermingle with slightly older substrates with pioneer vegetation as well as with substrates where the vegetation has had a long time to develop.

The formations and subformations in the High Altitude Dry Zone range from forest, woodland (steppe forest), scrub, scrubland and scrub steppe savanna to grass- and fernland (Table 16.1).

The most complex (structurally and floristically) vegetation of the High Altitude Dry Zone is an evergreen forest, dominated by *Scalesia cordata* (south Isabela) or *Scalesia microcephala* (north Isabela and Fernandina). *Opuntia saxicola* and *O. insularis* are common species in the zone and have similar geographical distributions in the low forest, woodland and scrub as *Scalesia cordata* and *S. microcephala*, respectively. It should be noted that *Scalesia cordata* and *S. microcephala* also occur as dominant species in the humid regions at much lower elevation on these two islands, but in the High Altitude Dry Zone the two species are found as dominants in a low forest or a scrub. The *Scalesia* species form a canopy at about only 3–4 m height; however, the appearance of the vegetation is that of a miniature forest. Vines and epiphytes are present, but occur in lower numbers than in the *Scalesia* forest in the humid regions. A more or less distinct shrub layer is composed of a number of species, some of which dominate in some areas while others dominate in other areas: *Zanthoxylum fagara*, *Opuntia* spp., *Baccharis gnidiifolia*,

*Tournefortia rufo-sericea*, *Duranta* spp., *Solanum erianthum*, *Psychotria rufipes*, *Darwiniothamnus* spp., *Cordia* spp., *Lippia rosmarinifolia*, *Croton scouleri*, *Macraea laricifolia*, etc.

The low forest or scrub dominated by these two species of *Scalesia* is well developed at the rim of the highest volcanoes (Fernandina Island and Volcan Cerro Azul, Volcan Darwin and Volcan Wolf on Isabela Island). Further down the high slopes of the volcanoes, *Scalesia cordata* and *S. microcephala* are in places characteristic members of woodland vegetation, in which the other woody species mentioned above also occur. A more or less seasonal ground cover, composed of rather few species, is present in both the low forest and the woodland.

Other types of scrub and scrubland occur in patches, mostly with *Macraea laricifolia* and *Lippia rosmarinifolia* as the characteristic woody species, and low-grown *Zanthoxylum fagara* and scattered individuals of *Opuntia* spp. as common constituents. In addition, *Polypodium tridens* is usually the most abundant species in the sparse ground cover, thereby linking this vegetation to the fernland found at these high altitudes. Fernland vegetation is widespread and is mostly dominated by *Polypodium tridens* and/or *Pteridium arachnoideum*, but with many other pteridophytes as common members too, such as other species of *Polypodium*, *Asplenium* spp. and *Lycopodium* spp.

A unique type of grassland is found at the summit of the highest volcanoes in the High Altitude Dry Zone. The grassland is dominated by *Pennisetum pauperum*, which forms the main part of a closed ground layer. Other species here are *Pteridium arachnoideum*, *Alternanthera* spp., *Commelina diffusa* and *Paspalum galapageium*. In places, the *Pennisetum* grassland turns into scrub steppe savanna with scattered woody plants, such as *Zanthoxylum fagara*, *Cordia* spp., *Scalesia* spp. and *Solanum erianthum* appearing in the grass cover.

The High Altitude Dry Zone on Fernandina and Isabela has not been explored as extensively as the other vegetation zones in the archipelago. Further studies might lead to a redefinition of some of the Galapagos vegetation zones as here presented. The close relationship between some of the formations of the pioneer vegetation and the Dry Lowland Zone, the High Altitude Dry Zone and the humid vegetation zones could be investigated.

## Aquatic vegetation

Freshwater aquatic vegetation is found on the large islands of Floreana, Isabela, Santa Cruz, San Cristóbal and Santiago, particularly in the humid vegetation zones, but temporary pools are also found in the Transition and even the Dry Lowland Zone. Springs, streams and temporary freshwater pools or small lakes are the habitats in which aquatic herbfield and fernfield vegetation (Table 16.1) occur, but springs and streams are rare in the archipelago, whereas temporary pools and small lakes are rather common.

The few springs and streams often have *Potamogeton pectinatus* as the characteristic species, but the temporary pools and small lakes display quite a variety of species. Usually the pools dry out in the cool season, at which time they are often covered with a number of herbs, such as *Heliotropium indicum* and *Ludwigia* spp. In the hot season, the areas fill up again and can be completely covered by floating *Azolla microphylla*. Larger temporary pools and small lakes generally have a fringe composed of both floating (*Azolla microphylla*, *Lemna aequinoctialis* and *Ludwigia peploides*) and rooted species (e.g. *Ludwigia leptocarpa*, *Eleocharis fistulosa* and other species of *Eleocharis*). In the temporary pools, the submerged plants usually found are *Najas guadalupensis* and algae (e.g. Characeae).

The springs, streams and temporary pools and small lakes contribute to the mosaic of vegetation in the landscape. The herbfield and grassland vegetation characteristic for a vegetation zone merge with the fringes of temporary pools and small lakes, so that the constituting species of the fringe vegetation of the pools vary between, for example, the *Scalesia* Zone and the Fern-Sedge Zone.

## Pioneer vegetation

The indigenous Galapagos flora is mainly composed of species with pioneer characteristics. Numerous indigenous species, including those that have evolved in the archipelago and are prominent members of the well-developed Galapagos vegetation formations, have presumably descended from pioneer ancestors (e.g. *Scalesia* spp.), and they have retained many traits characteristic for pioneers. Consequently, most of the pioneer species, which occur in habitats which have been formed recently by volcanic activity, upliftings or natural fires, also occur in vegetated areas on older substrates. In general, the vegetated areas on nearby older substrates provide the source areas for the colonising species, as only a few species seem to be colonising specialists with an exclusive occurrence in pioneer vegetation.

Pioneer vegetation is found at practically all elevations all over the archipelago, but most extensively on the islands which are most active volcanically. Because the local and microclimatic conditions differ at different elevations, and because the species composition of the source areas for pioneers differs at different elevations, the pioneer vegetation may display the first steps in a succession that over many years may lead to the formation of vegetation typical for the vegetation zone at a particular elevation.

Pioneer vegetation is found in a large range of habitats on lava (A'a and pahoehoe) and tephra (dust, ash, cinders, scoria, pumice, etc.) substrates, and in fumarole areas. The vegetation formations range from woodland, scrub and scrubland to cactus-, herb-, and sedgefield and fernland (Table 16.1).

Flows of A'a and pahoehoe lava are colonised by a number of species (Hamann, 1981; Huttel, 1995). The most conspicuous early pioneer species is *Brachycereus nesioticus*. This species seems to be a pioneer specialist on lava, where it is found scattered or forming striking cactusfields with numerous individuals. *Brachycereus* is not a regular member of other vegetation formations. A few other species, e.g. *Mollugo snodgrassii* and *Borreria ericaefolia*, are mainly found as early pioneers, too.

Depending on elevation and exposure, location in relation to older, vegetated areas, and to the age and weathering of the substrate, the vegetation formations on lava range from woodland to scrubland. Among the late pioneers are *Jasminocereus thouarsii*, *Darwiniothamnus* spp., *Scalesia* spp., *Lecocarpus* spp., *Macraea laricifolia*, *Chamaesyce viminea*, *Ipomoea habeliana* and *Polypodium tridens*. The two ferns, *Cheilanthes microphylla* and *Notholaena galapagensis*, mainly occur in pioneer vegetation. Each of the woody species just mentioned may characterise the vegetation, such as *Jasminocereus thouarsii* in woodland, *Macraea laricifolia* or *Scalesia* spp. in scrub, and *Lecocarpus* spp. in scrubland. The species composition in the formations on lava vary from location to location, because of the complement of species that characteristically form the older vegetation nearby, such as *Bursera graveolens*, *Opuntia* spp., *Lantana peduncularis*, *Cordia* spp., *Chiococca alba* and *Waltheria ovata* (Huttel, 1995).

Herbfield composed mainly of *Tiquilia* spp. is often found on ash and gravel plains, while fernland occurs on many different substrates, mainly at higher elevations. *Polypodium tridens* is often the dominant species on both lava and tephra. Fumarole areas are often surrounded by sedgefield vegetation composed by *Fimbristylis* spp. and *Cyperus* spp.

Bush fires are rather frequent in the Galapagos. Areas at higher altitudes, where fires completely destroy the vegetation, are commonly first recolonised by *Marchantia chenopoda* and then *Pteridium arachnoideum*, creating a fernland vegetation. Later on, other species, including woody plants, from the adjacent vegetation may invade and re-form the original vegetation formations of the area (Nowak *et al.*, 1990).

Information about the pioneer vegetation in the Galapagos is mostly found in the form of observations in relation to volcanic eruptions, to general studies of vegetation, or to floristic studies of little-explored areas. The only investigation carried out with a specific focus on pioneer vegetation is Huttel's work on the vegetation on lava flows (Huttel, 1995). Further studies on the pioneer vegetation and plants are needed in order to gain a deeper understanding of the ecology of the Galapagos vegetation.

## Aspects of vegetation dynamics and demography of woody plants (temporal patterns)

In the Galapagos, major natural events such as volcanic eruptions, fires and ENSO events change the vegetation over time, sometimes drastically. However, since the arrival of humans a few centuries ago, the Galapagos ecosystems have also been subjected to serious anthropogenic disturbances. All major disturbances, naturally-induced and human-induced, have a profound influence on the vegetation dynamics.

Changes in vegetation may be short-term or long-term and involve units at different levels, from individual plants to landscapes and vegetation regions. In the Galapagos, natural changes occur slowly in dry vegetation but relatively fast in humid vegetation, which is related to the biology and demography of the different species (Mueller-Dombois & Fosberg, 1998), and there is a marked difference in robustness and resilience between the plant species and vegetation of the dry zones and the humid zones (Hamann, 2001).

Vegetation changes over time have been studied in the Galapagos since 1966. For this purpose, several series of permanently marked vegetation quadrats in different vegetation types on different islands have been established, and repeated studies during several decades have generated much information on aspects of vegetation changes and on demography of selected woody species. The studies have focused on both natural and anthropogenic changes, and have often been carried out in connection with conservation management activities in the archipelago.

The mortality rates for different age groups in stands of selected taxa of *Scalesia* and *Opuntia*, and of *Bursera graveolens*, on the islands of Santa Cruz, Santa Fe and Pinta, are shown in Table 16.2. The average mortality per year of seedlings is generally high, while that of juveniles varies from high to low, which would be expected, considering the very variable conditions from year to year for plant establishment and growth. However, the average mortality for adult individuals displays a pattern: the *Scalesia pedunculata* stand has the highest mortality rate among the adults, indicating that 18% of the stand dies every year, i.e. there is a very rapid turnover in the humid *Scalesia* forest. *Scalesia helleri* ssp. *helleri* occurs only on Santa Fe, which only has Dry Lowland Zone vegetation (and Littoral Zone). *S. helleri* ssp. *helleri* shows a rather low average mortality of 7% per year among adults, indicating a much slower turnover in the stand than in *Scalesia pedunculata*. On Pinta, *S. baurii* ssp. *hopkinsii* presents a little higher average mortality among the adults at 10% per year, which might be related to the rather broad amplitude of ecological conditions in which it occurs, i.e. from the Dry Lowland Zone near the coast to the lower parts of the more humid *Zanthoxylum* Zone at higher elevations. The average annual mortality for adults of *Opuntia* spp. and *Bursera* ranges from 1% to 5%, which indicates a very slow turnover in the stands. Thus, there is a marked

Table 16.2 *Average mortality percent per year of age groups in stands of selected woody species on Santa Cruz, Santa Fe and Pinta. The average annual mortality over more than three decades of recording indicates that adult individuals, particularly of* Scalesia pedunculata, *have a very rapid turnover, while adults of* Opuntia *spp. and* Bursera graveolens, *which are typical dry zone plants, have a slow turnover in the stands.*

| | Adults | Juveniles | Seedlings | Time period |
|---|---|---|---|---|
| *Scalesia pedunculata* | 18 | 33 | 44 | 1970–2007[1] |
| *S. baurii* ssp. *hopkinsii* | 10 | 15 | 22 | 1970–2007 |
| *S. helleri* ssp. *helleri* | 7 | 14 | 19 | 1973–2007 |
| *Opuntia echios* var. *gigantea* | 5 | 2 | 26 | 1966–2007 |
| *O. echios* var. *barringtonensis* | 4 | 4 | 34 | 1972–2007 |
| *O. galapageia* var. *galapageia* | 3 | 3 | Ins. data[2] | 1970–2007 |
| *Bursera graveolens*, Santa Fe | 1 | 17 | Ins. data | 1972–2007 |
| *B. graveolens*, Pinta | 1 | 9 | 49 | 1970–2007 |

[1] No *Scalesia* individuals recorded in permanent quadrat since 2001.
[2] Insufficient data.

difference in average annual mortality between adults of *Scalesia pedunculata* on the one hand and on the other hand the Dry Lowland Zone species of *Opuntia* and *Bursera* as well as of *Scalesia helleri* ssp. *helleri*, while *Scalesia baurii* ssp. *hopkinsii* displays an intermediate position. The average annual mortality and the turnover among adults of these species appear to be in accordance with the life-cycle characteristics of the taxa, as well as being related to the different ecological conditions under which they occur.

## Disturbances, succession and regeneration

### *Naturally induced changes*

#### *Succession after volcanic eruption*

The higher parts of the westernmost island of Fernandina fall within the High Altitude Dry Zone, with vegetation formations ranging from forest to grassfield and fernland. Fernandina Island is one massive volcano with a central caldera, reaching 1494 m, and it is one of the most active of the Galapagos volcanoes; the last recorded eruption (minor) took place in April 2009. Hence, natural disturbances occur frequently and induce vegetation changes.

In 1968 a violent eruption took place on Fernandina that caused major changes in the vegetation on the caldera rim as well as in the caldera. Before this eruption, the whole rim of the caldera was covered with a low forest with *Scalesia microcephala* as the dominant species. Among the other characteristic woody plants were *Zanthoxylum fagara*, *Tournefortia rufo-sericea*, *Croton scouleri*, *Darwiniothamnus tenuifolius* and *Lippia rosmarinifolia*, and more scattered *Duranta repens*, *Opuntia insularis* and *Solanum erianthum* (Eliasson, 1972).

The 1968 eruption buried the vegetation on the western caldera rim in tephra. The post-eruption regeneration of vegetation on the caldera rim was studied over the following 9-year period (Hendrix, 1981; Hendrix & Smith, 1986). The vegetation covering about half of the caldera rim apparently survived the eruption, thereby forming a source for revegetation of the disturbed area on the western rim.

Only few species of herbs and grasses (e.g. *Aristida repens*) were able to colonise and establish themselves in gullies on the deep tephra. However, in places where the vegetation had been buried by just a shallow layer of tephra, a vigorous vegetative sprouting of shrubs and perennial herbs was recorded, resulting in a nearly complete vegetation cover by 1977 (Hendrix, 1981). In particular, *Solanum erianthum*, *Tournefortia rufo-sericea* and *Scalesia microcephala* regenerated very rapidly. The rate and pattern of the post-eruption regeneration of vegetation was determined by the different reproductive and growth strategies of the plant species, so that species reproducing by seed returned more slowly than species with vegetative reproduction. In addition, land iguanas facilitated seed dispersal of some species and appeared to be most important for vegetation regeneration in the deep, barren tephra (Hendrix, 1981; Hendrix & Smith, 1986).

Thus, the post-eruption regeneration of High Altitude Dry Zone vegetation on Fernandina showed a pattern comparable to that recorded from colonisation on lava flows at different altitudes and different islands by Huttel (1995). Two main factors seem to be important: the basic pioneering characteristics of most indigenous Galapagos plants and the presence of source areas of more or less undisturbed vegetation nearby. However, the time frame is presumably very different for colonisation of lava flows and regeneration of vegetation on tephra. Colonisation on lava flows depends on the very few microhabitats available for plant growth on the lava fields, and on the formation of new suitable microhabitats over time through weathering, accumulation of organic material, and so on. Thus, regeneration of vegetation on tephra obviously can be a fast process, while plant colonisation of lava flows is probably a very slow process lasting decades or hundreds of years, not least under the prevailing dry conditions.

(It would be interesting to obtain an estimate of the age of the different lava flows studied by Huttel (1995). It could provide an indirect estimate of the time frame for plant colonisation and succession, which then could be related to the differences in rainfall and the development stages of the vegetation in the different areas.)

### Scalesia *forest succession*

The genus *Scalesia* is endemic to the Galapagos and comprises 21 taxa, including 15 species and several subspecific taxa. The *Scalesia* species are all light-loving, erect, single-stemmed and non-deciduous woody plants, ranging in size from about 0.5 to 15 m in height. The *Scalesia* species grow very fast and have short life expectancies. Different species occupy different ecological niches in the islands, but generally the smaller, shrub-sized *Scalesia* species occur in various types of dry habitat, while the larger, tree-sized species occur in humid habitats (Eliasson, 1974, 1984, 1995; Hamann, 1979b, 1981, 1995; Lawesson, 1988; Itow & Mueller-Dombois, 1988; Adsersen, 1990; de Vries & Tupiza, 1990; Itow, 1992, 1995, 2003; Arsiniegas-Fuertes, 1996; Shimizu, 1997; Mueller-Dombois & Fosberg, 1998).

The three forest-forming species of *Scalesia* are *S. pedunculata* (Floreana, San Cristóbal, Santa Cruz and Santiago), *S. cordata* (southern Isabela) *and S. microcephala* (Fernandina and northern Isabela). The turnover in the *Scalesia* forest and the succession or regeneration after major disturbances is determined by both endogenous and exogenous factors. The forests are most commonly composed of even-aged, monospecific stands. Because of a lack of competing late-succession trees, the characteristics of these *Scalesia* species enable them to persist in the forest, even when the ecosystem matures. *Scalesia pedunculata* is the largest of the *Scalesia* species; it grows to a height of 15 m and is often dominant in the humid forest (Fig. 16.3). It displays a pattern of growth and survival not usually associated with forest trees: in just 2 to 5.5 years the trees grow to a height of 7–8 m and start flowering and fruiting. Several insects are recorded as pollinating the flowers, the most important being the endemic carpenter bee (*Xylocopa darwinii*). Darwin's finches may play a role in seed dispersal, but the achenes are small and are generally dispersed over short distances only, having no morphological adaptations for long-distance dispersal.

On Santa Cruz, most adult individuals of *Scalesia pedunculata* die when they are about 15–20 years old, and only few individuals survive several years more. On Isabela, adult individuals of *Scalesia cordata* are estimated to reach an age of about 60 years before they die (de Vries & Tupiza, 1990). Thus, the natural turnover in a *Scalesia* forest may differ from species to species and from one location or island to another, but when they have reached the senescent stage, they are predisposed to die. The natural succession or regeneration of the *Scalesia* forest may then follow different paths, determined by the occurrence or non-occurrence of major disturbances.

If a strong El Niño hits the forest while the *Scalesia* trees are in the senescent stage, almost all trees die at more or less the same time, displaying a synchronous stand-level dieback (Hamann, 1979b, 1986, 2001; Itow & Mueller-Dombois, 1988; Lawesson, 1988; Itow, 1992, 1995, 2003; Tye & Aldaz, 1999). The succession or regeneration of the *Scalesia* forest becomes self-cyclic, because new *Scalesia* trees are able to succeed the dead ones. Furthermore, they are shade-intolerant, so that not much regeneration of *Scalesia* takes place underneath a mature canopy; a new generation can only develop when the old one has died off. During the years before the dieback, a large bank of *Scalesia* seeds has usually accumulated in the soil, so seeds are available for regeneration. A strong El Niño event is often followed by a 1- to 2-year period of drought (La Niña). During this period, very little or no regeneration takes place, but as soon as a normal rainy season occurs, thousands of *Scalesia* seedlings appear. During the first year, there is an extremely high mortality among the seedlings (self-thinning), and during the next few years there is similarly a high mortality among the juveniles, as the plants grow and compete to become members of the canopy. But within just 4–5 years a new canopy is formed by even-aged trees that start flowering and fruiting. A rapid self-cyclic succession or regeneration takes place. If the *Scalesia* forest has reached the senescent stage, and no major disturbances occur, the old individuals die off gradually over a longer period of time, and no synchronous dieback occurs. In consequence, the canopy is thinned out, gaps occur and new *Scalesia* seedlings emerge. A gradual regeneration of the forest takes place.

Thus, two types of regeneration of the *Scalesia* forests in the Galapagos have been recorded, a synchronous and a continuous type. Shimizu (1997) suggested that the two types were characteristic for humid habitats and drier habitats, respectively. Shimizu found that the synchronous type fitted the pattern displayed in the *S. pedunculata* forest on Santa Cruz, while the continuous type of regeneration was observed on Isabela Island, where *S. cordata* on Volcan Sierra Negra and *S. microcephala* on Volcan Alcedo formed sparse forests with broken canopies. There, Shimizu (1997) found the *Scalesia* stands being heterogeneous in age structure, and synchronous regeneration did not take place, even after the 1982–83 El Niño.

Based on the observations covering 35 years it seems reasonable to suggest that both synchronous and continuous regeneration of *Scalesia* forest may take place after disturbances, in both humid and drier habitats, depending on the stage of development of the *Scalesia* trees and the strength of the disturbance. *Scalesia* stands of uneven age structure may well survive disturbances and regenerate continuously for a long time, because young and mature members of the stands survive while those that are senescent die. However, extreme disturbances may result in a stand-level dieback, such as the 5-year period of severe drought between 1935 and 1940 described by Kastdalen (1982), which killed almost all *Scalesia* individuals, and as it happened in the 1982–93 El Niño event. Such extreme, but rare, disturbances are then resetting the scene for a synchronous regeneration.

## *Human-induced changes*

As the resident human population in the Galapagos has grown sharply over the last 40–50 years, the anthropogenic impact on the biodiversity and ecosystems has increased dramatically, because of exploitation of natural resources, habitat destruction and fragmentation, invasive alien species, and human-made fires (Kastdalen, 1982; Adsersen, 1989; Nowak *et al.*, 1990; MacFarland & Cifuentes 1996; Hamann, 1997; Itow, 2003; Tye, 2003; Jäger *et al.*, 2007). In fact, the biological isolation of the Galapagos Islands has been breached, and many native ecosystems have been altered and modified. Examples of human-induced changes, with a main focus on disturbances caused by invasive species, are presented below.

## Impact on vegetation by invasive plants and animals

The most destructive force on isolated oceanic islands worldwide has been the direct destruction of native vegetation for farming, settlement and other purposes. However, in recent decades, this has been taken over in importance by the introduction of alien species, and the Galapagos is no exception (Macdonald *et al.*, 1989). Alien plants, animals and microorganisms have been introduced to both inhabited and uninhabited islands, purposely or accidentally. Their numbers are steadily increasing, and in many parts of the archipelago they have had serious impacts. Feral mammals such as goats, donkeys, pigs, horses and cattle destroy the native vegetation, and invasive plant species such as *Cinchona pubescens*, *Psidium guajava*, *Cedrela odorata*, *Rubus* spp., *Lantana camara*, *Passiflora* spp. and *Pennisetum purpureum* outcompete and displace native species and alter the ecosystems.

### *Life and death of a* Scalesia *stand on Santa Cruz*

On Santa Cruz a small section of *Scalesia* forest in the southwest part of the island has been studied from 1972 to 2007, through repeated investigations of the vegetation in a permanent quadrat. The original vegetation in this area was a semi-evergreen forest of the upper Transition Zone, with a canopy reaching about 8–12 m height. *Scalesia pedunculata* dominated the forest, which also included other indigenous trees such as *Psidium galapageium*, *Pisonia floribunda* and *Zanthoxylum fagara*. At the time the study began in 1972, the investigated stand of *Scalesia pedunculata* was mainly composed of juveniles, and during the following years many of them died, as some surviving individuals reached maturity and formed a closed canopy.

In 1982–3, the most intense ENSO event ever recorded in the Galapagos took place: Almost all the old *Scalesia* trees died during the 8-month period of El Niño heavy rains, as well as many of the *Psidium galapageium* trees. Afterwards, two years of drought (La Niña) followed, and no regeneration of *Scalesia* was recorded. However, after the next normal wet season in 1985, thousands of *Scalesia* seedlings emerged. The mortality among the seedlings was very high the first year (about 98%), but some survived to the juvenile stage, and it was expected that a self-cyclic regeneration of the forest was under way. But instead, the forest was changing dramatically: the El Niño–La Niña event triggered a massive invasion of alien plant species, in particular *Psidium guajava* and *Cedrela odorata*. Both these species had been present in low numbers in the forest before the ENSO, as well as being very common in the in the agricultural zone nearby, but after 1982–3 and continuing after 1997–8, these aggressive invaders outcompeted *Scalesia pedunculata*, primarily because they had survived the ENSO events and were able to form a shading canopy, which prevented seedlings and saplings of the shade-intolerant *Scalesia* from reaching maturity. Since 2003, no *Scalesia pedunculata* specimens have been recorded in the investigated plot, and by 2007 the vegetation had been transformed into a forest with *Cedrela odorata* forming the canopy and with an understorey of smaller *Psidium guajava* trees.

Thus, the combined effect of strong ENSO events and invasion of aggressive alien plant species effectively eliminated the original forest from this part of Santa Cruz Island. Fortunately, there are other areas in the Santa Cruz highlands where the *Scalesia* forest has not suffered such a massive takeover, although other invaders, such as *Pennisetum purpureum* and *Rubus niveus*, are invading the forest and gradually changing the ecosystem dynamics and thwarting its natural ability to regenerate.

### *Regeneration of dry vegetation after the elimination of feral goats*

The *Opuntia* cacti are a very conspicuous element in dry areas, from sea level to high altitudes on all islands and major islets. The genus is represented in the Galapagos by 14 endemic taxa (six species, plus several recognised varieties). The growth form ranges from prostrate to shrubby and arboreal, but most taxa grow as erect trees that with age develop large trunks without spines. Some may grow to 10 m height or more, but they grow very slowly; on Santa Cruz, the average growth of individuals of *Opuntia echios* var. *gigantea* is about 3–5 cm per year (measured in a permanent quadrat over 32 years). However, this average may be at the lower end of the scale, perhaps typical of very dry areas, because in wetter areas individuals have been recorded to grow around 10 cm per year.

Thus, in very dry areas it may take 50 years or more for *Opuntia* to reach maturity and start flowering and fruiting. But once the adult state is attained, the individuals may live another 150 years or more, so some of the very large individuals (Fig. 16.2) could easily be around 200 years old or even older. *Opuntia* cacti flower at the beginning of the hot season, around December/January, and are a primary source of food for many animals. Land iguanas and giant tortoises eat the pads, and several species of Darwin's finches feed on the flowers, fruits and seeds, and the animals disperse the seeds. Even though annual seed production may be regular and copious, the germination and establishment of *Opuntia* seedlings is sporadic and very dependent on favourable conditions that may not be present every year. Thus, natural regeneration and turnover in *Opuntia* populations may take decades, but as long-living xerophytes the *Opuntia* cacti are well adapted to survive the prevailing dry conditions over long periods. However, El Niño conditions have an adverse effect on *Opuntia*. The *Opuntia* cacti do not tolerate prolonged periods with excess water.

Santa Fe is a small, uninhabited island, which only has Littoral and Dry Lowland Zone vegetation. A small number of goats had been present on the island for perhaps 200 years before they were completely eliminated by the Galapagos National Park in 1971. It was expected that plants and vegetation on Santa Fe would gradually recover after the removal of the goats, and a long-term study was initiated in 1972, whereby the time frame and direction of the expected regeneration process could be investigated (Hamann, 2004).

The Dry Lowland Zone on Santa Fe is characterised by *Opuntia echios* var. *barringtonensis* and *Bursera graveolens*. In the long-term study on Santa Fe, the highest mortality for all life stages of *Opuntia* was recorded in association with the major El Niño events of 1982–3 and 1997–8 (Hamann, 2004). This is consistent with the general observations made during and after these two El Niño events elsewhere in the archipelago, when *Opuntia* populations throughout were decimated. The availability of water first caused the *Opuntia* cacti to become top-heavy with water stored in the pads, and when the soil became waterlogged for long periods the roots rotted; consequently, the large *Opuntia* died and fell over. At the same time, seedlings and juveniles were smothered by a heavy growth of vines and herbs.

During the first 5 years after removal of the goats, regeneration was beginning to occur on Santa Fe: *Scalesia helleri* ssp. *helleri*, which had suffered seriously from goat grazing (Eliasson, 1968), recovered rapidly and some recruitment of *Opuntia* and *Bursera* began as well, while the dominant shrub species *Cordia lutea*, *Encelia hispida* and *Lantana peduncularis* increased in numbers. But the El Niño events in 1982–3 and 1997–8 had a marked impact on the vegetation and the regeneration process. The incipient recovery of *Opuntia* and *Bursera* was halted, and even reversed in the case of *Opuntia* (Fig. 16.4). At the same time, the shrub species *Cordia lutea*, *Encelia hispida* and *Lantana peduncularis* were favoured by the El Niño conditions (Fig. 16.5) and continued to increase in numbers, so that the shrub layer of the dry forest and woodland formations on Santa Fe became more and more dense. By 2007, the vegetation on Santa Fe had changed, not in composition of species but in structure.

The strong impact of El Niño on the recovery process suggests that regeneration of *Bursera* and especially *Opuntia* on Santa Fe should be viewed over a much

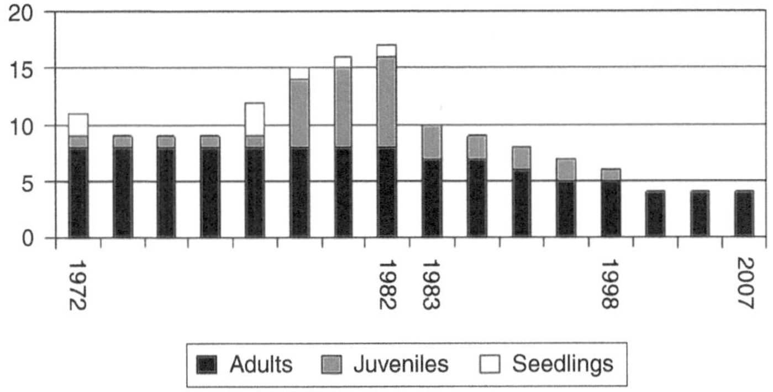


Figure 16.4 Number of individuals of *Opuntia echios* var. *barringtonensis* in permanent quadrats, Santa Fe, 1972–2007. During the first decade after the eradication of feral goats on Santa Fe in 1971, seedlings and juveniles appeared, and a recovery of the stands was noted. However, the El Niño events in 1982–3 and 1997–8 reversed the trend. In 2007, only a few adult individuals were still alive, and no new regeneration was recorded.

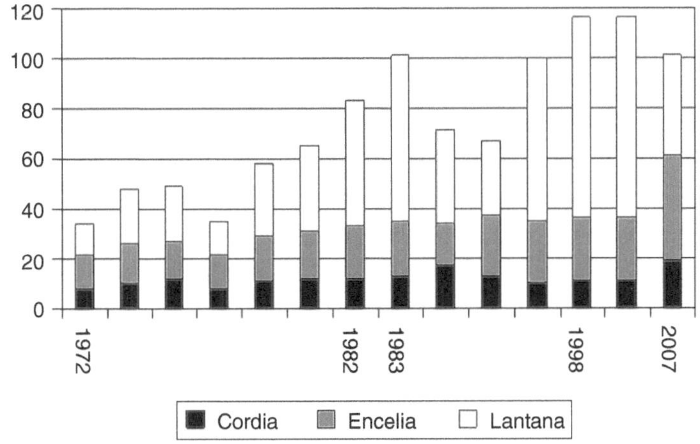


Figure 16.5 Number of individuals of *Cordia lutea*, *Encelia hispida* and *Lantana peduncularis* per 100 m$^2$ in permanent quadrats, Santa Fe. In contrast to *Opuntia*, the three most common shrub species in the permanent quadrats recovered well after the eradication of goats in 1971. The two El Niño events actually seemed to boost their regeneration. Although the number of individuals of the shrub species fluctuated from recording to recording, the general trend was an increase. Their numbers grew about three-fold during the 35 years of investigation.

longer time scale than the 35 years of the study period, in accordance with their estimated life expectancy of 150–200 years as adults. On such a time scale, the frequency and strength of El Niño events may determine whether the vegetation on Santa Fe will change more profoundly in structure as well as in relative abundance of the woody species. The recruitment and survival of *Opuntia* in particular may be hindered if El Niño events are to occur with increased frequency and strength. This could lead to a drastic reduction in the number of *Opuntia* individuals on the island, which would completely alter the aspect of the Santa Fe vegetation.

## Conservation

Plant conservation in the Galapagos is carried out at several levels: from single species to vegetation and ecosystems measures, and to the level of whole-island conservation. *In situ* plant conservation measures include protection of native species and vegetation, and control or elimination of invasive alien plants and animals. Over the last 50 years, the Galapagos National Park (GNP) and the Charles Darwin Foundation (CDF) have been able to remove some of the worst threats to the Galapagos plants and vegetation on both small and large islands. Carefully managed eradication campaigns targeted at feral goats and other introduced mammals have succeeded, particularly on the uninhabited islands. Thus, the future for many endangered species and ecosystems has been secured, and in many places noteworthy regeneration of vegetation has been recorded. At the same time, some of the critically endangered plant species and populations have been pulled back from the brink of extinction by focused conservation research and management. *Ex situ* plant conservation measures in the islands comprise growing and propagating rare and endangered species for habitat restoration or population enhancement. Outside the islands, such species may be conserved *ex situ* in seed and gene banks as the ultimate insurance against extinction.

Inhabited and uninhabited islands present both similar and very different conservation problems (MacFarland & Cifuentes, 1996; Tye *et al.*, 2002; Tye, 2003). On the inhabited islands, the growing number of human residents leads to many difficult conservation problems, including, for example, that caused by the continuous invasion of *Cinchona pubescens* from sources in the agricultural areas (Macdonald *et al.*, 1988; Jäger *et al.*, 2007). However, in the present context, the uninhabited island of Santiago will serve as an example of the different kinds of conservation activities that are of particular relevance for plants and vegetation. Over the last 50 years conservation actions on this island have been taken at all levels.

### *Single-species conservation:* Scalesia atractyloides

The endemic *Scalesia atractyloides* only occurs on the island of Santiago. In the early 1990s, it was believed to be extinct because of goat grazing and browsing. However, seeds had been collected for research purposes, and a small number of live plants were held in cultivation in the Botanic Garden, University of Copenhagen, Denmark. Consequently, *Scalesia atractyloides* plants were propagated in the

Botanic Garden in order to provide material for reintroduction of the species to the wild, should that become possible. However, thorough field investigations on Santiago showed that several small populations of both varieties of *S. atractyloides* were surviving, so reintroduction did not become necessary (Tye, 2003, 2006). But it did provide an example of the necessity for complementing *in situ* with *ex situ* measures. The main threat to the survival of *S. atractyloides*, as well as to all other species on Santiago, was grazing and browsing by alien, invasive mammals. Therefore, in order to exclude goats and protect the remaining small populations of *S. atractyloides*, some of these populations were fenced, at the same time as efforts continued to eradicate all goats, pigs and donkeys on Santiago.

During the same period the, the critically endangered *Calandrinia galapagosa* on San Cristóbal Island was saved: the main populations of the species were fenced and the feral goats in the area eliminated, while a major population of *Calandrinia* was enhanced by replanting *Calandrinia* individuals propagated *ex situ* (Arsiniegas-Fuertes, 1996; Mauchamp *et al.*, 1998; Tye, 2003).

## *Protection of vegetation on Santiago*

In the mid-1970s seven small fenced exclosures on Santiago were established to keep out goats, pigs and donkeys. These exclosures (covering less than 1 hectare) protected small samples of the Santiago vegetation, which served as small, living seed banks and were used for a comparative study of the impacts of goat grazing in the areas around the exclosures (Calvopiña & de Vries, 1975; de Vries & Calvopiña, 1977).

A major plant conservation effort was launched in 1993. Six areas (15.6 hectares in total) were selected in different vegetation zones and fenced. Thus, quite large representative samples of both endangered species and vegetation types from the Dry Lowland Zone to the humid areas in the highlands were protected against introduced herbivores. These exclosures actually contained 162 species out of the 318 species known to occur on the island (Blanton, 1995). In 1998, the set of exclosures was augmented by the expansion of the area of four of the existing ones and the construction of four new ones, three of them to protect additional remnants of *Scalesia* forest and one to protect the only population then known of *S. atractyloides* var. *atractyloides*. At the same time, all existing exclosures were repaired, as many of them had gaps so were no longer functioning to exclude goats.

## *Whole-island conservation*

To complement (where possible) conservation and management of species, populations and ecosystems, activities should also focus on whole-island conservation, in order to maintain natural and ecologically effective ecosystems.

Goats were introduced to Santiago in 1813, but when Charles Darwin visited the island in 1835 during his famous voyage around the world in HMS *Beagle*, he made no record of seeing goats. On the contrary, he found the upper regions of Santiago covered in green and flourishing vegetation (Darwin, 1839). Seventy years later, Alban Stewart (1911, 1915) described Santiago's dense forests, which in the upper parts were dominated by large *Scalesia pedunculata* trees. At those points in time, the vegetation presumably was relatively undisturbed by introduced mammals.

However, goats were again introduced to the island in 1906, and also later in the early part of the twentieth century. During the latter half of the century, the destructive impact of feral goats (plus pigs and donkeys) had become evident. In 1972 it was estimated that about 80–100 000 goats were present on the island, as well as several thousand pigs and hundreds of donkeys. In many places, dense forests had been transformed into woodland or herbaceous vegetation and only remnants of the *Scalesia* forest could be found. In 2003 the deterioration of vegetation had progressed even further. The distinctiveness of the plant formations was disappearing: forest dominated by *Psidium galapageium*, forest dominated by *Scalesia* and low forest dominated by *Zanthoxylum fagara* were all being transformed into the same depleted kind of vegetation, composed of a few resilient woody species, such as *Zanthoxylum fagara*, while the dominant vegetation type was open grassland. Because of the massive grazing and browsing by goats, pigs and donkeys, woody plants, epiphytes and vines were disappearing, seedlings were grazed and regeneration prevented, the soil seed bank was gradually depleted, short-lived herbs replaced perennials in the ground layer, fog-drip condensation decreased, and the habitats became more and more dry. Desertification and erosion had begun, and it could be foreseen that if great numbers of goats (in particular) were allowed to live on Santiago in the future, the vegetation would become more and more homogenised and less and less diverse (Fig. 16.6).

Fortunately, the plight of the Santiago ecosystems, plants and animals had long been recognised, and forceful action had been initiated. A major conservation management programme was initiated in 1997 by the CDF and the GNP aimed at eradicating all feral goats, donkeys and pigs on Santiago and northern Isabela, and in 2006 the goal was attained, a magnificent conservation result that has no parallel in scope, size and complexity anywhere in the world (F. Cruz, 2 April 2006, pers. comm.). The GNP has also successfully eradicated feral goats from Plaza Sur (in 1961), Santa Fe (1971), Rabida (1975), Española (1978), Marchena (1983) and Pinta (2000).

Prior to 2006, several hectares of remnants of *Scalesia pedunculata* forest in the Santiago highland had been fenced to protect the forest and allow it the possibility of regenerating, as mentioned above. According to F. Cruz of the CDF (2007, pers. comm.), the regeneration was already well under way a year later.

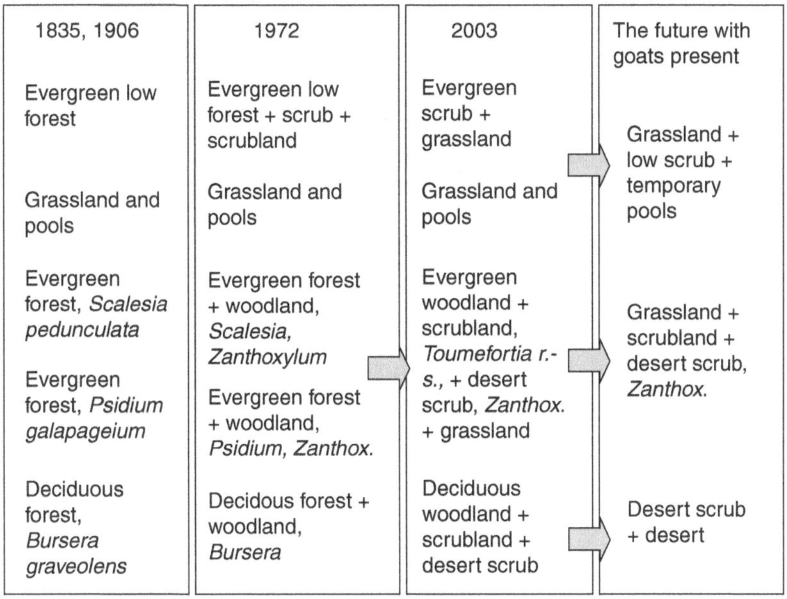


Figure 16.6 Suggested changes in vegetation formations on Santiago Island between Darwin's visit in 1835 and 2003. From bottom to top, the formations follow the increase in altitude and rainfall. Arrows indicate the suggested changes over time from distinct formations to less distinct and merged formations, at different altitudes. For example, evergreen *Psidium galapageium* forest and woodland as well as evergreen *Scalesia pedunculata* forest and woodland were still present at middle to high elevations in 1972, with both formations having *Zanthoxylum fagara* and *Tournefortia rufo-sericea* as prominent members. However, in 2003, these formations had merged into evergreen woodland, scrubland and desert scrub, and mixed with the now prevailing grassland. The surviving and now dominant constituents of these formations were *Zanthoxylum fagara* and *Tournefortia rufo-sericea*. If invasive mammals in great numbers were allowed to continue to live on Santiago, these formations could be predicted to change into scrub, scrubland and desert scrub with *Zanthoxylum fagara* as the main woody species, in addition to grassland. Thus, the vegetation formations became progressively poorer in species and less distinct because of the impact of thousands of goats and pigs and hundreds of donkeys. The degradation resulted in homogenisation and species depletion of the vegetation at all elevations.

By combining the protection of representative samples of vegetation through fencing, before species disappeared forever, with the execution of a major, systematic campaign against alien, invasive mammals, the plants and vegetation on Santiago have been saved for the future. Monitoring of the regeneration of the vegetation and ecosystems on Santiago has begun, and the final conservation action to be taken is to eradicate some introduced plant species that now regenerate as fast as the indigenous plants. When that happens, the goal of the whole island conservation approach will be attained, i.e. ensuring the existence of ecosystems without alien species and natural continuation and development of ecological and evolutionary processes in the ecosystems.

## *Strategies for future botanical research and plant conservation in the Galapagos*

Tye (2003) outlined the three themes of the Charles Darwin Foundation's approach to plant research and conservation in the Galapagos:

1 Ecology and conservation of native plants, with a main focus on the most endangered species and vegetation types. Key elements are inventories, monitoring, prioritisation and biological studies as a basis for providing management recommendations, so that species declines and habitat degradations can be reversed. Ultimately, the aim is to restore species and habitats to their original status.
2 Ecology and control of introduced plants, with a main focus on the worst invasive species. Again, inventories, monitoring, prioritisation and biological studies are key elements in order to attain results that can serve as basis for control or eradication of the introduced species.
3 Building the capacity for conservation, comprising all the functions needed to be able to attain the goals. This includes development of the Herbarium CDS and of a database of the Galapagos flora, as well as training Ecuadorian nationals in conservation biology, and producing research papers, other publications and presentations.

The Galapagos Islands is still the least altered tropical archipelago in the world (MacFarland & Cifuentes, 1996; Tye, 2003). The close collaboration between the Charles Darwin Foundation and the Galapagos National Park over 50 years has provided many impressive conservation results of importance to plants and vegetation, such as the complete eradication of invasive mammals on several islands, including Santiago. However, the Galapagos Islands are still under pressure, and there is a strong need for continuing along the lines described by Tye (2003). It is particularly important to obtain basic information on endangered species and vegetation types, as well as on alien, invasive species, in order to ensure that priorities can be set and efficient management activities be carried out. Plants and vegetation in the Galapagos can then be saved for the future.

## References

Adsersen, H. (1989). The Rare Plants of the Galapagos Islands and Their Conservation. *Biol. Conserv.* 47: 49–77.

Adsersen, H. (1990). Intra-archipelago distribution patterns of vascular plants in Galápagos. In J. E. Lawesson, O. Hamann, G. Rogers, *et al.* (eds.), 'Botanical Research and Management in Galápagos'. *Monogr. Syst. Bot. Mo. Bot. Gard.* 32: 67–78.

Arsiniegas-Fuertes, M. O. (1996). *Caracterizacion, evaluacion y conservacion de cuatro especies de plantas endemicas que se hallan en peligro de extincion en las Islas Galápagos: Calandrinia galapagosa, Lecocarpus darwinii, Scalesia helleri, Scalesia retroflexa.* Ingeniero Forestal thesis: Universidad Tecnica Luis Vargas Torres, Esmeraldas, Ecuador.

Atkinson, I. A. E. (1990). Mapping and classifying vegetation in the Galapagos Islands, some considerations. In J. E. Lawesson, O. Hamann, G. Rogers, *et al.* (eds.), 'Botanical Research and Management in Galápagos'. *Monogr. Syst. Bot. Mo. Bot. Gard.* 32: 101–115.

Blanton, C. (1995). *Informe de la Directora de la Estación Cientifica Charles Darwin, Noviembre de 1994–Marzo de 1995.* Charles Darwin Foundation, Galápagos, Ecuador.

Bowman, R. I. (1961). Morphological differentiation in the Galápagos finches. *Univ. Calif. Publ. Zool.* 58: 1–302.

Calvopiña, L. H. & de Vries, T. (1975). Estructura de la población de cabras salvajes (*Capras hircus* L.) y los daños causados en la vegetación de la isla San Salvador, Galápagos. *Revista Univ. Católica (Quito)* 3 (8): 219–214.

Darwin, C. (1839). *Journal of researches into the geology and natural history of the various countries visited by H.M.S. Beagle, Under the Command of Captain FitzRoy, R.N. from 1832 to 1836*, pp. 453–478. London.

de Vries, T. & Calvopiña, L. H. (1977). Papel de los chivos en los cambios de la vegetación de la isla San Salvador, Galápagos. *Revista Univ. Católica, Quito* 5 (16): 145–169.

de Vries, T. & Tupiza, A. (1990). The dynamics of *Scalesia cordata* (Asteraceae) in different habitats in Sierra Negra, Isabela, Galápagos. In J. E. Lawesson, O. Hamann, G. Rogers, *et al.* (eds.), 'Botanical Research and Management in Galápagos'. *Monogr. Syst. Bot. Mo. Bot. Gard.* 32: 137–147.

Eliasson, U. (1968). On the influence of introduced animals on the natural vegetation of the Galapagos Islands. *Notic. Galápagos* 11: 19–21.

Eliasson, U. (1972). Studies in Galápagos Plants XII. On the Vegetation of Fernandina before the Eruption in 1968. *Bot. Notiser* 125: 49–61.

Eliasson, U. (1974). Studies in Galápagos Plants XIV. The genus Scalesia Arn. *Opera Botanica* 36: 1–117.

Eliasson, U. (1984). Native climax forests. In R. Perry (ed.), *Key Environments. Galápagos*, pp. 101–114. Oxford, Pergamon Press.

Eliasson, U. (1995). Patterns of diversity in island plants. In P. M. Vitousek, L. L. Loope & H. Adsersen (eds.), *Islands, Biological Diversity and Ecosystem Function*, pp. 35–50. Berlin, Springer (Ecological Studies).

Geist, D. (1996). On the emergence and submergence of the Galapagos Islands. *Notic. Galápagos* 56: 5–9.

Hamann, O. (1974). Contributions to the Flora and Vegetation of the Galapagos Islands III. Five New Floristic Records. *Bot. Notiser* 127: 309–316.

Hamann, O. (1979a). On climatic conditions, vegetation types and leaf size in the Galapagos Islands. *Biotropica* 11: 101–122.

Hamann, O. (1979b). Dynamics of a stand of *Scalesia pedunculata* Hooker fil., Santa Cruz Island, Galápagos. *Bot. J. Linn. Soc.* 78: 67–84.
Hamann, O. (1981). Plant Communities of the Galapagos Islands. *Dansk Botanisk Arkiv* 34 (2): 1–163.
Hamann, O. (1986). The El Niño influence on the Galápagos vegetation. In G. Robinson & E. del Pino (eds.), *El Niño en las Islas Galápagos: el evento de 1982–1983*, pp. 299–330. Quito, Ecuador, CDF.
Hamann, O. (1995). Galapagos Islands, Ecuador. In *Centres of Plant Diversity. A Guide and Strategy for Their Conservation*, 2, pp. 556–564. Cambridge, UK, IUCN.
Hamann, O. (1997). Conservation of endangered plants of the Galápagos; integrating conservation actions. In D. H. Touchell & K. W. Dixon (eds.), *Conservation into the 21st Century*, pp. 137–150. Proceedings of the 4th International Botanic Gardens Conservation Congress, Perth, Western Australia.
Hamann, O. (2001). Demographic studies of three indigenous stand-forming plant taxa (*Scalesia*, *Opuntia* and *Bursera*) in the Galapagos Islands. *Biodivers. Conserv.* 10: 223–250.
Hamann, O. (2004). Vegetation changes over three decades on Santa Fe Island, Galápagos, Ecuador. *Nord. J. Bot.* 23: 143–152.
Hendrix, L. B. (1981). Post-eruption succession on Isla Fernandina, Galápagos. *Madroño* 28: 242–254.
Hendrix, L. B. & Smith, S. D. (1986). Post-eruption revegetation of Isla Fernandina, Galápagos: II. *Natl. Geogr. Res.* 2: 6–16.
Huttel, C. (1995). *Vegetación en coladas de lava, Islas Galápagos, Ecuador*, pp. 1–175. Fundación Charles Darwin para las Islas Galápagos & ORSTOM, Quito, Ecuador.
Itow, S. (1990). Herbaceous and ericaceous communities in the highlands of Santa Cruz, the Galapagos Islands. In J. E. Lawesson, O. Hamann, G. Rogers, *et al.* (eds.), 'Botanical Research and Management in Galápagos'. *Monogr. Syst. Bot. Missouri Bot. Gard.* 32: 47–58.
Itow, S. (1992). Altitudinal change in plant endemism, species turnover, and diversity on Isla Santa Cruz, the Galapagos Islands. *Pac. Sci.* 46: 251–268.
Itow, S. (1995). Phytogeography and ecology of *Scalesia* (Compositae), endemic to the Galapagos Islands. *Pac. Sci.* 49: 17–30.
Itow, S. (2003). Zonation pattern, succession process and invasion by aliens in species-poor insular vegetation of the Galapagos Islands. *Glob. Environ. Res.* 7: 39–58.
Itow, S. & Weber D. (1974). Fens and bogs in the Galapagos Islands. *Hikobia*: 39–52.
Itow, S. & Mueller-Dombois, D. (1988). Population structure, stand-level dieback and recovery of *Scalesia pedunculata* forest in the Galapagos Islands. *Ecol. Res.* 3: 333–339.
Jäger, H., Tye, A. & Kowarik, I. (2007). Tree invasion in naturally treeless environments: impacts of quinine (*Cinchona pubescens*) trees on native vegetation in Galápagos. *Biol. Conserv.* 140: 297–307.
Jørgensen, P. M. & León-Yanez, S. (eds.) (1999). *Catalogue of the Vascular Plants of Ecuador.* 1181 pp. St Louis, Missouri Botanical Garden Press.
Kastdalen, K. (1982). Changes in the biology of Santa Cruz Island between 1935 and 1965. *Notic. Galápagos* 35: 7–12.
Lawesson, J. E. (1988). Stand-level dieback and regeneration of forests in the Galapagos Islands. *Vegetatio* 77: 87–93.

Macfarland, C. & Cifuentes, M. (1996). Case study: Ecuador. In V. Dompka (ed.), *Human Population, Biodiversity and Protected Areas: Science and Policy Issues*, pp. 135–188. Washington, DC, American Association for the Advancement of Science, International Directorate.

Macdonald, I. A. W., Ortiz, L., Lawesson, J. E. & Nowak, J. B. (1988). The Invasion of Highlands in Galápagos by the Red Quinine-Tree *Cinchona succirubra*. *Environ. Conserv.* 15: 215–220.

Macdonald, I. A. W., Loope, L. L., Usher, M. B. & Hamann, O. (1989). Wildlife conservation and the invasion of nature reserves by introduced species: a global perspective. In J. A. Drake, H. A. Mooney, F. de Castri, *et al.* (eds.), *Biological Invasions: A Global Perspective*, SCOPE 37: 215–255. Chichester, UK.

Mauchamp, A., Aldaz, I., Valdebenito, J. & Ortiz, E. (1998). Threatened species: a re-evaluation of the status of eight endemic plants of the Galápagos. *Biodivers. Conserv.* 7: 97–107.

Mueller-Dombois, D. & Fosberg, F. R. (1998). *Vegetation of the Tropical Pacific Islands*, pp. 587–604. New York, Springer.

Nowak, J. B., Lawesson, J. E., Adsersen, H. & de Vries, T. (1990). A two-year study of post-fire vegetation dynamics on Southern Isabela, Galapagos Islands, Ecuador. In J. E. Lawesson, O. Hamann, G. Rogers, *et al.* (eds.), 'Botanical Research and Management in Galápagos', *Monogr. Syst. Bot. Mo. Bot. Gard.* 32: 123–136.

Shimizu, Y. (1997). Competitive relationships between tree species of *Scalesia* (*S. pedunculata, S. cordata, S. microcephala*) and introduced plants (*Cinchona succirubra, Psidium guajava, Lantana camara*) with reference to regeneration mechanism of Scalesia forests in the Galapagos Islands. *Reg. Views* 11: 23–172. Institute of Applied Geography, Komazawa University, Tokyo.

Snell, H. M. & Rea, S. (1999). The 1997–98 El Niño in Galápagos: can 34 years of data estimate 120 years of pattern? *Notic. Galápagos* 60: 11–20.

Stewart, A. (1911). Expedition of the California Academy of Sciences to the Galapagos Islands, 1905–1906. II. A Botanical Survey of the Galapagos Islands. *Proc. Calif. Acad. Sci.* Fourth series, 1: 7–288.

Stewart, A. (1915). Some observations concerning the botanical conditions on the Galapagos Islands. *Trans. Wisconsin Acad. Sci.* 18: 272–340.

Tye, A. (2003). *Plant Research for Conservation in Galápagos. Report for the years 1998–2003*, pp. 1–49. Charles Darwin Foundation for the Galapagos Islands, Puerto Ayora, Galápagos, Ecuador.

Tye, A. (2006). Restoration of the vegetation of the Dry Zone in Galápagos. *Lyonia* 9: 1–33.

Tye, A. & Aldaz, I. (1999). Effects of the 1997–98 El Niño Event on the Vegetation of Galápagos. *Notic. Galápagos* 60: 22–24.

Tye, A., Snell, H. L., Peck, S. B. & Adsersen, H. (2002). Outstanding terrestrial features of the Galapagos archipelago. In R. Bensted-Smith (ed.), *A Biodiversity Vision for the Galapagos Islands*, Charles Darwin Foundation and World Wildlife Fund, Puerto Ayora, Galápagos, Chapter 3: 1–11.

van der Werff, H. (1978). *The Vegetation of the Galapagos Islands*, pp. 1–102. PhD thesis: Rijksuniversiteit, Utrecht.

van der Werff, H. (1979). Conservation and vegetation of the Galapagos Islands. In D. Bramwell (ed.), *Plants and Islands*, pp. 391–404. London, Academic Press.

van der Werff, H. (1980). The vegetation types of Santa Cruz and Alcedo. *Notic. Galápagos* 31: 11–13.

van der Werff, H. (1990). Ferns as indicators of vegetation types in the Galapagos archipelago. In J. E. Lawesson, O. Hamann, G. Rogers, *et al.* (eds.), 'Botanical Research and Management in Galápagos'. *Monogr. Syst. Bot. Mo. Bot. Gard.* 32: 79–92.

van der Werff, H. H. & Adsersen, H. (1993). Dry coastal ecosystems of the Galapagos Islands. In E. van der Maarel (ed.), *Ecosystems of the World 2B. Dry Coastal Ecosystems*, pp. 459–475. London, Elsevier.

Wiggins, I. L. & Porter, D. M. (1971). *Flora of the Galapagos Islands.* Stanford, CA, Stanford University Press.

Wium-Andersen, S. & Hamann, O. (1986). Manglares de las Islas Galápagos. *Revista Geografica* 23: 101–122. Quito, Ecuador.

# 17

# New directions and challenges for the conservation of the flora of Madagascar

STUART CABLE

Madagascar, the fourth largest island in the world, is well known for its rich and diverse flora. Separated from Africa 160 million years ago (Ma), the flora evolved in isolation and today we estimate that there are 12 000–14 000 species of plants, of which over 90% are endemic. Similar diversity is seen in animals, with Madagascar home to 50% of the world's chameleon species and more primate taxa than any other country. Conservation International ranks Madagascar as a top 10 global biodiversity hotspot and the public perception of Madagascar's biological riches has been reinforced by media images of lush rainforests, dramatic pinnacles of limestone karst, exotic spiny forest and various charismatic plants and animals.

However, Madagascar has been considered a lost cause for conservation for decades. People only settled on the island about 1800 years ago (Burney *et al.*, 2004), but with severe consequences. The generally accepted view has been that 80% of the original forest cover has been cleared and we can only guess at the level of extinctions. The sub-fossil record shows the disappearance of such remarkable animals as dwarf hippos, giant ground-dwelling lemurs and elephant birds. We will never know how many plant species have become extinct, lost before they could be discovered by botanists. With a very poor and burgeoning rural population, subsisting on inefficient fire-based agricultural systems and charcoal for fuel, more than a third of Madagascar's remaining primary forest has been lost since 1975 and at current rates all remaining primary vegetation will have gone by 2067 (Moat & Smith, 2007). Up until the last few years the conservation community was largely powerless to stop the relentless loss, as contiguous forest has become fragmented, the fragments have disappeared and protected areas have been eroded by encroachment and fire.

*The Biology of Island Floras*, eds. David Bramwell and Juli Caujapé-Castells. Published by Cambridge University Press. 

The problems continue; but since 2003 a dramatic change has occurred, driven by Madagascar itself, which has precipitated renewed national and international optimism for the conservation of Madagascar's biological diversity. In September 2003 at the World Park Congress in Durban, former President Marc Ravalomanana declared Madagascar's intent to triple the area of protected land in Madagascar to over 60 000 $km^2$ (10% of the total land surface), in order to meet international obligations under the Convention on Biological Diversity and reach the targets of the Global Strategy for Plant Conservation. One of the newly created reserves is Makira Protected Area which, together with the contiguous Masoala National Park, contains 10% of the Madagascar's remaining humid forests and will conserve an estimated 50% of its biodiversity.

Ravalomanana was largely successful in raising most of the estimated $50 million needed to create the new protected areas, but more is needed to ensure the long-term survival of Madagascar's biodiversity. The World Conservation Society and Conservation International have recently demonstrated that the newly emerging carbon markets could fund conservation, with polluters paying for carbon sequestration by avoided deforestation. Despite recent political turmoil and some setbacks for conservation, there is still a momentum for designating and effectively managing new protected areas as well as working with local communities to guarantee their success. Major funders, such as the Critical Ecosystem Partnership Fund, have helped by funding numerous small-scale grassroots conservation and development projects and by bringing together the large conservation agencies and local NGOs to tackle the major issues.

## The Madagascar biodiversity hotspot

Madagascar is the fourth largest island in the world with a land area of 587 000 $km^2$, which is larger than France and twice the size of the United Kingdom. It measures 1600 km north to south and is up to 600 km wide from east to west. It is home to a striking variety of vegetation types with associated floras, from the spiny forest of the sub-arid southwest to the mega-diverse humid forests of the northeast. Madagascar is one of the most important global hotspots for biodiversity conservation (Myers *et al*., 2000) and the understanding of evolutionary processes (Myers & Knoll, 2001).

The island separated from Africa over 160 Ma and separated from India around 70 Ma (de Wit, 2003; Rakotosamimanana, 2003). This long isolation, combined with climatic and geological variation across the island, has resulted in high species richness and endemism. Gautier and Goodman (2003) provide a summary of the affinities and origin of the Malagasy flora. Madagascar separated from mainland Africa around the end of the Jurassic period and before the main radiation of the angiosperms.

Floristic affinities between Africa and Madagascar are probably secondary in origin (by land bridge or long-distance dispersal), with the strongest similarities seen in the flora of western Madagascar. Schatz (1996) reviewed the affinities with the Indo-Australo-Malaysian floras, which is reflected in the flora of eastern Madagascar. The Southern Hemisphere affinities are explained as Gondwanan relicts. The Malagasy flora also includes a strong pantropical element (28–36% of genera) as well as the palaeotropical element (12–14.5% of genera). Madagascar separated from India about 80 Ma, while India was still connected to South America (accounting for a neotropical element). The movement of plants may have been both ways, with taxa that evolved on Madagascar subsequently reaching other land masses, and taxa with high diversity on Madagascar having originated elsewhere. The potential for disjunct distributions is exemplified by the baobabs, which are often seen as a flagship Madagascan plant group. Of the eight species of *Adansonia*, one species is found across dryland Africa and the Mascarenes, one species is endemic to Australia and the remaining six species are endemic to Madagascar.

The Rubiaceae is the largest family of woody plants across the tropical floras. Davis *et al.* (2009) present a global assessment of the distribution, diversity and endemism of the Rubiaceae that clearly indicates the importance of Madagascar in terms of global biodiversity. Madagascar is ranked the tenth country for Rubiaceae species richness irrespective of land area, based on the total of 569 described species used in the analysis (not including up to 300 species that have yet to be described). Taking area into consideration, Madagascar is ranked the fourth country for species richness, after Venezuela, Columbia and Cuba. Madagascar also has the second highest total of endemic Rubiaceae species for any country, after New Guinea, with 520 endemics (91% of the Rubiaceae species).

## How many plant species are there?

Botanists have been studying the flora of Madagascar for over 350 years, yet there is still no accurate total for the number of species. Recent estimates vary between 10 000 and 14 000 species, which would represent about 3–5% of the estimated global total of 270 000 species (Global Strategy for Plant Conservation, 2002).

The first published list of plants from Madagascar was Etienne de Flacourt's *Histoire de la grande îsle Madagascar* (1658), which included notes on useful and medicinal plants. De Flacourt was a French apothecary who settled in Fort Dauphin in 1648 and began collecting plants and minerals. The first systematic collecting was by Aubert-Aubert du Petit-Thouars, an army captain who gave up his military career to study natural history. He was based on Mauritius for over 10 years but spent 6 months in Madagascar collecting plants. He published several pamphlets in the series *Histoire des végétaux recueillis dans les îsles australes d'Afrique* (1804–1806), describing and illustrating the plants he found. In 1822 he published the book *Histoire particulière des plantes orchidées recueillis sur les trois îsles australes d'Afrique, de France, de Boubon et de Madagascar,* which was the first systematic treatment of orchids for the region.

The nineteenth century then saw a succession of missionaries, diplomats, explorers and commercial collectors, risking life and limb adding new and exotic species to the known flora of Madagascar. Dorr (1997) gives concise biographies for these collectors, many with names familiar from plant species epithets, such as Diedrich Rutenberg, Johann Hildebrandt, Robert Lyall, Carl Hilsenberg and Wenceslas Bojer.

It was the British missionary, the Rev. Richard Baron, who published the first checklist of the plants of Madagascar, the *Compendium des Plantes Malagaches* (1900–1906, as discussed by Dorr, 1987). Baron's checklist includes over 4700 species and varieties in 152 families and 970 genera of vascular plants. Baron was a missionary who spent 35 years in Madagascar and was fluent in the Malagasy language. He wrote a Malagasy textbook on botany in 1882 and contributed botanical notes for Richardson's *A New Malagasy-English Dictionary* in 1885. Despite the Compendium and the textbook, Baron is best known for over 12 000 herbarium collections sent to Kew between 1880 and 1905. Among these collections were nearly 1000 new species that J. G. Baker described in the *Journal of the Linnean Society: Botany* between 1882 and 1890. Baron travelled to the north and east during a 5-month collecting trip, but most of his collections are from the Central Plateau as he lived in Fianarantsoa and near Antananarivo. Unfortunately for modern botanists he did not record detailed locality information. Baron also drew on the work of Henri Baillon, who published the *Liste des plantes de Madagascar* serially through the *Bulletin Mensuel de la Société Linnéenne de Paris* (1882–1895). Baillon also compiled, with Drake del Castillo, the six volumes of the *Histoire Naturalle des Plantes* (1886–1903) published as part of the 39 volumes of Alfred Grandidier's *Histoire Physique Naturelle et Politique de Madagascar (1875–1954*).

The twentieth century was dominated by French botanists, of whom the most important and prolific was Henri Perrier de La Bâthie. Originally visiting Madagascar as a prospector for a gold mining company, he was sent on several expeditions for the Muséum National d'Histoire Naturelle in Paris and then worked for the Service de Colonisation, collecting geological and soil samples. He collaborated with many of the colonial botanists whose names are now familiar as plant species epithets, including Karl Afzelius, Edmund Fris, Henri Humbert, Henri Poisson, René Viguier, Raymond Decary and Henri Jumelle. Perrier added hundreds of new species to the flora, but he is best known for his work writing half of the family treatments and directing the publication of the *Catalogue des Plantes de Madagascar* (1930–40). In 1936, Perrier estimated the vascular flora to total 7370 species.

Perrier's catalogue was superseded by the *Flore de Madagascar et des Comores*, edited by Henri Humbert from 1936 to 1967. Published by the Muséum National d'Histoire Naturelle, Paris, the Flora covers 80% of the families of vascular plants in 166 volumes. There are some large families that have not been completed, including Acanthaceae, Asclepiadaceae, Euphorbiaceae, Leguminosae, Poaceae and Rubiaceae. A third of all the volumes were published before 1960 and are now thought to be in need of revision (Gautier & Goodman, 2003). Humbert estimated the flora to total 7900 species, with a level of endemism of 81% (Humbert, 1962).

The Leguminosae (Du Puy *et al.*, 2002) and Palmae (Dransfield & Beentje, 1995) have been published in a separate series by the Royal Botanic Gardens, Kew (RBG Kew). The *Palms of Madagascar* set a new standard for modem taxonomic treatments of Madagascan plants, incorporating colour photographs, line drawings and maps for all species.

Visualising distributions through accurate maps is the key to identifying species and understanding floristic affinities and taxonomic relationships. Another important recent publication has been the *Generic Tree Flora of Madagascar* (Schatz, 2001), which revised and expanded Rene Capuron's *Essai d'introduction à l'etude de la flore forestière* (1957). Schatz calculated the tree flora as totalling 4220 species in 490 native genera and 107 plant families, and with species endemism of 96%.

The collecting effort has intensified over the past 40 years, leading to revised estimates for the size of the flora. Phillipson (1994) calculated the total number of native vascular plant species as 9345, with 81% endemism. Dejardin *et al.* (1973), Koechlin *et al.* (1974) and Schatz *et al.* (1996) all estimated the flora to be as high as 12 000 species, with endemism of 85%. Each new taxonomic treatment has seen large increases in the number of species known. Hedge *et al.* (1998) described 58 new species for the account of Lamiaceae in the *Flore de Madagascar et des Comores*, representing an increase of 70%. Research for *Orchids of Madagascar* (Hermans *et al.*, 2007) resulted in an increase of 30% in the total number of orchid species known, making the Orchidaceae the most diverse plant family in Madagascar with over 1000 species in 60 genera. Nearly 90% of the orchid species are endemic to Madagascar. The Leguminosae is another of the largest plant families known from Madagascar and Du Puy *et al.* (2002) treated 573 native species, of which 459 are endemic (80%) and 129 were new species (23%). They included a further 94 introduced and naturalised species (Labat and Moat, 2003).

Madagascar is home to a rich palm flora, which remained poorly known until publication of *The Palms of Madagascar* (Dransfield and Beentje, 1995), in which 171 species in 16 genera were described, about three times the number known from continental Africa. Their treatment included 70 species new to science, increasing the number of palm species known for Madagascar by 69%. Palms illustrate the uniqueness of the Madagascan flora and the challenges facing conservation. Of the palms listed 166 are endemic and many have very narrow distributions; 86 have IUCN Red List categories of Endangered, Critically Endangered or Extinct and the authors speculate that if their work had started 10 years later, many of their new species would never have been discovered as they would already be extinct (Dransfield & Beentje, 2003). Research into the palms of Madagascar has continued and many more new species have been discovered since *Palms of Madagascar* was published (see Chapter 7 in this volume).

The Rubiaceae is the largest family of woody plants in Madagascar, with 650 described species in 95 genera, representing 7–9% of the plant diversity of Madagascar (Davis & Bridson, 2003). Over 98% of species and 30% of genera are endemic to Madagascar. There are also up to 300 undescribed species known from herbarium collections, and new species are discovered on every collecting trip made by specialists. Recent taxonomic treatments have increased the known species for some of the larger genera by 25–50%. Davis *et al.* (2006), in their conspectus of the genus *Coffea*, list 59 species for Madagascar, an increase of 42 species from the previous taxonomic revision by Chevalier (1947). A further six species in the subgenus *Baracoffea* have just been published (Davis & Rakotonasolo, 2008). Many of the coffee species have narrow distributions and are vulnerable to extinction. The authors have applied IUCN red list criteria and categorise seven species as Critically Endangered, 28 species as Endangered, 8 species as Vulnerable and 8 species as Near Threatened. Only 8 species, out of 59, have wide ranges with good populations and are categorised as Least Concern. The final answer to how many plant species are known from Madagascar should be answered by Missouri Botanical Garden's Vahinala Project, which aims to make available online information on all native and naturalised vascular plant species known from Madagascar. It will summarise the available taxonomic literature and the specimen base for each taxon, and will be compiled as *The Catalogue of Vascular Plants of Madagascar*, an online database and eventually a book. The Missouri Botanical Garden leads the project in collaboration with other institutions and researchers. The anticipated end date was December 2010 for completing all species accounts with data on distribution, ecology and conservation. As of 24 October 2008, the catalogue contained information for 9756 accepted species and 7376 synonyms (www.eFloras.org). At current rates of discovery, the total number of species known from Madagascar will be 12 000–14 000 (Goodman & Benstead, 2005; Phillipson *et al.*, 2006) and with endemism > 90% (Schatz, 2001). The limiting factors are whether there are enough botanists to do the collecting, identification and taxonomic work and whether we can slow the rate of forest destruction and find and conserve plant species before they become extinct.

## The digital age

Producing a widely accessible list of known plant species for Madagascar is an obligation under Target 1 of the Global Strategy for Plant Conservation (2002). Most countries have signed the Convention on Biological Diversity, and the GSPC gives the Government of Madagascar and partner organisations targets to measure progress with conservation. The increasing availability of data online is significantly improving our prospects for achieving Target 1.

Missouri Botanical Garden's Tropicos database, containing over 170 000 botanical collection records for Madagascar, has been the single most important source of data for taxonomists and conservationists working on the flora for over 10 years. The herbaria of Paris and Kew are also making herbarium collection records available online. The study of the flora of Madagascar has undergone a revolution during the last couple of years with the global availability of online resources. The World Checklist Series by the Royal Botanic Gardens, Kew (http://apps.kew.org/wcsp) is a global synthesis of the taxonomy of the Monocotyledons and many Dicotyledon families, giving basic data on accepted names, synonyms and distribution. Eventually it will cover all species in combination with other major online databases such as Tropicos (see www.theplantlist.org/). However, the most important recent development facilitating work on the flora of Madagascar has been the Aluka Project funded by the Andrew W. Mellon Foundation and the Arcadia Fund. Aluka is an international project building an online digital library of scholarly resources from and about Africa (http://plants.jstor.org/). The key component of Aluka for botanists is the African Plants Initiative, a collaboration among more than 50 institutions in Africa, Europe and the USA, that has built a comprehensive online research tool aggregating scattered scientific resources about African plants. It contains over 34 000 high-resolution digital images of type specimens of Madagascan plant species. Other material includes photographs, drawings and digitised floras. The *Flore du Madagascar* is online at Missouri Botanical Garden's Tropicos portal. However, it is the images of the types that have become so useful for anyone wishing to identify Madagascan plants, whether based in Madagascar or herbaria around the world. The most complete herbarium collection of Madagascan plant species is that at the Muséum National d'Histoire Naturelle, Paris, which is in the process of digitising a significant proportion of its collection. The flora of Madagascar is now accessible to anyone with an internet connection.

## Seeing Red: assessing the extinction risk of species

The IUCN Red List of Threatened Species categorises species according to their risk of extinction. It is a searchable online database that currently contains assessments and supporting data for 44 834 species from around the world (www.iucnredlist.org/). Over 38% of these species are threatened with extinction (Vulnerable, Endangered or Critically Endangered) and 3246 (7%) are classified as Critically Endangered. As of June 2009, the IUCN Red List included 359 vascular plant species from Madagascar, with 278 listed as threatened with extinction (Vulnerable, Endangered or Critically Endangered) and 61 (17%) of these listed as Critically Endangered.

Randrianasolo *et al.* (2002) studied the application of IUCN categories and criteria to species in five genera of Anacardiaceae from Madagascar. They used GIS to create distribution maps for historic and recent occurrence data to determine the change in conservation status, area of occupancy and extent of occurrence for each species. They found that 80% of species had decreased in the number of subpopulations and the extent of occurrence over time. The study illustrated the importance of accurate taxonomy prior to a conservation assessment. Before the genus *Campnosperma* was revised, it was considered monospecific in Madagascar, with the single species covering 93 000 $km^2$ and including populations in at least seven protected areas. It would have been given a Least Concern rating under the IUCN ratings. After the revision, *Campnosperma* consisted of four species in Madagascar, two with very narrow distributions and all four categorised as being Endangered or Vulnerable. They also concluded that IUCN assessments can be misleading if based on collection data without any knowledge of the species in the field. They give *Campnosperma parvifolium* as an example. It is classified as vulnerable even though it is known from two large populations in well-protected reserves. Thirdly, they caution that specimen data might not be representative of species distributions due to poor sampling.

Callamander *et al.* (2007) reiterate the need for field observations to complement historical specimen data, particularly information on population size, sensitivity to disturbance and habitat threats. They demonstrate how application of IUCN Red List categories and criteria for a well-researched plant group (Pandanaceae in their study) can reveal centres of richness and rarity and gaps in Madagascar's protected area network. They assessed 91% of Panadanaceae species as threatened and identified 10 sites that are needed for the conservation of the 19 species that are not found within the current reserve network.

It should be of concern to conservationists that there are so few red-listed Madagascan plant species. It is clear that the lack of assessments does not reflect a healthy flora, just a lack of botanists to do assessments, a lack of data to do assessments accurately and a shortage of resources to add assessments to the IUCN database. Only 163 of the assessments currently listed on the IUCN database have been carried out during the last 10 years, yet many specialists routinely calculate IUCN Red List assessments as part of their taxonomic treatments.

Ultimately, the basis of success for plant conservation in Madagascar will be the ability of the national and international herbaria such as Park Tsimbazaza, Missouri Botanical Garden, Paris and the Royal Botanic Gardens, Kew, to systematically compile conservation assessments and feed them into the processes determining conservation policy. The necessary steps will include completing a working list of plant species, compiling accurate collection data and automating preliminary conservation assessments. Taxonomists can then concentrate on priority taxa for full IUCN assessments enhanced by field data. For discussion see Schatz (2002).

## Forests and grasslands, deforestation and fragmentation

Madagascar was settled by humans from 1800 years ago and there is a general consensus that forest covered up to 90% of the island at that time. This has given rise to the headline figure of 80% forest loss in Madagascar. There are a growing number of studies showing the negative effects of deforestation and forest fragmentation (e.g. Langrand & Wilme, 1997). A number of studies (e.g. Agarwahl *et al.*, 2005; Kull, 2000; Kull, 2002; McConnell, 2002; Réau, 2002) have begun to question the assumption that Madagascar was 90% forested prior to humans arriving and have reinterpreted current forest distribution based on patterns of historical land use. A key assumption has been that the central highlands were cleared of forest by slash-and-burn agriculture and by burning to promote new grass for zebu grazing. Grasslands cover > 75% of Madagascar, so understanding their history is important for calculating the overall total for forest loss. Bond *et al.* (2008) have questioned the assumption that the central highlands were once completely forested and that all grasslands indicate a degraded landscape. If the grasslands were created through human activity this would be apparent from a lower grass diversity compared to similar grasslands in neighbouring countries. The authors compared C4, or CAM (crassulacean acid metabolism), grasses in Madagascar and southern Africa and found that Madagascar has 88 C4 grass genera (including six endemic genera) or about 85–89% of the total found in southern Africa. The authors concluded that the C4 grass lineages in Madagascar and associated endemic plants and animals do not fit with the view that the grasslands are derived from degraded forests.

Harper *et al.* (2007) have enumerated forest loss since the 1950s. They suggest that, even with conservative estimates of forest cover, Madagascar had already lost more than half of its forests by 1950. Since 1950, forest cover had declined from 160 000 $km^2$ to 90 000 $km^2$ over 50 years, with annual rates of loss varying from 0.3% per year up to 1970, to 1.7% per year for 1970–1990 and 0.9% per year for 1990–2000. The humid, dry and spiny forests lost 43%, 41% and 28%, respectively, of their cover over that time. Fragmentation was dramatic, with core forest (> 1 km from a non-forest edge) decreasing from 90 000 $km^2$ to 20 000 $km^2$. Consiglo *et al.*

(2006) undertook a similar study focused on littoral forests and found that only 1.5% of the original cover remains today.

The Royal Botanic Gardens, Kew, has recently produced the Madagascar Vegetation Atlas (Moat & Smith, 2007), which mapped 14 vegetation types. The maps were based on satellite data and the enumerated vegetation types were ground-truthed by a team from RBG Kew, Missouri Botanical Garden and Park Tsimbazaza in Antananarivo. Funded by the Critical Ecosystem Partnership Fund, the new vegetation atlas will enable Madagascar to assess progress with its national conservation goals. The team calculated a 33% reduction in forests since the 1970s and current losses of 1500 km$^2$ per year. Their dismal prediction is that all primary vegetation will have gone by 2067.

## The Durban Vision and new protected areas

In September 2003, at the World Park Congress in Durban, former President Marc Ravalomanana declared Madagascar's intent to triple the area of protected land in Madagascar from 17 000 km$^2$ to over 60 000 km$^2$ (10% of the total land surface), in order to meet international obligations under the Convention on Biological Diversity and reach the targets of the Global Strategy for Plant Conservation. This goal has become known as the Durban Vision and is incorporated into Ravalomanana's National Environmental Action Plan (phase III). The process was nearly complete and Ravalomanana had been largely successful in raising most of the estimated $50 million needed to create the new protected areas when he lost the presidency in a *coup d'etat* early in 2009. The conservation community is currently working with the new Government of Madagascar to realise the Durban Vision, and the final refinement and legal designation of new protected areas will not be complete until at least 2012.

A number of technical groups were set up to help the Government of Madagascar identify new sites for protected area status. Claire Kremen led a team from the University of California, Berkeley, to analyse the available biological data, and their focus was on 2315 endemic species in six major taxonomic groups: ants, butterflies, frogs, geckos, lemurs and plants (Kremen *et al.*, 2008). The plant data used was a database of geo-referenced palm data derived from the latest taxonomic treatment of the family for Madagascar (Dransfield & Beentje, 2003), but the plan is to incorporate a wider taxonomic base through the new Rebioma web portal (www.rebioma.net). The problem for conservation in Madagascar is that, as well as high species richness, it has complex patterns of microendemism, making the design of protected area networks difficult. The aim of the Durban Vision is to provide a solution to maximise both the number of species protected and the proportion of their distribution within protected areas. The analysis by the Berkeley team identified several regions within the Central Plateau and littoral forests as priorities, even though these areas have low forest cover and have been historically neglected in favour of large blocks of humid forest.

Missouri Botanical Garden, with funding from the Critical Ecosystems Partnership Fund, selected 1250 plant species for full IUCN conservation assessments in order to identify priority areas for plant conservation. Their analysis identified 77 sites and they are currently working to establish seven of these as protected areas. National and local NGOs are playing a key role establishing the new protected areas and developing management plans in collaboration with local communities that incorporate appropriate sustainable development.

Rebioma (2009) report that by the end of 2008, around 38 000 $km^2$ of new protected areas had been identified and were at varying stages of planning. This will bring the total of protected areas in Madagascar to over 55 000 $km^2$, which is equivalent to 9.4% of the land surface.

## Seeing REDD: paying for protected areas

Deforestation produces 5–6 gigatonnes of carbon dioxide annually, which represents about 20% of global greenhouse gas emissions (Tollefson, 2008). REDD (reducing emissions from deforestation and forest degradation) is an emerging mechanism to use market incentives for funding forest conservation to reduce the emission of greenhouse gases resulting from deforestation and forest degradation. Proposed at the United Nations Framework Convention on Climate Change meeting in Bali in 2007, it was discussed as part of post Kyoto Protocol planning at the 15th Conference of the Parties (COP) in Copenhagen in December 2009. Although there was no overall climate deal agreed at COP, REDD is still on the agenda and could be funded through the new Copenhagen Green Climate Fund.

REDD has the potential for pouring billions of dollars from industrialised countries into forest conservation in developing countries. The Food and Agriculture Organization estimates that, between 1990 and 2005, deforestation in Madagascar released 583 million tonnes of carbon dioxide. Today that would be worth up to $17.5 billion dollars on European exchanges where carbon credits are traded at $10–30 per tonne of carbon dioxide. A number of key issues need to be resolved, including: determining the structure of funding mechanisms; calculating the actual cost of curbing deforestation; setting baseline levels to measure reduction in emissions (current deforestation rates against reduced deforestation rates); monitoring forest cover; engaging local communities; ensuring equitable benefits and preventing displacement of deforestation to adjacent areas.

A huge concern of developed countries interested in carbon trading is that a large number of carbon credits could swamp developing carbon markets. In

January 2008, the European Commission proposed to exclude forest credits from the EU's Emissions Trading Scheme, the world's largest carbon market, in order to protect efforts to reduce emissions in Europe. However, carbon trading on voluntary markets, such as the Chicago Climate Exchange, grew from around 4 million tonnes of $CO_2$ equivalent traded in 2004 to 20–50 million tonnes traded in 2006 (Carbonpositive, 2009), albeit at substantially lower prices than on the legally binding markets such as the EU's Emissions Trading Scheme. Despite the global recession, REDD continues to grow.

The Wildlife Conservation Society, Conservation International, WWF and the Government of Madagascar have developed a new model for financing forest conservation in Madagascar under the REDD framework. In June 2008, the Wildlife Conservation Society in collaboration with the Government of Madagascar launched the Makira Carbon Company and announced their agreement to market 9.5 million tonnes of carbon offsets over the 30 years. Carbon financing will help establish the new Makira Protected Area, manage its biodiversity for conservation and create sustainable development, with about half of the money raised going back to the local communities. Investors already include companies such as Mitsubishi Corporation and Johnson & Johnson, as well as the music groups Pearl Jam and Dixie Chicks. The Makira Protected Area covers nearly 350 000 ha of humid forests which, together with adjacent forests in the Masoala National Park, forms the largest tract of humid forest in Madagascar, covering over 600 000 ha, and accounts for 10% of the total of all remaining humid forest. It is home to over 50% of Madagascar's plant species, which represents over 2% of the world's total (www.wcs.org). Hannah *et al.* (2008) have estimated that the REDD mechanism could generate as much as US$72–144 million for forest conservation in Madagascar. The five REDD pilot projects in Madagascar are outlined in Ferguson (2009).

## Future challenges: species conservation, engaging communities, restoration and insurance against extinction

Madagascar is regarded as one of the most economically disadvantaged countries in the world, with a largely rural population dependent on subsistence agriculture, so it is easy to blame local people for species losses. However, the conservation and botanical communities have failed to tell local people about the global significance of the plant species around them and they have therefore been denied the opportunity to participate in or take responsibility for species conservation. Entrusting local communities may be the only option for the conservation of some species. Many threatened plant species have extremely restricted distributions and not all species can be conserved within the protected area network. With limited resources for conservation, decisions will be needed as to which species to keep and which to lose. Madagascar will need a methodology for choosing management priorities for threatened species, along the lines of the Project Prioritization Protocols (PPP) developed for conserving New Zealand's Red List species (Joseph *et al.*, 2009). PPP is a cost-effectiveness analysis that factors in the value of the species and the benefits, costs and likelihood of success of each management project. The projects that offer the greatest return on investment are ranked highest. The Critical Ecosystems Partnership Fund recently invested over $5 million to provide Madagascar with the capacity needed to manage biodiversity conservation and to facilitate the participation of local communities by funding small community projects that would otherwise find funding difficult. The most significant impact has been a dramatic increase in the role of national NGOs in achieving conservation and sustainable development at the local, regional and national levels (CEPF, 2006). Former President Ravalomanana also began implementing a new regional structure and a planning process focused on community resource management. People in Madagascar are generally sensitised to the need for forest conservation, and the ownership of forests and responsibility for their management has begun to be passed back to local communities. Elmqvist *et al.* (2007) have demonstrated that, for dry forests at least, the highest rates of deforestation occur in areas characterised by insecure property rights and that in areas where communities have ownership forests are stable or regenerating.

The link between the healthy ecosystems and quality of life is recognised throughout Madagascar in the development community (e.g. Zeller *et al.*, 1999) and numerous community-based forest restoration projects are starting up. Restoration is a developing discipline and the company Rio Tinto is setting the standard for mining companies in Madagascar with its plans to restore littoral forest and conserve endemic species in the Tolagnaro region (Lowry *et al.*, 2008) after titanium ore extraction. Rio Tinto, like other mining companies, has a legal obligation to restore forests after its operations and they have been experimenting with restoration for over 15 years. They have recently published their results from restoration (Vincelette *et al.*, 2007) and species reintroductions (e.g. Randriatafika *et al.*, 2007). They found that using quick-growing exotic species to establish a suitable microclimate for saplings of native species gave poor success rates, and they had better results when concentrating on native trees associated with nitrogen-fixing bacteria and mycorrhiza. The lack of topsoil was the limiting factor for successful restoration on the nutrient-poor sands around Tolagnaro. Volume 3 (2010) of the journal *Malagasy Nature* is entirely devoted to the Ambatovay Project's efforts to study and conserve the flora of an area of eastern humid forest within its nickel mining concession. If Madagascar is to restore natural vegetation on large areas of its denuded landscapes, it needs systematic ecological profiling of species, especially for trees, with information on soils, climate, germination and ecology. Taxonomists will have an important role to ensure that the right species are planted in the right places.

*Ex-situ* conservation offers insurance against extinction for dryland and temperate species. The Millennium Seed Bank Project of the Royal Botanic Gardens, Kew, working with the Silo National des Graines Forestieres (SNGF), have banked over 8 million seeds of 1000 plant species from Madagascar over the last 10 years. This represents 15–20% of the bankable flora. The seed collections are split between Madagascar and the UK to reduce the risk of them being destroyed, and the seeds are being used in taxonomic studies and to determine germination behaviour for potential reintroductions or restoration projects. Funds are being raised to continue the work for the next 10 years, banking a further 1500 species (Solofo Rakotoarisoa, pers. comm.). The focus will be on:

- endemic families and genera;
- characteristic and framework species of threatened vegetation (e.g. Didieraceae from spiny forest);
- culturally significant, charismatic and high-profile species (e.g. the baobabs);
- economically important species (e.g. yams, which are an important famine food);
- endangered species known from very few sites;
- endangered species not found in protected areas.

## Political turmoil and future challenges

Prior to 2003, the outlook for biodiversity conservation in Madagascar was bleak, but under former President Marc Ravalomanana (July 2002 to March 2009) conservation made huge progress with the tripling of protected areas to cover 10% of the country's surface. The Durban Vision was a key component of President Ravalomanana's five-year Madagascar Action Plan (2007–2012), which was lauded by the international community as an innovative programme for effective social and economic development. However, since the former mayor of Antananarivo, Andry Rajoelina, has come to power in what has been widely viewed as a *coup d'etat*, there is considerable confusion over the new regime's commitment to completing Ravalomanana's conservation initiatives. As of February 2011 the political uncertainty had not been resolved and the ensuing lawlessness has created the opportunity for illegal loggers to plunder national parks for *Dalbergia* and other valuable timbers, particularly in Marojejy and Masoala National Parks in the north (Ballet *et al.*, 2009). With major international donors unable to continue providing development funds and the collapse of revenues from the tourist industry, conservation has become a luxury that Madagascar cannot afford on its own.

While politicians and their policies come and go, over the long term the critical environmental issues remain the same. For Madagascar, these will be climate change, food security and reducing deforestation, soil erosion and destructive agricultural practices. If Madagascar is to avoid huge loss of biodiversity, then the conservation community must continue to develop and secure funds for innovative and long-term projects. The main challenges will be not only the management of protected areas over 10% of the land area, but also the restoration of natural forest cover and the stabilisation of ecosystems throughout the remaining 90% and, most importantly, gaining the support and commitment of the Malagasy people to conserve their unique flora.

## References


Agarwal, D. K., Silander, J. A., Gelfand, A. E., Dewar, R. E. & Mickelson, J. G. (2005). Tropical deforestation in Madagascar: analysis using hierarchical spatially explicit, Bayesian regression models, *Ecol. Model.* 185: 105–131.

Ballet, J., Lopez, P. & Rahaga, N. (2009). L'exportation de bois précieux (Dalbergia et Diospyros) 'illégaux' de Madagascar: 2009 et après? *Mad. Conserv. Dev.* 5 (2): 110–116.

Bond, W. J., Silander, J. A. Jr., Ranaivonasy, J. & Ratsirarson, L. (2008). The antiquity of Madagascar's grasslands and the rise of C4 grassy biomes. *J. Biogeogr.* 35: 1743–1758.

Burney, D. A., Burney, L. P., Godfrey, L. G., Jungers, W. L., Goodman, S. M., Wright, H. T. *et al.* (2004). A chronology for late prehistoric Madagascar. *J. Hum. Evol.* 47: 25–63.

Callamander, M. W., Schatz, G. E., Lowry II, P. P., Laivao, M. O., Raharimampionona, J., Andriambololonera, S. *et al.* (2007). Identification of priority areas for plant conservation in Madagsacar using Red List criteria: rare and threatened Pandanaceae indicate sites in need of protection. *Oryx* 41 (2): 168–176.

Carbonpositive (2009). www.carbonpositive.net

CEPF (2006). Assessing five years of CEPF Investment in the Madagascar and Indian Ocean Islands Biodiversity Hotspot. Critical Ecosystem Partnership Fund (/www.cepf.net).

Chevalier, A. (1947). Les caféiers du globe 3: Systématique des caféiers et faux-caféiers; maladies et insectes nuisibles. *Encycl. Biol.* 28: 1–352.

Consiglo, T. K., Schatz, G. E., McPherson, G., Lowry II, P. P., Rabenantoandro, J., Rogers, Z. S. *et al.* (2006). Deforestation and plant diversity of Madagascar's littoral forests. *Conserv. Biol.* 20: 1799–1803.

Davis, A. P. & Bridson, D. (2003). Introduction to the Rubiaceae. In S. M. Goodman & J. P. Benstead (eds.), *The Natural History of Madagascar*, pp. 431–434. Chicago, University of Chicago Press.

Davis, A. P. & Rakotonasolo, F. (2008). A taxonomic revision of the baracoffea alliance: nine remarkable *Coffea* species from western Madagascar. *Bot. J. Linn. Soc.* 158, 355–390.

Davis, A. P., Govaerts, R., Bridson, D. M. & Stoffelen, P. (2006). An annotated taxonomic conspectus of the genus *Coffea* (Rubiaceae). *Bot. J. Linn. Soc.* 152: 465–512.

Davis, A. P., Govaerts, R., Bridson, D. M., Ruhsam, M., Moat, J. & Brummitt, N. A. (2009). A global assessment of distribution, diversity, endemism, and taxonomic effort in the Rubiaceae. *Ann. Mo. Bot. Gard.* 96: 68–78.

Dejardin, J., Guillaumet, J.-L., & Mangenot, G. (1973). Contribution à la connaissance de l'élément non endémique de la flore malgache (végétaux vasculaires). *Candollea* 28: 325–391.

de Wit, M. J., (2003). Madagascar: heads it's a continent, tails it's an island. *Ann. Rev. Earth Planet. Sci.* 31: 213–248.

Dorr, L. J. (1987). Rev. Richard Baron's 'Compendium des Plantes Malagaches'. *Taxon* 36 (1): 39–46.

Dorr, L. J. (1997). *Plant Collectors in Madagascar and the Comoro Islands.* London, Royal Botanic Gardens Kew.

Dransfield, J. & Beenje, H. J. (1995). *The Palms of Madagascar.* London, Royal Botanic Gardens Kew and the International Palm Society.

Dransfield, J. & Beentje, H. J. (2003). Arecaceae, palms. In S. M. Goodman & J. P. Benstead (eds.), *The Natural History of Madagascar*, pp. 448–457. Chicago, Chicago University Press.

Du Puy, D., Labat, I.-N., Rabevohitra, J.-N., Villiers, R., Bosser, J.-F. & Moat, J. (2002). *Leguminosae of Madagascar.* London, Royal Botanic Gardens Kew.

Elqvist, T., Pyykonen, M., Tengo, M., Rakotondrasoa, F., Rabakonandrianina, E. & Radimijahy, C. (2007). Patterns of loss and regeneration of a tropical dry forest in Madagascar: the social institutional context. *PLoS One* 2 (5): e402.

Ferguson, B. (2009). REDD comes into fashion in Madagascar. *Madagascar Conservation and Development* 4 (2): 132–135.

Gautier, L. & Goodman, S. M. (2003). Introduction to the Flora of Madagascar. In S. M. Goodman & J. P. Benstead (eds.), *The Natural History of Madagascar*, p. 229. Chicago: University of Chicago Press.

Global Strategy for Plant Conservation (2002). *Global Strategy for Plant Conservation.* Quebec, Canada, Secretariat of the Convention on Biological Diversity.

Goodman, S. M. & Benstead, J. P. (2005). Updated estimates of biotic diversity and endemism for Madagascar. *Oryx* 39: 73–77.

Hannah, L., Radhika, D., Lowry II, P. P., Andelman, S., Andrianarisata, M., Andriamaro, L. *et al.* (2008). Climate change adaptation for conservation in Madagascar. *Biol. Lett.* 4, 5: 590–594.

Harper, G. J., Steininger, M. K., Tucker, C. I., Juhn, D. & Hawkins, F. (2007). Fifty years of deforestation and forest fragmentation in Madagascar. *Environ. Conserv.* 34: 325–333.

Hedge, I., Clement, R. A., Paton, A. J. & Phillipson, P. B. (1998). Lamiacées. In *Flore de Madagascar et des Comores.*

Hermans, J., Hermans, C., Du Puy, D., Cribb, P. J. & Bosser, I (2007). *Orchids of Madagascar*, 2nd edition. London, Royal Botanic Gardens Kew.

Humbert, H. (1962). *Histoire de l'exploration botanique a Madagascar.* AETFAT 1961: 127–144.

Joseph, L. N., Maloney, R. F. & Possingham, H. P. (2009). Optimal allocation of resources among threatened species: a project prioritization protocol. *Conserv. Biol.* 23: 328–338.

Koechlin, J., Guillaumet, J.-L. & Morat, P. (1974). *Flore et végétation de Madagascar.* Vaduz, Liechtenstein, J. Cramer Verlag.

Kremen, C., Cameron, A, Moilanen, A., Phillips, S. J., Thomas, C. D., Beentje, H. *et al.* (2008). Aligning conservation priorities across taxa in Madagascar with high-resolution planning tools. *Science* 320: 222–225.

Kull, C. A. (2000). Deforestation, erosion, and fire: degradation myths in the environmental history of Madagascar. *Environ. Hist.* 6: 421–450.

Kull, C. A. (2002). Madagascar's burning issue: the peristent conflict over fire. *Environment* 44 (3): 8–25.

Kull, C. A. (2004). *Isle of Fire: The Political Ecology of Landscape Burning in Madgascar.* Chicago, University of Chicago Press.

Labat, I-N. & Moat, J. (2003). Leguminosae (Fabaceae) In: Goodman, S. M. & Benstead, J. P. (eds.), *The Natural History of Madagascar*, pp. 346–373. Chicago, University of Chicago Press.

Langrand, O. & Wilmé, L. (1997). Effects of forest fragmentation on extinction patterns of the endemic avifauna on the Central High Plateau of Madagascar. In S. M. Goodman & B. D. Patterson (eds.), *Natural Change and Human Impact in Madagascar*, pp. 280–305. Washington DC, Smithsonian Institution Press.

Lowry, P. P., Randriatafika, F. & Rabenantoandro, J. (2008). Conservation status of vascular plant species from the QMMl Rio Tinto mining area at Mandena, Tolagnaro (Fort Dauphin) region, southeast Madagascar. *Mad. Conserv. Dev.* 3 (1): 55–63.

McConnell, W. (2002). Madagascar: Emerald isle or paradise lost? *Environment* 44: 10–22.

Moat, L. & Smith, P. P. (2007). *Atlas of the Vegetation of Madagascar.* London, Royal Botanic Gardens Kew.

Myers, N. & Knoll, H. A. (2001). The biotic crisis and the future of evolution. *Proc. Natl. Acad. Sci. USA* 98: 5389–5392.

Myers, N., Mittermeier, R. A., Mittermeier, C. G., da Fonseca, G. A. B. & Kent, J. (2000). Biodiversity hotspots for conservation priorities. *Nature* 403: 853–858.

Phillipson, P. B. (1994). Madagascar. In S. D. Davis, V. H. Heywood and A. C. Hamilton. *Centres of Plant Diversity: A Guide and Strategy for their Conservation. Volume 1: Europe, Africa, South West Asia and the Middle East.* Cambridge, UK, IUCN Publications Unit.

Phillipson, P. B., Schatz, G. E., Lowry, P. P. & Labat, J.-N. (2006). A catalogue of the vascular plants of Madagascar. In S. A. Ghanzan & H. J. Beentje (eds.) *Taxonomy and Ecology of African Plants: Their Conservation and Sustainable Use. Proceedings of 17th annual AETFAT Congress*, pp. 613–627. London, Royal Botanic Gardens Kew.

Rakotosamimanana, B. (2003). Foreword. In S. M. Goodman & J. P. Benstead (eds.), *The Natural History of Madagascar*, p. xi. Chicago, University of Chicago Press.

Randrianasolo, A, Miller, J. S. & Consiglio, T. K. (2002). Application of IUCN criteria and Red List categories to species of five Anacardiaceae genera in Madagascar. *Biodivers. Conserv.* 11 (7): 1289–1300.

Randriatafika, F., Rabenantoandro, J., Birkinshaw, C. & Vincelette, M. (2007). Biology, ecology, risk of extinction and conservation strategy for *Eligmocarpus cynometroides* (Fabaceae): a priority species at Petriky. In *Biodiversity, Ecology, and Conservation of Littoral Forest Ecosystems in southeastern Madagascar, Tolagnaro (Fort Dauphin).*

Réau, B. (2002). Burning for zebu: the complexity of deforestation issues in western Madagascar. *Norsk Geografisk Tidsskrift* 56: 219–229.

Rebioma (2009). Madagascar Digital Conservation Atlas, Final Report 2009. www.rebioma.net.

Schatz, G. E. (1996). Malagasy/Indo-Australo-Malesian phytogeographic connections. In W. R. Lourenço (ed.), *Biogeography of Madagascar.* Paris, Editions ORSTOM.

Schatz, G. E. (2001). *Generic Tree Flora of Madagascar.* London, Royal Botanic Gardens Kew and St Louis, Missouri Botanical Garden.

Schatz, G. E. (2002). Taxonomy and herbaria in service of plant conservation: lessons from Madagascar's endemic families. *Ann. Mo. Bot. Gard.* 89: 145–152.

Schatz, G. E., Lowry II, P. P., Lescot, M., Wolf, A-E., Andriambololonera, S., Raharimalala, V. *et al.* (1996). Conspectus of the vascular plants of Madagascar: a taxonomic and conservation electronic database. In L. J. G. van der Maesen, X. M. van der Burgt & J. M. van Medenbach de Rooy (eds.), *The Biodiversity of African Plants*, pp. 10–17. Wageningen: Kluwer Academic.

Tollefson, J. (2008). Save the trees. *Nature* 452: 8–9.

Vincelette, M., Rabenantoandro, J., Randrihasipara, L., Randriatafika, F. & Ganzhorn, J. U. (2007). Results from 10 years of restoration experiments in the southeastern littoral forests of Madagascar. In J. U. Ganzhorn, S. M. Goodman & M. Vincelette (eds.), *Biodiversity, Ecology, and Conservation of Littoral Forest Ecosystems in southeastern Madagascar, Tolagnaro (Fort Dauphin)*, pp. 337–354. Washington, DC, Smithsonian Institution.

Zeller, M., Lapenu, C., Minten, B., Ralison, E., Randrianaivo, D. & Randrianarisoa, C. (1999). Pathways of rural development in Madagascar: an empirical investigation of the critical triangle between environmental sustainability, economic growth and poverty alleviation. *Q. J. Int. Agric.* 2.

# 18

# Climate change and island floras

DAVID BRAMWELL

Island plants and their habitats will be affected in the future not just by climate change but by a whole series of factors that make up the ongoing process of global change. While much of the focus in recent years has been on the impacts of climate change, these do not operate, now, nor will they in the future, in isolation but closely interact with human population changes and alterations in disturbance regimes. This inevitably leads to an impoverishment of biodiversity and loss or fragmentation of habitats.

Climate change and land-use modification are both key drivers of biodiversity loss and, when their effects are examined separately rather than together, we are likely to underestimate the extent of projected loss of biodiversity.

In the specific case of islands the threats posed by climate change are considerable (Bramwell, 2010). Islands, however, make up an important component of the world's biodiversity 'hotspots' (Myers *et al.*, 2000). The 34 Conservation International areas now defined as biodiversity hotspots are home to over 50% of the world's plant species and 42% of all terrestrial vertebrate species with very high numbers of endemics, yet their combined area of remaining natural habitat covers only 2.3% of the Earth's land surface.

Islands are of particular importance in a biodiversity context as they cover about 5% of the Earth's land surface but have more than 35% of the world's vascular plant species including about 50 000 endemics, of which some 20 000 are already threatened, without taking into account the future effects of climate change. At the same time, about 15% of the world's mammals, amphibians and birds are exclusively insular in distribution and 35% of all threatened bird species are island endemics (Pimm *et al.*, 2006).

*The Biology of Island Floras*, eds. David Bramwell and Juli Caujapé-Castells. Published by Cambridge University Press. 

Of the 34 Conservation International biodiversity hotspots 14 are islands, island archipelagos or have an important insular component and 6 others include offshore islands within their limits. These include such biodiversity centres as New Zealand, Polynesia/Micronesia, Madagascar and the Indian Ocean islands, the Caribbean Islands, the Philippines (with over 7000 islands), the Mediterranean Basin including Macaronesia, the Juan Fernandez Islands and the Galapagos Islands, among others. In the last two cases the high insular endemicity means that they generally have little in common with the adjacent continents in terms of their unique biodiversity.

Island organisms have always been especially vulnerable to human activities, from the destruction of natural vegetation for agricultural and urban exploitation, to the more subtle threat of competition from invasive, introduced alien species. Since the arrival of humans on islands, the native species have been up against it, with classic examples such as the extinction of flightless birds, including the dodo, through over-hunting. Now, because of human-induced climate change they face further threatening challenges including a rise in sea level of, according to recent estimates, between 2 and 15 metres in the next 100 years. Although generally the lowest islands are not the most rich in endemic biodiversity, it would be a pity to lose their unique contribution to our knowledge of colonisation and evolution.

## The 'island syndrome' and conservation

Posing the question of what makes insular organisms and, in this particular case, insular plants so vulnerable, the special conditions of evolution and isolation must be considered. Almost 35 years ago Carlquist (1974) defined a series of characteristics common to many insular organisms as an 'island syndrome'. This syndrome includes such phenomena as woodiness and longevity in generally herbaceous groups of plants and giantism in animals, diminished efficiency of dispersal in plants and the loss of flight capacity in birds and insects. The obvious tendency for island organisms to be poor competitors in the face of invasive species is also considered to be one of the syndrome's characteristics. The island syndrome includes adaptation to narrow, specific ecological niches through what Carlquist (1974) considers to be an irreversible specialisation, including the need to adapt to new pollinators, especially as many islands do not have rich pollinating insect faunas. Barratt (1998), in a review of reproductive biology of island plants, cites as examples the depauperate native insect fauna of the Hawaii archipelago with only six moth species, two butterflies and two bumblebees, and of the Galapagos Islands with only one pollinating bee species.

The strong trend to obligate outcrossing through dioecy and gynodioecy, the so-called 'escape from homozygosity' (Carlquist, 1974) after passing through initial genetic bottlenecks, also seems to qualify as part of the island syndrome. In an early paper on island colonisation, Baker (1955) pointed out the advantages of being hermaphrodite and self-compatible for the colonisation of islands remote from continental sources (Baker's Law, or Baker's Rule). As Barratt (1998) suggests, Baker's Rule seems to be supported in general by the evidence, from various sources such as the endemic floras of New Zealand, Hawaii and the Galapagos Islands. The post-colonisation, insular evolution of outcrossing is demonstrated by the endemic flora of Hawaii where Sakai *et al.* (1995a, 1995b) show that of about 290 original colonists only about 10% were dimorphic but that in the endemic flora almost 15% of species are dioecious and 21% are sexually dimorphic, apparently the highest proportions known from any flora. As an example of the development of outcrossing mechanisms, the situation in Hawaiian *Bidens* species is significant, 50% of the endemic species are gynodioecious but this mechanism to avoid inbreeding is not found in any of the 200 or so *Bidens* species and their relatives occurring outside the Hawaiian archipelago.

A recent paper by Crawford *et al.* (2008) on Canary islands *Tolpis* species indicates that the endemic taxa are incompletely self-incompatible or, as the authors term it, 'pseudo-self compatible' but the only non-endemic species examined, the Mediterranean *Tolpis barbata*, which also occurs in the Canaries, is described as highly autogamous, so here again there is probably a trend towards the development of a post-colonisation outcrossing mechanism in the endemic species.

A further contributor to the 'island syndrome', according to Carlquist (1974), is the phenomenon of adaptive radiation, a speciation process with a marked tendency to enhance floristic richness and lead to the evolution of specialised niche-specific species with narrow ecological and geographical ranges, weak intrinsic isolating mechanisms and poor dispersal capacity. Generally, selection in this process is for immediate fitness but, as Givnish (1998) points out, in changing situations such as new competition or unstable environmental conditions the specialised characteristics arising from adaptive radiation can make the resulting species particularly prone to extinction.

From a conservation point of view it is informative to consider some of the island syndrome characteristics in the context of climate change. It is frequently stated that the isolated position of islands has made them living laboratories for understanding the adaptation and evolution of species. Islands house unique biodiversity in terms of both ecosystems and species. Isolation, however, has also made them extremely susceptible, resulting in some of the highest extinction rates of all. They can also be highly climate-sensitive, for example, coral bleaching on atoll reefs due to the long-term warming of surface waters has become more widespread around both Pacific and Caribbean islands since the 1990s.

Other possible climate change concerns include increased extinction rates of high elevation species that have limited opportunities for migration. Changes in coastal ecosystems such as mangroves, dunes and halophyte communities, and declines in forest cover due to floods, droughts, or increased incidence of pests, pathogens, or fire will affect island endemics. In addition it is probable that disturbance to natural ecosystems caused by increases in the frequency or intensity of hurricanes and other major climatic incidents would generally favour the intrusion of invasive species, for example in the Caribbean islands. In addition, the unique cloud forests located on many islands of higher altitude occupy a narrow geographical and climatic niche where a relatively small change in temperature or precipitation patterns could possibly cause forest zones to shift upwards enough to be eliminated altogether (Loope & Giambelluca, 1998). Certainly there is no doubt that islands are very vulnerable and the synergy between climate change itself and the ongoing destruction of habitat due to human activities will, in the future, cause a massive increase in the numbers of threatened island plants and will lead to the extinction of many of them.

## Response to climate change

In order to consider how plants might respond to climate change we have to take into account the fact that ecosystems are made up of a whole series of interacting individual species of many types of different organisms, each with its own particular eco-climate envelope. Modelling of the climate envelope can be useful in predicting alterations in distribution under climate change conditions. Generally, however, as a result of climate change each organism, including plants, will have basically three options: to migrate, to adapt individual climatic envelopes to changing local conditions or to become extinct.

Plant response to climate change in the past, at least from what the palaeobotanical evidence tells us, is one of migration to areas of a more suitable climate. This seems to have been, for example, the principal response in recent interglacial periods. Obviously, therefore, migration is an issue to take into account for future response to climate change, but species will migrate at different rates depending on a number of independent factors such as generation times, dispersal mechanisms and availability of suitable migration routes, for example. In the case of islands, finding suitable climatic, edaphic and topographical conditions within a migration distance also means re-overcoming such phenomena as genetic bottlenecks in order to establish new and genetically viable populations with optimum rates of population growth. These are the principal requirements for survival. The availability of specific niches, competition with broad climate envelope species such as invasive weeds and encountering potentially new pathogens and predators, will also affect the capacity of plants to re-establish after migration. Finding the right habitat is not just a question of finding a suitable temperature and water regime; it is obviously a much more complicated issue. Recent studies unfortunately tend to suggest that migration will not be able to keep pace with the rate of climate change (Etterson & Shaw, 2001; Neilson *et al.*, 2005; Huntley, 1991).

Turning to the alternative of adaptation: natural adaptation is not generally a rapid process; even our accelerated, artificial plant breeding processes for annual crop species need various generations to produce genetically stable cultivars. Climate change presents a major challenge for adapting organisms, principally because of the diversity of factors that may be modified in the local environment by changing climatic conditions. Species will, in many cases, be pushed outside their natural eco-climate envelope and have to adapt physiologically to higher temperatures and modified water regimes as well as increasing competition from invasive species. Temperature changes will also affect flowering and fruiting periods, and the loss of synchronous relationships with specialist pollinators and seed dispersers can disrupt dispersal and pollination efficiency. At the same time, many plant species depend, for passing periods of adversity, on the availability of seed reserves in the soil seed bank to rebuild populations. In turn the survival of seeds in the soil depends on factors such as humidity and microbial activity. Therefore, we must ask the question: what will happen to microbial activity as the soil warms? The evidence points to an initial increase in activity until the individual species get to their climate envelope limits. This will affect both the beneficial, mycorrhiza type of organisms and also the microbial seed-infecting pathogens. Changes in both temperature and moisture in the soil may also have an adverse effect on seed dormancy so further diminishing the soil seed bank's capacity for regeneration. Thus, if plants are to stay where they are and adapt to changing local conditions they may be faced with having to respond to the simultaneous selection pressures for widely differing new traits, not just to the pressure of adaptation to a single factor. Bradshaw and McNeilly (1991), however, as a result of their intensive research on plant adaptation, make the following comment: 'Although some evolution in relation to climatic change may take place, we must not therefore presume that its power will be unlimited in relation to adaptation to future climatic change. As a result, and because the natural migration of species is now very restricted, we should expect considerable numbers of extinctions.'

The characteristics of the 'island syndrome' do not, therefore, seem to help either adaptation or migration. Insular woodiness and the tendency towards longevity, implying longer generation times, is not conducive to rapid adaptation to changing climatic and ecological conditions. In fact, Carlquist (1974) argues that a moderate, rather humid, stable climate with relatively long growing seasons is what has favoured the development of insular woodiness. If climate change, as predicted by many models, implies rapidly increasing temperature and aridity and intensification of extreme conditions, the elements required for survival of woody insular endemics will be disrupted, leading to a new wave of extinctions. Obviously on isolated islands any migration can only be local and whole climatic zones with their endemic flora may disappear as the possibility to migrate becomes exhausted by long generation times, ecological limitations or lack of physical space. A second characteristic of the island syndrome, loss of dispersability also limits the possibilities of migration. The change in breeding systems suggested by Carlquist (1974) as a feature of the island syndrome also has a restricting effect on migration and adaptation. Dioecy, for example, has major disadvantages for migration (Baker, 1955) and for adaptation to climate change where the symbiosis with pollinators can be critically affected by alterations to pollinator distribution.

In general terms, therefore, it can be said that the island syndrome creates a critical balance between insular endemics and their environment that can easily be disrupted by climate change and does not help the prospects for their *in situ* conservation in the face of rapidly changing conditions. Even though the numbers of endemics are relatively manageable, in many cases conservationists, in these circumstances, will still be faced with what Vane-Wright *et al.* (1991) described as the 'agony of choice'. It seems obvious that up to now humans have not been very good at the sustainable management and conservation of the world's islands and their ecosystems, even in stable climatic conditions (Diamond, 2005). How, therefore, are we going to cope with *in situ* conservation in very delicate and precarious climate change situations on islands where limited resources for human survival may well lead to the further sacrifice of sustainability and conservation? Thus the real agony of choice is where best to use the limited conservation resources available and how to manage them effectively. In these special circumstances, where some 15% of the world's natural plant resources are under threat, biodiversity conservation cannot depend on waiting to see if climate change can be managed *in situ*, especially as, for most islands, even the most sophisticated climate change models are not good at predictions (Hijman & Grahams, 2006).

There is, therefore, a major challenge for conservationists and particularly in the case of islands for *ex situ* conservation centres. In stable conditions, the ideal system is a combination of *in situ* conservation with an 'insurance policy' back-up of seed banks and *ex situ* collections. However, in view of the problems that *in situ* conservation already faces on islands and elsewhere through human activity, with the added implications of global change: can major reserves and protected areas be managed or even moved to compensate for such change? The World Wide Fund for Nature report from 2003 (WWF, 2004) shows the extent to which protected areas may be disrupted by climate change and suggests that a major re-evaluation of conservation policies is needed. The reliance on *in situ* conservation alone in the wake of climate change without the *ex situ* insurance policy will surely lead to numerous extinctions, especially in the case of biodiversity-rich islands.

## The challenge for conservation

The challenge for conservation is all about how to contribute to a species-based insurance policy in the coming years.

The first major step is to redefine the priorities. Current thinking on endangered species is anchored to the assignment of IUCN Red List categories. However, although many of the world's islands already have red lists, the principles for assessing the conservation status and extinction risk for species have been based on past and current areas of distribution and on the fluctuation of numbers of individuals and populations. With climate change, however, island plants will be faced with a whole new series of threats, meaning that current concepts of what are threatened species and how they are defined will need reconsideration and considerable revision.

In the context of target 8 of the Global Strategy for Plant Conservation (CBD/BGCI, 2002), the increasing importance of *ex situ* conservation collections in botanic gardens must be recognised, not only as repositories for living collections but also through their capacity for seed storage. Many gardens around the world have developed effective seed banks for conserving the germplasm of wild plant species, and more regional and local seed banks on islands should be established in response to target 8.

The vegetation and plant community level of challenge for island ecosystems is, however, basically the same for all ecosystems. How can the erosion and destruction of the natural environment be slowed down and even halted in the future?

Initially this must be promoted through education and the communication the real value of natural diversity in the face of global change, especially on islands where tourism and tourist development often enter into direct conflict with biodiversity conservation. In response, island biologists must become involved in networking the development of public interfaces on biodiversity and climate change. The world's botanic gardens, for example, receive several hundred million visitors a year and are, therefore, the primary interface between botanical science, environmental issues and the general public and should be striving to communicate with visitors on biodiversity and the climate change issue.

At the same time, much more accurate and refined climate change models are needed on a local or regional scale because corrective measures for the conservation of floras and ecosystems will have to be taken at these levels. This implies that the role of specialist field botanists will become more and more important for monitoring changes in vegetation distribution, species composition in plant communities, species distribution and so on on a local scale. With the trends away from traditional biological science in universities, however, botanic gardens are often the only centres with personnel capable of identifying and monitoring local threatened species and their ecosystems.

Climate change is opening up new fields for research in reproductive biology, molecular population genetics, the ecology of colonisation of new habitats and so on but 'classical taxonomy', inventories and basic classification will still be needed in order to learn as much as possible about biodiversity before it is too late. New Guinea, Madagascar and New Caledonia can be cited as examples of islands where probably thousands of species are still undescribed.

Indeed, in the face of global change, the human race is faced with a great challenge. The biodiversity essential for the Earth's future wellbeing is disappearing rapidly, not just on islands but throughout the planet and, as E. O. Wilson has pointed out, it is up to the human race to prevent massive rapid extinctions of thousands of other organisms caused by its own folly.

## References

Baker, H. G. (1955). Self-compatibility and establishment after 'long distance' dispersal. *Evolution* 9: 347–349.

Barratt, S. (1998). The reproductive biology and genetics of island plants. In P. R. Grant (ed.), *Evolution on Islands*, pp. 18–34. Oxford, Oxford University Press.

Bradshaw, A. D. & McNeilly, T. (1991). Evolutionary response to global climatic change. *Ann. Bot.* 67: 5–14.

Bramwell, D. (2010). Island hot spots: the challenge of climate change. In D. M. Williams & S. Knapp (eds.), *Beyond Cladistics: The Branching of a Paradigm*, pp. 91–100. Berkeley, CA, University of California Press.

CBD/BGCI (2002). *Global Strategy for Plant Conservation*. Secretariat of the Convention on Biological Diversity. Gland, Switzerland, WWF.

Carlquist, C. (1974). *Island Biology*. New York, Columbia University Press.

Crawford, D. J., Archibald, J. K., Stroermer, D., Mort, M. E., Kelly, J. K. & Santos, A. (2008). A test of Baker's Law: breeding systems and the radiation of *Tolpis* (Asteraceae) in the Canary Islands. *Int. J. Plant Sci.* 169 (6): 782–791.

Diamond, J. M. (2005). *Collapse: How Societies Choose to Fail or Succeed*. New York, Viking Press.

Etterson, J. R. & Shaw, R. (2001). Constraint to adaptive evolution in response to global warming. *Science* 5 (249): 151–154.

Givnish, T. J. (1998). Adaptive plant evolution on islands: classic patterns, molecular data, new insights. In P. R. Grant (ed.), *Evolution on Islands*, pp. 281–304. Oxford, Oxford University Press.

Hijmans, R. J. & Graham, C. H. (2006). The ability of climate envelope models to predict the effect of climate change on species distributions. *Glob. Change Biol.* 12: 2272–2281.

Huntley, B. (1991). How plants respond to climate change: migration rates, individualism and the consequences for plant communities. *Ann. Bot.* 67: 15–22.

Loope, L. L. & Giambelluca, T. W. (1998). Vulnerability of island tropical montane cloud forests to climate change with special reference to East Maui Hawaii. *Climatic Change* 39: 503–517.

Myers, N., Mittermeier, R. A., Mittermeier, C. G., da Fonseca, G. A. B. & Kent, J. (2000). Biodiversity hotspots for conservation priorities. *Nature* 403: 853–858.

Neilson, R. P., Pitelka, L. F., Solomon, A. M., Nathan, R., Midgley, G. F., Fragoso, J. M. *et al.* (2001). Forecasting regional to global plant migration in response to climate change. *Bioscience* 55 (9): 749–759.

Pimm, S., Raven, P., Peterson, A., Sekercioglu, C. H. & Erlich, P. R. (2006). Human impacts on rates of recent, present and future bird extinctions. *Proc. Natl. Acad. Sci. USA* 103: 10941–10946.

Sakai, A., Wagner, W. L., Ferguson, D. M. & Herbst, D. R. (1995a). Biogeographical and ecological correlates of dioecy in the Hawaiian Flora. *Ecology* 76: 2530–2543.

Sakai, A., Wagner, W. L., Ferguson, D. M. & Herbst, D. R. (1995b). Origins of dioecy in the Hawaiian Flora. *Ecology* 76: 2517–2529.

Vane-Wright, R., Humphries, C. J. & Williams, P. H. (1991). What to protect? Systematics and the agony of choice. *Biol. Conserv.* 55: 235–254.

WWF (2004). *How Effective are Protected Areas? A Preliminary Analysis of Forest Protected Areas.* Report prepared for the Seventh Conference of parties to the Convention on Biological Diversity. Gland, Switzerland, WWF.

# 19

# Conservation status of endemic plants on Isla del Coco, Costa Rica: applying IUCN Red List criteria on a small island

JENNIFER L. TRUSTY, HERBERT C. KESLER, JORGE RODRÍGUEZ AND JAVIER FRANCISCO-ORTEGA

One in six plant species grows on oceanic islands, yet one in every three endangered plants is an insular endemic (World Conservation Monitoring Centre, 1992). Habitat loss and fragmentation, together with the negative effect of exotic species, imperil island native plants (Davis *et al.*, 1986; Frankel *et al.*, 1995). As a preliminary step in the conservation of endemic island plants, an extinction risk assessment provides a framework for identifying and prioritising threatened species (Rodrigues *et al.*, 2006). The World Conservation Union (IUCN) has recognised the rapid loss of the world's biodiversity and has created a system to assess the threat of extinction faced by each species. This system, established in 1963, is known as the IUCN Red List (IUCN website, 2004). Under the IUCN Red List criteria, hundreds of thousands of species have been assessed and over 12 000 species are considered threatened with extinction (IUCN, 2003). Island plants account for a large number of IUCN threatened species and seven of the top 10 most threatened floras are thought to be on islands (Walter & Gillett, 1998). Red Lists using the system implemented by the IUCN in 2001 have been published for a small number of large islands or archipelagos such as the Canary Islands (Bañares *et al.*, 2004), the Falkland Islands (Broughton & McAdam, 2002), the Galapagos Islands (Lawesson, 1990) and New Caledonia (Jaffre *et al.*, 1998). Conservation assessments of plants from small- or single-island floras using the IUCN system are uncommon despite their importance to global biodiversity (but see Bacchetta & Pontecorvo, 2005; Gray *et al.*, 2005; Kingston & Waldren, 2005).

The high global extinction threat for many island endemic species reflects the geographic isolation and small area of most islands at the global scale. In order for the extinction threat designation of island plants to provide meaningful information at

*The Biology of Island Floras*, eds. David Bramwell and Juli Caujapé-Castells. Published by Cambridge University Press. 

the within-island/archipelago level, the IUCN Red List criteria should be employed at the regional scale.

Ranking at both the global and regional scales is recommended to promote conservation actions at the national or regional level while providing valuable information to the global listing (Gärdensfors *et al.*, 2001; Milner-Gulland *et al.*, 2006). Species found on a single island create a unique difficulty to the use of regional Red List criteria. The IUCN recommends that all species whose distributions are completely within a geopolitical region use the global ranking 'without modification' (Gärdensfors *et al.*, 2001). The result of this recommendation is that all species restricted to small areas (such as small islands) are considered to be threatened with extinction by the IUCN even when local conservationists feel that 'this [use of small area of occupancy] does not necessarily reflect the extinction threat to these taxa' (Waldren *et al.*, 1995). In order to convey regional information, previous extinction threat estimates for islands have included creating an additional ranking system (Kingston & Waldren, 2005) or have modified the IUCN system (Keith, 1998). In this chapter, we use the IUCN Red List criteria to assess the extinction threat for the 36 endemic plant taxa found on the small Pacific island of Isla del Coco, Costa Rica. Rather than creating a new system or using expert opinion to determine regional status, we suggest augmenting the IUCN global rank with the best available estimate of total population size for all island endemic species. Based on the use of IUCN Red List criteria and population size, this chapter provides the first extinction threat assessment for Isla del Coco endemic plant species. Finally, we highlight the need for immediate conservation action and suggest conservation actions that will help to ensure the long-term survival of these rare plants.

## Isla del Coco: a case study

Isla del Coco (also known as Cocos Island) is a small, 24 km$^2$ volcanic island located at 5° 32′ 57″ N and 86° 59′ 17″ W in the Pacific Ocean (Fig. 19.1). The island is located about 500 km (300 miles) west of Costa Rica and 680 km (350 miles) northeast of the Galapagos Islands. The highest point on the island reaches 630 m (2275 ft) at Cerro Iglesias. The island was claimed by Costa Rica in 1869 but remains almost uninhabited to this day.

Isla del Coco was designated as a Costa Rican National Park by Executive Decree in 1978. In 1991, the park's limits were extended to include the marine ecosystems up to a distance of 15 km around the island. At this time, the entire park was declared a zone of absolute protection, and no resources could be extracted. In 1995, both the marine and terrestrial regions were designated as National Conservation Areas and in 1997 Isla del Coco was placed on the

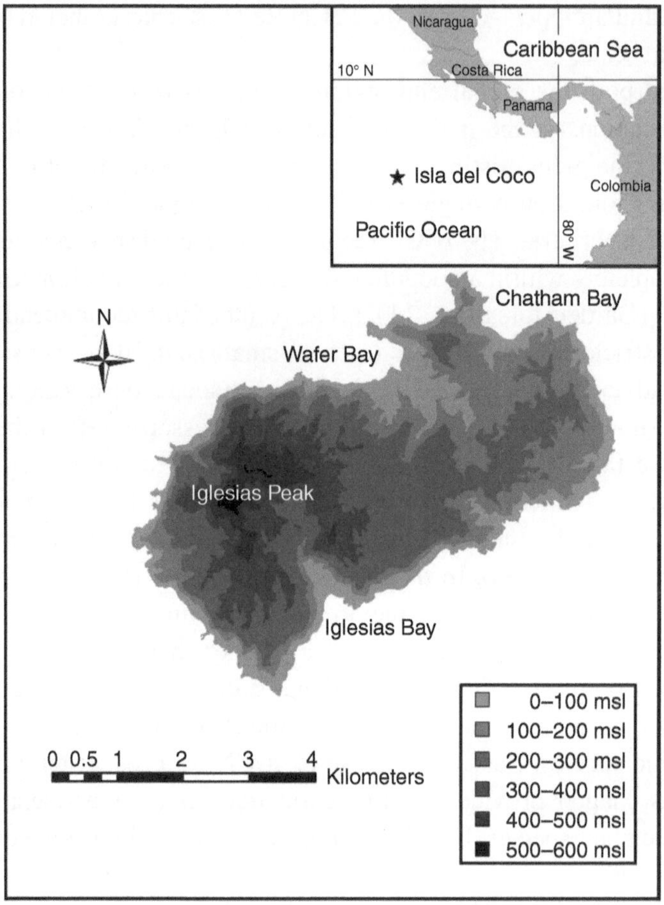


Figure 19.1 Location and topography of Isla del Coco, Costa Rica.

World Natural Heritage List by the World Conservation Union (IUCN). In addition, in 1998 Cocos was designated as a Wetland of International Importance by the RAMSAR Convention.

## Flora and vegetation

The coastline of Cocos is irregular, with cliffs rising almost vertically out of the sea to heights of up to 200 m. Beyond the narrow shore, green mountain ranges rise up in all directions, separated by steep ravines. This imposing geography has protected the island from human colonisation and provides a home to its unique flora and fauna. There are seven major vegetation types on Isla del Coco: bayshore, coastal cliff, riparian, low elevation humid forest, high elevation cloud forest, landslide and islet (Trusty *et al.*, 2006).

Recent floristic work on the island identified a total of 263 species for the island. Of these, 192 species are considered native and 71 are considered introduced (Trusty *et al.*, 2006). Thirty-two species and four subspecies are endemic to Isla del Coco (Table 19.1).

## Threats to Isla del Coco native and endemic plant species

The motivation for assessing the extinction risk for plants in a fully legally protected area is due to the observed, suspected and future impacts of introduced species on Isla del Coco. Although the island is protected from extensive development or habitat destruction by humans, the flora and fauna of Isla del Coco have to survive in a habitat that continues to be altered by introduced animals and plants.

In July 1793 James Colnett, captain of the English whaler *Rather*, introduced pigs (*Sus scrofa*) and goats (*Capra hircus*) onto the island (Lievre, 1893). Today, the feral pig is still a prominent feature of the island and is one of the main threats for the native plants of Cocos. The current introduced fauna of the island consists of feral pigs, feral cats (*Felis domesticus*), goats, white-tailed deer (*Odocoileus virginianus*) and two species of rat (*Rattus rattus*) and (*R. norvegicus*) (Montoya, 1990; Gómez, 2004).

The Cocos feral pig population is estimated at 400–500 individuals (Sierra, 2001a). Research has shown that areas rooted by the feral pig population have a rate of soil erosion nearly 100 times greater than unrooted areas and that 10–20% of the total island surface is rooted annually by these animals (Sierra, 2001b). The impact of over 200 years of feral pig inhabitation on the native vegetation of Isla del Coco is not known, but recent pig exclusion fencing has coincided with visible differences in understorey cover and diversity (pers. obs.). The changes to soil properties caused by rooting, which affect plant germination and establishment on the island, can be inferred. Our research on Isla del Coco has documented the rarity of regeneration for a number of tree and shrub species, as noted in Appendix 19.1. In addition, six of ten understorey shrub species have population estimates less than 100 and one understorey species has not been found in over 100 years. Park officials have approved the eradication of feral pigs contingent on obtaining a long-term funding source (Montoya, 1990). More recently, the IUCN Invasive Species Specialist Group is investigating ways to eradicate all introduced mammal species on Isla del Coco (Species Survival Commission, 2004). In the meantime, conservation actions must be undertaken to protect the rare endemic species on Isla del Coco.

August Gissler and his Cocos Island Agricultural Company cultivated a large area in Wafer Bay with bananas, beans, cacao, coffee, corn, avocado, achiote, sugar cane, oranges, limes, pineapples, tobacco, cherimoya, almonds and yucca (Jinesta,

Table 19.1 *Extinction risk assessment for Isla del Coco endemic plant species. All area estimations are in square kilometres. Criterion B estimates area while criterion D is determined by population size. Plant names in bold are epiphytic or islet species.*

| Taxon | Population size | EOO ($km^2$) | AOO ($km^2$) | IUCN B1 | IUCN B2 | IUCN D | IUCN global |
|---|---|---|---|---|---|---|---|
| *Acalypha pittieri* | Not found | DD | DD | EX | EX | EX | **EX** |
| *Ardisia cuspidate* | > 10 000 | 23.67 | 19.8 | CR | EN | VU | **CR** |
| *Asplenium barclayanum* | 150–250 | 3.81 | > 1 | CR | CR | EN | **CR** |
| *Asplenium delicatulum var. cocosensis* | < 50[a] | 25.36 | DD | CR | DD | CR | **CR** |
| *Bertiera angustifolia* | < 50[a] | 25.36 | DD | CR | DD | CR | **CR** |
| *Cecropia pittieri* | > 5 000 | 25.36 | 17.8 | CR | EN | VU | **CR** |
| ***Chloris paniculata*** | > 1 000 | 25.36 | > 5 | CR | CR | VU | **CR** |
| *Cyathea alfonsiana* | > 20 000 | 23.67 | 20 | CR | EN | VU | **CR** |
| *Cyathea nesiotica* | > 750 | 25.36 | 4 | CR | CR | VU | **CR** |
| *Cyathea notabilis* | > 1 500 | 23.67 | 8.1 | CR | CR | VU | **CR** |
| ***Elaphoglossum cocosense*** | > 2 000 | 16.23 | 11.2 | CR | EN | VU | **CR** |
| *Elaphoglossum incognitum* | > 500 | 21.97 | 13.8 | CR | EN | VU | **CR** |
| *Elaphoglossum reptans* | > 250 | 1.59 | 1.02 | CR | CR | VU | **CR** |
| ***Epidendrum cocoense*** | > 4 000 | 25.36 | 22.3 | CR | EN | VU | **CR** |
| ***Epidendrum insulanum*** | > 1 000 | 16.23 | 18.2 | CR | EN | VU | **CR** |
| ***Epidendrum jimenezii*** | Not found | DD | DD | EX | EX | EX | **EX** |
| *Eugenia cocosensis* | > 1 000 | 25.36 | 17.8 | CR | EN | VU | **CR** |
| *Eugenia pacifica* | < 50[a] | 25.36 | DD | CR | CR | CR | **CR** |
| *Hoffmannia nesiota* | < 50[a] | 1.59 | 1.02 | CR | CR | CR | **CR** |
| *Hoffmannia piratarum* | < 100 | 23.67 | 8.1 | CR | CR | CR | **CR** |
| ***Huperzia brachiata*** | Not found | DD | DD | EX | EX | EX | **EX** |
| ***Huperzia pittieri*** | > 250 | 21.97 | 25.3 | CR | EN | VU | **CR** |
| *Hymenophyllum cocosensis* | > 10 000 | 21.97 | 18.4 | CR | EN | VU | **CR** |
| *Hypolepis lellingeri* | > 250 | 25.36 | 13.4 | CR | EN | VU | **CR** |
| *Kyllinga nudiceps* | > 1,000 | 0.01 | > 1 | CR | CR | VU | **CR** |
| *Marcgravia waferi* | > 20,000 | 23.67 | 20 | CR | EN | VU | **CR** |
| *Miconia acanthotheca* | > 1,000 | 23.67 | 19.3 | CR | EN | VU | **CR** |
| *Miconia sp. nov.* | < 50[a] | 25.36 | DD | CR | DD | CR | **CR** |
| *Pilea gomeziana* | > 1,000 | 25.36 | 10.1 | CR | EN | VU | **CR** |
| *Psychotria cocosensis* | > 5,000 | 19.10 | 15.3 | CR | EN | VU | **CR** |
| *Saccoloma elegans var. cuneata* | > 500 | 25.36 | 13.4 | CR | EN | VU | **CR** |
| *Sacoglottis holdridgei* | > 50,000 | 25.36 | 22.4 | CR | EN | VU | **CR** |
| *Terpsichore cocosence* | > 2,500 | 21.97 | 17.3 | CR | EN | VU | **CR** |
| *Thelypteris calypso* | > 250 | 25.36 | 4 | CR | CR | EN | **CR** |
| *Thelypteris cocos* | > 1,000 | 25.36 | 21.3 | CR | EN | VU | **CR** |
| ***Trichomanes capillaceum var. cocos*** | > 20,000 | 25.36 | 22.4 | CR | EN | VU | **CR** |
| ***Trichomanes collariatum var. alvaradoi*** | >10,000 | 25.36 | 17.8 | CR | EN | VU | **CR** |

[a] Less than five individuals of this species were found by the authors

1937). Today, avocado, bananas, cacao and coffee remain from Gissler's fields. In addition, the park service maintains several tropical fruits and vegetables in cultivation for their own consumption. Although 71 non-native plant species were identified on the island, the only introduced plant that has escaped into a number of forested areas of the island is the shade-loving coffee plant (*Coffea arabica* L.; Trusty *et al.*, 2006). Currently, this species is still limited to areas near trails on the northern part of the island. Coffee has been found to be an invasive species in many tropical island ecosystems and its removal has been recommended (Pacific Island Ecosystems at Risk, 2002; Space *et al.*, 2003; Trusty *et al.*, 2006).

The introduction of non-native invasive species can rapidly alter island ecosystems. Many islands have encountered the extinction of native plant and animal species due to invasive exotics (Baeza *et al.*, 2002; Léon de la Luz *et al.*, 2003). As of yet, the majority of Isla del Coco has not been greatly affected by invasive exotic plant species but, as visitation and tourism increase, the number of introduced or exotic species on Isla del Coco is likely to grow. Our hope is that this threat assessment will bring international recognition to the floristic diversity found on this small island and will help to mobilise efforts to conserve this biodiversity for the future.

## Materials and methods

### *Survey data*

The data on Isla del Coco plant distributions were compiled from herbarium records and through field collection. Plant collections of all endemic species were located in the Museo Nacional de Costa Rica (CR), the herbaria of the Universidad de Costa Rica (USJ), the Instituto Nacional de Biodiversidad (INB), the Gray Herbarium at Harvard (GH), the United States National Herbarium (US) and the British Museum (BM). Loan requests for Isla del Coco collections were made to the California Academy of Sciences (CAS), the Brooklyn Botanic Garden (BKL) and the United States National Herbarium and included many type collections. The nomenclature and status of endemic species used in this study follows Trusty *et al.* (2006).

Distribution and/or abundance data located on the specimen label was recorded and included all of the endemic taxa known from the island. In addition, three collecting trips were made between 2001 and 2002: Trusty (July 2001), Trusty and Kesler (January–March 2002) and Trusty and Kesler (November–December 2002). Global positioning system (GPS) coordinates were taken at the collection site for each new collection. The local abundance of each specimen was recorded and the altitudinal distribution of each species was determined. Complete sets of these collections can be found in CR and Fairchild Tropical Botanic Garden (FTG). The location data recorded on the historical collections and the above information made during recent collection is the basis for the measurement of the extent of occurrence (EOO) of each taxon (IUCN, 2004). Three endemic species, *Acalypha pittieri* Pax and K. Hoffm., *Epidendrum jimenezii* Hagsater and *Huperzia brachiata* (Maxon) Holub, were not relocated in the 2001–2 collecting trips. Epiphytic orchid and fern species and plants restricted to islets located during surveys were not considered in decline in this extinction assessment due to the lack of direct evidence for negative impacts of introduced species on their habitat quality or quantity.

### *IUCN data*

There are five decision rules (criteria A–E) used to determine extinction risk under the IUCN system: A = rate of population decline; B = distributional range in conjunction with population structure and decline; C = population size in conjunction with population structure and decline; D = population size only and E = quantitative probability of extinction. Designation using the IUCN Red List rule B is dependent on having a measurement of the geographical area occupied by the species being evaluated. There are two accepted measurements, the first, known as extent of occurrence (EOO), is the area of a minimum convex polygon that incorporates all of the known populations of the species. The second accepted measurement, known as area of occupancy (AOO), is determined by placing a grid over a map of the known populations and counting all of the occupied grid cells. The grid size is not standardised in the IUCN systems but it is suggested that the grid scale should be appropriate to the relevant biological aspects of the taxon, the nature of threats and the available data (IUCN, 2001).

A topological map of Isla del Coco was scanned and georeferenced using ArcView 8.3 (ESRI, Redlands, California). The 100 m elevation lines were digitised as polygons. The elevation polygons were converted to coverages in ArcInfo 8.3. The GPS coordinates of all endemic species were overlaid onto this map to verify altitudinal distributions. The EOO was calculated by determining the minimum convex polygon than contained the distribution of that species. The EOO of the entire island is 22.5 km$^2$.

In order to measure geographic range by AOO, we used ArcInfo to calculate the area for each 100 m rise in elevation. This data was used to gain an estimate of potential available habitat determined by the established elevational range for each species. This second methodology is equivalent to using an infinitely small grid size.

The population size of each taxon was determined by direct observation and count (for species whose numbers were fewer than 20) or by extrapolation of subjective density estimates made while in the field.

## Results

The global IUCN Red List categorisation for the endemic plant taxa of Isla del Coco is shown in Table 19.1. The final assessment category was determined using the highest threat category met by a single criterion, as recommended by the IUCN (IUCN, 2004). All terrestrial taxa evaluated under IUCN criterion B were designated by subrules a (i.e. known to exist at a single location) and biii (i.e. continuing decline, observed, inferred or projected, in the area extent and quality of habitat).

All taxa designated as vulnerable using criterion D were designated by D2 (i.e. population with an AOO of less than 20 $km^2$ and fewer than five locations). Due to the absence of any information on previous population sizes, we were unable to use criteria A or C which estimate the degree of population reduction, decline or fluctuation for most taxa evaluated. In addition, the demographic data necessary to estimate the quantitative extinction risk (criterion E) for Isla del Coco species is not available.

## Discussion

According to the IUCN rules, the final categorisation is the strongest designation given among criteria (IUCN, 2004). In our assessment of Isla del Coco endemic plants, we found that all Isla del Coco terrestrial species were considered CR at the global level. The risk designation of Isla del Coco plants was influenced by the geographic range limits set by the IUCN criterion B1. The geographic size of the study area becomes critical in the designation of extinction threat. Palmer *et al.* (1997), in their risk assessment of the British flora, state that 'in a small country, the [IUCN] limits form a greater proportion of the total territory, the smaller the country, the greater the number of species which are likely to be threatened'. Small islands are an extreme example of the importance of the area threshold for the IUCN Red List criteria. Criterion B1 in the IUCN system has set size-dependent decision rules for the extent of occurrence (EOO) of a taxon estimated to less than 100 $km^2$ for a Critically Endangered (CR) designation, less than 500 $km^2$ for Endangered (EN) and less than 20 000 $km^2$ for Vulnerable (VU). This criterion also has subrules that must be met on the number of populations and evidence of decline or extreme fluctuation. For any single island smaller than 100 $km^2$ that is experiencing a decline in habitat quantity or quality (such as Isla del Coco), every plant species that is present as a single population on the island is therefore designated as Critically Endangered at the global level. Although it is known that many island species have high extinction risks, the numerically rare species found on this small island cannot be distinguished from numerous taxa because all have the same designation of CR. In order to have an effective regional conservation assessment, the IUCN criteria must reflect the difference in rarity and risk within the island itself (Keith, 1998).

In contrast to EOO, an alternative to geographic range measurement is the grid-based measurement called 'area of occupancy' or AOO, used by IUCN criterion B2. The IUCN (2001) recognises that the measurements of geographic range measured as AOO are dependent on scale and suggest that the most appropriate scale should be determined by the 'relevant biological aspects of the taxon'. Problems using the AOO methodology were exposed by Hartley and Kunin (2003) who noted that definitions of rarity based on the AOO threshold for rarity can be met as long as the AOO is calculated at a high-resolution grid. They found that the AOO methodology results in a failure of the risk assessment as species with only coarse-scale information available will appear less rare due to a large grid size, compared to those measured on a finer scale (Hartley & Kunin, 2003). This problem is compounded on islands, whose irregular shapes and high percentage of coastline incorporates more water in large grids than in small, inflating the overall potential habitat. Like many other IUCN extinction risk assessments, we did not assess the flora of Isla del Coco using a set grid and therefore did not have the necessary presence/absence data to determine AOO. Herbarium data used to determine AOO is also dependent on the grid size and, in addition, is often incomplete, which leads to higher threat categorisation (Willis *et al.*, 2003; Golding, 2004). Instead we relied upon the base of species-specific knowledge gained during our field surveys. We found that the majority of species distributions responded strongly to the elevational gradient within Isla del Coco. In order to determine the 'AOO' for each endemic plant in this study we used the elevational distribution of each taxon on the island as a surrogate for grid-based sample location points. We determined the actual area at each elevation using the digitised topographic map and compared these 'modified AOO' geographic distributions to those following the IUCN EOO methodology. This methodology yielded the following designations: three EX, ten CR, 21 EN and two data-deficient (DD) Isla del Coco endemic species (Table 19.1).

We feel that IUCN AOO designation better reflects the true extinction risk for the Isla del Coco endemic plant taxa and provides a better basis for the prioritisation of conservation activities. Unfortunately, the AOO grid-based data collection recommended by the IUCN requires a larger investment of time and effort by specialists to gain the required fine-scale knowledge of the species locations and distributions when compared to the EOO methodology. Often, the ruggedness of island habitats makes many areas inaccessible for vegetation surveys and may not be a viable option in many isolated and uninhabited environments.

In order to gain the information necessary to compare extinction risk at the global and regional scales without the added time and cost of a full grid-based survey, we recommend that the global IUCN designation be supplemented with the most

Figure 19.2 Rare Isla del Coco endemic plant species. Clockwise from top left: *Miconia* sp. B, *Elaphoglossum reptans*, *Asplenium barclayanum* and *Hoffmannia piratarum*.

current population size estimate for each island species. This small piece of information allows for the quick within-region ranking of extinction threat at nearly any scale. Specifically, on small islands where the global threat is high, this information quickly identifies species that are numerically rare and those that are in need of updated survey efforts. When extinction threat is supplemented by population size, Isla del Coco has 11 taxa that are numerically rare and deserve immediate conservation action; nine species have fewer than 50 known plants and two have fewer than 250 (Table 19.1; Fig. 19.2).

## The reality of small-island plant risk

Many island plants are at high risk of extinction (Waldren *et al.*, 1995; IUCN, 2003). This risk is due to the same factors that imperil continental species: loss and destruction of habitat and introduced species (Cronk & Fuller, 1995). Within a given area, extinction risk will not be the same for all species. Many aspects of a species' biology will modify its vulnerability to extinction threats. Details such as habitat specificity, habitat availability in time and space, and pollinator and dispersal requirements are of crucial importance in determining the size and frequency of a given population (Hartley & Kunin, 2003). The most relevant causes of extinction are due to stochastic and demographic processes on small populations (Soulè, 1987). Therefore, an extinction risk assessment for single, geographically restricted populations is best when based on an estimate of population size and degree of threat measured at both the global and regional scales. In addition, it is necessary to incorporate an understanding of the species' life history traits that are influencing mortality and recruitment. We believe that the use of the IUCN Red List criteria is a warranted and extremely important tool for the majority of the world's species but feel that the designation of island and endemic species should provide meaningful ranking information at the regional scale. Finding a compromise between the ease of data collection and designation of extinction risk and the reliability and accuracy of the designation for single-island systems is both necessary and urgent.

Isla del Coco is a small isolated island, but it is not alone. The United Nations Environment Programme Island Directory has over 500 named islands less than 100 $km^2$ (UNEP, 2004). Although many small islands do not have endemic plant species, there are notable exceptions (Table 19.2). In addition, many large and remote archipelagos such as French Polynesia, the Republic of Kiribati, the Philippines and the Federated States of Micronesia are composed of hundreds or thousands of small islands. Many of the small islands in these archipelagos are not well known floristically and are likely to have endemic species. All endemics will not be widespread, even within a large island chain. For example, in the well-studied Hawaiian archipelago, Price (2004) found that 79% of all endemic species were only found on a single island. In order to assess extinction risk for plants on islands, island size can be important but knowledge of the population sizes and habitat threats for each endemic plant species is necessary.

## Conclusions

We propose caution in the strict use of the global Red List criteria for species whose populations are restricted to areas less than 100 $km^2$. Field data on the population size and density is necessary to make an accurate comparative assessment for species in these geographically restricted areas. We recommend that any species (and especially those on islands) that is assessed as CR using the IUCN methodology should include the best estimate possible of total population size. This recommendation will help distinguish those plants that are numerically rare from those that are limited in distribution due to natural phenomena such as island or ecological endemism. Although extinction risk assessments are not intended to be used to determine conservation priorities, they were originally designed as a hierarchy to explicitly indicate rank. Adding a regional risk assessment rank based on population size to the global IUCN rank, as we recommend, allows for the

Table 19.2 *Selected small islands containing endemic plant species.*

| Island | Main archipelago system | Ocean | Area ($km^2$) | Number of endemic species | Reference |
|---|---|---|---|---|---|
| Ascension | St. Helena | Antlantic | 94 | 6 | Cronk (1980), Procter & Fleming (1999) |
| Bermuda | Bermuda | Atlantic | 54 | 17 | Procter & Fleming (1999) |
| Bonin Islands/ Ogasawara-Gunto | Bonin Islands | Pacific | 80 | 140 | Kobayashi & Ono (1978), Ono (1998) |
| Clarion | Revillagigidos Islands | Pacific | 24 | 3 | Levin & Moran (1989) |
| Eiao | Marquesas Islands | Pacific | 4.3 | 5 | UNEP (2004) |
| Fatu Hiva | Marquesas Islands | Pacific | 77 | 6 | Davis *et al.* (1986) |
| Futuna | Wallis and Futuna | Pacific | 80 | 3 | St. John (1977) |
| Henderson | Pitcairn Island Group | Pacific | 36 | 5 | Waldren *et al.* (1995) |
| Kazan Retto | Minami-Iwojima Islands | Pacific | 4 | 4 | Ono & Okutomi (1983) |
| Lord Howe | Australia | Pacific | 15.4 | 105 | Rodd & Pickard (1983), Flora of Australia (1994) |
| Macquarie Island | Australia | Pacific | 11 | 3 | Flora of Australia (1993) |
| Mangareva/Gambier | New Caledonia | Pacific | 25 | 11 | Florence *et al.* (1995) |
| Masafuera | Juan Fernandez Islands | Pacific | 50 | 35 | Stuessy *et al.* (1998) |
| Masatierra | Juan Fernandez Islands | Pacific | 48 | 55 | Stuessy *et al.* (1998) |

Table 19.2 (*cont.*)

| Island | Main archipelago system | Ocean | Area ($km^2$) | Number of endemic species | Reference |
|---|---|---|---|---|---|
| Nihoa | Hawaiian Islands | Pacific | 0.7 | 3 | Price (2004) |
| Norfolk Island | Australia | Pacific | 40 | 40 | Flora of Australia (1994) |
| Philip Island | Norfolk Island | Pacific | 2.5 | 3 | Melville (1969), Flora of Australia (1994) |
| Pitcairn | Pitcairn Island Group | Pacific | 4.5 | 9 | Waldren *et al.* 1995 |
| Rapa (iti) | Austral Islands | Pacific | 40 | 67 | Florence (1997) |
| Raoul Island | Kermadec Islands | Pacific | 29 | 17 | Sykes (1977), Sykes & West (1996) |
| Savage Islands | Macaronesia | Atlantic | 2.73 | 4 | Press & Short (1994) |
| San Benedicto | Revillagigidos Islands | Pacific | 20 | 2 | Levin & Moran (1989) |
| Ua Pu | Marquesas Islands | Pacific | 98 | 5 | Davis *et al.* (1986) |
| Ullung Island | Ullung Island | Pacific | 73 | 37 | Sun & Stuessy (1998) |

rapid identification of species that need immediate conservation action to survive (Possingham *et al.*, 2002).

Although the example of plant populations restricted to a single 24 $km^2$ oceanic island may seem to be an extreme case, human-induced destruction of native plant habitat is rapidly reducing the geographic size of continental plant populations. In essence, this fragmentation and loss of habitat is creating 'islands' within continental regions. Although the biological characteristics that make a plant species rare or common in addition to its overall distribution are important, extinction is a reality for plants with small and shrinking population sizes. There is real urgency in our need to accurately assess extinction threat for island and mainland taxa that is comparable among species at both the global and regional levels.

## Acknowledgements

This manuscript represents part of the doctoral dissertation of Jennifer Trusty submitted in partial fulfilment of the graduate requirements of Florida International University. This research was funded by a United States Environmental Protection Agency STAR Fellowship to J. Trusty, an Organization for Tropical Studies post-course grant to J. Trusty, a Florida International University Tropical Biology grant to J. Trusty and by the generosity of the Cocos Island National Park Service in providing transportation to Isla del Coco and room and board during field research on the island for J. Trusty and H. Kesler. All plant material was collected following national and international permit regulations. CITES permits were obtained for the collection of tree fern and orchid taxa collected on Cocos. This is contribution number 186 of the Tropical Biology Program of Florida International University.

## Appendix 19.1 Conservation recommendations for Isla del Coco rare plant species

### Acalypha pittieri *Pax & K. Hoffm.*

EX

*Description.* A small monoecious shrub with pubescent branches and leaves with white flowers. Extinction threat is due to feral pig rooting and stochastic processes on small populations.

*Distribution.* Rare, near the coast in Wafer Bay. This species has not been collected from Cocos Island since 1902.

*Conservation recommendations.* The long period of time since the last collection of this species makes a thorough, gridded survey the first priority. If this species is located, it is recommended that individuals be fenced for protection and that *ex situ* collections from seeds or cuttings be initiated immediately.

### **Asplenium barclayanum** *C. D. Adams (Aspleniaceae)*

EN (> 250 individuals)

*Description.* A small lithophilic pendent epiphyte with dark green fronds. Reproducing by numerous spores but with a very discrete insular microhabitat. No known herbivores. Extinction threat due to stochastic processes on small population sizes, the presumed lack of seed bank, and specific habitat and moisture requirements.

*Distribution.* Locally frequent but only found in two locations on Isla del Coco: on the wet volcanic rock cliffs near the pools of the large waterfalls in Wafer and Iglesias Bays. Both adult and juvenile plants were observed.

*Conservation recommendations.* This species should be propagated in an *ex situ* conservation programme using *in vitro* methods. In addition, these populations should be monitored on frequent (yearly) basis to determine the demographic rates and longevity of plants.

### **Asplenium delicatulum** *C. Presl var.* **cocosensis** *A. Rojas & Trusty (Aspleniaceae)*

CR (> 50 individuals)

*Description.* A small lithophilic epiphyte. Reproducing by numerous spores but with a very discrete insular microhabit. No known herbivores. Extinction threat due to stochastic processes on small population sizes, the presumed lack of seed bank, and specific moisture requirements.

*Distribution.* Found only near the waterfall in Iglesias Bay.

*Conservation recommendation.* This species should be propagated in an *ex situ* conservation programme using *in vitro* methods. Additional surveys should be undertaken to locate new populations and current population should be monitored yearly to collect demographic data.

### **Bertiera angustifolia** *Benth. (Rubiaceae)*

CR (> 50 individuals)

*Description.* A small tree with white flowers and dark purple fruits found in the understorey of moist *Sacoglottis holdrigei* forest. Only three observed individuals in two locations. All individuals were reproductive adults but no recruitment was observed.

*Distribution.* Rare in the closed canopy forest between 250 and 450 msl.

*Conservation recommendations.* As the recruitment of this species is most likely negatively affected by herbivores and/or pig rooting, an area around adult plants should be fenced to promote natural regeneration. In addition, the seeds of this species should be collected for *ex situ* propagation and viability testing. Finally, additional surveys should be employed to locate any additional individuals.

### **Elaphoglossum reptans** ***A. Rojas***

EN (> 250 individuals)

*Description.* A small, terrestrial, rhizomatous fern with glabrous fronds. Reproducing by numerous spores and rhizomatously but with a very discrete insular microhabitat. Extinction threat due to the presumed lack of seed bank, elevational requirements and feral pig rooting.

*Distribution.* Rare but locally common at a single location within a small area of cloud forest above 400 msl on Cerro Iglesias.

*Conservation recommendations.* Although feral pig rooting is less common in the cloud forest (Sierra 2001b), the small area of overall potential habitat makes this species vulnerable to infrequent pig disturbances. *Ex situ* collection and propagation is recommended.

### **Epidendrum jimenezii** ***Hágsater (Orchidaceae)***

EX

*Description.* An epiphytic orchid with one or two white flowers. There is some question that this species is simply a rare hybrid between *E. cocoense* and *E. insulanum* (Trusty & Blanco, 2006). This species was not found in the 2003–4 surveys and was last collected on the island in 1973. The putative cause for the extinction threat to this epiphytic species is stochastic events on small population sizes.

*Distribution.* Rare, found at below 100 msl near Wafer Bay.

*Conservation recommendations.* To confirm its presence or extinct status, a gridded survey should be attempted within the island to ensure complete survey coverage.

### **Eugenia pacifica** *Benth (Myrtaceae)*

EN (> 250 individuals)

*Description.* A small understorey shrub with large leaves and dark purple fruits. A single individual of this species was observed during the surveys and only about 5 fruits were observed. No regeneration was observed. Extinction threat due to feral pig rooting.

*Distribution.* Rare in the understorey of mid-elevation *Sacoglottis holdrigei* forest.

*Conservation recommendations.* Due to the inconspicuous nature of this species, further survey efforts will be required to locate additional individuals. Individuals should be fenced to encourage regeneration and monitored yearly. In addition, the seeds of this species should be collected for *ex situ* propagation and viability testing.

### **Hoffmannia nesiota** *Donn. Sm. (Rubiaceae)*

EN (> 250 individuals)

*Description.* A small shrub with chartaceous leaves and red fruits. The main extinction threat is due to feral pig rooting.

*Distribution.* Rare, found in Wafer and Iglesias Bays. Endemic to Isla del Coco.

*Conservation recommendations.* An area around adult plants should be fenced to promote natural regeneration. Seeds of this species should be collected for *ex situ* propagation and viability testing. In addition, a monitoring programme should be initiated to elucidate phenology and demographic parameters in this species.

### **Hoffmannia piratarum** *Standl. (Rubiaceae)*

EN (> 100 individuals)

*Description.* A suffrutescent herb with red fruits. Extinction threat due to feral pig rooting and anthropogenic disturbance near the Genio River.

*Distribution.* Rare and found in a single location along the riparian corridor of the Genio River in Wafer Bay.

*Conservation recommendations.* As the recruitment of this species is most likely negatively affected by herbivores and/or pig rooting, an area around adult plants should be fenced to promote natural regeneration. In addition, the seeds of this species should be collected for *ex situ* propagation and viability testing. Finally, yearly demographic monitoring should be initiated.

**Huperzia brachiata** ***(Maxon) Holub (Lycopodiaceae)***

EX

*Description.* A pendent epiphyte with very narrow stems. The putative cause for the extinction threat to this epiphytic species is stochastic events on small population sizes.

*Distribution.* Not collected since 1997. Previously found in closed canopy forest.

*Conservation recommendations.* This distinctive species was not found during the 2003–4 surveys. To confirm its presence or extinct status, a gridded survey should be attempted within the island to ensure complete coverage.

**Huperzia pittieri** ***(H. Christ) Holub (Lycopodiaceae)***

EN (> 250 individuals)

*Description.* A pendent epiphyte with thick stems and yellow-green leaves. The main extinction threats to this epiphytic species are stochastic events on small population sizes and overcollection.

*Distribution.* Infrequently found in the closed canopy forest above 200 msl.

*Conservation recommendations.* This species should be propagated in an *ex situ* conservation programme using *in vitro* methods. In addition, populations should be monitored yearly.

**Hypolepis lellingeri** ***A. Rojas***

EN (> 250 individuals)

*Description.* A large terrestrial fern found in openings of humid *Sacoglottis holdridgei* forest. Extinction threat due to feral pig rooting.

*Distribution.* Infrequently found in forest gaps and trail edges between 0 and 250 msl.

*Conservation recommendations.* This species should be propagated in an *ex situ* conservation programme. In addition, information on the ecological requirements and demography of this species could be gained through yearly monitoring and surveys for additional individuals.

**Miconia** ***sp. B Almeda ined. (Melastomataceae)***

CR (> 50 individuals)

*Description.* A small shrub (up to 1.5 m tall) with narrow, glossy leaves. Only two known collection locations of this plant are known, one with fewer than five individuals. Extinction threat due to feral pig rooting and stochastic processes on small populations.

*Distribution.* Rare, on Cerro Iglesias and also on ridge tops near Iglesias Bay.

*Conservation recommendations.* As the recruitment of this species is most likely negatively affected by herbivores and/or pig rooting, an area around adult plants should be fenced to promote natural regeneration. In addition, the seeds of this species should be collected for *ex situ* propagation and viability testing. Additional surveys and frequent monitoring of populations is also recommended.

### **Thelypteris calypso** *L. D. Gómez (Thelypteridaceae)*

EN (> 250 individuals)

*Description.* A medium-sized terrestrial fern found in partial openings within low-elevation humid forest. Extinction threat due to feral pig rooting.

*Distribution.* Infrequently found in canopy openings and trail edges between 0 and 100 msl.

*Conservation recommendations.* As the persistence of this species is most likely negatively affected by pig rooting it is recommended that it be placed in *ex situ* conservation collection and known individuals should be monitored on a frequent (yearly) basis.

## References

Bacchetta, G. & Pontecorvo, C. (2005). Contribution to the knowledge of the endemic vascular flora of Iglesiente (SW Sardinia – Italy). *Candollea* 60: 481–501.

Baeza, C. M., Stuessy, T. F. & Marticorena, C. (2002). Notes on the Poaceae of the Robinson Crusoe (Juan Fernández) Islands, Chile. *Brittonia* 54: 154–163.

Bañares, Á., Blanca, G., Güemes, J., Moreno, J. C. & Ortiz, S. (eds.) (2004). *Atlas y Libro Rojo de la Flora Vascular Amenazada de España: Taxones Prioritarios.* Madrid, Organismo Autónomo de Parques Nacionales.

Broughton, D. A. & McAdam, J. H. (2002). A red data list for the Falkland Islands vascular flora. *Oryx* 36: 279–287.

Cronk, Q. B. C. (1980). Extinction and survival of the endemic vascular flora of Ascension Island. *Biol. Conserv.* 17: 207–219.

Cronk, Q. B. C. & Fuller, J. (1995). *Plant Invaders.* London, Chapman & Hall.

Davis, S. D., Droop, S. J. M., Gregerson, P., Henson, L., Leon, C. J., Vila-Lobos, J. L. *et al.* (1986). *Plants in Danger. What Do We Know?* Gland, Switzerland, International Union for Conservation of Nature and Natural Resources.

Flora of Australia (1993). Vol. 50. *Oceanic Islands 2*. Canberra, Australian Governmental Publishing Service.

Flora of Australia (1994). Vol. 49. *Oceanic Islands 1*. Canberra, Australian Governmental Publishing Service.

Florence, J. (1997). *Flore de la Polynésie française*. Volume 1. Collection faune et flore tropicales 34. Paris, Editions de l'ORSTROM.

Florence, J., Waldren, S. & Chepstow-Lusty, A. J. (1995). The flora of the Pitcairn Islands: a review. *Biol. J. Linn. Soc.* 56: 79–119.

Frankel, O. H., Brown, A. H. D. & Burdon, J. J. (1995). *The Conservation of Plant Biodiversity*. Cambridge, Cambridge University Press.

Gärdenfors, U., Hilton-Taylor, C., Mace, G. M. & Rodriguez, J. P. (2001). The application of IUCN red list criteria at regional levels. *Conserv. Biol.* 15: 1206–1212.

Golding, J. S. (2004). The use of specimen information influences the outcomes of Red List assessments: the case of southern African plant specimens. *Biodivers. Conserv.* 13: 773–780.

Gómez, J. R. (2004). Estudio denso-poblacional de los reodores introducidos y su impacto sobre la fauna nativa en la Isla del Coco. Master's thesis: Universidad Nacional de Costa Rica.

Gray, A., Pelembe, T. & Stroud, S. (2005). The conservation of the endemic vascular flora of Ascension Island and threats from alien species. *Oryx* 39: 449–453.

Hartley, S. & Kunin, W. E. (2003). Scale dependency of rarity, extinction risk and conservation priority. *Conserv. Biol.* 17: 1559–1570.

IUCN (2001). 2001 Categories and Criteria (version 3.1). www.redlist.org/info/categories_criteria2001.html. Accessed 21 April 2004.

IUCN (2003). Release of the 2003 IUCN Red List of threatened species. www.icun.org/themes/ssc/RedList2003/English/newsreleaseEN.htm. Accessed 21 April 2004.

IUCN (2004). www.iucnredlist.org/info/introduction.html. Accessed 21 April 2004.

Jaffre, T., Bouchet, P. & Veillon, J.-M. (1998). Threatened plants of New Caledonia: is the system of protected areas adequate? *Biodivers. Conserv.* 7: 109–135.

Jinesta, R. (1937). *La Isla del Coco*. San José, Imprenta Falco Hermanos.

Keith, D. A. (1998). An evaluation and modification of the World Conservation Union Red List criteria for classification of extinction risk in vascular plants. *Conserv. Biol.* 12: 1076–1090.

Kingston, N. & Waldren, S. (2005). A conservation appraisal of the rare and endemic vascular plants of Pitcairn Island. *Biodivers. Conserv.* 14: 781–800.

Kobayashi, S. & Ono, M. (1978). A revised list of vascular plants indigenous and introduced to the Bonin (Ogasawara) and Volcano (Kazan) Islands. *Ogasawara Res.* 13: 1–55.

Lawesson, J. E. (1990). Threatened plant species and priority plant conservation sites in the Galapagos Islands. *Mon. Syst. Bot. Mo. Bot. Gard.* 32: 153–167.

León de la Luz, J. L., Rebman, J. P. & Oberbauer, T. (2003). On the urgency of conservation in Guadelupe Island, Mexico: Is it a lost paradise? *Biodivers. Conserv.* 12: 1073–1082.

Levin, G. & Moran, R. (1989). *The Vascular Flora of Isla Socorro, Mexico*. San Diego, San Diego History Museum Press.

Lievre, D. (1893). Una isla desierta en el Pacifico. La Isla del Coco (America). In *Los Viajes de Cockburn y Lievre por Costa Rica* (1963), pp. 103–134. San José, Editorial Costa Rica.

Melville, R. (1969). The endemics of Phillip Islands. *Biol. Conserv.* 1: 170–172.

Milner-Gulland, E. J., Kreuzberg-Mukhina, E., Brebot, B., Ling, S., Bykova, E., Abdusalamov, I. *et al.* (2006). Application of IUCN red listing criteria at the regional and national levels: a case study from Central Asia. *Biodivers. Conserv.* 15: 1873–1886.

Montoya, J. M. (1990). Plan de Majejo Parque Nacional Isla del Coco. Sistema de Parques y Reservas Marinas (SIPAREMA), Servicio de Parque Nacionales (SPN), Ministerio de Recursos naturales, Energia y Minas (MINEMEM). Documento de Trabajo ver. 14.

Ono, M. (1998). Conservation of the endemic vascular plant species of the Bonin (Ogasawara) Islands. In T. F. Stuessy & M. Ono (eds.), *Evolution and Speciation of Island Plants*. Cambridge: Cambridge University Press.

Ono, M. & Okutomi, K. (eds.) (1983). *Endemic Species and Vegetation of the Bonin Islands*. Kanagawa, Aboc-sha Company.

Pacific Island Ecosystems at Risk (2002). http://hear.org/pier_v3.3.htm. Accessed 1 April 2005.

Palmer, M. A., Hodgetts, N. G., Wigginton, M. J., Ing, B. & Stewart, N. F. (1997). The application to the British flora of the World Conservation Union's revised Red List criteria and the significance of Red Lists for species conservation. *Biol. Conserv.* 82: 219–226.

Possingham, H. P., Andelman, S. J., Burgman, M. A., Medillín, R. A., Master, L. L. & Keith, D. A. (2002). Limits to the use of threatened species lists. *Trends Ecol. Evol.* 17: 503–507.

Press, J. R. & Short, M. J. (eds.) (1994). *Flora of Madeira*. Koenigstein, Koeltz Scientific Books.

Price, J. P. (2004). Floristic biogeography of the Hawaiian Islands: influences of area, environment and paleogeography. *J. Biogeogr.* 31: 487–500.

Procter, D. & Fleming, L. V. (eds.) (1999). *Biodiversity: The UK Overseas Territories*. Peterborough, UK, Joint Nature Conservancy Committee.

Rodd, A. N. & Pickard, J. (1983). Census of the vascular flora of Lord Howe Island. *Cumminghamia* 1: 267–280.

Rodrigues, A. S. L., Pilgrim, J. D., Lamoreux, J. F., Hoffman, M. & Brooks, T. M. (2006). The value of the IUCN Red List for conservation. *Trends Ecol. Evol.* 21: 71–76.

Sierra, C. (2001a). El cerdo cimarrón (*Sus scrofa*, Suidae) en la Isla del Coco, Costa Rica: composicíon de su dieta, estado reproductivo y genética. *Rev. Biol. Trop.* 49: 1147–1157.

Sierra, C. (2001b). El cerdo cimarrón (*Sus scrofa*, Suidae) en la Isla del Coco, Costa Rica: escarbaduras, alteraciones al suelo y erosión. *Rev. Biol. Trop.* 49: 1159–1170.

Soulé, M. E. (1987). *Viable Populations for Conservation*. Cambridge, Cambridge University Press.

Space, J. C., Waterhouse, B. M. & Miles, J. E. (2003). *Report to the Republic of Palau on invasive plant species of environmental concern*. Honolulu, USDA Forest Service.

Species Survival Commission (2004). Saving 'Treasure Island' from the ravages of invasive species. *Species: Newslett. Species Surv. Comm.* 41: 9.

St John, H. (1977). Additions to the flora of Futuna Island, Horne Islands. *Phytologia* 36: 367–373.

Stuessy, T. F., Crawford, D. J., Marticorena, C. & Rodriguez, R. (1998). Island biogeography of the Juan Fernandez archipelago. In: T. F. Stuessy & M. Ono (eds.), *Evolution and Speciation of Island Plants*. Cambridge, Cambridge University Press.

Sun, B.-Y. & Stuessy, T. F. (1998). Preliminary observations on the evolution of endemic angiosperms of Ullung Island, Korea. In: T. F. Stuessy & M. Ono (eds.), *Evolution and Speciation of Island Plants*. Cambridge, Cambridge University Press.

Sykes, W. R. (1977). Kermadec Islands flora: an annotated checklist. Department of Scientific and Industrial Research Bulletin 219.

Sykes, W. R. & West, C. J. (1996). New records and other information on the vascular flora of the Kermadec Islands. *New Zeal. J. Bot.* 34: 447–462.

Trusty, J. L. & Blanco, M. A. (2006). The orchids of Isla del Coco. *Orchids* (November): 826–831.

Trusty, J. L., Kesler, H. C. & Haug, G. (2006). Vascular flora of Isla del Coco. *Proceedings of the California Academy of Sciences* 57: 247–355.

UNEP (2004). United Nations Environment Programme Island Directory. http://islands.unep.ch/isldir.htm. Accessed 22 November 2005.

Waldren, S., Florence, J. & Chepstow-Lusty, A. J. (1995). Rare and endemic vascular plants of the Pitcairn Islands, south-central Pacific Ocean: a conservation appraisal. *Biol. Conserv.* 74: 83–98.

Walter, K. S. & Gillett, H. J. (eds.) (1998). *1997 IUCN Red List of threatened plants.* Compiled by the World Conservation Monitoring Centre. Gland: IUCN – The World Conservation Union.

Willis, F., Moat, J. & Paton, A. (2003). Defining a role for herbarium data in Red List assessments: a case study of *Plectranthus* from eastern and southern tropical Africa. *Biodivers. Conserv.* 12: 1537–1552.

World Conservation Monitoring Centre (1992). *Global Biodiversity: Status of the Earth's Living Resources.* London, Chapman & Hall.

# 20

# Botanic gardens and the conservation of island floras

SARA OLDFIELD

Initiatives to conserve the world's threatened flora have developed over the past 40 years as awareness of the damage to ecosystems and species has grown. Botanic gardens have been actively involved in various aspects of plant conservation, helping to develop policy, implement legislation, survey and protect species in the wild and maintain rare and threatened species as living collections or in seed banks. Plant conservation on islands is a particular imperative. Islands generally have rich endemic floras that are increasingly fragile as a consequence of threats of habitat modification and destruction for agriculture, forestry, mineral extraction, tourism; over-exploitation of species; invasive species and increasingly climate change. The work of botanic gardens plays an essential role in the conservation of island floras.

Islands vary markedly in their age, size, geology, relief, climate and distance from nearest land mass – all factors that influence the degree of plant endemism. Island endemics are generally of two types: relict species or palaeo-endemics that were formerly more widespread when an island was attached to a larger land mass; and neo-endemics that have evolved as new species from the spores and seeds of so-called founder species carried by the wind and sea. In total it is estimated that there are between 50 000 and 52 000 island plant endemics of which about 20 000 are estimated to be threatened (Bramwell, 2002). Comparable data on the threat status of individual island floras is in urgent need of updating but is summarised in Table 20.1. Local capacity to conserve endemic and threatened island plants varies considerably from island to island and increasingly regional and international collaboration is required to prevent the extinction of island floras.

*The Biology of Island Floras*, eds. David Bramwell and Juli Caujapé-Castells. Published by Cambridge University Press. 

Table 20.1 *The number of botanic gardens, plant diversity and threatened plants, by island.*

| Continent | Island | Size of island ($km^2$)[1] | No. of BGs[2] | Approx. no. of endemics[3] | % endemics[3] | No. of species threatened[4] |
|---|---|---|---|---|---|---|
| Antarctica | French Southern Territories | 439 781 | 0 | 11 | 22 | |
| | Falkland Islands (Malvinas) | 12 200 | 0 | 14 | 8.5 | 6 |
| Africa | Ascension Island | 94 | 0 | 11 | 44 | 5 |
| | Cape Verde | 4 033 | 0 | 86 | 11.1 | 1 |
| | Comoros | 2 235 | 0 | 136 | 18.9 | 4 |
| | Madagascar | 587 041 | 3 | 5 000–8 000 | 68.4 | 306 |
| | Mauritius | 2 040 | 2 | 329 | 37.5 | 294 |
| | Reunion | 2 510 | 4 | 485 | 63 | 47 |
| | Rodrigues | 104 | 0 | 48 | 33.1 | 19 |
| | Saint Helena | 121 | 1 | 50 | 8.6 | 40 |
| | Sao Tome and Principe | 964 | 0 | 134 | 15 | 3 |
| | Seychelles | 451 | 1 | 250 | 15.2 | 78 |
| Asia | Andaman and Nicobar Islands | 8 120 | 0 | 225 | 9.9 | 136 |
| | British Indian Ocean Territory | 60 | 0 | 0 | 0 | 1 |
| | Caroline Islands | 1 170 | 0 | 267 | 26.9 | 4 |
| | Maldives | 298 | 0 | 5 | 1.8 | |
| | Phillipines | 300 000 | 12 | 3 500 | 39.3 | 360 |
| | Socotra | 3 625 | 0 | 215 | 31.6 | 131 |
| | Sri Lanka | 65 610 | 6 | 900 | 28 | 455 |
| | Taiwan | 34 507 | 4 | 892 | 25 | 325 |
| Europe | Azores | 2 235 | 0 | 55 | 9.2 | 30 |
| | Balearic Islands | 5 014 | 1 | 94 | 6.5 | 60 |
| | Canary Islands | 7 273 | 2 | 500 | 25.0 | 382 |

Table 20.1 (*cont.*)

| Continent | Island | Size of island ($km^2$)[1] | No. of BGs[2] | Approx. no. of endemics[3] | % endemics[3] | No. of species threatened[4] |
|---|---|---|---|---|---|---|
| | Cyprus | 9 251 | 0 | 88 | 5.2 | 51 |
| | Faeroe Islands | 1 399 | 0 | 1 | 0.4 | |
| | Iceland | 103 000 | 2 | 1 | 0.3 | 1 |
| | Ireland | 81 638 | 16 | 0 | 0 | 1 |
| | Malta | 316 | 1 | 5 | 0.5 | 15 |
| | Madeira | 828 | 5 | 145 | 18.1 | 86 |
| | Sardinia | 24 090 | 0 | 27 | 1.35 | 18 |
| | Sicily | 25 708 | 2 | 41 | 1.67 | 28 |
| | UK | 244 820 | 116 | 16 | 1 | 18 |
| Americas | Anguilla | 102 | 1 | 1 | 0.3 | |
| | Antigua and Barbuda | 442 | 1 | NA | [0.7] | 3 |
| | Aruba | 193 | 0 | 25 | 5.4 | |
| | Bahamas | 13 878 | 3 | 115 | 9.4 | 31 |
| | Barbados | 431 | 3 | 5 | 0.8 | 2 |
| | Bermuda | 53 | 1 | 15 | 9 | 10 |
| | Cayman Islands | 260 | 1 | 19 | 3.4 | 13 |
| | Cuba | 110 861 | 12 | 3 229 | 49.6 | 888 |
| | Dominica | 754 | 2 | 1 | 0.8 | 57 |
| | | | | 1 | | |
| | Dominican Republic | 48 442 | 1 | 1 800 | [0.8] | 136 |
| | Galapagos Islands | 7 844 | 1 | 229 | 42.2 | 150 |
| | Grenada | 344 | 1 | 4 | 0.4 | 8 |
| | Guadeloupe | 1 628 | 2 | 26 | 1.6 | 26 |
| | Haiti | 27 751 | 2 | 1 800 | 31.8 | 100 |
| | Hawaii | 16 641 | 23 | NA | NA | 743 |
| | Jamaica | 11 100 | 6 | 906 | 27.4 | 744 |

| | | | | | | |
|---|---|---|---|---|---|---|
| | Juan Fernandez | 93 | 0 | 118 | 80.3 | 94 |
| | Martinique | 1 128 | 2 | 30 | 1.9 | 44 |
| | Montserrat | 102 | 1 | 2 | 0.3 | 2 |
| | Netherlands Antilles | 800 | 0 | NA | NA | 2 |
| | Puerto Rico | 9 104 | 4 | 235 | 9.4 | 223 |
| | St Kitts and Nevis | 261 | 1 | NA | NA | 4 |
| | St Lucia | 620 | 0 | 11 | 1.1 | 6 |
| | St Vincent and the Grenadines | 389 | 1 | NA | NA | 9 |
| | Trinidad and Tobago | 5 128 | 2 | 226 | 9.3 | 21 |
| | Turks and Caicos Islands | 417 | 0 | 9 | 2 | 2 |
| | British Virgin Islands | 153 | 1 | NA | NA | 14 |
| | US Virgin Islands | 346 | 1 | 27 | 3.0 | 16 |
| Oceania | American Samoa | 199 | 1 | 10 | 2.2 | 9 |
| | Antipodes | 21 | 0 | 1 | 1.6 | 1 |
| | Auckland Islands | 625 | 0 | 6 | 3.2 | 0 |
| | Campbell Islands | 114 | 0 | 3 | 1.3 | 0 |
| | Chatham Islands | 1 235 | 0 | 40 | 13.3 | 16 |
| | Christmas Island | 135 | 0 | 15 | 3.9 | 14 |
| | Cook Islands | 240 | 0 | 3 | 1.1 | 12 |
| | Fiji | 18 274 | 2 | 812 | 49.9 | 74 |
| | French Polynesia | 4 167 | 0 | NA | NA | 187 |
| | Gambier Islands | 25 | 0 | 11 | 4.4 | 25 |
| | Guam | 541 | 0 | NA | NA | 18 |
| | Kermadec Islands | 33.5 | 0 | 23 | 11.8 | 7 |
| | Kiribati | 726 | 0 | 2 | 3.3 | |
| | Lord Howe Island | 13 | 0 | 74 | 19.5 | 73 |
| | Marquesas Islands | 1 275 | 0 | 103 | 70.1 | 58 |
| | Marshall Islands | 181 | 0 | 4 | 3.6 | |
| | Micronesia | 702 | 0 | NA | NA | 4 |

Table 20.1 (*cont.*)

| Continent | Island | Size of island ($km^2$)[1] | No. of BGs[2] | Approx. no. of endemics[3] | % endemics[3] | No. of species threatened[4] |
|---|---|---|---|---|---|---|
| | Nauru | 21 | 0 | 1 | 1.9 | |
| | New Caledonia | 18 575 | 0 | 2 480 | 80.2 | 480 |
| | New Zealand | 268 680 | 19 | 1 942 | 81.9 | 211 |
| | Niue | 260 | 0 | 1 | 0.6 | 1 |
| | Norfolk Island | 39 | 1 | 48 | 27.6 | 41 |
| | North Marianas Islands | 1 007 | 0 | 81 | 25.7 | 11 |
| | Palau | 459 | 0 | NA | NA | |
| | Pitcairn Islands | 47 | 0 | 14 | 18.4 | 360 |
| | Solomon Islands | 28 896 | 1 | 30 | 0.9 | 42 |
| | Tokelau | 10 | 0 | 0 | 0 | |
| | Tonga | 748 | 0 | 25 | 5.4 | 2 |
| | Tuvalu | 26 | 0 | NA | NA | |
| | Vanuatu | 12 200 | 0 | 50 | 5 | 26 |
| | Wallis and Futuna Islands | 264 | 0 | 5 | 2 | |
| | Western Samoa | 2 831 | 1 | 117 | 16.9 | 18 |

[1] Davis *et al.* (1986)
[2] BGCI GardenSearch Database
[3] World Conservation Monitoring Centre (1992)
[4] Walter & Gillett (1998)

## The threats

Factors that make plants on islands particularly vulnerable to extinction include their natural rarity, with limited natural distributions and population sizes making the species more sensitive to over-exploitation, habitat clearance and stochastic events. Evolution in the absence of predators also makes island species particularly vulnerable to introduced grazing animals. The impacts of various threats to the small South Atlantic island of St Helena have been well documented (e.g. Cronk, 1989, 1993, 2000; Procter & Fleming, 1999). Goats were introduced by Portuguese sailors as long ago as 1513, and within 75 years had formed large herds. By the time botanists arrived on the island in the early nineteenth century the flora had been devastated and the true size of the indigenous flora remains unknown. Additional threats to the flora of St Helena have included felling of timber and the development of plantations of New Zealand flax, *Phormium tenax*, originally used to make rope and now an invasive species. The flora of St Helena consists of around 580 higher plant species of which 50 are endemic and 40 are globally threatened.

Further north in the Atlantic, the island flora of Macaronesia is of outstanding global significance. Included in the Mediterannean Basin Hotspot (Mittermeir *et al.*, 2004) the flora of the Canary Islands, Madeira, the Azores and Cape Verde Islands includes around 4170 species. The Canary Islands alone have around 1200 indigenous plant species (see Table 20.1) of which around 500 are endemic and 380 are globally threatened. The main threats are considered to be tourism and residential developments, overgrazing, off-road vehicles and the spread of invasive plants. Natural habitats that are under threat on the Canary Islands include the laurel forests, which are of great ecological importance both in controlling erosion and in influencing local rainfall. On Tenerife, the laurel forests have been reduced by more than 90%. Clearance for agriculture has been the chief cause of forest loss.

The Caribbean Islands are also considered to be a global biodiversity hotspot (Mittermeir *et al.*, 2004). The islands support about 11 000 indigenous plant species and 182 endemic plant genera (Maunder *et al.*, 2008). Within the region, threats to the island floras includes centuries of colonial agriculture and wild harvesting of valuable timbers such as mahogany, *Swietenia mahogani*, and lignum vitae, *Guiacuam sanctum*. More recently the development of the islands for tourism has been a major factor in the decline of plant diversity. Cuba has a particularly rich flora with 7020 vascular plant species of which 50% are endemic to the island. The main threats to the flora are habitat loss, fires, agricultural and forestry development and mining. Recently 1414 taxa have been evaluated using the IUCN Red List categories and criteria, including 1089 endemic plant species. Of the recorded endemic species 21 are extinct, and 1006 are threatened with extinction on a global scale (Iturralde *et al.*, 2005). With changing climatic conditions the wild populations of these globally threatened species are likely to be placed under more severe stress. The 70 threatened species that are confined to high altitude cloud forest are potentially most at risk as temperatures increase.

Lying in the Pacific Ocean over 2000 miles from the nearest continental land mass, the islands of Hawaii are the most isolated high islands in the world. The neo-endemics have adapted to a wide variety of different habitats. In total, about 1200 plant species are native to Hawaii with more than 90% endemism. Approximately half of these naturally rare species are threatened with extinction and over 100 species are known from 20 or less remaining individuals (Sugii & Lamoureaux, 2004). Already 55 of the islands' plant species are recorded as extinct and a further 42 probably no longer exist in the wild. Deforestation, historical over-exploitation of species such as the native sandalwood, the effects of fires, overgrazing by introduced livestock and competition with invasive plant species have been the major threats to the endemic flora. Modification of the vegetation began around 2000 BP with the arrival of Polynesian settlers, and speeded up dramatically with the arrival of Europeans in the nineteenth century. Now adding to these threats is the over-arching threat of global climate change.

Situated in the Indian Ocean, 400 km from the southeastern coast of Africa, Madagascar is the world's fourth largest island. Up to 90% of Madagascar's plant species are endemic. For woody plants, the degree of endemism is even higher, with 96% of Madagascar's trees and shrubs found only on the island. Basic threats to the endemic flora of the island result from the poverty of the majority of Malagasy people, with almost 80% of the country's inhabitants dependent on subsistence agriculture. Clearance of forest to produce rice and other subsistence crops continues. Burning of the vegetation to produce fresh pasture for herds of zebu cattle is another major threat together with timber extraction, mining and over-collection of valuable species both for export and local use.

## Conservation solutions

The high concentration of endemic and threatened plant species places a significant conservation responsibility on island administrators. Frequently, as in so-called small island developing states (SIDS) that are mainly tropical, local conservation resources are limited, given the scale of the conservation task. A legal framework for biodiversity conservation is important on islands as part of government policy, and nowadays integrated solutions are increasingly called for that take into account the livelihood needs of local people. Botanic gardens on islands play an important role in local conservation, helping to develop and support government policy, informing local people and visitors about the importance and fragility of island floras and carrying out integrated conservation action. Table 20.1 shows which islands have at least one botanic garden.

Generally, protection of wild plants in their natural habitats is considered the best conservation option, combined with actions to tackle the predominant threats, but for the rare and fragile island floras *ex situ* conservation by botanic gardens is also invaluable. With climate change providing an uncertain future, maintaining *ex situ* collections is becoming critically important as a form of biodiversity insurance policy.

The 'Viera y Clavijo' Botanic Garden in the Canary Islands is one of the world's leading botanic gardens dedicated to plant conservation. For the past 30 years, the Viera y Clavijo has been actively involved in both *ex situ* and *in situ* conservation as well as helping to develop global plant conservation policy. One of the many species that the Viero y Clavijo is helping to conserve is *Anagyris latifolia*, a leguminous shrub considered to be Critically Endagered by IUCN and classified as Endangered by the Canarian Government. Populations are now very fragmented and composed of a small number of individuals. The decline of the plant is thought to be due to the economic importance of the species as a forage plant and destruction of the laurel forest in the Canary Islands (González-Pérez *et al.*, 2009). The species occurs within protected areas and in *ex situ* collections at Viero y Clavijo. Reinforcement of natural populations is recommended for this species (Navarro Déniz *et al.*, 2007).

One particular aim of the "Viera y Clavijo" Botanic Garden has been to use the knowledge of the local flora to secure a series of nature reserves on the island of Gran Canaria to cover as many as possible of the local vegetation types and endangered species. Sites for subsidiary local reserves for single local endemics have been recommended throughout the islands. Threatened plants are maintained by the Garden in both living collections and in the seed bank. The seed bank, initially established in 1983, now serves as the Macaronesian Islands Regional Seed Bank (BASEMAC) operating with partner centres on Madeira and on Faial in the Azores (Bramwell, 2007).

Frequently botanic gardens from different parts of the world work together to develop conservation initiatives for island floras. The small Macaronesian volcanic island of Madeira has around 130 native plant species that are unique to its flora. Madeira Botanical Garden works in conjunction with the Madeira Natural Park authorities in monitoring species and habitats in need of preservation. It works with Viera y Clavijo in seed conservation and also collaborates with the Conservatoire Botanique National (CBN) in Brest, France, in finding solutions for the conservation of Madeira's threatened plant species. The Brest CBN specialises in the conservation of highly threatened island plant species from around the world. Macaronesian plants that the Conservatoire has helped to save include

*Normania tryphilla*, an endemic Madeiran plant long presumed to be extinct until its re-discovery in 1991, *Monizia edulis*, an arborescent species with edible carrot-like roots also from Madeira and *Euphorbia stygiana* from the Azores. Another species that the CBN has helped to conserve is *Lysimachia minoricensis*. This plant, endemic to Menorca, was discovered in 1878, brought into cultivation in 1926 in the botanic garden at Barcelona, became extinct in the wild through unknown causes, was located in the Berlin-Dahlem Botanic Garden in 1974 and then extensively propagated by CBN so that seed could be returned to Menorca.

There are numerous other examples of the rescue of endangered plant species on islands by botanic gardens. Hawaii has 23 botanic gardens. The National Tropical Botanical Garden (NTBG), headquartered on Kauai, was established by the United States Government in 1964 as a privately supported research and education institution. There are now three gardens on the island of Kauai managed by NTBG, together with one on Maui and three nature reserves. Concentrated conservation efforts are a top priority for the Critically Endangered Hawaiian plant species that have fewer than 50 individuals remaining in the wild. These species are included in Hawaii's so-called Genetic Safety Net (GSN) list. Collection of GSN plants from the wild takes place according to strict guidelines to ensure representative genetic sampling of populations and careful location mapping for the individuals in their natural habitats. Where possible, seed is collected for long-term *ex situ* conservation. Seeds and/or cuttings are collected from the wild material for propagation within the gardens (Burney *et al.*, 2005). NTBG works closely with other botanic gardens, voluntary conservation groups, US government agencies and private landowners in its plant conservation efforts. The Lyon Arboretum has, for example, played a major role in propagating GSN plants using micropropagation techniques (Sugii & Lamoureaux, 2004).

NTBG has over 100 endangered species held in its collections and aims to reintroduce species to the wild wherever possible. Among these are the endemic *Pritchardia* palms. *Pritchardia* is a genus of about 27 species found in the Pacific with 22 species in the Hawaiian archipelago. Each Hawaiian species is restricted to a single island and only two of these survive as reproducing wild populations. NTBG aims to maintain all the Hawaiian species in *ex situ* collections with up to 50 individuals per species, where possible, to ensure genetic representation of the remaining wild species.

Reintroduction of *Pritchardia* palms to their original wild locations or to sites close by is now being carried out by NTBG. The Critically Endangered *P. limahuliensis* is, for example, being planted at the Limahuli Preserve, a site adjacent to the Limahuli Botanic Garden that is managed as part of an integrated conservation programme by NTBG. The Preserve is in the upper part of the Limahuli Valley, an area of remnant natural vegetation rich in biodiversity. Threats to the endemic flora include overgrazing by feral pigs and goats. These are now excluded from the Preserve by fencing and the native vegetation is being restored using a mix of native trees and shrubs. The regeneration of *P. limahuliensis* has been limited, mainly because of seed predation by rats and pigs, but now this species is being restored in its protected native habitat using nursery-grown material.

Another example of integrated *in situ* and *ex situ* conservation is provided by the work of the Norfolk Island Botanic Gardens situated on the small South Pacific island between Australia, New Zealand and New Caledonia. This Botanic Garden is managed in conjunction with the Norfolk Island National Park which comprises 462 hectares and Phillip Island (190 hectares) some 6 km to the south of Norfolk Island. The Botanic Garden concentrates on the native and endemic plants of the Norfolk group. Many of Norfolk's plants are under pressure from changing land uses, competition with invasive plants and grazing. Specific plants receiving specialised *in situ* attention are *Hibiscus insularis, Wikstroemia australis* (locally known as kurrajong), *Achyranthes arborescens, Meryta latifolia* and *Euphorbia norfolkiana*. With the exception of the *Wikstroemia*, these and many other species have been planted into the Botanic Garden and planted into many of the rehabilitation areas in the national park. In addition to these rehabilitation areas, a major invasive plant control programme is under way in areas of prime remnant vegetation and is playing a significant part in salvaging the forest health and quality.

Conserving the threatened flora of Madagascar is a major undertaking given the size of the island, the diversity of the flora and the increasing rural poverty. International collaboration is essential. The island has three small botanic gardens, two in the capital Antananarivo and one in the arid southwest region in Tulear. Missouri Botanical Garden (MBG) in the US provides extensive support for botanical conservation on the island. For over 30 years, botanists from MBG have been involved in writing the flora of Madagascar, conducting botanical inventories, training local botanists and conservationists and, more recently, collaborating in community-based conservation. In Madagascar, MBG now has a team of over 50 Malagasy people dedicated to studying and conserving the unique and highly threatened flora. The Fairchild Tropical Botanic Garden also supports community-based plant conservation in Madagascar and the Royal Botanic Gardens, Kew, is actively involved in conservation on the island. Kew's work involves mapping the flora to assist in conservation planning and seedbanking. Since 2000, the Millennium Seed Bank has collected seed from over 1000 plant species and has provided advice to local partners on all aspects of the seed conservation process. Kew has supplied equipment to the Silo National des Graines Forestieres (SNGF), the national seed bank. The use of seed in restoration of Madagascar's degraded flora is an important consideration.

## Strengthening international collaboration

Botanic gardens work together in a variety of ways to support the conservation of plant diversity. At a national level, for example, the 11 botanic gardens of Cuba form a collaborative network. Regionally the 38 botanic gardens of the Caribbean Islands worked together to produce the Conservation Action Plan for Botanic Gardens of the Caribbean Islands (Burbidge & Wyse Jackson, 1998).

At present the main international framework for plant conservation is the Global Strategy for Plant Conservation (GSPC) of the Convention on Biological Diversity (CBD). This Strategy was adopted by over 180 countries in 2002. The ultimate goal of the GSPC is to halt the current and continuing loss of plant diversity, with 16 outcome targets to be met by 2020. Botanic gardens have been closely involved in the development and adoption and implementation of the GSPC. This has been facilitated by the work of Botanic Gardens Conservation International (BGCI), an organisation that aims to mobilise skills and resources worldwide in support of plant conservation.

Individual botanic gardens have been actively involved in supporting national implementation of the GSPC in some island states. In the Seychelles, for example, the National Botanic Gardens worked with the Plant Conservation Action Group to develop a national response to the GSPC (Plant Conservation Action Group and Botanical Garden Section, 2005). With regard to the conservation of threatened plants this report notes that 53% of the endemic flowering plants of the granitic islands are threatened. About 50% of the endemics have been successfully propagated *ex situ*. Further work is needed to prioritise species for protection and recovery. Research collaborations with international partners is proposed for work leading to reintroduction of threatened species in the wild.

Primary responsibility for GSPC implementation is at the national level, but most of the 16 targets of the GSPC have an international organisation designated to help facilitate implementation and monitor progress. BGCI acts as the lead facilitating agency for GSPC target 8 on *ex situ* plant conservation and restoration and also GSPC target 14 on education and public awareness – both of which are clearly key to the work of botanic gardens around the world. The GSPC has also been widely embraced by the botanic garden community. Actions that support the GSPC complement the International Agenda for Botanic Gardens in Conservation (Wyse Jackson & Sutherland, 2000), a document that was taken into account when developing the GSPC. The International Agenda sets out a framework for the many actions that botanic gardens can take to conserve plants, regardless of their size or location.

GSPC target 8 calls for at least 75% of threatened plant species in accessible *ex situ* collections, preferably in the country of origin, and at least 20% of them available for recovery and restoration programmes. BGCI and Bioversity International are lead facilitating agencies for this target. BGCI's PlantSearch Database is being used as a tool to monitor progress towards target 8. This database was launched as an online searchable database on BGCI's website (www.bgci.org) in 2003. The database currently holds records for approximately 180 000 taxa, provided by nearly 700 botanic gardens. The plant records are linked to databases on the threat status and uses of plants – the 1997 and 2006 IUCN Red Lists of threatened plant species. *Anagyris latifolia*, the Canary Island endemic shrub, for example, is currently recorded in six botanic gardens. Table 20.2 indicates the degree to which *Pritchardia* spp. are found in botanic garden collections as recorded in the PlantSearch Database.

The ability to cross-reference the species in PlantSearch with current Red List data shows which threatened plant species are in cultivation as part of documented *ex situ* collections. At present, nearly 12 000 globally threatened species are recorded in the database. Using the 1997 IUCN Plant Red List as a baseline, this suggests that over 30% of globally threatened plants were recorded in *ex situ* collections by the end of 2009. Thus relatively good progress is being made for target 8 but recovery and restoration efforts will certainly need to be increased and island floras will remain a priority.

Target 5 of the GSPC calls for 50% of the most important areas for plant diversity to be protected. Around 70 countries have taken steps to identify important plant areas (IPAs) at a national level. With regard to protection, the scope of the target goes beyond formally protected areas and the challenge will be to find ways for local communities to be involved in sustainable management of areas rich in plant diversity. Missouri Botanical Garden's work in Madagascar supporting plant red listing, IPA identification and community-based conservation initiatives is an important example of support for target 5. Complementing target 5, GSPC target 7 calls for 60% of the world's threatened species to be conserved *in situ*. This again is of relevance for the work of botanic gardens on islands.

## Looking ahead

Island floras have long exerted a fascination for botanists and are generally relatively well known. The threats are understood and frameworks, policies and techniques exist to conserve threatened plant species worldwide. Nevertheless, conservation resources and actions need to be urgently increased. The Caribbean, for example, although recognised as a global biodiversity hotspot, has limited resources to conserve the rich endemic island floras. As noted by Maunder *et al.* (2008), 'the botanical infrastucture of the region is profoundly under invested'. The same is considered to be true for the Pacific island floras. Wagner *et al.* (2005) point out that:

Table 20.2 Pritchardia *species in botanic garden collections.*

| Species | Conservation status | No. of PlantSearch garden records |
|---|---|---|
| *Pritchardia affinis* | Critically Endangered | 19 |
| *Pritchardia aylmer-robinsonii* | Critically Endangered | 4 |
| *Pritchardia forbesiana* | Endangered | 8 |
| *Pritchardia glabrata* | Endangered | 2 |
| *Pritchardia hardyi* | Critically Endangered | 4 |
| *Pritchardia kaalae* | Critically Endangered | 5 |
| *Pritchardia lanigera* | Endangered | 8 |
| *Pritchardia lowreyana* | Vulnerable | 6 |
| *Pritchardia munroi* | Critically Endangered | 4 |
| *Pritchardia napaliensis* | Critically Endangered | 4 |
| *Pritchardia remota* | Endangered | 15 |
| *Pritchardia schattaueri* | Critically Endangered | 5 |
| *Pritchardia thurstonii* | Lower Risk: Near Threatened | 12 |
| *Pritchardia viscosa* | Critically Endangered | 5 |
| *Pritchardia woodfordiana* | Data Deficient | 1 |

In general, preservation programs throughout the Pacific are severely inadequate, with poor policies and few designated protected areas … In many cases where land has been set aside as a natural preserve or national park, the plants most at risk are located outside the protected area … Ecological and botanical research is urgently needed to provide a solid foundation for preservation and management programs.

*(Wagner et al., 2005)*

Increasingly conservation planning needs to take into account global climate change. Supporting *ex situ* conservation is likely to become a growing imperative in response to climate change. It is anticipated that the pace of plant extinctions will increase dramatically as global temperatures increase and that this will be amplified on islands where there are few migration options at higher altitudes or latitudes. Botanic gardens and associated seed banks are well placed to identify, store and propagate wild plant material. Potentially the horticultural knowledge and skills of botanic gardens will be very important in helping to adapt to climate change, both in growing plant species in new locations and in restoring vegetation with potentially new species combinations. There is a high degree of uncertainty but, with so many island plant species already at risk, an immediate priority is to ensure that as many as possible are secure in *ex situ* collections.

## References

Bramwell, D. (2002). How many plant species are there? *Plant Talk* 28: 32–34.

Bramwell, D. (2007). The response of botanic gardens to climate change. *BGjournal* 4 (2): 3–8.

Burbidge, B. & Wyse Jackson, P. (1998). *Action Plan for Botanic Gardens of the Caribbean Islands*. London, Botanic Gardens Conservation International.

Burney, D. A., Bender, D., Perlman, S., Salvador, C. & Wichman, C. (2005). The National Tropical Botanical Garden: creating a genetic safety net for Hawaii's vanishing flora. *BGjournal* 2 (1): 17–19.

Cronk, Q. (1989). The past and present vegetation of St. Helena. *J. Biogeogr.* 16: 47–64.

Cronk, Q. (1993). Extinction and conservation in the St Helena flora: the paleobiological and ecological background. *Boletim do Museu Municipal do Funchal* 2: 69–76.

Cronk, Q. (2000). *The Endemic Flora of St Helena*. Oswestry, UK, Anthony Nelson.

Davis, S. D., Droop, S. J. M., Gregerson, P., Henson, L., Leon, C. J., Vila-Lobos, J. L. *et al.* (1986). *Plants in Danger: What Do We Know?* Gland, Switzerland, International Union for Conservation of Nature and Natural Resources.

González-Pérez, M. A., Sosa, P. A. & Batista, F. J. (2009). Genetic variation and conservation of the endangered endemic *Anagyris latifolia* Brouss. Ex Willd. (Leguminosae) from the Canary Islands. *Plant Syst. Evol.* 279: 59–68.

Iturralde, R., Berazaín, F., Lara, J. & Torres, L. (2005). *Lista roja de la flora vascular cubana*. Gijón, Cuba, Jardín Botánico Atlántico.

Maunder, M., Levia, A., Santiago-Valentín, E., Stevenson, D. W., Acevedo-Rodríguez, P., Meerow, A. *et al.* (2008). Plant conservation in the Caribbean Island biodiversity hotspot. *Bot. Rev.* 74: 197–207.

Mittermeir, R. A., Robles Gil, P., Hoffmann, M., Pilgrim, J., Brooks, T., Goettsch Mittermeier *et al.* (2004). *Hotspots Revisited*. CEMEX. Available from Chicago, University of Chicago Press.

Navarro Déniz, J., Naranjo Suarez, J. & Oliva Tejera, F. (2007). *Flora de las Islas Canarias en Peligro Crítico. Top 100*. Las Palmas de Gran Canaria, Jardín Botánico Canario 'Viera y Clavijo'.

Plant Conservation Action Group and Botanical Garden Section (MENR) (2005). *Seychelles National Strategy for Plant Conservation 2005–2100.*

Procter, D. & Fleming, L. V. (eds.) (1999). *Biodiversity: The UK Overseas Territories*. Peterborough, UK, Joint Nature Conservancy Committee.

Sugii, N. & Lamoureux, C. (2004). Tissue culture as a conservation method: an empirical view from Hawaii. In E. O. Guerrant, K. Havens & M. Maunder (eds.), *Ex Situ Plant Conservation: Supporting Species Survival in the Wild*. Washington and London, Island Press.

Wagner, W. L., Mix, D. & Price, J. (2005). Pacific oceanic islands. In G. A. Krupnik & W. J. Kress (eds.), *Plant conservation: A Natural History Approach*. Chicago, University of Chicago Press.

Walter, K. S. & Gillett, H. (eds.) (1998). *1997 IUCN Red List of Threatened Plants*. Gland, Switzerland and Cambridge, UK, IUCN.

World Conservation Monitoring Centre (1992). *Global Biodiversity: Status of the Earth's Living Resources*. London, Chapman & Hall.

Wyse-Jackson, P. S. & Sutherland, L. A. (2000). *International Agenda for Botanic Gardens in Conservation*. London, Botanic Gardens Conservation International.

# 21

# The hazardous future of island floras

VERNON H. HEYWOOD

## Introduction: the changed context

When I contributed the final chapter to *Plants and Islands* (Bramwell, 1979), my conclusions about the future of island organisms and communities were described by the editor as 'rightly pessimistic'. Now 30 years later, it is perhaps even more difficult to be optimistic about their future. This is despite the fact that the intervening years have seen the coming into force of international treaties and agreements such as CITES, the Convention on Biological Diversity, the Climate Convention, the Migratory Species Convention, the Leipzig Report and the International Treaty on Plant Genetic Resources for Food and Agriculture, as well as the publication of a series of major reports and assessments such as the Agenda 21, the Global Biodiversity Assessment and the Millennium Ecosystem Assessment. The period has also witnessed major advances in our understanding of island floras, their nature, origin, evolution, biogeography, the threats to them and conservation responses, as the preceding chapters in this volume will testify.

There are many reasons for being seriously concerned about the prospects for island floras, not the least being that, writing this in 2010, the International Year of Biodiversity, it is patently obvious that our combined efforts to reduce the loss of biodiversity have not been successful and that many of the goals we have set ourselves, whether they be the Millennium goals, those of the Global Strategy for Plant Conservation or national or regional targets (Butchart *et al.*, 2010), have not been reached. Some of us would argue that the achievement of these goals never was a likely prospect, given the impossibility of coordinating the countless initiatives that have been undertaken across the globe and the major discrepancies between the capacity of the developed and the developing world to respond.

*The Biology of Island Floras*, eds. David Bramwell and Juli Caujapé-Castells. Published by Cambridge University Press. 

The basic reason for our failure to meet our targets has been the inescapable fact that not enough effort and resources have been applied at national and global levels to address the problems. Although this may seem paradoxical, given the many new initiatives that have been generated as a direct consequence of setting these targets and goals, it can be explained by our having not having understood the amount of effort and resources needed to meet them. There has also been a failure by the public, and the governments they elect, to acknowledge fully the vital importance of maintaining biodiversity for our survival and to allocate the necessary resources to ensure this. Although we now understand much better the drivers of biodiversity loss (Millennium Ecosystem Assessment, 2005), we have underestimated the difficulties of controlling the relentless forces of destruction. As I have stressed elsewhere (Heywood, 2010), we are not winning the argument for biodiversity conservation, a concept that many of the public do not really understand. Moreover, in the last few years the focus is moving away to climate change, although of course biodiversity and climate change are intimately interrelated.

A substantial rethinking of our approaches to conservation has taken place in recent years as a response to the worldwide loss, conversion, fragmentation and degradation of habitats, largely as a result of human action, and the need to combine conservation with sustainability of resources. There is now also an increasing appreciation of the importance of agricultural biodiversity and the wider landscape in our conservation planning and of the need to take fully into account the interests and concerns of local populations and to involve them in the management of biodiversity. One of the most significant changes in our perception of conservation has been the growing realisation of the extent and importance of dynamism in ecosystems and populations. It has in fact been suggested that 'one does not wish to conserve vegetation, which is a thing, but rather one is attempting to conserve a dynamic' (Pickett *et al.*, 1992). This has led to the notion that conservation is essentially managing dynamic change in species and ecosystems. There has also been a growing realisation of the importance of ecosystems as producers of goods and services and the consequent need to maintain ecosystem health.

While politicians and conservationists have been digesting the implications of these changes, growing understanding of the probable consequences of global change, and in particular climate change, has wrong-footed the conservation community. The tipping point, perhaps, was the publication of a series of reports – the *Stern Review on the Economics of Climate Change* (Stern, 2007), the IPCC Reports (IPCC, 2007) and *Confronting Climate Change: avoiding the unmanageable and managing the unavoidable* (Scientific Expert Group, 2007) – which, together with other findings in the literature, combined to present a picture of large, serious and damaging climatic impacts on our way of life and on biodiversity in the short, medium and long term. While there has been widespread acceptance of the main thrust of these reports and of their implications for biodiversity, although there are still many disagreements on the detail and major gaps in our knowledge (Schiermeier, 2010), I would argue that these documents have plunged biodiversity conservation into an even deeper crisis, which calls for an even more radical reappraisal of our current planning and strategies. It has also led to a drastic reduction in our planning horizons: no longer can we usefully consider distant scenarios but have to focus on the next 10 to 50 years during which critical actions will have to be taken to avoid irreversible changes.

At the same time, conservation biology, since it evolved in the 1980s, has fed on a variety of other areas of biology, notably ecology, demography, population biology, population genetics, biogeography, landscape ecology, environmental management and economics (Heywood & Iriondo, 2003) and has developed a powerful battery of methodologies, models and techniques, such as:

- conservation genetics (Avise & Hamrick 1996);
- the use of gap analysis;
- the use of systematic conservation planning, including complementarity techniques in reserve selection;
- measures for assessing the effectiveness of reserves;
- methods for predicting species' distributions and patterns of richness;
- metapopulation ecology dynamics and biology;
- fragmentation biology and patch dynamics;
- incorporation of phylogenetic and molecular methods into conservation assessment;
- use of DNA bar-coding in identifying populations at risk;
- applications of spatial analysis and phylogeographic methods for a better understanding of diversity patterns and of what to conserve;
- application of biodiversity informatics;
- novel methods for assessing extinction risks.

Many of these were designed to make conservation choices and actions better informed and more effective, and it is legitimate to question what impact this impressive array of approaches has had on the biodiversity crisis. In effect, they have only been applied to conservation practice on a very limited scale, largely because of the costs and difficulty of implementing them. Also, as Martín *et al.* (2010) comment, 'Regardless of the approach used, these measures are ineffective if not accompanied by political actions that legitimise and enforce them...' and, as we have seen, such commitments often fall short of what is needed.

## Island biodiversity and its conservation

Turning our focus specifically to islands, there is abundant evidence to suggest that island floras are even more vulnerable than they were 30 years ago. Most of the predictions I made in 1979 have, sadly, become true. As discussed below, there has been a massive expansion of travel and tourism to many islands, with even ecotourism posing a threat to areas such as the Galapagos where annual tourist numbers have grown from 41 000 to more than 160 000 since 1991, while the local population has grown by 4% every year to reach more than 40 000; agricultural expansion has continued – a topical example being the massive clearance of forests in parts of Southeast Asia such as Indonesia to plant African oil palm, partly in response to the demand for biofuel; and the number and extent of plant invasions on some islands has increased.

On the positive side, many new protected areas have been established – for example, in 2003 the government of Madagascar announced its intention to triple the country's protected area coverage from 5 million to 15 million acres (from 20 234 to 60 700 $km^2$) or about 10% of the country's total land surface (see Chapter 17 in this volume). Recent studies have addressed the design of this new network in order to optimise biodiversity conservation, including the recognition of key biodiversity areas (KBAs) as sites of global importance for biodiversity conservation, characterised by the presence of globally significant populations or assemblages of threatened or restricted range species (Langhammer *et al.*, 2007), 80 priority sites for plant conservation (PAPC) (MBG, 2005), and important plant areas (IPAs) (Plantlife, 2010). In addition, Rogers *et al.* (2010) developed a framework incorporating human-related threats, ecosystem services and biological measures to identify priorities within the developing protected area network. Sixteen key biodiversity sites were shown to be especially important for both biodiversity and ecosystem services.

Overall, as indicated in the 2003 United Nations List of Protected Areas (Chape *et al.*, 2003), the number of protected areas in the 'mixed island ecosystems' biome is 3425 and the percentage of the biome that is protected went up from 16.32% in 1997 to nearly 30% in 2003. Caution is needed, however, in interpreting these figures as many of these areas are not adequately protected or managed (UN Economic and Social Council, 1999). On the other hand, oceanic islands are underrepresented worldwide in protected areas.

Another example of a major breakthrough has been in Socotra which is home to 825 plant species of which 307 (37%) are endemic to the archipelago. In 1992, a Conservation Zoning Plan was developed which focused on preserving existing traditional uses and practices and forms the basis of the master plan for development of the archipelago for biodiversity conservation, environmental management

Table 21.1 *Plant micro-reserves (PMR) in Crete.*

| Species | PMR locations and area covered |
|---|---|
| *Androcymbium rechingeri* | Elafonisi islet (2 ha) |
| *Anthemis glaberrima* | Agria Gramvousa islet (4.5 ha) |
| *Bupleurum kakiskalae* | Linoseli (1 ha)* |
| *Cephalanthera cucullata* | Koustogerako – mountain range of Lefka Ori (12 ha) |
| *Hypericum aciferum* | Fournoti (6.5 ha) |
| *Nepeta sphaciotica* | Svourichti summit at Lefka Ori mountain (4.8 ha)* |
| *Phoenix theophrasti* | Chrisoskalitissa (2.2 ha) |

* The only known populations in the world.

and development. Its preparation involved extensive consultation with local community leaders throughout the islands of Socotra. The Zoning Plan created several key zones distinguished according to their mode of usage and introduced National Park Zones that cover more than 72% of the island's surface, and nature sanctuaries with the highest level of protection. The Socotra Archipelago Biosphere Reserve was designated as a UNESCO MAB Reserve in 2003 and, in 2008, the Socotra archipelago, Yemen, on the recommendation of IUCN was placed on the World Heritage List, for its outstanding biodiversity.

Another significant initiative was the establishment of a series of plant micro-reserves on the islands of Menorca and Crete, as part of the EU-LIFE programme (EC, 2008). Under the LIFE project 'Conservation of areas with threatened plant species in Minorca (Spain)' which ran from 2001 to 2004. In it, the government of Menorca (*Consell Insular*) developed a set of comprehensive actions to recover the plant species and priority habitats protected by the Habitats Directive, including the drafting of a network of 24 plant micro-reserves. Since 2004, LIFE has been supporting the CretaPlant project (a pilot network of plant micro-reserves in Western Crete (Greece) which was developed by the Mediterranean Agronomic Institute of Chania (MAICh) and the National and Kapodistrian University of Athens (NKUA), with the aim of adapting the PMR concept to the province of Chania in Western Crete; see Table 21.1 for details).

## Lack of baseline data

During the past 30 years, major efforts have been made to inventory and study the biodiversity of many islands or island complexes, as several chapters in this volume attest. There are still, however, serious lacunae in our knowledge of even the most basic data for many islands or island groups. As a study on climate change and biodiversity in Melanesia by the Indo-Pacific Conservation Alliance notes, there is 'currently a significant dearth of several kinds of data for many islands or island groups such as Melanesia that are important to understand climate change and how to adapt to it. Such data gaps include information on baseline rainfall, temperatures, biogeographical patterns, species physiology and ecological requirements of most marine and terrestrial species. This information is necessary to build better climate models that will allow more accurate regional projections and forecast potential impacts on biota and ecosystem services' (IPCA, 2010; see also James, 2008).

In Madagascar, although many plant species are seriously threatened, formal assignment to the new IUCN threat categories has been undertaken for only a small fraction of the flora. Similarly, in an assessment of climate change and biodiversity in Melanesia, James (2008) notes that, currently, less than 1% of the flora is listed as Vulnerable or above for New Guinea, Vanuatu and the Solomon Islands while, for the better documented floras of Fiji and New Caledonia, 6% of native species are listed as Threatened. It is, therefore, a matter of urgency to improve our knowledge of both the floristic and ecological composition of such islands and the threat status of the endemic species and establish baselines for monitoring. In addition, as in other types of habitat, we need more studies on the range of genetic variation, physiological tolerance and dispersal capacity of island species if we are to be able to make realistic assessments of the possible effects of climate change and other aspects of global change on islands within Melanesia and other island groups and of the adaptation capacity of individual species or their ability to migrate.

## Impacts of global change

Islands, like other areas, are being subjected to a whole series of changes – collectively known as global change – and climate change is just one of these components and not by itself likely to be the main driver of plant and animal biodiversity losses, but rather in combination with existing stresses, such as habitat loss, overuse of resources, and invasive species that will amplify and accelerate these losses. Chazal and Rounsevell (2009), for example, argue that if we examine the climate change and land-use change in isolation, we risk making inadequate estimates of their projected impacts on biodiversity. This is particularly relevant when using bioclimatic modelling to project the future migrations of plants in response to climate change and the nature and state of the 'new' habitats that develop.

In some islands, climate change is not projected to be the major cause of future biodiversity loss but rather changes in other factors such as disturbance regime, impacts of alien invasive species, etc. For example, a study by Gritti *et al.* (2006) on the vulnerability of ecosystems in five of the main islands in the Mediterranean basin (Mallorca, Corsica, Sardinia, Crete and Lesvos) to climate change and invasion by exotic plants showed that the main factors promoting invasions is habitat disturbance, while the effects of climate change alone are likely to be negligible. According to their simulations, almost all the ecosystems will be dominated by invasive aliens in the longer term. On the other hand, experimental work undertaken on the invasibility of ecosystems at the establishment phase, using *Ailanthus altissima*, *Carpobrotus* spp. and *Oxalis pes-caprae* as exemplars, at more than 200 locations in Crete, Lesvos, Sardinia, Porquerolles, Mallorca and Menorca, showed that, with the exception of *Oxalis*, they showed widespread resistance to invader establishment (Vilà *et al.*, 2008). These results are only preliminary and there are other factors involved, as the authors indicate.

Likewise, climate change is not considered to be the primary threat to biodiversity in the Canary Islands, although predicted changes in the wind direction and the consequent changes in temperature and precipitation will seriously affect the remaining laurel forests which will be reduced in extent or possibly displaced to favourable areas (Del Arco, 2008). The lower-zone *Euphorbia balsamifera* and *E. canariensis* scrub communities are expected to expand upwards to some degree, subject to the limits imposed by increased urbanisation. The high altitude ecosystems are likely to suffer from the effects of rising temperatures and some species such as *Bencomia montana* and *Rhamnus integrifolia* with small populations will be at risk, as will several rupicolous taxa such as various species of *Aeonium* which may not be able to migrate to other suitable habitats in a short time scale. The coastal vegetation and dunes will be seriously affected by rising sea levels. Another risk from climate change is the expansion of existing invasive species and the introduction of new ones. An example is the African grass, *Pennisetum setaceum*, which is already established in arid zones of the islands and is expected to extend its range (García-Gallo *et al.*, 1999).

In Madeira the most characteristic vegetation is the laurel forest, which is rich in endemic species and today covers some 15 000 ha or 16% of the island. These laurel forests are better conserved than those in the Canary Islands but are the communities most vulnerable to climate change, especially in the intensity of the north trade winds. Other threats are from species which, although not invasive until now, are able to benefit from the changing climate and are beginning to spread and invade the native forests.

The effects of global change on islands will in some cases be catastrophic, as in low-lying islands and coastal areas which are already under threat from rising sea levels caused by melting glaciers and thermal expansion of seawater (IPCC, 2001; Mimura *et al.*, 2007). In most cases they will require changes in our conservation strategies. Because of their frequently high biotic diversity and the damage caused by anthropogenic interventions, including habitat conversion for agriculture, tourism and development, and their susceptibility to predation and invasion by exotic organisms, global change will exacerbate an already critical situation. The biodiversity of small island states will be especially at risk through the effects of climate change on the extent, location and health of ecosystems and habitats, and the IPCC report expects a wide range of changes, including alterations in population size, species distribution and composition. There is a possibility of 'ecosystem collapse', for example in the case of mangroves, should they be forced to migrate from the shore inland.

## Population changes

Population change refers to both changes in the pattern of distribution of human populations and to demographic growth. Large-scale migrations of human populations can be caused by social, economic, political and health factors. War and civil conflict can leave large areas of land devastated or unusable and can cause large human migrations, thus affecting the natural and agroecosystems involved and their biodiversity. According to the 2008 edition of the UN *World Population Prospects*, in mid-2008, world population stood at 6.7 billion, up from 6.0 billion in 1999. The next milestone, 7 billion, will likely be passed in 2011 or 2012 and the 9 billion milestone by 2050. This demographic growth has also affected islands, thus adding pressure through increased demand on resources. One of the consequences of climate change and environmental degradation is the displacement of peoples from their homes. The inhabitants of some Pacific islands already have to face the possibility of permanent relocation to other islands (Loughry & McAdam, 2008). Population movements are another consequence of demographic growth, with high population growth leading to migration from smaller outer islands to larger islands and from rural areas to towns, especially national capitals (World Bank, 2006). The inhabitants of the Carteret Islands of Papua New Guinea started to evacuate to Bougainville in 2009, as a result of rising sea levels, loss of mangroves and other factors. It is probable that these islands and the Pacific nations of Tuvalu and Kiribati may disappear altogether by the middle of the twenty-first century as a result of the rise in sea level.

The problem of environmental refugees – people fleeing excruciating, often fatal, environmental conditions – is one of the most challenging aspects of global change. As climate change accelerates and leads to more intense storms, flooding, heat waves and droughts, many island communities will likely be affected – for example, worldwide, nearly 200 million people live in coastal flood zones that are at risk – and a sea level rise of 0.5 m would flood many coastal communities, especially in the poorest developing countries, such as the Andaman Islands and the Maldives, leading to an increase in the number of environmental refugees. The IPCC (2007) has suggested that the number of environmental refugees will increase by 200 million by the middle of this century. Their effects on biodiversity could be serious in that they will move into territories not able to support or feed them without large-scale disruption.

### Impacts of tourism

Another form, albeit temporary, of population migration is annual tourism. Tourism is an important incentive for the maintenance of biodiversity for many islands. Tourism may be used to justify conservation efforts since biodiversity is an obvious attraction for visitors who are interested in the wildlife and natural landscapes. For many small-island economies tourism is often one of the few forms of development available, which focuses attention on how tourism can be made sustainable. The Caribbean has been identified as the most tourist-dependent region in the world (Gayle & Goodrich, 1993) and it has been estimated that the industry contributes 15% of the region's national income, or gross domestic product (GDP), and much higher fractions for some islands – more than two-thirds in several of the smaller countries. In 2004, the tourism industry in the Caribbean had a turnover of $28 billion industry and employed 2.4 million people.

The Mediterranean is the leading tourist destination in the world with over 12 million tourists visiting the Mediterranean islands each year. The most robust recent growth in tourism has been among the islands of the Caribbean, Mediterranean and Pacific (Apostolopoulos & Gayle, 2002).

The increase in tourism has led to massive urban and tourist-related development with accompanying infrastructural effects such as irrigation, drainage, desalinisation, large-scale transport infrastructures and so on. This is especially accentuated in coastal areas such as parts of the Mediterranean (Berardo, 1999) and on islands, and coupled with increased coastal urbanisation (Plan Bleu, 2001) leads to the phenomenon known as 'coastalisation' – the concentration of population and economic activities in coastal spaces. This has inevitably led to environmental degradation such as pollution and coastal erosion and to an impoverishment of biodiversity and loss or fragmentation of habitats, many of which are home to fragile ecosystems. These side effects of tourism, combined with other threats to coastal ecosystems such as overfishing, reef destruction caused by the use of dynamite and cyanide poisoning, illegal dumping of waste and uncontrolled sewage outlets, can constitute a major threat to the survival of some elements of island biodiversity. The effects vary from region to region, as noted by the Global Development Research Center (GDRC, 2007), with nations such as Papua New Guinea still able to sustain relatively intact forests while, in other nations, the natural habitats have been almost entirely destroyed.

As a means of assessing tourism impact, McElroy and de Albuquerque (1998, 1999) developed the tourism penetration index (TPI) which is constructed from three variables: per capita visitor spending, daily visitor density per 1000 population and number of hotel rooms per km$^2$. McElroy (2002) applied it to a global sample of 47 islands which included 21 in the Caribbean, 15 in the Pacific, five in the Indian Ocean (Comoros, Maldives, Mauritius, Reunion and Seychelles), and three each in the Mediterranean (Balearics, Cyprus and Malta) and Atlantic (Bermuda, Canaries and Cape Verde). The 'most penetrated' islands form a subgroup of 11 internationally visible, highly developed destinations characterised by average per capita visitor spending over $8500 and an average daily visitor density close to 200 tourists per 1000 residents. Tourists thus represent the rough equivalent of a 20% increase in the daily population. Their insular landscapes are crowded, on average, with nearly 50 hotel rooms per km$^2$ of area. This cluster comprises four popular Caribbean resort areas – Aruba, British Virgins (BVI), Caymans and St Maarten – plus Bermuda and Canaries in the Atlantic, the Balearics and Malta in the Western Mediterranean, and Hawaii, Guam and Northern Marianas in the Pacific. As McElroy (2002) comments, 'Such islands literally define the postwar pleasure periphery for North America, Western Europe and Japan.' The TPI can be used as an early warning system to help state decision makers assess the overall pressure of tourism on fragile sociocultural and ecological assets. McElroy (2002, 2003) proposes that the siting and design of facilities and access nodes for tourism must fit within the natural insular buffering systems to maintain the terrestrial–marine landscape, scenic amenities and biodiversity.

Island tourism can also act as a means of transporting new species and diseases. This is especially a problem in small island states. The isolation that was partly the reason for the creation of high biodiversity often found on such islands can be opened up by tourism, and the prime tourist habitats such as beaches, sand dunes and coral reefs risk being damaged by visitors and by the infrastructure that is developed to cater for them (Sheldon, 2005).

In an island biogeographical approach to the interactions between island tourism and biodiversity in the Caribbean and Pacific, Hall (2010) found that few of the countries examined have anywhere near the recommended percentage of area protected, with marine ecosystems being least conserved.

## Threats from invasions

Biological invasions are one of the main drivers of biodiversity change and it is well known that islands are often much more susceptible to invasion by alien species than mainland sites, despite the fact that the invasives have the additional hazard of having necessarily to cross water to reach the islands.

In the last 30 years, there has been ample evidence of the spread of alien invasive species on islands (see Chapter 15). For example, the fire tree, *Morella* (*Myrica*) *faya*, which began to spread in the Hawaiian Islands in the 1920s shortly after its introduction for reforestation and was declared a noxious weed in 1944, has continued its spread in many areas in Hawaii, Maui, Lanai, Oahu and Kauai and now occupies vast extensions forming a dense monospecific cover, reaching as high as 2134 m on Maui. According to a report in 2003 (Starr *et al.*, 2003), its eradication in some areas at that time did not seem feasible. The South African succulents *Carpobrotus edulis* and *C. acinaciformis* are among the most harmful alien invasive species in the Mediterranean basin and on islands such as Menorca, Mallorca and Porquerolles and pose a serious threat to the native vegetation.

The prospect of accelerated climate change has now added a new dimension to the threat from alien invasive organisms. There is good evidence that changes in climatic conditions affect the rate and extent of biological invasions (Dukes & Mooney, 1999; Thuiller *et al.*, 2007), although these changes interact synergistically with other drivers of global change, which requires adopting a more integrated approach if we are to gain a better understanding of the consequent effects (Thuiller *et al.*, 2007; Walther *et al.*, 2009). The challenges for biodiversity conservation on islands are difficult and complex and not as self-evident as might appear. We cannot predict accurately which species will invade nor what impacts they will have on the ecosystems or their functioning, although in recent years there have been considerable advances in modelling the potential distribution of alien invasive species (Gallien *et al.*, 2010). However, more work is needed to test the ability of bioclimatic models to forecast the effects of climate change or the spread of invasive organisms (Jeschke & Strayer, 2008). A possible new hazard – the potential risk of introduced biofuel crops becoming invasive – was assessed by Buddenhagen *et al.* (2009) who, using a widely accepted weed risk assessment system, found that a comprehensive list of 40 biofuel crops proposed for Hawaii were two to four times more likely to be naturalised or invasive there, compared with a sample of introduced non-biofuel species. The authors consider that their results are applicable to other vulnerable island ecosystems.

The costs of control or eradication of invasive species is a major concern and not easily affordable on some islands. For example, the eradication from parts of the Galapagos Islands of donkeys and goats that cause soil degradation, so as to protect the fragile ecosystems, endemic species and the local tourist economy, is reported to cost more than $8 million annually (Murphy & Cheesman, 2006). The eradication of *Carpobrotus edulis* and *C. acinaciformis* in various parts of the Mediterranean, notably Mallorca and Menorca (Spain), incurs annual costs of hundreds of thousands of euros and has been included in *LIFE Nature* projects. Some idea of the scale of the operations involved can be obtained from the campaign to eradicate *Carpobrotus* undertaken in Menorca from 2002 to 2005: 233 785 $m^2$ of *Carpobrotus* were eliminated, representing the removal of 832 148 kg of biomass and involving 9041 hours of work.

Moore *et al.* (2010) explore the issue of how much effort should be invested in quarantine to reduce the risk of an invasive species arriving, versus surveillance, looking for the invasive species on an island with the view of eradicating it before it gets out of control. As they note, 'preventing invasion events on islands through the dual management actions of off-island quarantine and on-island surveillance provides a potentially effective means of minimising the future impact of new, or recently eradicated, invasives'. One approach to limiting the introduction and spread of invasives is the use of voluntary codes such as the St Louis Voluntary Codes of Conduct which were developed following a workshop on 'Linking ecology and horticulture to prevent plant invasions' held in 2001 at Missouri Botanical Garden (Fay *et al.*, 2001) and in Chicago in 2002 (Fay *et al.*, 2002) and the Code of Conduct on Horticulture and Invasive Alien Plants commissioned by the Council of Europe (Heywood & Brunel, 2009).

## The impacts of climate change on islands

The impacts of climate change on islands have attracted considerable attention recently, as discussed in Chapter 18. This is not surprising in view of the large number of species that are endemic to islands – estimated at some 70 000 by Caujapé-Castells *et al.* (2010) – even though many of the islands have only relatively small floras, with the consequence that the actual and potential effects of climate change on global biodiversity may be correspondingly large. In a recent global assessment of endemism and species richness across islands and mainland regions, Kier *et al.* (2009) proposed using an index combining both endemism and species richness – 'endemism richness' – to help in assessing the relative conservation value of islands and mainlands.[1] The survey of 90 biogeographical regions showed that endemism richness of plants on islands exceeds that of mainland regions by a factor of 9.5.

Projection of the impacts of climate change are still at an early stage of development and, while bioclimatic modelling has made great advances in the past few years, many problems remain regarding scale of sampling and ability of models to reflect habitat data (Gallien *et al.*, 2010). The application of bioclimatic modelling to islands has so far been very limited and is complicated by the unreliability or incompleteness of the necessary baseline data such as site-specific climate and species information, as already mentioned. In the case of small islands, climate models may not even 'see' them because of the scale of modelling employed.

[1] Endemism richness is calculated by giving each species the same value that is equally distributed across its range so that when 100% of a species' range falls into a mapping unit (e.g. a grid cell or biogeographic region), its entire value is attributed to this area. When only half its distribution area falls into a mapping unit, only 50% of its value is attributed to the mapping unit, and so on. By summing up the fraction of range equivalents for all species within a mapping unit, both endemism and species richness are reflected. For details see Kier & Barthlott (2001).

The impacts of climate change on islands will vary according to location, size, whether continental or oceanic, degree of habitation, and so on. The most vulnerable areas are likely to be coastal regions and mountain ecosystems. Climate change, sea level rise and more frequent extreme weather events such as hurricanes will have repercussions on coastal development, water supply, energy, agriculture and health, among other sectors.

The following are some of major impacts that will affect tropical islands such as Hawaii: stronger winds and changed direction; increased frequency and severity of storms; drought; heavier rainfall; reduction in water resources; sea level rise; deep incursions from strong ocean surges; shoreline erosion; saltwater intrusion; sea surface temperature rise; ocean acidification; and higher temperatures (State of Hawaii, 2008).

The anticipated climate changes will accelerate the erosion of coastal beaches, land and protective mangroves.

Not only will biodiversity be affected but also food security. As a report on the Caribbean and climate change notes,

> Food security is also of concern due to the vulnerability and limited scale of Caribbean agriculture, already facing uncertain impacts from temperature and precipitation changes. Many islands, including Barbados, Jamaica, and Puerto Rico, are highly dependent on imported food and agricultural products, and very susceptible to changes in world food prices. Such prices may spike upwards as climate change exacerbates droughts and floods in the world's major agricultural producing regions.
>
> *(Bueno et al., 2008)*

Montane ecosystems are particularly susceptible to climate change, largely as a consequence of the specialised types of habitat that are occupied by alpine/montane species which often grow in small, isolated populations that are less able to survive because they lack the genetic variability needed to adapt to changing conditions, and because of the limited migration possibilities open to them. The GLORIA (Global Observation Research Initiative in Alpine Environments) project is a global set of sites from the poles to the tropics that have been set aside for the long-term surveillance of the migration of plant species attributed to climate change. Recognition of the importance of the mountain regions of Europe led to the development of GLORIA-Europe, the European dimension of the project. It is estimated that alpine species that occur in Europe's high mountains account for some 20–25% of Europe's total plant diversity (Grabherr, 2009; Nagy & Grabherr, 2009) with notable concentrations in the Alps and in the Appenines and Apuan Alps, Italy, the French and Spanish Pyrenees, Sierra Nevada, Baetic and Subbaetic sierras of Spain, the Lefka Auri (White Mountains), Crete, Mount Olympus and the mountains of southern and central Greece, and the Troödos mountains, Cyprus.

## The problems of small island states

Small island developing states (SIDS), according to the United Nations Department of Economic and Social Affairs, are low-lying coastal countries that share similar sustainable development challenges, including small population, limited resources, remoteness, susceptibility to natural disasters, vulnerability to external shocks and excessive dependence on international trade. They are especially vulnerable to natural and human actions and by their very nature the amount of land available for habitation, building infrastructure, food production and industrial development is constrained in extent, making them economically, socially and physically vulnerable. Unable to produce all the goods and services needed to satisfy the domestic economy, they have to rely largely on imports (Tompkins *et al.*, 2005). As with many other kinds of island, they often rely on tourism to boost the economy and generate the foreign exchange necessary for imports. Small island states vary considerably in terms of their biodiversity: some of them have very high levels of endemism (Table 21.2), such as coral reefs with the highest known diversity among marine ecosystems, while others such as low reef islands have low biodiversity and low levels of endemism (IPCC, 2001). They are especially susceptible to the impacts of accelerated climate change and other aspects of global change as discussed below.

According the IPCC Working Group report on adaptation and vulnerability (IPCC, 2001), small island states:

> … urgently need sensible predictive information and tools for minimising the likely outcomes of climate change and sea-level rise because it is now generally accepted that small islands are intrinsically more vulnerable than larger countries. The outcome of climate change on human and natural systems in small island states will depend on changes induced by climate itself, feedback mechanisms, human adaptive capacity, and the resilience of biophysical and human systems.

Many small island states are especially vulnerable to rises in sea level which will impact both human populations and natural ecosystems. According to a World Bank report (Vergara, 2005), anticipated sea level rises in the Caribbean will threaten aquifer-based freshwater supplies through saline intrusion in many of the smaller islands as well as leading to flooding of coastal zones. This is a major concern, given that over 50% of the people in most Caribbean states reside within 2 km of the coast. Thus resources critical to island and coastal populations – including beaches, wetlands, fresh water, fisheries, coral reefs and atolls and wildlife habitat – are all at risk. If we know, however, what the existing risks are and put in place monitoring systems to track changes, island governments will be in a better position to prepare for future hazards.

Table 21.2 *Biodiversity status for selected small island states, 1990.*[1]

| Country | Known mammal species | Known endemic mammal species | Known bird species | Known endemic bird species | Known plant species | Known endemic plant species |
|---|---|---|---|---|---|---|
| Cuba | 31 | 12 | 342 | 22 | 6 004 | 3,229 |
| Dominican Republic | 20 | — | 254 | 0 | 5 000 | 1,800 |
| Fiji | 4 | 1 | 109 | 26 | 1 307 | 760 |
| Haiti | 3 | 0 | 220 | 0 | 4 685 | 1,623 |
| Jamaica | 24 | 3 | 262 | 25 | 2746 | 923 |
| Mauritius | 4 | 2 | 81 | 9 | 700 | 325 |
| Palau | 214 | 57 | 708 | 80 | 10 000 | — |
| Solomon Islands | 53 | 19 | 223 | 44 | 2 780 | 30 |
| Trinidad and Tobago | 100 | 1 | 433 | 1 | 1 982 | 236 |

[1] Extracted from IPCC (1998).

## Long-distance dispersal

In assessing the impacts of global change on biodiversity, one of the critical questions is to determine which species will be able to adapt to the new conditions and survive, which will be able to migrate and track their climate envelopes, which will succumb and, in the case of islands in particular, what is the source of the migrants? A key factor is the dispersal capacity of individual species. On islands, in attempting to assess the likely migration of species in the face of accelerated climate change, one of the critical factors that needs to be examined more closely is long-distance dispersal, a topic that has attracted much attention from Darwin onwards (Carlquist, 2009). As Pearson and Dawson (2005) note, there has been renewed interest in recent years in the potential role of rare long-distance dispersal (LDD) events as drivers of rapid plant migrations (for a review see Nathan *et al.*, 2003). While it is possible that low-probability LDD events may allow propagules of some species to arrive within the short time scale envisaged, there is no guarantee that this will lead to successful establishment or spread, as this will depend on habitat availability and its permeability, life history and biological characteristics (Higgins & Richardson, 1999) and degree of saturation of the island in terms of species richness, as discussed by Sax and Gaines (2008).

## Novel conservation approaches

The role of protected areas in the face of climate change is coming under close scrutiny. Although protected areas will continue to be a major component of any conservation strategy, there is increasing evidence that they will be unable to prevent climate change pushing species beyond their boundaries (Kharouba & Kerr, 2010). The fact that their boundaries are fixed means that, while some species will be able to adapt to the new climate and other conditions that will ensue, many species may have to migrate to track their climatic envelope within and outside the present boundaries, while others will succumb. This, combined with the inward migration of other species and the possible increase in pioneer or invasive species, will lead to a large turnover of species and the formation of novel or non-analogue assemblages of species. Various solutions for adapting protected areas to climate change have been proposed (Halpin, 1997; Shafer, 1999; Hannah & Salm, 2003; Lovejoy, 2006; Ervin *et al.*, 2010), including the extension of current protected areas into areas projected to be climatically suitable, the establishment of patches or mosaics, and the creation of migration corridors. On the other hand, Pearson and Dawson (2005) caution against the justification of large-scale corridors on grounds of climate change when migration along corridors by standard dispersal mechanisms is unlikely to keep pace with projected change for many species. They suggest that artificial translocation may provide a more effective and efficient solution but, as discussed below, this approach is controversial.

In the case of smaller islands at least, such options for adapting protected areas in response to climate change are more limited and, as we have noted already, the likelihood is that there will be an increase in the spread of alien invasive species and the successful colonisation by new ones which will further complicate future planning.

Faced with such challenges, a number of novel conservation approaches have been proposed such as human-aided translocation of species populations as a means of countering biodiversity loss from global change. This technique, known as assisted migration (McLachlan *et al.*, 2007) or assisted colonisation[2] (Hunter 2007; Hoegh-Guldberg *et al.*, 2008), is being suggested for situations where the rate of change, the existence of obstacles or barriers or the lack of continuous suitable habitat are considered likely to prevent natural migration. Moving species into new environments is a complex and contentious issue and may involve considerable risks (Ricciardi & Simberloff, 2009; Richardson *et al.*, 2009). It is also potentially costly and is a venture that needs to be subject to careful cost–benefit analysis and perhaps used, only in exceptional circumstances, for species judged locally or globally of high priority that are otherwise unlikely to survive climate change. Its role

[2] Hunter uses the term *assisted colonization* in contrast to *assisted migration* 'because many animal ecologists reserve the word *migration* for the seasonal, round-trip movements of animals … and because the real goal of translocation goes beyond assisting dispersal to assuring successful colonization, a step that will often require extended husbandry'.

in conserving island species endangered by climate change needs to be assessed critically.

## The hazardous future

Even if we exclude stochastic factors such as the devastating 2004 tsunami that cost over 225 000 human lives and caused major ecological damage in coastal regions of Sri Lanka and Indonesia, the future for island biodiversity and plant life in particular looks bleak in the light of growing human populations, increasing tourist numbers, continuing deforestation and loss of mangroves, a spread of biological invasions, and the interacting impacts of accelerated climate change. A recent review of the conservation of oceanic island floras estimated that between 3500 and 6800 of the estimated 70 000 insular endemic plant species might be highly threatened (CR and EN *sensu* IUCN) and, of these, between 2000 and 2800 in critical danger of extinction (Caujapé-Castells *et al.*, 2010). What then can be done to reduce the effects of all these factors? Some of the actions needed are:

- more effective planning of conservation actions, including the design of protected areas and adaptation measures;
- much greater effort in devising means of conserving species outside protected areas in the wider landscape;
- more effective and realistic use of targets and effective monitoring of progress in achieving them;
- intelligent priority setting – of target areas, ecosystems and species;
- filling in the gaps in our baseline data;
- more extensive and effective monitoring;
- much more study of the ecophysiological tolerance of species;
- finer-scale climatic modelling;
- much more application of bioclimatic modelling and broadening the range of factors included in the models so as to help make projections of the likely impacts of climate change on ecosystems and species;
- formulation and adoption of appropriate preventive measures based on scientific risk assessments (such as weed risk assessments, WRA) and import pathway analyses, using climate change scenarios, where appropriate; and
- strengthening of quarantine measures and codes of conduct to regulate or control the entry and or spread of alien invasive species.

(See also Caujapé-Castells *et al.* (2010) who give a detailed analysis of the threats to insular endemics and the scientific and political conservation needs of nine focal archipelagos.)

In addition to these scientific and technical issues, the future of island biodiversity will depend very largely on political, financial and socioeconomic factors (Heywood, 2006; Caujapé-Castells *et al.*, 2010). It will require much greater commitment by governments to biodiversity conservation and allocation of resources.

Islands are an endless source of inspiration and discovery. Many of Darwin's ideas about the island biogeography are still being extended and clarified in the light of new observations and the application of modern approaches such as genetic analysis, cytogenetics, molecular systematics and phylogeography, which were not available to him (Carlquist, 2009). We also need to remember that islands are home to many people. As the Millennium Ecosystem Assessment (2005) notes, 'The coastal, marine, and inland ecosystems of islands provide valuable regulating, provisioning, and cultural services to more than 500 million people.' It is not just the future of island biodiversity that is at risk but the future of those who both depend on it and help maintain it.

## References

Apostolopoulos, Y. & Gayle, D. J. (eds.) (2002). From MIRAB to TOURAB? Searching for sustainable development in the Caribbean, Mediterranean and Pacific. In *Island Tourism and Development: Caribbean, Pacific and Mediterranean*, pp. 3–14. Santa Barbara, CA, Greenwood Press.

Avise, J. C. & Hamrick, J. L. (eds.) (1996). *Conservation Genetics*. New York, Chapman & Hall.

Berardo, C. (1999). Spatial dynamics of Mediterranean coastal regions. *J Coastal Conserv.* 5: 105–112.

Bramwell, D. (ed.) (1979). *Plants and Islands*. London, Academic Press.

Buddenhagen, C. E., Chimera, C. & Clifford, P. (2009). Assessing biofuel crop invasiveness: a case study. *PLoS ONE* 4(4): e5261. doi:10.1371/journal.pone.0005261.

Bueno, R., Herzfeld, C., Stanton, E. A. & Ackerman, F. (2008). *The Caribbean and Climate Change: The Costs of Inaction*. Stockholm Environment Institute–US Center Global Development and Environment Institute, Tufts University.

Butchart, S. H. M., Walpole, M., Collen, B. *et al.* (2010). Global biodiversity: indicators of recent declines. *Science Express*. www.sciencexpress.org /29 April 2010 /Page 5/10.1126/science.1187512.

Carlquist, S. (2009). Darwin on island plants. *Bot. J. Linn. Soc.* 161: 20–25.

Caujapé-Castells, J., Tye, A., Crawford, D. J., Santos-Guerra, A., Sakai, A., Beaver, K. *et al.* (2010). Conservation of oceanic island floras: present and future global challenges. *Perspect. Plant Ecol. Evol. Syst.* 12: 107–129.

Chape, S., Blyth, S., Fish, L., Fox, P. & Spalding, M. (2003). *2003 United Nations List of Protected Areas*. Gland, Switzerland and Cambridge, UK, IUCN and Cambridge, UK, UNEP-WCMC.

de Chazal, J. & Rounsevell, M. D. A. (2009). Land-use and climate change within assessments of biodiversity change: a review. *Glob. Environ. Change* 19: 306–315.

del Arco, M. (2008). Consecuencias del cambio climático sobre la flora y vegetación canaria. In J. M. Méndez Pérez & M. Vázquez Abeledo (eds.), *El Cambio Climático en Canarias*. Academia Canaria de Ciencias. Serie Monografías N° 1: 79–100.

Dukes, J. S. & Mooney, H. A. (1999). Does global change increase the success of biological invaders? *Trends Ecol. Evol.* 14: 135–139.

EC (European Commission) (2008). *LIFE and Endangered Plants: Conserving Europe's Threatened Flora*. Luxembourg, Office for Official Publications of the European Communities, 0207.

Ervin, J., Mulongoy, K. J., Lawrence, K., Game, E., Sheppard, D., Bridgewater, P. *et al.* (2010). *Making Protected Areas Relevant: A Guide to Integrating Protected Areas into Wider Landscapes, Seascapes and Sectoral Plans and Strategies*. CBD Technical Series No. 44. Montreal, Canada, Convention on Biological Diversity.

Fay, K., Fay, K. C. and Associates (2001). Linking Ecology and Horticulture to Prevent Plant Invasions. Proceedings of the Workshop at the Missouri Botanical Garden, St Louis, MO, 1–4 December 2001. Available at: www.centerforplantconservation.org/invasives/DownloadPDF/Proceedings – FINAL.pdf

Fay, K., Fay, K. C. and Associates (2002). Linking Ecology and Horticulture to Prevent Plant Invasions II Proceedings of the Workshop II at the Chicago Botanic Garden, Chicago, IL, 31 October 2002. Available at: www.centerforplantconservation.org/invasives/DownloadPDF/CBG–Proceedings.pdf

Gallien, L., Münkemüller, T., Albert, C. H., Boulangeat, I. & Thuiller, W. (2010). Predicting potential distributions of invasive species: where to go from here? *Diversity and Distributions* 16: 33–342.

García-Gallo, A., Wildpret de la Torre, W., Rodríguez Delgado, O., PÉREZ DE PAZ, P. L., León Arencibia, M. C., Suárez Rodríguez, C. *et al.* (1999). El xenófito *Pennisetum setaceum* en las Islas Canarias (Magnoliophyta, Poaceae). *Vieraea* 27: 133–158.

Gayle, D. J. & Goodrich, J. N. (eds.) (1993). *Tourism, Marketing and Management in the Caribbean*. London and New York, Routledge.

GDRC (2007). *An Overview of Major Environmental Threats Encountered by Small Island Developing States*. The Global Development Research Center. www.gdrc.org/oceans/10things/index.html

Grabherr, G. (2009). Biodiversity in the high ranges of the Alps: ethnobotanical and climate change perspectives. *Glob. Environ. Change* 19: 167–172.

Gritti, E. S., Smith, B. & Sykes, M. T. (2006). Vulnerability of Mediterranean Basin ecosystems to climate change and invasion by exotic plant species. *J. Biogeogr.* 33: 145–157.

Hall, C. M. (2010). An island biogeographical approach to island tourism and biodiversity: an exploratory study of the Caribbean and Pacific Islands. *Asia-Pac. J. Tourism Res.* 15 (3): 383–399.

Halpin, P. N. (1997). Global climate change and natural-area protection: management responses and research directions. *Ecol. Appl.* 7: 828–843.

Hannah, L. & Salm, R. (2003). Protected areas and climate change. In L. Hannah & T. Lovejoy (eds.), *Climate Change and Biodiversity: Synergistic Impacts*, pp. 91–100. Washington, DC, Conservation International.

Heywood, V. H. (2006). Island ecosystems – priorities for conservation. In Pinheiro de Carvalho, *et al.* (eds.), *II Symposium Island Ecosystems*, pp. 85–99, University of Madeira, Centre for Macaronesian Studies.

Heywood, V. H. (2010). Developing new biodiversity conservation strategies in response to global change. *Bollettino dei Musei e degli Istituti Biologica dell'universita di Genova* 72: 95–122.

Heywood, V. H. & Brunel, S. (2009). *Code of Conduct on Horticulture and Invasive Alien Plants.* Nature and Environment No. 155. Strasbourg, Council of Europe Publishing.

Heywood, V. H. & Iriondo, J. M . (eds.) (2003). Plant conservation: old problems, new perspectives. In *Plant Conservation Biology: Emerging Tools and Strategies. Biological Conservation* 113: 321–335.

Higgins, S. I. & Richardson, D. M. (1999). Predicting plant migration rates in a changing world: The role of long-distance dispersal. *Am. Nat.* 153: 464–475.

Hoegh-Guldberg, O., Hughes, L., McIntyre, S., Lindenmayer, D. B., Parmesan, C., Possingham, H. P. *et al.* (2008). Assisted colonization and rapid climate change. *Science* 321: 345–346.

Hunter, M. L. (2007). Climate change and moving species: furthering the debate on assisted colonization. *Conserv. Biol.* 21: 1356–1358.

IPCA (2010). Indo-Pacific Conservation Alliance. IPCA joint study yields important data about global warming in the Pacific. www.indopacific.org/news.asp (accessed 26 April 2010).

IPCC (1998). *The Regional Impacts of Climate Change: An Assessment of Vulnerability.* A Special Report of IPCC Working Group II (R. T. Watson, M. C. Zinyowera & R. H. Moss, eds.) Cambridge, Cambridge University Press.

IPCC (2001). *Climate Change 2001:Working Group II: Impacts, Adaptation and Vulnerability. Chapter 17: Small Island States.* IPCC. Cambridge, Cambridge University Press.

IPCC (2007). *Climate Change 2007 – The Physical Science Basis. Contribution of Working Group I to the Fourth Assessment Report of the IPCC; Climate Change 2007 – Impacts, Adaptation and Vulnerability. Contribution of Working Group II to the Fourth Assessment Report of the IPCC; Climate Change 2007 – Mitigation of Climate Change. Contribution of Working Group III to the Fourth Assessment Report of the IPCC.* Cambridge, Cambridge University Press.

James, S. A. (2008). Climate change impacts on native plant communities in Melanesia. CCBM Paper 8. *Bishop Mus. Tech. Rep.* 42 (8).

Jeschke, J. M. & Strayer, D. L. (2008). Usefulness of bioclimatic models for studying climate change and invasive species. *Ann. N.Y. Acad Sci.* 1134: 1–24.

Kharouba, H. M. & Kerr, J. T. (2010). Just passing through: global change and the conservation of biodiversity. *Biol. Conserv.* 143: 1094–1101.

Kier, G. & Barthlott, W. (2001). Measuring and mapping endemism and species richness: a new methodological approach and its application on the flora of Africa. *Biodivers. Conserv.* 10: 1513–1529.

Kier, G., Kreft, H., Lee, T. M., Jetz, W., Ibisch, P.., Nowicki, C. *et al.* (2009). A global assessment of endemism and species richness across island and mainland regions. *Proc. Natl. Acad. Sci. USA* 106(23): 9322–9327.

Langhammer, P. F., Bakarr, M. I., Bennun, L. A., Brooks, T. M. *et al.* (2007). *Identification and gap analysis of Key Biodiversity Areas: targets for comprehensive protected area systems.* Gland, Switzerland, IUCN.

Loughry, M. & McAdam, J. (2008). Kiribati – relocation and adaptation. *Forced Migr. Rev.* 31 (Climate Change and Displacement): 51–52.

Lovejoy, T. E. (2006). Protected areas: a prism for a changing world. *Trends Ecol. Evol.* 21: 329–333.

Martín, J. L., Cardoso, P., Arechavaleta, M. *et al.* (2010). Using taxonomically unbiased criteria to prioritize resource allocation for oceanic island species conservation. *Biodivers. Conserv.* 19: 1659–1682.

McElroy, J. (2002). The impact of tourism in small islands: a global comparison. In F. di Castri & V. Balaji (eds.), *Tourism,Biodiversity and Information*, pp. 151–167. Leiden, Backhyus Publishers.

McElroy, J. L. (2003). Tourism development in small islands across the world. *Geografiska Annaler* Series B *Human Geography* 85 (4) (special issue: Nature-Society Interactions on Islands): 231–242.

McElroy, J. & Albuquerque, K. (1998). Tourism penetration index in small Carribean islands. *Ann. Tourism Res.* 25: 145–168.

McElroy, J. & Albuquerque, K. (1999). Measuring tourism penetration in small islands. *Pac. Tourism Rev.* 3: 161–169.

McLachlan, J. S., Hellmann, J. J., & Schwartz, M. W. (2007). A framework for debate of assisted migration in an era of climate change. *Conserv. Biol.* 21: 297–302.

Millennium Ecosystem Assessment (2005). Island systems. In R. Hassan, R. Scholes & N. Ash (eds.), *Ecosystems and Human Well-being: Current State and Trends: Findings of the Condition and Trends Working Group.* Washington, DC, Island Press.

Mimura, N., Nurse, L., McLean, R. F., Agard, J., Briguglio, L., Lefale, P. *et al.* (2007). Small islands. In M. L. Parry, O. F. Canziani, J. P. *et al.* (eds.), *Climate Change 2007: Impacts, Adaptation and Vulnerability. Contribution of Working Group II to the Fourth Assessment Report of the Intergovernmental Panel on Climate Change*, pp. 687–716. Cambridge, Cambridge University Press.

MBG (2005). Missouri Botanical Garden. *Assessment of Priority Areas for Plant Conservation* (*CEPF Final Project Completion Report*) www.cepf.net.

Moore, J. L., Rout, T. M., Hauser, C. E., Moro, D., Jones, M., Wilcox, C. *et al.* (2010). Protecting islands from pest invasion: optimal allocation of biosecurity resources between quarantine and surveillance. *Biol. Conserv.* 143: 1068–1078.

Murphy, S. T. & Cheesman, O. D. (2006). The Aid Trade International Assistance Programs as Pathways for the Introduction of Invasive Alien Species: A Preliminary Report. Washington, DC, World Bank.

Nagy, L. & Grabherr, G. (2009). *The Biology of Alpine Habitats.* Oxford, UK, Oxford University Press.

Nathan, R., Perry, G., Cronin, J. T., Strand, A. E. & Cain, M. L. (2003). Methods for estimating long-distance dispersal. *Oikos* 103: 261–273.

Pearson, T. G. & Dawson, T. P. (2005). Long-distance plant dispersal and habitat fragmentation: identifying conservation targets for spatial landscape planning under climate change. *Biol. Conserv.* 123: 389–401.

Pickett, S. T. A., Parker, V. T., & Fiedler, P. (1992). The new paradigm in ecology: implications for conservation biology above the species level. In P. L. Fiedler & S. Jain (eds.), *Conservation Biology: The Theory and Practice of Nature Conservation, Preservation and Management*, pp. 65–88. New York, Chapman & Hall.

Plan Bleu (2001). *Urban Sprawl in ihe Mediterranean Region*. Blue Plan Regional Activity Centre, Sophia Antipolis, France.

Plantlife (2010). *Important Plant Areas around the World*. Successes and case studies in implementing the Global Strategy for Plant Conservation (GSPC). Salisbury, UK, Plantlife.

Ricciardi, A. & Simberloff, D. (2009). Assisted colonization is not a viable conservation strategy. *Trends Ecol. Evol.* 24: 248–253.

Richardson, D. M., Hellmann, J. J., McLachlan, J. S. *et al.* (2009). Multidimensional evaluation of managed relocation. *Proc. Natl. Acad. Sci. USA* 106: 9721–9724.

Rogers, H. M., Glew, L., Honzák, M. & Hudson, M. D. (2010). Prioritizing key biodiversity areas in Madagascar by including data on human pressure and ecosystem services. *Landscape Urban Plan.* 96: 48–56.

Sax, D. F. & Gaines, S. D. (2008). Species invasion and extinction: the future of native biodiversity on islands. *Proc. Natl. Acad. Sci. USA* 105 (Suppl 1): 11490–11497.

Schiermeier, Q. (2010). The real holes in climate science. *Nature* 463: 284–287.

Scientific Expert Group on Climate Change (2007). *Confronting Climate Change: Avoiding the Unmanageable and Managing the Unavoidable* (R. M. Bierbaum, J. P. Holdren, M. MacCracken, R. H. Moss and P. H. Raven, eds.). Report for Sigma Xi and the United Nations Foundation, Research Triangle Park, NC and Washington, DC.

Shafer, C. L. (1999). National park and reserve planning to protect biological diversity: some basic elements. *Landscape Urban Plan.* 44: 123–153.

Sheldon, P. J. (2005). *The Challenges to Sustainability in Island Tourism*. School of Travel Industry Management, University of Hawaii, Occasional Paper 2005–01.

Starr, F., Starr, K. & Loope, L L. (2003). *Morella faya* Firetree Myricaceae. www.hear.org/Pier/pdf/pohreports/morella_faya.pdf (accessed 5 May 2010).

State of Hawaii (2008). *Studying the Impacts of Global Climate Change in Hawaii*. ORMP (Ocean Resources Management Plan) Working Group. http://hawaii.gov/dbedt/czm/ormp/reports/wg_proposal_climate_change.pdf

Stern, N. (2007). *The Economics of Climate Change. The Stern Review*. Cambridge, UK, Cambridge University Press.

Thuiller, W., Richardson, D. M. & Midgley, G. F. (2007). Will climate change promote alien plant invasions? In W. Nentwig, W. (ed.), *Biological Invasion*. pp. 197–211. Berlin and Heidelberg, Springer Verlag.

Tompkins, E.L., Nicholson-Cole, S.A., Hurlston, L.-A. *et al.* (2005). *Surviving climate change in small islands: A guidebook*. Tyndall Centre for Climate Change Research, University of East Anglia, Norwich.

UN Economic and Social Council (1999). *Progress in the Implementation of the Programme of Action for the Sustainable Development of Small Island Developing States*. Commission on Sustainable Development. E/CN.17/1999/6/Add.5.

Vergara, W. (2005). *Adapting to Climate Change: Lessons Learned, Work in Progress, and Proposed Next Steps for the World Bank in Latin America*. Washington, DC, World Bank, Latin America and Caribbean Region.

Vilà, M., Siamantziouras, A-K. D., Brundu, G., Camarda, I., Lanbdon, P., Médail, F. *et al.* (2008). Widespread resistance of Mediterranean island ecosystems to the establishment of three alien species. *Divers. Distrib.* 14: 839–851.

Walther, G.-R., Roques, A., Hulme, P. E., Sykes, M. T., Pyšek, P., Kühn, I. *et al.* (2009). Alien species in a warmer world: risks and opportunities. *Trends Ecol. Evol.* 24: 686–693.

World Bank (2006). *The Pacific Infrastructure Challenge: A Review of Obstacles and Opportunities for Improving Performance in the Pacific Islands*, January 2006. Washington, DC, World Bank. http://siteresources.worldbank.org/INTPACIFICISLANDS/Resources/PacificReportFinal.pdf (accessed 6 April 2010).

# Index